Arabidopsis

Annual Plant Reviews

A series for researchers and postgraduates in the plant sciences. Each volume in this annual series will focus on a theme of topical importance and emphasis will be placed on rapid publication.

Arabidopsis

Edited by

MARY ANDERSON
Director
Arabidopsis Stock Centre
University of Nottingham

and

JEREMY ROBERTS
Reader in Plant Biology
University of Nottingham

CRC Press

First published 1998
Copyright © 1998 Sheffield Academic Press

Published by
Sheffield Academic Press Ltd
Mansion House, 19 Kingfield Road
Sheffield S11 9AS, England

ISBN 1-85075-890-5
ISSN 1460-1494

Published in the U.S.A. and Canada (only) by
CRC Press LLC
2000 Corporate Blvd., N.W.
Boca Raton, FL 33431, U.S.A.
Orders from the U.S.A. and Canada (only) to CRC Press LLC

U.S.A. and Canada only:
ISBN 0-8493-9732-4
ISSN 1097-7570

Printed on acid-free paper in Great Britain by
Bookcraft Ltd, Midsomer Norton, Bath

British Library Cataloguing-in-Publication Data:
A catalogue record for this book is available from the British Library

Library of Congress Cataloging-in-Publication Data:
Arabidopsis / edited by Mary Anderson and Jeremy Roberts.
     p.   cm. -- (Annual plant reviews ; v. 1)
  Includes bibliographical references (p.) and index.
  ISBN 0-8493-9732-4 (alk. paper)
  1. Arabidopsis.   I. Anderson, Mary (Mary Louise)   II. Roberts, J.
A. (Jeremy A.)   III. Series.
QK495.C9A685 1998
583'.64--dc21

98-10358
CIP

Preface

In its natural environment, *Arabidopsis thaliana* (thale cress, mouse-eared cress) is a rather insignificant weed, but do not be misled by appearances; take *Arabidopsis* into a laboratory environment—its new niche—and it blossoms as one of the most useful systems for investigating many important facets of plant development. The true power of this diploid, rapid-cycling, prolific, seed-producing plant stems from the fact that it can be readily mutagenised, that large populations of material can be screened easily for mutant phenotypes and that it has a small and simply organised genome. These key attributes have led to the use of *Arabidopsis* as a mutation machine and the emergence of a multi-national concerted effort to generate both a highly detailed genetic map and a complete physical map of the genome as a prelude to its complete sequencing.

This volume discusses recent advances in *Arabidopsis* research at the genome level, including the construction of the physical map, sequencing of the genome, and strategies for structure–function analysis. The power of mutagenesis as a tool to gain insights into plant developmental processes is illustrated in a range of stages in the life cycle of *Arabidopsis*, including embryogenesis, vegetative development, flowering, reproduction and cell death. In addition, the control of metabolism, secretion and biological rhythms is examined and the ways in which development is regulated by such stimuli as plant hormones and light are evaluated.

There is no doubt that *Arabidopsis* research has started to unravel many of the biological processes that are central to plant development. What is particularly exciting is that this information can be transferred to other plant systems through the identification of othologous genes. Hence the impact of *Arabidopsis* in answering central, unanswered questions relating to species of particular agricultural or horticultural significance is coming to fruition.

Contributors

Dr Malcolm J. Bennett Department of Biological Sciences, University of Warwick, Coventry, CV4 7AL, UK

Dr Alice Y. Cheung Department of Biochemistry and Molecular Biology, University of Massachusetts, Amherst, MA 01003, USA

Dr Christopher S. Cobbett Department of Genetics, The University of Melbourne, Parkville, Victoria 3052, Australia

Dr Liam Dolan Department of Cell Biology, John Innes Centre, Colney, Norwich, NR4 7UH, UK

Dr Susannah Gal Department of Biological Sciences, The State University of New York, Binghamton, NY 13902-6000, USA

Dr Jérôme Giraudat Institut des Sciences Végétales, CNRS, 1 Avenue de la Terrasse, 91198 Gif-sur-Yvette Cedex, France

Dr John Gray Department of Agronomy, 205 Curtis Hall, University of Missouri, Columbia, MO 65211, USA

Dr Guri S. Johal Department of Agronomy, 205 Curtis Hall, University of Missouri, Columbia, MO 65211, USA

Dr Joe Kieber University of Illinois at Chicago, Department of Biological Sciences (M/C 567), Molecular Biology Research Facility, 900 South Ashland Avenue, Chicago, IL 60607, USA

Dr Ilha Lee The Salk Institute for Biological Studies, Plant Biology Laboratory, 10010 North Torrey Pines Road, La Jolla, CA 92037, USA

Dr Rob Martienssen Cold Spring Harbor Laboratory, 1 Bungtown Road, Cold Spring Harbor, NY 11724, USA

Dr Andrew J. Millar Department of Biological Sciences, University of Warwick, Coventry, CV4 7AL, UK

Dr Peter Morris Department of Biological Sciences, Heriot-Watt University, Riccarton, Edinburgh, EH14 4AS, UK

Dr François Parcy The Salk Institute for Biological Studies, Plant Biology Laboratory, 10010 North Torrey Pines Road, La Jolla, CA 92037, USA

Dr Andy Pereira DLO-Centre for Plant Breeding and Reproduction Research (CPRO-DLO), PO Box 16, Droevendaalsesteeg 1, NL-6700AA, Wageningen, The Netherlands

Dr Renate Schmidt Max-Delbrück-Laboratorium in der Max-Planck-Gesellschaft, Carl-von-Linné-Weg 10, 50829 Köln, Germany

Dr Willem J. Stiekema DLO-Centre for Plant Breeding and Reproduction Research (CPRO-DLO), PO Box 16, Droevendaalsesteeg 1, NL-6700AA, Wageningen, The Netherlands

Dr Ramón A. Torres Ruiz Institut für Genetik, Technische Universität München, Lichtenbergstrasse 4, 85747 Garching, Germany

Dr Detlef Weigel The Salk Institute for Biological Studies, Plant Biology Laboratory, 10010 North Torrey Pines Road, La Jolla, CA 92037, USA

Dr Garry C. Whitelam Department of Biology, University of Leicester, Leicester, LE1 7RH, UK

Dr Hen-Ming Wu Department of Biochemistry and Molecular Biology, University of Massachusetts, Amherst, MA 01003, USA

Contents

7 Embryogenesis   223
R. A. TORRES RUIZ

8 Patterns in vegetative development   262
R. MARTIENSSEN and L. DOLAN

9 Genetic control of floral induction and floral patterning 298
I. LEE, D. WEIGEL and F. PARCY

10 Light regulation and biological clocks 331
G. C. WHITELAM and A. J. MILLAR

11 Programmed cell death in plants

J. GRAY and G. S. JOHAL

1 The *Arabidopsis thaliana* genome: towards a complete physical map

Renate Schmidt

1.1 Introduction

Arabidopsis thaliana (thale cress, *Arabidopsis*) is ideally suited for the molecular genetic analysis of many plant processes and consequently has become an important model organism for plant molecular biology. Its small size, short generation time and prolific seed production have enabled numerous genes controlling developmental or metabolic processes to be revealed by mutational analyses (Meyerowitz and Somerville, 1994).

The *Arabidopsis* nuclear genome is characterised by a low level of repetitive sequences (Leutwiler *et al.*, 1984; Pruitt and Meyerowitz, 1986) and its haploid size is estimated to be approximately 100–120 Mb (Megabase pairs), considerably smaller than the genomes of nearly all other well-studied plant systems.

The construction of complete physical maps of the five *Arabidopsis* chromosomes is an important goal, since a complete genomic map will provide, for the first time, the basis for understanding the organisation of a plant genome. For example, the chromosome maps will allow the study of the physical linkage of genes and the distribution of repetitive elements. Furthermore, centromeric and sub-telomeric sequences can be identified. Where the physical map has been generated with the help of molecular mapped markers, it can readily be compared to the genetic map, hence the relationship of genetic and physical distance in different areas of the genome can be determined. Most importantly, a physical map is also an efficient means for map-based cloning experiments and it provides a framework for genomic sequencing programmes. The many advantages of having a physical map thus outweigh the laborious process required to generate it.

A variety of approaches, both random and directed, can be used to generate a physical map of an organism. Random strategies include restriction mapping of clones and fingerprinting analysis. The latter has been very successful for the generation of contiguous regions of overlapping cosmid clones (cosmid contigs) for the *Caenorhabditis elegans* genome (Coulson *et al.*, 1986), which is of similar size to that of *Arabidopsis*. A drawback of random approaches is that it is difficult to follow the actual progress of the mapping. The number of clones needing to be analysed to achieve a certain theoretical coverage of a given genome can be calculated, but it is very hard to measure accurately and take into account the influence of chimaeric clones as well as the under-representation of certain genomic regions in the clone libraries, factors which have a great impact on the completeness of the map.

Using a directed and map-based strategy, the whole genome does not need to be analysed at once, rather it is possible to focus on a particular chromosome or genomic region. Useful tools for such a directed approach are chromosome-specific clone libraries or clone libraries that have been greatly enriched for sequences from a particular chromosome. However, the small size of the *Arabidopsis* chromosomes and their relatively homogeneous length means that flow-sorting of chromosomes is unlikely to be useful in the generation of chromosome-specific libraries.

Nevertheless, very extensive molecular-marker maps have been developed for many organisms, including *Arabidopsis*. These can be exploited for physical mapping experiments. Molecular markers with a known genetic-map position serve to identify and anchor clones on the genetic map. Providing a large number of mapped markers and clone libraries with large insert sizes are available, clones will not only be anchored by a single marker, but by two or more markers in the same region of the genome. Hence, large contiguous regions of overlapping clones (contigs) that share common markers can be assembled. The genome size of an organism, the redundancy and the average insert size of the clones, as well as the number of anchors utilised, all determine the size and the number of contigs that can be assembled using such a marker content mapping approach. The resulting contigs can then be joined in chromosome walking experiments to establish ultimately a single contig for each of the chromosomes.

The specific topics discussed in this chapter include the characteristics of the *Arabidopsis* genome and its known repetitive elements; the available resources for genome mapping experiments, including molecular mapped markers and libraries of clones containing large inserts of *Arabidopsis* DNA; the status of the physical mapping experiments and the first results on the organisation of the *Arabidopsis* genome. Furthermore, the way in which data for *Arabidopsis* might be utilised for related plant species will be discussed.

1.2   The *Arabidopsis* genome

1.2.1   Genome size

Measurements of the *Arabidopsis* genome size have been performed with a number of different techniques, such as microspectrophotometric measurements of Feulgen-stained nuclei (Bennett and Smith, 1976), DNA reassociation kinetics (Leutwiler *et al.*, 1984), quantitative genome blot hybridisations (Pruitt and Meyerowitz, 1986), electron microscopic measurements of the volume of chromosomes (Heslop-Harrison and Schwarzacher, 1990), flow cytometry (Arumuganathan and Earle, 1991) and sampling of genomic clone libraries

(Hauge *et al.*, 1991). These diverse methods have resulted in different genome-size estimates, ranging from 50 to 200 Mb. The observed differences can, however, largely be accounted for by the use of different internal standards in the individual experiments. Although an accurate size estimate of the *Arabidopsis* genome will depend on the availability of complete chromosome maps, and ultimately the nucleotide sequence, it is now assumed that the genome encompasses approximately 100–120 Mb. Hence, *Arabidopsis* has one of the smallest genomes observed in higher plants. Plant genomes with a DNA content more than 1000-fold larger have been observed (Bennett and Smith, 1976).

1.2.2   Karyotype

As early as 1907, a chromosome number of 2n = 10 was established for *Arabidopsis thaliana* (Laibach, 1907). The small sizes of the *Arabidopsis* chromosomes, which range in size from 1.5 μm to 2.8 μm (Schweizer *et al.*, 1987), make cytological studies difficult; nevertheless in 1963 a first clear description of the karyotype was given (Steinitz-Sears, 1963). Cytological observations using trisomic lines identified two different chromosomes that carry a nucleolus organiser region (NOR) (Sears and Lee-Chen, 1970). Subsequently, the Giemsa C-banding technique for staining heterochromatin was used to characterise the *Arabidopsis* chromosomes further (Ambros and Schweizer, 1976). Schweizer *et al.* (1987) attempted to correlate the cytological and molecular chromosome data to the genetic linkage map (Koornneef *et al.*, 1983) and assigned the two NORs to the smallest chromosomes, chromosomes 2 (1.5 μm) and 4 (2.1 μm). Chromosome 1 was found to be the largest chromosome (2.8 μm) whereas the sizes of chromosomes 3 and 5 were 2.2 and 2.4 μm, respectively. The relative lengths of the five *Arabidopsis* chromosomes can also be determined using synaptonemal complex preparations. These studies confirm that the chromosomes carrying the NORs are the smallest chromosomes, accounting for 16 and 17% of the total complement. The other three chromosomes have relative lengths of 19, 22.5 and 25% (Albini, 1994).

1.2.3   Tandemly repeated sequences

Consistent with the small genome size, only a few repetitive sequences are found in the *Arabidopsis* genome. Reassociation kinetic studies have shown that only 10% of the genome consists of rapidly annealing sequences (Leutwiler *et al.*, 1984). A distinct Giemsa C-band was found at the centromere of each of the chromosomes, approximately 12.5% of total chromosome length being banded (Ambros and Schweizer, 1976). Schweizer *et al.* (1987) noted that this amount of constitutive C-heterochromatin correlated well with the proportion of very rapidly reannealing sequences, 10–14%, as revealed by reassociation kinetic studies (Leutwiler *et al.*, 1984). Fluorescent *in situ* hybridisation (FISH) experiments did indeed show that two families of tandemly repeated elements (Martinez-Zapater *et al.*, 1986; Simoens *et al.*, 1988) co-localise with the centromeric hetero-chromatin (Maluszynska and Heslop-Harrison, 1991; Murata *et al.*, 1994), al-though one of them is primarily found on one pair of chromosomes (Bauwens *et al.*, 1991). These two families of tandemly repeated sequences share sequence similarity and are characterised by monomer lengths of approximately 180 and 500 bp (base pairs) respectively (Simoens *et al.*, 1988). It could be shown that the 180 bp repeat is present in arrays of 50 kb (kilobase pairs) or longer (Martinez-Zapater *et al.*, 1986; Murata *et al.*, 1994). Both families show meth-ylation at CpG sites (Martinez-Zapater *et al.*, 1986; Simoens *et al.*, 1988; Murata *et al.*, 1994) and with G+C contents of 36 (180 bp) and 38% (500 bp; Simoens *et al.*, 1988), have a lower than average (41%; Leutwiler *et al.*, 1984) G+C content. An *Alu*I repeated element with a monomer length of approximately 160 bp also has a similarly low G+C content (32–36%; Simoens *et al.*, 1988), however, it does not show the same centromeric location (Bauwens *et al.*, 1991). Together, the three different repeated sequence families account for approximately 2% of the total genome. The highest copy number, 3100–6000, is observed for the 180 bp family (Martinez-Zapater *et al.*, 1986; Simoens *et al.*, 1988).

The *Arabidopsis* telomeric sequences were the first to be cloned from a higher eukaryotic organism (Richards and Ausubel, 1988). The telomeres are primarily composed of a 7 bp long repeat (5'-CCCTAAA-3') arranged in tandem and are heterogeneous in size. As observed in other organisms, degenerated telomere sequence motifs have also been found in *Arabidopsis* at non-telomeric locations. Simoens *et al.* (1988) noted that the 500 bp *Hind*III repeat contained a 190 bp domain with telomere-similar sequences and indeed screening for sequences corresponding to degenerated telomeric motifs led to the independent isolation of the 500 bp repeat (Richards *et al.*, 1991). Restriction fragment length polymorphism (RFLP) mapping studies of sequences adjacent to these tandem repeats established a map position linked to the centromere of chromosome 1.

The basic repeat unit of the 18S–25S rDNA sequences is 10.0–10.5 kb long and present in 570–750 copies (Pruitt and Meyerowitz, 1986; Copenhaver and Pikaard, 1996a). The nucleotide sequences of the rRNA genes, the intergenic region and the internal transcribed spacers have been determined (Gruendler *et al.*, 1989; Unfried *et al.*, 1989; Unfried and Gruendler, 1990). Length heteroge-neity in the intergenic region has been also documented (Gruendler *et al.*, 1991; Luschnig *et al.*, 1993). Several hundred copies of the basic repeat unit of the 18S–25S rDNA sequences are arranged in simple head-to-tail arrays at two simi-larly sized NORs (Copenhaver and Pikaard, 1996a). FISH experiments also re-vealed two rDNA loci (Bauwens *et al.*, 1991; Maluszynska and Heslop-Harrison, 1991) and the use of trisomic lines unequivocally showed that the loci were located on chromosomes 2 and 4 (Maluszynska and Heslop-Harrison, 1991), which is consistent with earlier cytological studies (Schweizer *et al.*, 1987). Genetic and physical mapping showed that the NORs adjoin the telomeres of either chromosome 2 or chromosome 4 (Copenhaver and Pikaard, 1996b).

The genes encoding the 5S ribosomal RNA are also organised in tandem arrays with a repeat length of 497 bp, but they are unlinked to the 18S–25S rDNA loci (Campell *et al.*, 1992). Approximately 1000 copies are present in the haploid genome.

1.2.4 Dispersed repetitive elements

Transposable elements are a prominent class of dispersed repetitive elements in most other plant species and represent a considerable proportion of the genome. The *del2* element in lily, for example, is present in 240 000 copies (reviewed in Smyth, 1991). In *Arabidopsis*, however, the situation is quite different: *Athila* is the only described retroelement that is present in more than 100 copies (Péllissier *et al.*, 1995; Thompson *et al.*, 1996a). Far lower copy numbers have been observed for the other known transposable elements or their derivatives: *Ta1* and its related elements (Voytas and Ausubel, 1988; Konieczny *et al.*, 1991), *Tat1* (Peleman *et al.*, 1991) and *Tag1* (Tsay *et al.*, 1993).

A number of different middle repetitive elements from *Arabidopsis* which hybridise to between 20 and 150 genomic fragments have been described. So far no homology to known genes or proteins has been found (Schmidt *et al.*, 1995; Thompson *et al.*, 1996a; Thompson *et al.*, 1996b).

1.3   Genetic maps

Because of the small size of mature plants, a short generation time and prolific seed production from a single plant, *Arabidopsis* is ideally suited for large mutagenesis screens. Ionising radiation and chemicals such as EMS are being extensively used to mutagenise *Arabidopsis*. The development of efficient transformation systems based on *Agrobacterium tumefaciens* has allowed the exploitation of T-DNA as an efficient insertion mutagen (reviewed in Feldmann *et al.*, 1994). Transposable elements have also been used to tag genes in *Arabidopsis*. These insertional mutagenesis experiments (Aarts *et al.*, 1993; Bancroft *et al.*, 1993; Long *et al.*, 1993) have predominantly used transposons from maize, although at least one mutant has been caused by the endogenous transposable element *Tag1* (Tsay *et al.*, 1993).

Using these various techniques, a large number of mutants has been generated. One compilation revealed that more than 1000 genetic loci have been identified by mutation in *Arabidopsis* (Meinke *et al.*, 1995). The spectrum of mutants include, for example, plants that are impaired in metabolic processes, plants with altered morphology and mutants that show an altered response to plant pathogens (Meyerowitz and Somerville, 1994). A particularly extensive collection has been assembled for embryo and seedling lethal mutants (Franzmann *et al.*, 1995).

The first genetic-linkage studies were carried out in the 1960s and established six (Rédei and Hirono, 1964) and four (McKelvie, 1965) linkage groups. The use of trisomics allowed the correlation of these linkage groups with particular chromosomes (Lee-Chen and Bürger, 1967; Lee-Chen and Steinitz-Sears, 1967; Koornneef and van der Veen; 1983). Koornneef *et al.* (1983) established the first comprehensive genetic map, which assigned 76 loci to the five chromosomes, and gave positions for four of the five centromeres. Because of a lack of suitable telotrisomic lines, the centromere on chromosome 4 could not be mapped. The location of each of the five centromeres was established using a system that allows genome-wide analysis of recombination (Copenhaver *et al.*, 1997). To date, about 370 mutants have been placed on the genetic map (Meinke *et al.*, 1995; http://mutant.lse.okstate.edu/genepage/genepage.html).

The first molecular-marker maps generated for the *Arabidopsis* chromosomes utilised RFLP markers. Two maps, containing approximately 100 markers each, were constructed using independent mapping populations (Chang *et al.*, 1988; Nam *et al.*, 1989). The integration of these two maps with the genetic map (Koornneef *et al.*, 1983) was possible because mapped mutations were included in the crosses used to generate the mapping populations and their meiotic seg-regation was followed alongside the segregation of the RFLP markers (Chang *et al.*, 1988; Nam *et al.*, 1989). This integrated map was calculated using the JOINMAP computer package (Stam, 1993) and also included data for 123 additional RFLP markers (Hauge *et al.*, 1993).

Recombinant inbred (RI) populations have also been established: one derived from a cross between the ecotype Wassilewskija and the marker line W100, which carries nine phenotypic markers (Reiter *et al.*, 1992), and the other from a cross between the two ecotypes, Columbia and Landsberg *erecta* (Lister and Dean, 1993). The progeny of a particular cross is self-pollinated to provide the F2 generation. Individual F2 plants are then self-pollinated to generate the F3 generation. The process of self-pollination is continued by single-seed descent until at least the F8 generation. The RI lines are then virtually homozygous and individual lines can be replicated indefinitely without disturbing their genetic fidelity. The inexhaustible seed supply enables the RI lines to be distributed widely. This has resulted in the considerable advantage of new markers being mapped predominantly using these populations. A total of 672 markers has been assigned to the five chromosomes using one set of RI lines (Lister and Dean, 1993; compiled in Anderson, 1997; http://nasc.life.nott.ac.uk/new_ri_map.html). More than 27% of all the markers used have been placed on the largest chromosome, chromosome 1. As the genetic map encompasses approximately 550 cM (centiMorgan), this is, on average, a marker every 0.8 cM. Not all RFLP markers that have been described for *Arabidopsis* (Chang *et al.*, 1988; Nam *et al.*, 1989; Hauge *et al.*, 1993; McGrath *et al.*, 1993) have been mapped on these RI lines, therefore, far more than 672 molecular markers are currently available for mapping in *Arabidopsis*.

The majority of the 672 markers are RFLP markers (Chang *et al.*, 1988; Nam *et al.*, 1989; McGrath *et al.*, 1993; Liu *et al.*, 1996), however, polymerase chain reaction (PCR)-based markers have also been developed, as have amplified fragment length polymorphisms (AFLPs) (Vos *et al.*, 1995), cleaved amplified polymorphic sequences (CAPS) (Konieczny and Ausubel, 1993; http://genome-www.stanford.edu/Arabidopsis/aboutcaps.html), microsatellite markers (Bell and Ecker, 1994; http://cbil.humgen.upenn.edu/~atgc/SSLP_info/SSLP.html) and randomly amplified polymorphic DNA markers (RAPDs) (Williams *et al.*, 1990; Reiter *et al.*, 1992). AFLPs are especially useful if large numbers of polymorphic molecular markers are required, while microsatellite and CAPS markers are often used to map newly identified mutant loci. This is an efficient alternative to using mapping strains that carry multiple morphological markers on each chromosome (Koornneef and Hanhart, 1983; Koornneef *et al.*, 1987; Franzmann *et al.*, 1995).

1.4 Physical mapping

1.4.1 *Physical mapping in* Arabidopsis *using cosmid libraries*

The fingerprinting technique developed for *C. elegans* cosmids (Coulson *et al.*, 1986) has also been used in *Arabidopsis* to identify overlapping cosmids. An analysis of 20 000 clones representing eight genome equivalents was performed and 750 cosmid contigs with an average size of 120–130 kb were constructed. It was calculated that the 750 contigs correspond to 91–95% of the genome, hence most of the contigs should overlap or at least be in close proximity (Hauge *et al.*, 1991; Goodman *et al.*, 1995). Yeast artificial chromosome (YAC) clones were used in an attempt to join the contigs efficiently. This strategy was attempted as it had been very successful in linking the cosmid contigs that were established in *C. elegans* (Coulson *et al.*, 1988, 1991). Hybridisations of YAC clones to ordered arrays of cosmid clones representative of the 750 cosmid contigs from *Arabidopsis* were performed. Using non-chimaeric YACs from one particular area of chromosome 4, as probes to the ordered arrays, revealed six cosmid contigs corresponding to two regions, both approximately 300 kb in size. The identified cosmid contigs spanned approximately 50% of the analysed regions. It was also shown that the distance between adjacent cosmid contigs in the two regions was less than 100 kb (Thompson *et al.*, 1996c). This analysis confirms that the cosmid contigs do indeed cover a large part of the *Arabidopsis* genome and that it should be possible to bridge most of the gaps between adjacent cosmid contigs with clones containing large inserts of *Arabidopsis* DNA, such as YACs or bacterial artificial chromosomes (BACs) Shizuya *et al.*, 1992).

However, in most of the other experiments aimed at linking the cosmid contigs, random YACs derived from the two *Arabidopsis* YAC libraries EG and EW (Grill and Somerville, 1991; Ward and Jen, 1990; see below) were used.

Although joins between cosmid contigs could be established, the strategy was not as fruitful as for the *C. elegans* genome project. This was because of the high frequency of chimaeric YAC clones and because many of the clones used for the hybridisations carried repetitive sequences (Goodman *et al.*, 1995).

1.4.2 Large insert size clone libraries

There are several different clone libraries that are suitable for physical mapping. Insert fragments much larger than 40 kb can be cloned and maintained in yeast and bacterial artificial chromosomes (Burke *et al.*, 1987; Shizuya *et al.*, 1992) and in P1 clones (Sternberg, 1990). Libraries of clones carrying large inserts are usually gridded in microtiter plates (96 or 384 wells), thereby assigning unique co-ordinates to every clone. The libraries can therefore easily be transferred between different laboratories and mapping results directly compared.

Cloning vectors carrying yeast sequences that function as telomeres, centromeres and selectable markers can be exploited to generate clones carrying large insert fragments as artificial chromosomes in yeast (Burke *et al.*, 1987). Clones that harbour more than a megabase of insert DNA have been obtained and stably maintained. These large inserts make YAC clones particularly suitable for genome analysis. This versatility is demonstrated by the generation of YAC contig maps of entire chromosomes (Chumakov *et al.*, 1992; Foote *et al.*, 1992).

A number of YAC libraries have been constructed from high molecular weight genomic *Arabidopsis* DNA. Four different clone libraries have been made from DNA of the ecotype Columbia (Ecker, 1990; Ward and Jen, 1990; Grill and Somerville, 1991; Creusot *et al.*, 1995) and are widely used for physical-mapping experiments (Table 1.1). Additionally, YAC libraries have been constructed from DNA of the ecotpye Landsberg *erecta* and the *abi-1* mutant (Grill and Somerville, 1991).

All of the clones in the YAC libraries made from Columbia DNA contain insert sizes that are on average larger than the average size of the cosmid contigs that have been generated using the fingerprinting approach. Nevertheless, the average insert sizes of the clones derived from the four libraries show considerable differences. While clones from the EG (Grill and Somerville, 1991) and EW (Ward and Jen, 1990) libraries both have an average insert size of approximately 160 kb, the clones that were generated using *Eco*RI partially digested DNA from the yUP (Ecker, 1990) and the CIC (Creusot *et al.*, 1995) YAC libraries are on average 250 and 420 kb in size, respectively (Figure 1.1). Three of the libraries (EG, EW and yUP) contain between 2200 and 2300 clones each, while the CIC library consists of only 1152 clones; together these libraries represent 8.5 genome equivalents (Schmidt *et al.*, 1996, 1997).

Large differences between the libraries are observed on comparing the frequencies of various repetitive sequences (compiled in Goodman *et al.*, 1995).

Table 1.1 Libraries containing large inserts of *Arabidopsis* DNA

Library	No. of clones	Vector	DNA fragmentation method	Average insert size
CIC YAC	1152	pYAC4	*Eco*RI partial	420 kb
EG YAC	~2300	pYAC41	*Bam*HI partial	160 kb
EW YAC	~2200	pYAC3	Random shear	160 kb
yUP YAC	~2300	pYAC4	*Eco*RI partial	250 kb
TAMU BAC	~12000	pBeloBACII	*Hin*dIII partial	100 kb
IGF BAC	10752	pBeloBAC-Kan	*Eco*RI partial	100 kb
P1	10080	pAd10*sac*BII	*Sau*3A partial	80 kb

The lowest frequencies of clones carrying rDNA and chloroplast DNA sequences were found in the EW library, which was constructed from randomly sheared DNA. In contrast to this, the EG library, made from *Bam*HI partially digested DNA, showed the highest number of clones carrying these repetitive sequences. The 180 bp *Hin*dIII repeated sequences, however, were much more prevalent in the EW library than in the other three YAC libraries. The observed differences in the frequencies of different repetitive elements in the various libraries can at least partly be accounted for by the different methods that were used to cleave the *Arabidopsis* genomic DNA before cloning.

Considerable differences between the four YAC libraries were also found when the frequency of chimaeric clones was investigated. An analysis of 1179 clones was performed to give an estimate of the rate of chimaerism and 186 clones (15.8%) were found to be chimaeric (Schmidt *et al.*, 1997). Since not all 1179 clones could be fully analysed, this is a minimal estimate. The fewest chimaeric clones were found in the EW (1.7%) and the CIC (3.5%) libraries while 4- to 20 fold higher frequencies were found in the yUP (15.4%) and EG (33.2%) libraries, respectively (Schmidt *et al.*, 1997; Figure 1.1). Interestingly, a considerable proportion of the chimaeric clones in the yUP and EG libraries consisted of clones carrying single-copy sequences in addition to unlinked repetitive sequences, such as chloroplast DNA and rDNA (Schmidt *et al.*, 1996, 1997). Because of the high frequency of chimaeric clones in the YAC libraries it is important not to rely on results that have been obtained with a single YAC clone, instead mapping results need to be established with at least two independent YAC clones.

YAC clones carrying tandemly repeated sequences were found to exhibit a high degree of instability (Schmidt *et al.*, 1994). Since this phenomenon is observed in all four YAC libraries tested, regions of the genome that carry such sequences need to be mapped using different techniques. Tandemly repeated sequences with high copy numbers have been found thus far at the telomeres, the centromeres and at the two NORs in *Arabidopsis*.

The apparent drawbacks of YAC libraries, i.e. the high frequency of chimaeric clones, the instability of tandemly repeated sequences and the generally low clone redundancy of the libraries, led to alternative cloning systems in *Escherichia coli* being exploited, particularly as the transformation of *E. coli* by electroporation is more efficient than yeast spheroplast transformation.

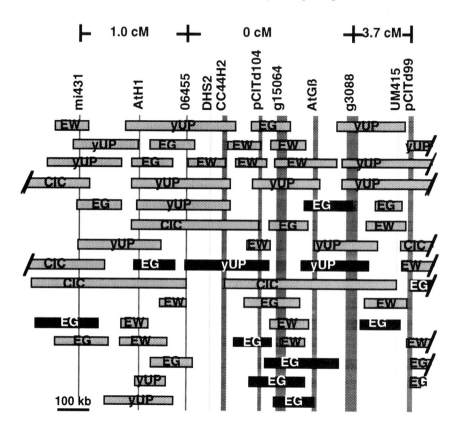

Figure 1.1 Representation of a YAC contig. A section of a YAC contig located on the long arm of chromosome 4 is shown (redrawn after Schmidt *et al.*, 1995, 1996). The YACs shown in the contig originate from four different YAC libraries: CIC (Creusot *et al.*, 1995); EG (Grill and Somerville, 1991); EW (Ward and Jen, 1990); yUP (Ecker, 1990). All YACs are drawn to scale and chimaeric YACs are represented as black boxes. The markers that were used to identify and position the YAC clones are shown at the top of the drawing. Taking into account marker hybridisation data and the sizes of all YACs, the YACs could be positioned relative to each other and to the markers. A tentative location of the markers within the YACs is given by lines crossing the YACs. The approximate sizes of the markers (in kb) are reflected by the thickness of the lines. This representation of a YAC contig allows the physical distances between the markers used for the mapping experiments to be estimated. Recombinant inbred lines (Lister and Dean, 1993) were used for the genetic mapping experiments. The map distances given were calculated with the MAPMAKER programme (version 3.0) using the Kosambi mapping function.

A cloning system that allows the cloning of inserts as large as 95 kb was derived from the bacteriophage P1 (Sternberg, 1990; Pierce and Sternberg, 1992). High molecular weight *Arabidopsis* DNA prepared from nuclei was partially digested with *Sau*3A and cloned in the pAd10*sac*BII P1 vector (Pierce *et al.*, 1992). The resulting library consisted of 10080 clones with an average insert size of approximately 80 kb (Liu *et al.*, 1995), representing about eight genome equivalents (Table 1.1).

An F-factor-derived cloning system has been used to clone high molecular weight DNA. Inserts as large as 300 kb have been obtained for human DNA (Shizuya *et al.*, 1992). The resulting BACs are maintained as single-copy plasmids in *E. coli* and can easily be isolated using standard plasmid techniques.

Using BAC vectors, clone libraries from *Arabidopsis* high molecular weight DNA have been constructed (Table 1.1). The TAMU BAC library initially contained 3948 clones with an average insert size of 100 kb, thus covering approximately four genome equivalents (Choi *et al.*, 1995). The library presently consists of approximately 12 000 clones. For its construction, high molecular weight DNA isolated from nuclei was partially digested with *Hin*dIII and cloned into pBeloBACII. The library has been analysed for the frequency of several repetitive sequences. Approximately 1.9, 1.7 and 0.7% of the clones carried rDNA, chloroplast DNA and the 180 bp *Hin*dIII repeated DNA sequences, respectively (http://cbil.humgen.upenn.edu/~atgc/physical-mapping/BAC_data/repBACs.html and http://http.tamu.edu:8000/~creel/TOC.html).

The IGF BAC library consists of 10 752 clones, which exhibit a mean insert size of 100 kb (Altmann *et al.*, 1997). This library has been established by cloning *Eco*RI partially digested DNA into a derivative of pBeloBAC. Since the high molecular weight DNA was prepared from root suspension cultures only 20 clones, less than 0.2% of the total, have been found to contain chloroplast DNA in this library. The frequency of clones carrying rDNA and the 180 bp *Hin*dIII repeated sequences was 10.5 and 2.8%, respectively (http://cbil.humgen.upenn.edu/~atgc/physical-mapping/BAC_data/repBACs.html).

All these libraries are important resources for mapping experiments in *Arabidopsis*. The large insert sizes of the YAC clones are especially advantageous for building large contigs quickly, while the high redundancy of the BAC and P1 libraries are important for generating reliable contigs as substrates for genomic sequencing approaches.

1.4.3 Generation of insert fragments as probes in chromosome walking experiments

End fragments from insert sequences directly adjacent to the vector sequences can be used as probes in chromosome walking experiments to reveal overlapping clones. Different techniques are being exploited to generate such end fragments. The PCR is the basis for inverse PCR (IPCR) (Ochman *et al.*, 1988), thermal asymmetric interlaced PCR (TAIL PCR) (Liu and Whittier, 1995) and vectorette

PCR (Riley *et al.*, 1990), techniques that can be used to isolate both end fragments adjoining the vector sequences.

Plasmid rescue of insert fragments adjacent to one of the YAC vector ends exploits the fact that these vectors carry an origin of replication and an antibiotic resistance gene functional in *E. coli*. However, for pYAC4 and related vectors, which have been used to construct the *Arabidopsis* YAC libraries, it is only possible to adopt this strategy for the cloning of insert DNA sequences that are adjacent to the left arm of the YAC vector (Schmidt and Dean, 1993). Using this method, insert fragments adjacent the right end of BAC vector sequences can also be rescued (Woo *et al.*, 1994).

1.4.4 Screening of large insert size clone libraries

Large insert size clone libraries are screened using colony hybridisations or PCR. Colony hybridisations are a highly reliable method for identifying clones corresponding to a particular probe, especially if the screening is carried out with duplicate filters or with filters on which all clones have been plated twice. While the screening of P1 and BAC libraries can be carried out using standard *E. coli* colony hybridisation techniques, the screening of YAC libraries requires the removal of yeast cell walls prior to the colony hybridisation experiments (Coulson *et al.*, 1988; Schmidt and Dean, 1995). To minimise the number of filters that need to be screened, the libraries can be plated at high density using 96- or 384-prong replicators. If hybridisation techniques are employed, no sequence information about the probe is required.

Despite the high reliability of colony hybridisation results, the results are often confirmed using Southern blot analyses. Such experiments also show whether or not the clones span the entire probe. This information can be used to position the clones relative to each other and to the probes (Figure 1.1). Southern blot analyses of cut clone DNA can also reveal whether a particular probe identifies different members of a multigene family (Schmidt *et al.*, 1996, 1997).

If nucleotide sequence information for the sequence to be detected is available, the libraries can also be screened by PCR. To minimise the number of PCR reactions required to identify a clone containing a particular sequence unambiguously, pooling strategies are employed. The library to be screened is divided into pools, each consisting of several hundred clones derived from several microtiter plates (superpool). For each of these different superpools, pools are prepared to allow the identification of a particular clone, if all of them are screened. Pools can, for example, be prepared from clones corresponding to half a microtiter plate, the particular rows in the microtiter plates and the particular columns.

For the PCR screening of a library, a first set of PCR reactions is performed on all superpools. If a PCR fragment of the correct size is found in a particular superpool, a second set of PCR reactions is then carried out on all the pools that correspond to this superpool. The plate, column and row number of a positive clone can be determined unequivocally if PCR fragments of the correct size are found in one of the pools corresponding to half a microtiter plate, one of the row pools and one of the column pools. DNA is then isolated from the putative positive clone and tested individually to confirm the result. Using such a scheme only 30 PCR reactions are needed to establish and confirm the co-ordinate of an individual clone in the CIC–YAC library (Creusot *et al.*, 1995).

A method combining pooling and the dot hybridisation strategy has been developed for the P1 library: pools are prepared from all clones in the different microtiter plates and from all clones in particular well positions, then DNA from all these pools is isolated and dotted onto membranes, which can then be used in hybridisation experiments. The candidate clones revealed in such a primary screening are then analysed in a secondary screening experiment. This approach minimises the number of filters that needs to be screened (Liu *et al.*, 1995).

1.4.5   Status of mapping using BAC and YAC libraries

For the construction of a physical map of the *Arabidopsis* genome based on YAC clones, a map-based approach, rather than a random strategy, was chosen. Such an approach allowed the sharing of the mapping work between different laboratories.

RFLP markers (Chang *et al.*, 1988; Nam *et al.*, 1989; McGrath *et al.*, 1993; Liu *et al.*, 1996) were used as probes in hybridisation experiments to identify and anchor YAC clones on the different chromosomes. When the sequence information for microsatellite markers (Bell and Ecker, 1994), CAPS markers (Konieczny and Ausubel, 1993) and, most importantly, expressed sequence tags (ESTs) (Höfte *et al.*, 1993; Newman *et al.*, 1994) was available PCR-based screening experiments of the YAC libraries were also carried out.

The first experiments used 125 RFLP markers distributed over the five chromosomes and, mainly, the EG YAC library (Grill and Somerville, 1991). Despite the relatively low number of anchors and the comparatively small average insert size found for YAC clones in the EG YAC library, it could be estimated that the YAC contigs established by the 125 markers together accounted for approximately 30% of the *Arabidopsis* genome (Hwang *et al.*, 1991). Although the majority of the YAC clones were anchored by a single RFLP marker, 12 YAC contigs spanned multiple markers. In total, 296 YAC clones were identified and positioned.

To assess the feasibility of long-range walking experiments, a region encompassing 10 cM on chromosome 4, located between RFLP markers m210 and m226, was analysed. Two contiguous regions of approximately 1000 and 1300 kb could be built around seven RFLP markers; however, 60 end probes from 47 YAC clones derived from the EG, EW and yUP libraries were necessary to establish these contigs (Schmidt and Dean, 1993).

It can therefore be concluded that marker content mapping experiments of YAC clones efficiently allows coverage of large parts of the genome with YAC contigs, while chromosome walking experiments are comparatively much more labour intensive: gain from the latter in terms of genome coverage is rather limited.

The excellent genome coverage that can be achieved in marker content mapping experiments has been calculated by Ewens *et al.* (1991): for a genome of 100 Mb, the use of 500 anchors and 2300 clones with an average insert size of 250 kb would result in the construction of 180 contigs, with these contigs spanning 87% of the genome and having an average size of 550 kb. Furthermore, it was predicted that most of the contigs would be overlapping and that only 68 gaps with an average size of merely 200 kb would remain. At such a stage of a project it is useful to implement chromosome walking experiments since the efficiency of linking contigs by using more markers is diminished because most of the markers will be positioned in existing contigs rather than linking two adjacent contigs. Furthermore, because of the small size of the majority of the remaining gaps, most of them can easily be closed by using chromosome walking experiments.

For genome mapping in *Arabidopsis*, more than 672 markers (Chang *et al.*, 1988; Nam *et al.*, 1989; McGrath *et al.*, 1993; Liu *et al.*, 1996; compiled in Anderson, 1997) and two YAC libraries, CIC and yUP, are currently available, which have an average insert size as described by Ewens *et al.* (1991) or an even larger one. Therefore using all these resources should result in very good coverage (>87%) of the *Arabidopsis* genome with YAC contigs.

Many of the tools, markers and libraries used in genome mapping experiments are available through the *Arabidopsis* Biological Resource Center in Ohio (ABRC) (http://aims.cps.msu.edu/aims). Most importantly, the information that has been generated is present on a number of Web pages on the Internet (links to all relevant information are provided by URL http://genome-www.stanford.edu/Arabidopsis). Furthermore, much of the data has been submitted to the *Arabidopsis thaliana* database (AtDB) (http://genome-www.stanford.edu/Arabidopsis). This ensures the accessibility of the information obtained by mapping efforts.

Chromosome 1 is the largest *Arabidopsis* chromosome (Albini, 1994). The yUP and CIC YAC libraries have mainly been screened with markers mapping to this chromosome. The use of more than 200 markers and more than 130 YAC end fragments identified and positioned approximately 600 YAC clones on this chromosome and, so far, 33 YAC contigs have been established (Dewar *et al.*, 1996; http://cbil.humgen.upenn.edu/~atgc/physical-mapping/physmaps.html).

The YAC contig map of chromosome 2 is almost complete. Using 130 probes in total, and both marker hybridisation and chromosome walking experiments, 73 CIC YAC clones were identified, which together span the majority of the chromosome. The repetitive nature of some areas of chromosome 2 and the lack of certain sequences in the CIC library did not allow the construction of a single YAC contig; nevertheless, the clones could be arranged into only four contigs. The total size of the four contigs has been estimated to be approximately 13.3 Mb (Zachgo *et al.*, 1996; http://weeds.mgh.harvard.edu/goodman/c2.html).

Although initially 60% of chromosome 3* was estimated to be represented in 42 YAC contigs (reviewed in Goodman *et al.*, 1995; http://cbil.humgen.upenn. edu/~atgc/physical-mapping/physmaps.html), the number of YAC contigs on this chromosome has been substantially reduced by mapping efforts in France and Japan (D. Bouchez, personal communication; Sato *et al.*, 1997). In most cases the CIC YAC library was screened, but in areas where CIC YAC clones could not be identified, the yUP YAC library was also used. Utilising 261 probes, 12 YAC contigs were established that encompass 22.6 Mb and comprise 168 YAC clones (D. Bouchez, personal communication).

A total of 563 YAC clones, derived from all four YAC libraries and made from DNA of the Columbia ecotype, were positioned on chromosome 4 by using 141 markers and chromosome walking experiments. Four contigs were established, which represent more than 90% of the genetic map and together span approximately 17 Mb (Schmidt *et al.*, 1996; http://genome-www.stanford. edu/Arabidopsis/JIC-contigs.html or http://nasc.life.nott.ac.uk/JIC-contigs/JIC-contigs. html; Figure 1.2).

In addition, for the YAC contig map of chromosome 5*, the YAC clones from the EG and EW libraries have been screened, as well as the larger insert size clones of the CIC and yUP libraries. By using 142 molecular markers more than 600 YAC clones were arranged into 31 contigs, which together span approximately 25 Mb (Schmidt *et al.*, 1997; http://genome-www.stanford.edu/ Arabidopsis/JIC-contigs.html or http://nasc.life.nott.ac.uk/JIC-contigs/JIC-contigs. html). A comprehensive effort to link the remaining YAC contigs with BACs successfully reduced the number of contigs to 15 (C. Dean, personal communication). An effort in Japan used marker hybridisation experiments in combination with chromosome walking experiments to establish YAC contigs on chromosome 5. This approach established 10 YAC contigs comprising 127 CIC YAC clones, which together cover 26.8 Mb and equate to the majority of this chromosome (Sato *et al.*, 1997; http://www.kazusa.or.jp/arabi).

Approximately 65 YAC contigs have been established for the *Arabidopsis* genome. Many of the established YAC contigs are in excess of 1 Mb and some of the contigs are even larger than 5 Mb. So far, for none of the chromosomes could a YAC contig spanning the entire chromosome be established; nevertheless, very high coverage with few YAC contigs has been achieved for most of the chromosomes. Only on chromosome 1 do more than 30 YAC contigs remain to be joined; however, one can estimate that at this stage of the project most of these contigs are in close proximity or are overlapping. This has convincingly

---

* The YAC contig maps for chromosomes 3 and 5 have been updated (Camilleri *et al.*, 1998); http//genome-www.stanford.edu/Arabidopsis/Chr3-INRA; http//www.kazusa.or.jp/arabi

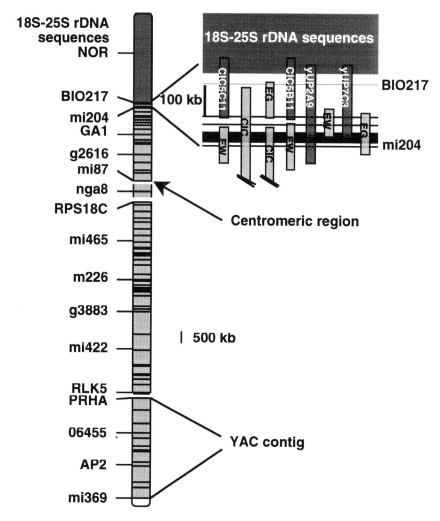

Figure 1.2 YAC contigs covering chromosome 4. The four YAC contigs covering the majority of chromosome 4 are shown as light grey boxes (redrawn after Schmidt *et al.*, 1995). The extent of the contigs is drawn to scale. All the genetically mapped markers that were used to identify and position the YACs on chromosome 4 are shown and their relative positions within the YAC contigs are indicated by lines crossing the chromosome. The designations for some of these markers are shown next to the chromosome map. The sizes of the gaps between the YAC contigs are not known, hence the gaps are not drawn to scale. A tentative location of the centromeric region is indicated. The 18S–25S rDNA sequences have been positioned distal to marker BIO217 and all known chromosome 4 markers. A section of the YAC contig covering this chromosomal region (redrawn after Schmidt *et al.*, 1996) is shown on the right-hand side. The approximate size of the NOR, represented by a dark grey box on the left-hand side, has been determined by two-dimensional mapping techniques involving a combination of pulsed-field and conventional gel electrophoresis (Copenhaver and Pikaard, 1996a).

been demonstrated on chromosome 5, since most of the 31 YAC contigs established by marker hybridisations could be successfully joined by using chromosome walking experiments and the linking of YAC contigs with BAC clones.

For genomic sequencing efforts, BAC clones are a more amenable and reliable substrate than YAC clones because of their smaller insert size and lower rate of chimaerism. Hence, large-scale mapping efforts using BAC clones are currently being carried out, either directed at certain areas of the genome or at the genome as a whole. For example, it is not expected that the chromosome 1 YAC contigs in chromosome walking experiments will be linked using YAC end fragments, instead the emphasis is on mapping BAC clones to this chromosome. This effort is focused primarily on regions of the chromosome currently being sequenced and so far more than 600 BAC clones have been placed in different areas of chromosome 1 (K. Dewar, personal communication; http://cbil.humgen. upenn.edu/~atgc/physical-mapping/physmaps.html). Using YAC sequences as probes to the BAC libraries, a large number of BACs have also been positioned on chromosome 4 (Bancroft and Bevan, 1997). These approaches are being complemented by a random strategy. Contigs located around the genome have been established using end fragments from several hundred BAC clones. All BAC efforts have so far established 359 BAC contigs, among which 157 putative overlaps exist. It has been predicted that these contigs should equate to a genome coverage of approximately 90% (Mozo *et al.*, 1997; http://194.94.225.1/private_ workgroups/pg_101/bac.html).

1.4.6 *Mapping sequences onto the YAC contig maps*

RFLP mapping can readily be used to map any sequence of interest to the five *Arabidopsis* chromosomes. However, this approach is limited if several hundred sequences need to be mapped. Because of the extensive coverage of the *Arabidopsis* chromosomes with YAC contigs, YAC mapping presents a promising alternative. Currently, the CIC YAC library is most useful for such an approach because contigs consisting of CIC YAC clones have been established for the majority of the genome. This library consists of only 1152 clones and shows a low frequency of chimaeric clones. The presence of chimaeric clones in the library means, however, that a reliable positioning of a sequence is achieved only with at least two YAC clones mapping to the same area identified. With the increasing coverage of the genome with BAC contigs, BAC libraries might in future also be exploited to establish map positions for sequences of interest.

Taking advantage of the YAC mapping approach, T-DNA and transposon-flanking sequences have successfully been mapped (Aarts *et al.*, 1995; Osborne *et al.*, 1995; Schmidt *et al.*, 1995; Long *et al.*, 1997). The versatility of this approach for large numbers of sequences has clearly been demonstrated by the mapping of several hundred EST sequences using either a PCR-based approach (Agyare *et al.*, 1997; http://genome-www.stanford.edu/Arabidopsis/EST2YAC.

html) or a hybridisation strategy (M. Stammers *et al.*, personal communication; http://genome-www.stanford.edu/Arabidopsis/EST2CIC.html).

1.5   Map-based cloning

Numerous genes controlling developmental and metabolic processes have been revealed by mutational analyses in *Arabidopsis* (Meyerowitz and Somerville, 1994). However, unless the mutation has been caused by the insertion of a known sequence, the isolation of the mutant loci is severely hampered because of the lack of knowledge about the biochemical nature of the gene products encoded by the mutant loci. In such cases, the cloning has to rely solely on the mutant phenotypes and the genetic-map positions. The high number of molecular markers which have been mapped onto the five chromosomes, the small genome size of approximately 100–120 Mb and the relatively low frequency of repetitive sequences facilitate map-based cloning experiments. Consequently, many such experiments have been initiated and several have been completed (e.g. Arondel *et al.*, 1992; Giraudat *et al.*, 1992). These experiments generate local physical maps at various regions around the genome and integrate the classical genetic with the molecular genetic and physical maps.

The general strategy for map-based cloning experiments in *Arabidopsis* is to generate or utilise a YAC contig covering the locus of interest, in combination with fine mapping of the locus within that YAC contig. The mutant locus to be cloned is initially mapped relative to other visible or molecular markers. The F2 progeny from a cross between two parents that are polymorphic at the DNA level is analysed with markers flanking the locus of interest to select recombination events close to the mutant locus. Subsequently, all available molecular markers mapping to the interval between the flanking markers are used to establish the genotypes of the selected recombinants. The analyses of the recombination break points in the various recombinants determines the map position of the locus of interest with respect to all the molecular markers.

YAC clones corresponding to the RFLP markers flanking the locus of interest are then identified by using the RFLP markers as probes on YAC libraries, if no YAC mapping information is found in the databases for these particular markers. In many cases, YAC clones will be found that span the flanking markers as well as the locus of interest. However, if the region of interest has not previously been covered, chromosome walking experiments have to be initiated. In this case, end fragments of YAC inserts are generated from YAC clones corresponding to the RFLP markers flanking the locus of interest. The end fragments are then hybridised to the YAC libraries to find overlapping YAC clones. To determine the direction of the walks initiated from the clones corresponding to the two flanking markers, the recombinants are analysed using the YAC end fragments as RFLP markers. Once the walks are orientated, they are extended towards the mutant locus in order to build a contiguous region covering the locus of interest.

As soon as the locus of interest has been mapped to a YAC or even a small part of it, DNA from this clone is subcloned into vectors, which can be used in plant transformation experiments. It is also possible to use the YAC as a probe to a library of *Arabidopsis* DNA cloned in a plant transformation vector. The resulting subclones are used in transformation experiments with the mutant *Arabidopsis* line to identify subclones that complement the mutant phenotype. The final identification of the open reading frame encoded by the locus of interest is achieved through additional rounds of complementation with smaller subclones in conjunction with sequence and transcript analyses.

The map-based cloning experiments aimed at cloning the *ABI3* and *ETR1* genes used cosmid libraries to construct contigs spanning the locus of interest (Giraudat *et al.*, 1992; Chang *et al.*, 1993). It was unnecessary to establish large YAC contigs spanning several hundred kb, since RFLP markers cosegregating or extremely closely linked with the locus of interest were available. In other experiments, an RFLP marker closely linked to the locus of interest could be identified and a YAC clone found that spanned the RFLP marker and the locus of interest, obviating the need for extensive chromosome walking efforts (Arondel *et al.*, 1992; Leyser *et al.*, 1993; Lukowitz *et al.*, 1996). However, in other areas of the genome, contigs encompassing more than 500 kb had to be established (Weigel *et al.*, 1992; Putterill *et al.*, 1993; Pepper *et al.*, 1994; Grant *et al.*, 1995; Sakai *et al.*, 1995; Vijayraghavan *et al.*, 1995; Busch *et al.*, 1996; Chapple *et al.*, 1996; Hardtke and Berleth, 1996; Li *et al.*, 1996; Macknight *et al.*, 1997).

The RFLP markers PG11 and m600 located on chromosome 4 were starting points for map-based cloning experiments aimed at cloning the *abi1* and *rps2* loci. Coincidentally, these loci were found to be in the immediate vicinity of each other (Bent *et al.*, 1994; Leung *et al.*, 1994; Meyer *et al.*, 1994; Mindrinos *et al.*, 1994).

The resources of the BAC libraries have also been exploited for map-based cloning experiments. For the isolation of the *LSD1* gene, for example, BACs were used in conjunction with YACs (Dietrich *et al.*, 1997).

Map-based cloning experiments are now greatly facilitated by the availability of YAC and/or BAC contigs for the majority of the genome and the wealth of molecular markers. If, however, only sparse marker and contig coverage is found in a region of interest, AFLP marker technology (Vos *et al.*, 1995) can be exploited to generate the necessary closely linked markers. These can then be used to identify the corresponding YAC or BAC clones (Cnops *et al.*, 1996).

1.6 Chromosome organisation

The organisation of *Arabidopsis* chromosomes 2 and 4 has been studied utilising the YAC contig maps. Both chromosomes carry an NOR, which cannot be covered by YAC contigs because of the instability of tandemly repeated sequences in YAC clones. However, YAC clones were found that contained rDNA sequences as well as single-copy sequences, allowing identification of the sequences adjacent to the rDNA loci. These results positioned one of the NORs distal of marker ve012 and all other known markers on chromosome 2, and the other NOR distal of marker BIO217 and all other known markers on chromosome 4 (Figure 1.2). This is in excellent agreement with data from physical and genetic mapping experiments that have identified the NORs in very close physical and genetic proximity to the telomeres of chromosomes 2 and 4, respectively (Copenhaver and Pikaard, 1996b). Hence, the localisation of the NORs with their associated telomeres provides end points for the genetic and physical maps of these two chromosomes.

The centromeric regions of chromosomes 2 and 4 (Figure 1.2) were positioned by identifying YAC clones that hybridised to the tandemly repeated 180 bp repeated DNA sequence (Martinez-Zapater *et al.*, 1986; Simoens *et al.*, 1988), which has been shown to co-localise with the heterochromatin surrounding the centromeres of all five *Arabidopsis* chromosomes (Bauwens *et al.*, 1991; Maluszynska and Heslop-Harrison, 1991; Murata *et al.*, 1994). However, owing to the high frequency of repetitive elements in these areas of the chromosomes, it has not been possible to bridge the centromeric regions entirely with YAC clones (Schmidt *et al.*, 1995; Zachgo *et al.*, 1996). Nevertheless, mapping of the rDNA loci and the repeated sequences flanking the centromeres resulted in the integration of the physical and cytogenetic maps of these chromosomes (Schmidt *et al.*, 1995; Zachgo *et al.*, 1996).

The high density of YAC clones and markers on chromosome 4 could be exploited to position all 563 YAC clones relative to each other and to the markers by taking into account marker and end-fragment hybridisation data as well as the sizes of all YAC clones. This analysis allowed the estimation of physical distances between the majority of chromosome 4 markers (Figure 1.1). As most of the markers used for the YAC mapping experiments have also been mapped genetically, a detailed comparison of physical to genetic distances could be carried out for this chromosome (Figure 1.1). This analysis showed the frequency of recombination to vary significantly, with relative hot and cold spots occurring along the whole of chromosome 4 (Schmidt *et al.*, 1995). A detailed analysis of one area of chromosome 5 also showed great variation when physical distances were compared to genetic distances, ranging from 100 to 720 kb/cM (Chapple *et al.*, 1996). The average value that could be established for the four chromosome 4 contigs was 185 kb/cM (Schmidt *et al.*, 1995), very similar to the average values of 190 and 200 kb/cM that have been found in two areas of chromosome 5 (Putterill *et al.*, 1993; Chapple *et al.*, 1996). In different regions of chromosome 1, average values of 200 and 400 kb/cM were observed (Vijayraghavan *et al.*, 1995; Hardtke and Berleth, 1996).

The clone contig maps can also be exploited to study the distribution of repetitive elements in a particular genomic region or chromosome. Various dispersed repetitive elements were analysed, only one of which, representing a diverged copy of the *Athila* retrotransposon (Péllissier *et al.*, 1995; Thompson *et al.*, 1996a), showed homology to known genes or proteins. The different repetitive elements were hybridised to YAC clones representing the majority of chromosome 4 and were found in a complex arrangement across the centromeric region but nowhere else on the chromosome (Schmidt *et al.*, 1995; Thompson *et al.*, 1996a). Interestingly, the analysis of sequences adjacent to the 180 bp repeated sequence, which co-localise with the paracentromeric heterochromatin, identified middle-repetitive sequences and, in most of the cases, *Athila* sequences (Péllissier *et al.*, 1996), confirming the centromeric distribution of such sequences. YAC hybridisation experiments suggest that at least some of the repetitive sequences are not only clustered in the centromeric region of chromosome 4 but also in at least some of the other centromeric regions (Péllissier *et al.*, 1996; Thompson *et al.*, 1996a; Thompson *et al.*, 1996b). This has been corroborated by *in situ* hybridisation experiments for at least one of the repetitive elements (Thompson *et al.*, 1996b).

Despite the successful use of repeated sequences in FISH experiments, it has only recently been possible to exploit this technique for the detection of low-copy DNA sequences in *Arabidopsis* (Murata and Motoyoshi, 1995). Resolution is limited by the small sizes of *Arabidopsis* chromosomes. However, the use of extended DNA fibres has dramatically increased the resolution of this method. It is now possible to establish the order of partially overlapping cosmid clones as well as the extent of the overlap. The degree of stretching of the DNA fibres was found to be approximately 3.27 kb/μm (Fransz *et al.*, 1996).

The instability of tandemly repeated sequences in most large insert clones means that FISH experiments will be a very important tool for studying regions of the genome that carry such sequences. This has been elegantly demonstrated by using YACs and repeated sequences, in addition to single-copy sequences, in FISH experiments for the construction of a pachytene map of an entire *Arabidopsis* chromosome arm (Fransz *et al.*, 1997). For the analysis of highly repeated DNA regions, such as the NORs and the centromeric regions, pulsed field gel electrophoresis experiments are a powerful tool and Southern blot analysis of two-dimensional gels has successfully been used to reveal the fine structure and size of such areas (Copenhaver and Pikaard, 1996a; Round *et al.*, 1997).

1.7 Comparative mapping

Because of the high conservation of gene sequences during evolution, it is possible to utilise RFLP markers derived from one species for genetic mapping experiments in closely related species. For example, for the closely related genera *Arabidopsis* and *Brassica* conservation in protein coding regions is on average 86% (Lydiate *et al.*, 1993).

Using the same set of RFLP markers for genetic mapping experiments in related species allows the construction of comparative genetic maps of these species. To reveal the degree of genome colinearity, comparative genetic mapping experiments have also been carried out in the family Brassicaceae. A first indication for conserved linkage arrangements was found by mapping a small number of genes in both *Arabidopsis thaliana* and *Brassica rapa* (Teutonico and Osborn, 1994). More extensive comparative mapping experiments using *A. thaliana* and *B. oleracea* revealed conserved regions extending from 3.7 to 49.6 cM; however, evidence for extensive rearrangements, translocations and inversions has also been found (Kowalski *et al.*, 1994). An analysis of 11 DNA fragments derived from a 1.5 Mb contig of *A. thaliana* chromosome 5 was used for fine-scale comparative mapping experiments with *B. nigra*. In three regions of *B. nigra* a colinear arrangement of the *Arabidopsis* probes could be established, consistent with the mainly triplicated nature of the *B. nigra* genome (Lagercrantz and Lydiate, 1996). However, in one of the *B. nigra* regions, a large chromosomal inversion was detected (Lagercrantz *et al.*, 1996). Mapping a set of five genes located on a 15 kb segment of *A. thaliana* chromosome 3 in *B. rapa*, *B. oleracea* and *B. nigra* showed that the gene sequences were clustered on a single linkage group in each of the three *Brassica* species. However, since almost no recombination within a particular gene cluster was observed, the gene order could not be determined and compared to the arrangement in *Arabidopsis*. The highly complex nature of the *Brassica* genomes was further unveiled by the presence of multiple copies of the *A. thaliana* gene cluster in the different *Brassica* genomes (Sadowski *et al.*, 1996).

Genome colinearity over short genetic distances was thus established by analysing different species of the Brassicaceae. Experiments aimed at establishing short-range colinearity between *Arabidopsis* and species belonging to different families and even taxa are now underway (Paterson *et al.*, 1996).

The information gained from comparative genetic mapping experiments, however, is limited to the gross chromosomal organisation. It now needs to be tested whether local gene order is conserved in a similar way to gene order over large chromosomal regions since this may open up map-based cloning to many more plant species, especially crop plants with large genomes. The genetic mapping of agronomically interesting loci would then be carried out in the species with the large genome and the cloning could be performed in *Arabidopsis*, for which clone contig maps are available. Furthermore, it could be determined whether the conservation of sequences between closely related species is restricted to gene sequences or whether it also extends to intergenic regions.

1.8   Conclusion

As part of the *Arabidopsis* genome project, several clone libraries carrying large inserts of *Arabidopsis* genomic DNA have been established using various vectors. Furthermore, more than 100 molecular mapped markers are now available for each of the *Arabidopsis* chromosomes. Exploiting these resources, clone contig maps have been constructed which cover the majority of the *Arabidopsis* genome.

YAC clones have been efficiently used to build large contigs that are anchored to the genetic map of the five chromosomes. However, because of their large size and high frequency of chimaerism, YAC clones are not as amenable for large-scale genomic sequencing experiments as BAC clones. Hence, BAC clones are increasingly being used to establish high-density physical maps, which are crucial for the ongoing effort to determine the entire nucleotide sequence of the *Arabidopsis* genome.

The extensive coverage of the genome with contigs allows the clone libraries to be used efficiently to map sequences to the *Arabidopsis* chromosomes. This strategy offers an important alternative to RFLP mapping experiments.

The organisation of the *Arabidopsis* genome has also been studied using the YAC contig maps, revealing that several classes of middle-repetitive elements are clustered in centromeric regions. The genetic and physical maps can readily be integrated and a comparison of physical and genetic distances on chromosome 4 showed considerable variation in the frequency of recombination in different regions of the chromosomes. Furthermore, for two of the chromosomes an integration of the cytogenetic and physical maps was achieved. Comparative mapping studies are also increasingly being used to exploit the information generated in the *Arabidopsis* genome project for the study of closely related genomes.

Efficient accessibility of the mapping information is guaranteed owing to the compilation of data on a large number of Web pages on the Internet as well as in AtDB. Hence, in the future, gene-cloning experiments should be greatly facilitated and no longer be a major limiting factor in the dissection of complex plant processes.

Acknowledgements

I thank Drs David Bouchez (INRA, Versailles, France), Caroline Dean (John Innes Centre, Norwich, UK) and Ken Dewar (University of Pennsylvania, Philadelphia, USA) for communicating data prior to publication. Drs John Chandler and David Flanders are gratefully acknowledged for commenting on the manuscript.

References

Aarts, M.G., Dirkse, W.G., Stiekema, W.J. and Pereira, A. (1993) Transposon tagging of a male sterility gene in *Arabidopsis*. *Nature*, **363** 715-17.

Aarts, M.G.M., Corzaan, P., Stiekema, W.J. and Pereira, A. (1995) A two-element *Enhancer-Inhibitor* transposon system in *Arabidopsis thaliana. Mol. Gen. Genet.*, **247** 555-64.

Agyare, F.D., Lashkari, D.A., Lagos, A., Namath, A.F., Lagos, G., Davis, R.W. and Lemieux, B. (1997) Mapping expressed sequence tag sites on yeast artificial chromosome clones of *Arabidopsis thaliana* DNA. *Genome Res.*, **7** 1-7.

Albini, S.M. (1994) A karyotype of the *Arabidopsis thaliana* genome derived from synaptonemal complex analysis at prophase I of meiosis. *Plant J.*, **5** 665-72.

Altmann, T., Mozo, T., Willmitzer, L., Meier-Ewert, S. and Lehrach, H. (1997) Physical mapping of the *Arabidopsis thaliana* genome using bacterial artificial chromosomes (BAC), in *Plant Genome V*, San Diego, CA. Abstract P207.

Ambros, P. and Schweizer, D. (1976) The Giemsa C banded karyotype of *Arabidopsis thaliana. Arabidopsis Inf. Serv.*, **13** 167-71.

Anderson, M. (ed.) (1997) The latest RI map using Lister and Dean RI lines. *Weeds World*, **4** 1-10.

Arondel, V., Lemieux, B., Hwang, I., Gibson, S., Goodman, H.M. and Somerville, C.R. (1992) Map-based cloning of a gene controlling omega-3 fatty acid desaturation in *Arabidopsis. Science*, **258** 1353-54.

Arumuganathan, K. and Earle, E.D. (1991) Nuclear DNA content of some important plant species. *Plant Mol. Biol. Rep.*, **9** 208-18.

Bancroft, I. and Bevan, M. (1997) The EU *Arabidopsis* genome sequencing programme, in *8th International Meeting on Arabidopsis Research*, Madison, Wisconsin, Abstract 12-4.

Bancroft, I., Jones, J.D.G. and Dean, C. (1993) Heterologous transposon tagging of the *DRL1* locus in *Arabidopsis. Plant Cell*, **5** 631-38.

Bauwens, S., Van Oostveldt, P., Engler, G. and Van Montagu, M. (1991) Distribution of the rDNA and three classes of highly repetitive DNA in the chromatin of interphase nuclei of *Arabidopsis thaliana. Chromosoma*, **101** 41-48.

Bell, C.J. and Ecker, J.R. (1994) Assignment of 30 microsatellite loci to the linkage map of *Arabidopsis. Genomics*, **19** 137-44.

Bennett, M.D. and Smith, J.B. (1976) Nuclear DNA amounts in angiosperms. *Phil. Trans. Roy. Soc. Lond.*, **274** 227-74.

Bent, A.F., Kunkel, B.N., Dahlbeck, D., Brown, K.L., Schmidt, R., Giraudat, J., Leung, J. and Staskawicz, B.J. (1994) *RPS2* of *Arabidopsis thaliana*: a leucine-rich repeat class of plant disease resistance genes. *Science*, **265** 1856-60.

Burke, D.T., Carle, G.F. and Olson, M.V. (1987) Cloning of large segments of exogenous DNA into yeast by means of artificial chromosome vectors. *Science*, **236** 806-12.

Busch, M., Mayer, U. and Jürgens, G. (1996) Molecular analysis of the *Arabidopsis* pattern formation of gene *GNOM*: gene structure and intragenic complementation. *Mol. Gen. Genet.*, **250** 681-91.

Camilleri, C., Lafleuriel, J., Macadré, C., Varoquaux, F., Parmentier, Y., Picard, G. , Caboche, M. and Bouchez, D. (1998) A YAC contig map of *Arabidopsis thaliana* chromosome 3. *Plant J.* (in press).

Campell, B.R., Song, Y., Posch, T.E., Cullis, C.A. and Town, C.D. (1992) Sequence and organization of 5S ribosomal RNA-encoding genes of *Arabidopsis thaliana. Gene*, **112** 225-28.

Chang, C., Bowman, J.L., DeJohn, A.W., Lander, E.S. and Meyerowitz, E.M. (1988) Restriction fragment length polymorphism linkage map for *Arabidopsis thaliana. Proc. Natl Acad. Sci. USA*, **85** 6856-60.

Chang, C., Kwok, S.F., Bleecker, A.B. and Meyerowitz, E.M. (1993) *Arabidopsis* ethylene-response gene *ETR1*: similarity of product to two-component regulators. *Science*, **262** 539-44.

Chapple, R.M., Chaudhury, A.M., Blömer, K.C., Farrell, L.B. and Dennis, E.S. (1996) Construction of a YAC contig of 2 megabases around the *MS1* gene in *Arabidopsis thaliana. Aust. J. Plant Physiol.*, **23** 453-65.

Choi, S., Creelman, R.A., Mullet, J.E. and Wing, R.A. (1995) Construction and characterization of a bacterial artificial chromosome library of *Arabidopsis thaliana. Plant Mol. Biol. Rep.*, **13** 124-28.

Chumakov, I., Rigault, P., Guillou, S., Ougen, P., Billaut, A., Guasconi, G., Gervy, P., LeGall, I., Soularue, P., Grinas, L., Bougueleret, L., Bellanné-Chantelot, C., Lacroix, B., Barillot, E., Gesnouin, P., Pook, S., Vaysseix, G., Frelat, G., Schmitz, A., Sambucy, J.-L., Bosch, A., Estivill, X., Weissenbach, J., Vignal, A., Riethman, H., Cox, D., Patterson, D., Gardiner, K., Hattori, M., Sakaki, Y., Ichikawa, H., Ohki, M., Le Paslier, D., Heilig, R., Antonarakis, S. and Cohen, D. (1992) Continuum of overlapping clones spanning the entire human chromosome 21q. *Nature*, **359** 380-86.

Cnops, G., den Boer, B., Gerarts, A., Van Montagu, M. and Van Lijsebettens, M. (1996) Chromosome landing at the *TORNADO1* locus using an AFLP-based strategy. *Mol. Gen. Genet.*, **253** 32-41.

Copenhaver, G.P. and Pikaard, C.S. (1996a) Two-dimensional RFLP analyses reveal megabase-sized clusters of rRNA gene variants in *Arabidopsis thaliana*, suggesting local spreading of variants as the mode for gene homogenization during concerted evolution. *Plant J.*, **9** 273-82.

Copenhaver, G.P. and Pikaard, C.S. (1996b) RFLP and physical mapping with an rDNA-specific endonuclease reveals that the nucleolus organizer regions of *Arabidopsis thaliana* adjoin the telomeres on chromosomes 2 and 4. *Plant J.*, **9** 259-72.

Copenhaver, G.P., Browne, W. and Preuss, D. (1997) Assaying genome-wide recombination and centromere functions with *Arabidopsis* tetrads. *Proc. Natl Acad. Sci. USA*, **95** 247-52.

Coulson, A., Sulston, J., Brenner, S. and Karn, J. (1986) Toward a physical map of the genome of the nematode *Caenorhabditis elegans*. *Proc Natl Acad. Sci. USA*, **83** 7821-25.

Coulson, A., Waterston, R., Kiff, J., Sulston, J. and Kohara, Y. (1988) Genome linking with yeast artificial chromosomes. *Nature*, **335** 184-86.

Coulson, A., Kozono, Y., Lutterbach, B., Shownkeen, R., Sulston, J. and Waterston, R. (1991) YACs and the *C. elegans* genome. *BioEssays*, **13** 413-17.

Creusot, F., Fouilloux, E., Dron, M., Lafleuriel, J., Picard, G., Billault, A., Le Paslier, D., Cohen, D., Chabouté, M.-E., Durr, A., Fleck, J., Gigot, C., Camilleri, C., Bellini, C., Caboche, M. and Bouchez, D. (1995) The CIC library: a large insert YAC library for genome mapping in *Arabidopsis thaliana*. *Plant J.*, **8** 763-70.

Dewar, K., Dunn, P.J., Li, Y.-P., Kim, C.J., Bouchez, D. and Ecker, J.R. (1996) Genetic and physical mapping of *Arabidopsis thaliana* chromosome 1, in *7th International Conference on Arabidopsis Research*. Norwich, UK. Abstract S45.

Dietrich, R.A., Richberg, M.H., Schmidt, R., Dean, C. and Dangl, J. (1997) A novel zinc finger protein is encoded by the *Arabidopsis LSD1* gene and functions as a negative regulator of plant cell death. *Cell*, **88** 685-94.

Ecker, J.R. (1990) PFGE and YAC analysis of the *Arabidopsis* genome. *Methods*, **1** 186-94.

Ewens, W.J., Bell, C.J., Donnelly, P.J., Dunn, P., Matallana, E. and Ecker, J.R. (1991) Genome mapping with anchored clones: theoretical aspects. *Genomics*, **11** 799-805.

Feldmann, K.A., Malmberg, R.L. and Dean, C. (1994) Mutagenesis in *Arabidopsis*, in *Arabidopsis* (eds. E.M. Meyerowitz and C.R. Somerville), Cold Spring Harbor Laboratory Press, Cold Spring Harbor, New York, pp. 137-72.

Foote, S., Vollrath, D., Hilton, A. and Page, D. (1992) The human Y chromosome: overlapping DNA clones spanning the euchromatin region. *Science*, **258** 60-66.

Fransz, P.F., Alonso-Blanco, C., Liharska, T.B., Peeters, A.J.M., Zabel, P. and de Jong, J.H. (1996) High-resolution physical mapping in *Arabidopsis thaliana* and tomato by fluorescence in situ hybridization to extended DNA fibers. *Plant J.*, **9** 421-30.

Fransz, P., Armstrong, S. and Jones, G. (1997) The short arm of chromosome 4 of *Arabidopsis thaliana* reconstructed by FISH using repeats, single copy DNA clones and YACs, in *Plant and Animal Genome V*, San Diego, CA, Abstract P215.

Franzmann, L.H., Yoon, E.S. and Meinke, D.W. (1995) Saturating the genetic map of *Arabidopsis thaliana* with embryonic mutations. *Plant J.*, **7** 341-50.

Giraudat, J., Hauge, B.M., Valon, C., Smalle, J., Parcy, F. and Goodman, H.M. (1992) Isolation of the *Arabidopsis* ABI3 gene by positional cloning. *Plant Cell*, **4** 1251-61.

Goodman, H.M., Ecker, J.R. and Dean, C. (1995) The genome of *Arabidopsis thaliana*. *Proc. Natl Acad. Sci. USA*, **92** 10831-35.

Grant, M.R., Godiard, L., Straube, E., Ashfield, T., Lewald, J., Sattler, A., Innes, R.W. and Dangl, J.L. (1995) Structure of the *Arabidopsis RPM1* gene enabling dual specificity disease resistance. *Science*, **269** 843-46.

Grill, E. and Somerville, C. (1991) Construction and characterization of a yeast artificial chromosome library of *Arabidopsis* which is suitable for chromosome walking. *Mol. Gen. Genet.*, **226** 484-90.

Gruendler, P., Unfried, I., Poitner, R. and Schweizer, D. (1989) Nucleotide sequence of the 25S-18S ribosomal gene spacer from *Arabidopsis thaliana*. *Nucl. Acids Res.*, **17** 6395-96.

Gruendler, P., Unfried, I., Pascher, K. and Schweizer, D. (1991) rDNA intergenic region from *Arabidopsis thaliana*: structural analysis, intraspecific variation and functional implications. *J. Mol. Biol.*, **221** 1209-22.

Hardtke, C. and Berleth, T. (1996) Genetic and contig map of a 2200 kb region encompassing 5.5cM on chromosome 1 of *Arabidopsis thaliana*. *Genome*, **39** 1086-92.

Hauge, B.M., Hanley, S., Giraudat, J. and Goodman, H.M. (1991) Mapping the *Arabidopsis* genome, in *Molecular Biology of Plant Development* (eds. G.I. Jenkins and W. Schuch), The Company of Biologists, Cambridge, pp. 45-56.

Hauge, B.M., Hanley, S.M., Cartinhour, S., Cherry, J.M., Goodman, H.M., Koornneef, M., Stam, P., Chang, C., Kempin, S., Medrano, L. and Meyerowitz, M. (1993) An integrated genetic/RFLP map of the *Arabidopsis thaliana* genome. *Plant J.*, **3** 745-54.

Heslop-Harrison, J.S. and Schwarzacher, T. (1990) The ultrastructure of *Arabidopsis thaliana* chromosomes, in *Fourth International Conference on Arabidopsis Research*, Vienna. Abstract p. 11.

Höfte, H., Desprez, T., Amselem, J., Chiapello, H., Rouzé, P., Caboche, M., Moison, A., Jourjon, M.-F., Charpenteau, J.-L., Berthomieu, P., Guerrier, D., Giraudat, J., Quigley, F., Thomas, F., Yu, D.-Y., Mache, R., Raynal, M., Cooke, R., Grellet, F., Delseny, M., Parmentier, Y., de Marcillac, G., Gigot, C., Fleck, J., Philipps, G., Axelos, M., Bardet, C., Tremousaygue, D. and Lescure, B. (1993) An inventory of 1152 expressed sequence tags obtained by partial sequencing of cDNAs from *Arabidopsis thaliana*. *Plant J.*, **4** 1051-61.

Hwang, I., Kohchi, T., Hauge, B.M., Goodman, H., Schmidt, R., Cnops, G., Dean, C., Gibson, S., Iba, K., Lemieux, B., Arondel, V., Danhoff, L. and Somerville, C. (1991) Identification and map position of YAC clones comprising one-third of the *Arabidopsis* genome. *Plant J.*, **1** 367-74.

Konieczny, A. and Ausubel, F. (1993) A procedure for quick mapping of *Arabidopsis* mutants using ecotype specific markers. *Plant J.*, **4** 403-10.

Konieczny, A., Voytas, D.F., Cummings, M.P. and Ausubel, F.M. (1991) A superfamily of *Arabidopsis thaliana* retrotransposons. *Genetics*, **127** 801-809.

Koornneef, M. and Hanhart, C.J. (1983) Linkage marker stocks of *Arabidopsis thaliana*. *Arabidopsis Inf. Serv.*, **20** 89-92.

Koornneef, M. and van der Veen, J.H. (1983) Trisomics in *Arabidopsis thaliana* and the location of linkage groups. *Genetica*, **61** 41-46.

Koornneef, M., van Eden, J., Hanhart, C.J., Stam, P., Braaksma, F.J. and Feenstra, F.J. (1983) Linkage map of *Arabidopsis thaliana*. *J. Hered.*, **74** 265-72.

Koornneef, M., Hanhardt, C.J., van Loenen-Martinet, E.P. and van der Veen, J.H. (1987) A marker line, that allows the detection of linkage on all Arabidopsis chromosomes. *Arabidopsis Inf. Serv.*, **23** 46-50.

Kowalski, S.P., Lan, T.-H., Feldmann, K.A. and Paterson, A.H. (1994) Comparative mapping of *Arabidopsis thaliana* and *Brassica oleracea* chromosomes reveals islands of conserved organization. *Genetics*, **138** 499-510.

Lagercrantz, U. and Lydiate, D. (1996) Comparative genome mapping in Brassica. *Genetics*, **144** 1903-10.

Lagercrantz, U., Putterill, J., Coupland, G. and Lydiate, D. (1996) Comparative mapping in *Arabidopsis* and *Brassica*, fine scale genome collinearity and congruence of genes controlling flowering time. *Plant J.*, **9** 13-20.

Laibach, F. (1907) Zur Frage nach der Individualität der Chromosomen im Pflanzenreich. *Beih. Bot. Zentralbl.*, **22** 191-200.

Lee-Chen, S. and Bürger, D. (1967) The location of linkage groups on the chromosomes of *Arabidopsis* by the trisomic method. *Arabidopsis Inf. Serv.*, **4** 4-5.

Lee-Chen, S. and Steinitz-Sears, L.M. (1967) The location of linkage groups in *Arabidopsis thaliana*. *Can. J. Genet. Cyt.*, **9** 381-84.

Leung, J., Bouvier-Durand, M., Morris, P.C., Guerrier, D., Chefdor, F. and Giraudat, J. (1994) *Arabidopsis* ABA response gene *ABI1*: features of a calcium-modulated protein phosphatase. *Science*, **264** 1448-52.

Leutwiler, L.S., Hough-Evans, B.R. and Meyerowitz, E.M. (1984) The DNA of *Arabidopsis thaliana*. *Mol. Gen. Genet.*, **194** 15-23.

Leyser, H.M., Lincoln, C.A., Timpte, C., Lammer, D., Turner, J. and Estelle, M. (1993) *Arabidopsis* auxin-resistance gene *AXR1* encodes a protein related to ubiquitin-activating enzyme E1. *Nature*, **364** 161-64.

Li, J., Nagpal, P., Vitart, V., McMorris, T.C. and Chory, J. (1996) A role for brassinosteroids in light-dependent development of *Arabidopsis*. *Science*, **272** 398-401.

Lister, C. and Dean, C. (1993) Recombinant inbred lines for mapping RFLP and phenotypic markers in *Arabidopsis thaliana*. *Plant J.*, **4** 745-50.

Liu, Y.-G. and Whittier, R.F. (1995) Thermal asymmetric interlaced PCR: automatable amplification and sequencing of insert end fragments from P1 and YAC clones for chromosome walking. *Genomics*, **25** 674-81.

Liu, Y.-G., Mitsukawa, N., Vazquez-Tello, A. and Whittier, R.F. (1995) Generation of a high-quality P1 library of *Arabidopsis* suitable for chromosome walking. *Plant J.*, **7** 351-58.

Liu, Y.-G., Mitsukawa, N., Lister, C., Dean, C. and Whittier, R.F. (1996) Isolation and mapping of a new set of 129 RFLP markers in *Arabidopsis thaliana* recombinant inbred lines. *Plant J.*, **10** 733-36.

Long, D., Martin, M., Sundberg, E., Swinburne, J., Puangsomlee, P. and Coupland, G. (1993) The maize transposable element system *Ac/Ds* as a mutagen in *Arabidopsis*: identification of an *albino* mutation induced by *Ds* insertion. *Proc Natl Acad. Sci. USA*, **90** 10370-74.

Long, D., Goodrich, J., Wilson, K., Sundberg, E., Martin, M., Puangsomlee, P. and Coupland, G. (1997) *Ds* elements on all five *Arabidopsis* chromosomes and assessment of their utility for transposon tagging. *Plant J.*, **11** 145-48.

Lukowitz, W., Mayer, U. and Jürgens, G. (1996) Cytokinesis in the *Arabidopsis* embryo involves the syntaxin-related KNOLLE gene product. *Cell*, **84** 61-71.

Luschnig, C., Bachmair, A. and Schweizer, D. (1993) Intraspecific length heterogeneity of the rDNA-IGR in *Arabidopsis thaliana* due to homologous recombination. *Plant Mol. Biol.*, **22** 543-45.

Lydiate, D., Sharpe, A., Lagercrantz, U. and Parkin, I. (1993) Mapping the *Brassica* genome. *Outlook Agric.*, **22** 85-92.

Macknight, R., Bancroft, I., Page, T., Lister, C., Schmidt, R., Love, K., Westphal, L., Murphy, G., Sherson, S., Cobbett, C. and Dean, C. (1997) *FCA*, a gene controlling flowering time in *Arabidopsis*, encodes a protein containing RNA-binding domains. *Cell*, **89** 737-45.

Maluszynska, J. and Heslop-Harrison, J.S. (1991) Localization of tandemly repeated DNA sequences in *Arabidopsis thaliana*. *Plant J.*, **1** 159-66.

Martinez-Zapater, J.M., Estelle, M.A. and Somerville, C.R. (1986) A highly repeated DNA sequence in *Arabidopsis thaliana*. *Mol. Gen. Genet.*, **204** 417-23.

McGrath, J.M., Jancso, M.M. and Pichersky, E. (1993) Duplicate sequences with a similarity to expressed genes in the genome of *Arabidopsis thaliana*. *Theor. Appl. Genet.*, **86** 880-88.

McKelvie, A.D. (1965) Preliminary data on linkage groups in Arabidopsis. *Arabidopsis Inf. Serv.*, **1** (suppl.) 79-84.

Meinke, D., Caboche, M., Dennis, E., Flavell, R., Goodman, H., Jürgens, G., Last, R., Martinez-Zapater, J., Mulligan, B., Okada, K. and Van Montagu, M. (1995) *The Multinational Coordinated Arabidopsis thaliana Genome Research Project. Progress Report: Year Five*, National Science Foundation, Arlington, VA.

Meyer, K., Leube, M.P. and Grill, E. (1994) A protein phosphatase 2C involved in ABA signal transduction in *Arabidopsis thaliana. Science*, **264** 1452-55.

Meyerowitz, E.M. and Somerville, C.R. (eds.) (1994) *Arabidopsis*, Cold Spring Harbor Laboratory Press, Cold Spring Harbour, New York.

Mindrinos, M., Katagiri, F., Yu, G.L. and Ausubel, F.M. (1994) The *A. thaliana* disease resistance gene RPS2 encodes a protein containing a nucleotide-binding site and leucine-rich repeats. *Cell*, **78** 1089-99.

Mozo, T., Lehrach, H., Meier-Ewert, S., Willmitzer, L. and Altmann, T. (1997) Physical mapping of the *Arabidopsis thaliana* genome using bacterial artificial chromosomes (BAC), in *8th International Meeting on Arabidopsis Research*, Madison, Wisconsin, Abstract 12-2.

Murata, M. and Motoyoshi, F. (1995) Floral chromosomes of *Arabidopsis thaliana* for detecting low-copy DNA sequences by fluorescence in situ hybridization. *Chromosoma*, **104** 39-43.

Murata, M., Ogura, Y. and Motoyoshi, F. (1994) Centromeric repetitive sequences in *Arabidopsis thaliana. Jpn. J. Genet.*, **69** 361-70.

Nam, H.-G., Giraudat, J., den Boer, B., Moonan, F., Loos, W.D.B., Hauge, B.M. and Goodman, H. (1989) Restriction fragment length polymorphism linkage map of *Arabidopsis thaliana. Plant Cell*, **1** 699-705.

Newman, T., de Bruijn, F.J., Green, P., Keegstra, K., Kende, H., McIntosh, L., Ohlrogge, J., Raikhel., N., Somerville, S., Thomashow, M., Retzel, E. and Somerville, C. (1994) Genes galore: a summary of methods for accessing results from large-scale partial sequencing of anonymous Arabidopsis cDNA clones. *Plant Physiol.*, **106** 1241-55.

Ochman, H., Gerber, A.S. and Hartl, D.L. (1988) Genetic application of an inverse polymerase chain reaction. *Genetics*, **120** 621-23.

Osborne, B.I., Wirtz, U. and Baker, B. (1995) A system for insertional mutagenesis and chromosomal rearrangement using the Ds transposon and Cre-*lox. Plant J.*, **7** 687-701.

Paterson, A.H., Lan, T.-H., Reischmann, K.P., Chang, C., Lin, Y.-R., Liu, S.-C., Burow, M.D., Kowalski, S., Katsar, C.S., DelMonte, T.A., Feldmann, K.A., Schertz, K.F. and Wendel, J.F. (1996) Toward a unifying genetic map of higher plants, transcending the monocot-dicot divergence. *Nature Genet.*, **14** 380-82.

Peleman, J., Cottyn, B., Van Camp, W., Van Montagu, M. and Inzé, D. (1991) Transient occurence of extrachromosomal DNA of an *Arabidopsis thaliana* transposon-like element, *Tat 1. Proc. Natl Acad. Sci. USA*, **88** 3618-22.

Péllissier, T., Tutois, S., Deragon, J.M., Tourmente, S., Genestier, S. and Picard, G. (1995) *Athila*, a new retroelement from *Arabidopsis thaliana. Plant Mol. Biol.*, **29** 441-52.

Péllissier, T., Tutois, S., Tourmente, S., Deragon, J.M. and Picard, G. (1996) DNA regions flanking the major *Arabidopsis thaliana* satellite are principally enriched in *Athila* retroelement sequences. *Genetica*, **97** 141-51.

Pepper, A., Delaney, T., Washburn, T., Poole, D. and Chory, J. (1994) *DET1*, a negative regulator of light-mediated development and gene expression in Arabidopsis, encodes a novel nuclear-localized protein. *Cell*, **78** 109-16.

Pierce, J.C. and Sternberg, N. (1992) Using the bacteriophage P1 system to clone high molecular weight (HMW) genomic DNA, in *Methods in Enzymology*, Vol. 216 (ed. R. Wu), Academic Press, San Diego, pp. 549-74.

Pierce, J.C., Sauer, B. and Sternberg, N. (1992) A positive selection vector for cloning high molecular weight DNA by the bacteriophage P1 system: improved cloning efficacy. *Proc. Natl Acad. Sci. USA*, **89** 2056-60.

Pruitt, R.E. and Meyerowitz, E.M. (1986) Characterization of the genome of *Arabidopsis thaliana. J. Mol. Biol.*, **187** 169-84.

Putterill, J., Robson, F., Lee, K. and Coupland, G. (1993) Chromosome walking with YAC clones in *Arabidopsis*: isolation of 1700 kb of contiguous DNA on chromosome 5, including a 300 kb region containing the flowering-time gene *CO. Mol. Gen. Genet.*, **239** 145-57.

Rédei, G.P. and Hirono, Y. (1964) Chromosome studies in *Arabidopsis thaliana. Arabidopsis Inf. Serv.*, **1** 9-10.

Reiter, R.S., Williams, J.G.K., Feldmann, K.A., Rafalski, J.A., Tingey, S.V. and Scolnik, P.A. (1992) Global and local genome mapping in *Arabidopsis thaliana* by using recombinant inbred lines and random amplified polymorphic DNAs. *Proc. Natl Acad. Sci. USA*, **89** 1477-81.

Richards, E.J. and Ausubel, F.M. (1988) Isolation of a higher eukaryotic telomere from *Arabidopsis thaliana*. *Cell*, **53** 127-36.

Richards, E.J., Goodman, H.M. and Ausubel, F. (1991) The centromere region of *Arabidopsis thaliana* chromosome 1 contains telomere-similar sequences. *Nucl. Acids Res.*, **19** 3351-57.

Riley, J., Butler, R., Ogilvie, D., Finniear, R., Jenner, D., Powell, S., Anand, R., Smith, J.C. and Markham, A.F. (1990) A novel, rapid method for the isolation of terminal sequences from yeast artificial chromosome (YAC) clones. *Nucl. Acids Res.*, **18** 2887-90.

Round, E.K., Flowers, S.K. and Richards, E.J. (1997) *Arabidopsis thaliana* centromere regions—genetic-map positions and repetitive DNA structure. *Genome Res.*, **7**, 1045-53.

Sadowski, J., Gaubier, P., Delseny, M. and Quiros, C.F. (1996) Genetic and physical mapping in *Brassica* diploid species of a gene cluster defined in *Arabidopsis thaliana. Mol. Gen. Genet.*, **251** 298-306.

Sakai, H., Medrano, L. and Meyerowitz, E.M. (1995) Role of *SUPERMAN* in maintaining *Arabidopsis* floral whorl boundaries. *Nature*, **378** 199-203.

Sato, S., Kotani, H., Nakamura, Y., Asamizu, E., Kaneko, T., Fukami, M., Miyajima, N. and Tabata, S. (1997) Progress of the *Arabidopsis* genome sequencing project at Kazusa DNA research institute, in *8th International Meeting on Arabidopsis Research*, Madison, Wisconsin, Abstract 12-25.

Schmidt, R. and Dean, C. (1993) Towards construction of an overlapping YAC library of the *Arabidopsis thaliana* genome. *BioEssays*, **15** 63-69.

Schmidt, R. and Dean, C. (1995) Hybridization analysis of YAC clones. *Meth. Mol. Cell. Biol.*, **5** 309-18.

Schmidt, R., Putterill, J., West, J., Cnops, G., Robson, F., Coupland, G. and Dean, C. (1994) Analysis of clones carrying repeated DNA sequences in two YAC libraries of *Arabidopsis thaliana* DNA. *Plant J.*, **5** 735-44.

Schmidt, R., West, J., Love, K., Lenehan, Z., Lister, C., Thompson, H., Bouchez, D. and Dean, C. (1995) Physical map and organization of *Arabidopsis thaliana* chromosome 4. *Science*, **270** 480-83.

Schmidt, R., West, J., Cnops, G., Love, K., Balestrazzi, A. and Dean, C. (1996) Detailed description of 4 YAC contigs representing 17Mb of chromosome 4 of *Arabidopsis thaliana* ecotype Columbia. *Plant J.*, **9** 755-65.

Schmidt, R., Love, K., West, J., Lenehan, Z. and Dean, C. (1997) Description of 31 YAC contigs spanning the majority of *Arabidopsis thaliana* chromosome 5. *Plant J.*, **11** 563-72.

Schweizer, D., Ambros, P., Gründler P. and Varga F. (1987) Attempts to relate cytological and molecular chromosome data of *Arabidopsis thaliana* to its genetic linkage map. *Arabidopsis Inf. Serv.*, **25** 27-34.

Sears, L.M.S. and Lee-Chen, S. (1970) Cytogenetic studies *in Arabidopsis thaliana. Can. J. Genet. Cyt.*, **12** 217-23.

Shizuya, H., Birren, B., Kim, U.-J., Mancino, V., Slepak, T., Tachiiri, Y. and Simon, M. (1992) Cloning and stable maintenance of 300-kilobase-pair fragments of human DNA in *Escherichia coli* using an F-factor-based vector. *Proc. Natl Acad. Sci. USA*, **89** 8794-97.

Simoens, C.R., Gielen, J., Van Montagu, M. and Inzé, D. (1988) Characterization of highly repetitive sequences of *Arabidopsis thaliana*. *Nucl. Acids Res.,* **16** 6753-66.

Smyth, D.R. (1991) Dispersed repeats in plant genomes. *Chromosoma,* **100** 355-59.

Stam, P. (1993) Construction of integrated genetic linkage maps by means of a new computer package: JOINMAP. *Plant J.,* **3** 739-44.

Steinitz-Sears, L.M. (1963) Chromosome studies *in Arabidopsis thaliana*. *Genetics,* **48** 483-90.

Sternberg, N. (1990) A bacteriophage P1 cloning system for the isolation, amplification and recovery of DNA fragments as large as 100kbp. *Proc Natl Acad. Sci. USA,* **87** 103-107.

Teutonico, R.A. and Osborn, T.C. (1994) Mapping of RFLP and qualitative trait loci in *Brassica rapa* and comparison to the linkage maps of *B. napus, B. oleracea,* and *Arabidopsis thaliana*. *Theor. Appl. Genet.,* **89** 885-94.

Thompson, H.L., Schmidt, R. and Dean, C. (1996a) Identification and distribution of seven classes of middle-repetitive DNA in the *Arabidopsis thaliana* genome. *Nucl. Acids Res.,* **24** 3017-22.

Thompson, H.L., Schmidt, R., Brandes, A., Heslop-Harrison, J.S. and Dean, C. (1996b) A novel repetitive sequence associated with the centromeric regions of *Arabidopsis thaliana* chromosomes. *Mol. Gen. Genet.,* **253** 247-52.

Thompson, H.L., Schmidt, R. and Dean, C. (1996c) Analysis of the occurrence and nature of repeated DNA in an 850kb region of *Arabidopsis thaliana* chromosome 4. *Plant Mol. Biol.,* **32** 553-57.

Tsay, Y.-F., Frank, M.J., Page, T., Dean, C. and Crawford, N.M. (1993) Identification of a mobile endogenous transposon in *Arabidopsis thaliana*. *Science,* **260** 342-44.

Unfried, I. and Gruendler, P. (1990) Nucleotide sequence of the 5.8S and 25S rRNA genes and of the internal transcribed spacers from *Arabidopsis thaliana*. *Nucl. Acids Res.,* **18** 4011.

Unfried, I., Stocker, U. and Gruendler, P. (1989) Nucleotide sequence of the 18S rRNA gene from *Arabidopsis thaliana* Col0. *Nucl. Acids Res.,* **17** 7513.

Vijayraghavan, U., Siddiqi, I. and Meyerowitz, E. (1995) Isolation of an 800 kb contiguous DNA fragment encompassing a 3.5 cM region of chromosome 1 in *Arabidopsis* using YAC clones. *Genome,* **38** 817-23.

Vos, P., Hogers, R., Bleeker, M., Reijans, M., van de Lee, T., Hornes, M., Frijters, A., Pot, J., Peleman, J., Kuiper, M. and Zabeau, M. (1995) AFLP: a new technique for DNA fingerprinting. *Nucl. Acids Res.,* **23** 4407-14.

Voytas, D.F. and Ausubel, F.M. (1988) A copia-like transposable element family in *Arabidopsis thaliana*. *Nature,* **336** 242-44.

Ward, E.R. and Jen, G.C. (1990) Isolation of single-copy-sequence clones from a yeast artificial chromosome library of randomly-sheared *Arabidopsis thaliana* DNA. *Plant Mol. Biol.,* **14** 561-68.

Weigel, D., Alvarez, J., Smyth, D.R., Yanofsky, M.F. and Meyerowitz, E.M. (1992) *LEAFY* controls floral meristem identity in Arabidopsis. *Cell,* **69** 843-59.

Williams, J.G.K., Kubelik, A.R., Livak, K.J., Rafalski, J.A. and Tingey, S.V. (1990) DNA polymorphisms amplified by arbitrary primers are useful as genetic markers. *Nucl. Acids Res.,* **18** 6531-35.

Woo, S.-S., Jiang, J., Gill, B.S., Paterson, A.H. and Wing, R.A. (1994) Construction and characterization of a bacterial artificial chromosome library of *Sorghum bicolor*. *Nucl. Acids Res.,* **22** 4922-31.

Zachgo, E.A., Wang, M.L., Dewdney, J., Bouchez, D., Camilleri, C., Belmonte, S., Huang, L., Dolan, M. and Goodman, H.M. (1996) A physical map of chromosome 2 of *Arabidopsis thaliana*. *Genome Res.,* **6** 19-25.

2 Unravelling the genome by genome sequencing and gene function analysis

Willem J. Stiekema and Andy Pereira

2.1 Introduction

Arabidopsis thaliana (*Arabidopsis*) is an excellent model organism for the molecular genetic dissection of numerous complex plant biological processes. This is illustrated by the identification of thousands of mutants, involved in plant growth and development, addressing fundamental questions in plant physiology, biochemistry, cell biology and pathology. The identification of genes and their associated functions in biological processes of *Arabidopsis* contributes to a molecular understanding of analogous processes in a wide range of economically important plant species.

The basic structure of a plant genome will be described by the complete sequence of *Arabidopsis*, for which the low copy regions are expected to be complete by 2003 (Bevan *et al.*, 1997). This will be the reference point for future studies on plant gene function, expression and evolution. As summarised by the 1995 Multinational Coordinated *Arabidopsis thaliana* genome research project (http://www.nsf.gov:80/bio/pubs/nsf9643/start.htm), the estimated genome size of 100 Mb is predicted to accommodate approximately 20 000 genes, but the available characterised mutant collection is around 1000, suggesting that only about 1 in 20 genes will display an obvious mutant phenotype. This is consistent with the mutation frequencies observed in general random insertional mutagenesis experiments with T-DNA or transposons. Higher mutation frequencies can be obtained with specific and detailed screening methods, including the selection of conditional mutants. Approximately 500 mutants showing a lethal phenotype have been described in *Arabidopsis*. For most other non-plant model organisms, about 1 in 3 genes are required for viability (Miklos and Rubin, 1996), which suggests that in plants other mechanisms might be involved that minimise the recovery or selection of vital genes, and different methods for gene analysis are required to close the phenotype gap.

Nucleotide sequence information may lead to the identification of the protein product of a gene and suggest the general class to which the protein belongs, e.g. regulatory protein, enzyme, etc. The actual function of what the protein really does in the living cell can be described only by experimentation. In other genome projects, e.g. *Drosophila*, gene disruptions are an integral component of gene function analysis (Spradling *et al.*, 1995), but the discovery of numerous phenotypically silent genes induced the development of new techniques for their analysis. A number of strategies to discover the functions of genes are described here. Their systematic implementation to whole genome analysis, as done now in yeast (Oliver, 1996) will help to decipher the function of individual genes and the regulatory networks embodying biological function.

2.2   Genome sequencing

2.2.1   Genomics

The haploid genome of *Arabidopsis* contains approximately 100 Mb (megabases) of low copy DNA sequences (Stammers *et al.*, 1996) arranged on five chromosomes. Unlike other plant species it accommodates a remarkably low number of repeated sequences. Only approximately 10% of the genome has been shown to be highly repetitive while another 10% occurs as moderately repetitive DNA (Meyerowitz, 1992). Moreover, *Arabidopsis* contains a very low methylation status of only 4.6% (Leutwiler *et al.*, 1984).

Physical maps, in yeast artificial chromosome (YAC) and cosmid clones of *Arabidopsis* chromosome 4 (Schmidt *et al.*, 1995) and chromosome 2 (Zachgo *et al.*, 1996), are available while similar maps for the other three chromosomes are under construction (see chapter 1). Four YAC contigs of chromosome 4 have been assembled, covering approximately 17 Mb of mostly unique and low level repetitive DNA, which allows a straightforward approach to sequence the protein-encoding DNA of this chromosome. The most abundant highly repetitive 180 base pair *Arabidopsis* tandem repeated sequence, which is present 4000–6000 times across the genome, appears to be exclusively located at the centromeric region of chromosome 4. The boundaries of this centromeric region have been determined. There is also in this region a 500 base pair sequence, which is repeated 500 times, copies of 5S rDNA and six other highly repeated sequence, which could not be detected elsewhere on chromosome 4. At one of the telomeric ends of this chromosome a nucleolus organiser region (NOR) of about 3.5 Mb has been mapped and this contains tandemly repeated rDNA units.

These features, in addition to a suite of YAC, bacterial artificial chromosome (BAC), cosmid, lambda and cDNA libraries freely available to the scientific community, make *Arabidopsis* an excellent candidate for complete sequence analysis of its genome, with chromosome 4 as an appropriate beginning. To define the structure of transcribed genes, projects to identify the sequences of cognate mRNAs have also been initiated.

2.2.2   Genomic DNA sequencing

2.2.2.1   The ESSA project: 1993–1996
In 1993 a number of European laboratories set out on a collaborative *Arabidopsis*

genome sequencing project known as ESSA (European Scientists Sequencing *Arabidopsis*). The ESSA project was a pilot project that laid the foundation for sequencing the entire genome, which has been a global undertaking. The ESSA project resembled the Yeast Genome Sequencing Network and consisted of a trans-European network of laboratories contributing to the sequencing effort co-ordinated by Mike Bevan (John Innes Centre, Norwich, UK). Besides genomic sequencing, ESSA was also involved in random and cognate cDNA sequencing. The genomic sequencing network was organised in a number of teams respon-sible for DNA coordination, DNA informatics, joint sequencing of chromosome 4 and individual chromosome sequencing focused on specific genomic loci (Table 2.1). The DNA coordinators were responsible for the collection and pro-duction of the materials necessary for the joint sequencing team. YAC clones, which were part of the constructed YAC contig of chromosome 4 covering the *FCA* (Macknight *et al.,* 1997) and *AP2* regions, were subcloned into overlapping cosmids. These cosmid clones were analysed by digestion with two restriction enzymes and distributed to the laboratories that took part in joint sequencing of the *FCA* and *AP2* region. Two high quality *Arabidopsis* BAC clone libraries (Choi *et al.,* 1995; Altmann, unpublished) are now available to the network and are routinely used to obtain sequence-ready contigs spanning chromosome 4.

In a typical sequencing strategy, selected cosmid and BAC clones were sheared into 0.8 to 1.5 kb DNA fragments and shotgun cloned into *E. coli* plasmid vectors. The clones obtained were randomly sequenced and the raw sequence data assembled into a final cosmid or BAC insert sequence using DNA assembly programs such as the Staden package. Data generated by the joint sequencers was subsequently sent to the DNA informatics coordinator at the Münich Institute for Protein Sequences (MIPS). MIPS was responsible for setting up and operating an electronic network linking all participants to a central node, as it did for the Yeast Genome Sequencing Network (Goffeau *et al.*, 1996). MIPS assembled and analysed the genomic sequences generated, and was responsible for the quality control of the sequences submitted. It is generally accepted that to identify potential coding regions at least 99.99% accuracy must be achieved (Dujon, 1996). For this purpose the submitted data was system-atically analysed for transmission errors, vector contamination, artifactual internal repeats and correspondence with the provided cosmid or BAC clone restriction maps. The accuracy of the individual sequencing laboratories was monitored and discrepancies detected by analysis of overlapping genomic DNA sequences and comparison with verified sequences. MIPS also examined the submitted sequence for homologues of predicted open reading frames (ORFs), introns, tRNA, rRNAs, regulatory elements and repeats. In addition, compari-sons to existing cDNA and EST (expressed sequenced tag) databases were made. MIPS will also establish an *Arabidopsis* genome database that contains all genomic DNA sequences obtained and its related information resulting from

Table 2.1 Achievements of the ESSA consortium

DNA coordination and bio-informatics	Region	Distribution or analysis
M. Bevan, John Innes Centre, Norwich, UK	Coordination	
I. Bancroft, John Innes Centre, Norwich, UK	FCA region	1 830 000
H.W. Mewes, MIPS, Martinsreid, Germany	FCA region	Analysis
Keygene, Wageningen, The Netherlands	AP2 region	300 000
Universiteit van Gent, Belgium	AP2 region	Analysis

FCA region

Laboratory	Base pairs
G. Murphy, John Innes Centre, Norwich, UK	381 185
W.J. Stiekema, CPRO-DLO, Wageningen, The Netherlands	325 709
T. Pohl, GATC, Konstanz, Germany	181 071
R. Wambutt, AGON, Berlin, Germany	159 290
N. Terryn, Rijksuniversiteit van Gent, Belgium	89 300
T. Kavanagh, Trinity College, Dublin, Ireland	85 994
M. Kreis, Université de Paris-Sud, Paris, France	85 220
K.-D. Entian, SRD, Oberursel, Germany	77 324
R. James, University of East Anglia, Norwich, UK	72 411
P. Puigdomenech, CSIC, Barcelona, Spain	71 358
M. Rieger, Biotechnologische und Molekularbiologische Forschung, Wilhemsfeld, Germany	70 600
P. Hatzopoulos, Agricultural University of Athens, Greece	65 956
B. Obermaier, MediGene, Martinsreid-München, Germany	52 623
A. Düsterhöft, Qiagen, Hilden, Germany	44 733
J. Jones, Sainsbury Laboratory, Norwich, UK	43 339
K. Palme, Max-Planck-Institut, Köln, Germany	33 600
V. Benes, EMBL, Heidelberg, Germany	29 600
M. Delseny, Université de Perpignan, Perpignan, France	19 400
Total	1 841 645

AP2 region

Laboratory	Base pairs
N. Terryn, Rijksuniversiteit van Gent, Belgium	143 000
Keygene, Wageningen, The Netherlands	156 839
Total	299 839

Individual regions

Laboratory	Region	Base pairs
N. Terryn, Rijksuniversiteit van Gent, Belgium	Pfl/ACC, chr.1	56 354
F. Schöffl, Universität Tubingen, Germany	HSTF, chr.1	24 840
M. Delseny, CNRS, Perpignan, France	EM, chr.3	28 964
R. Mache, Université J. Fourier, Grenoble, France	GAPA, chr. 3	75 286
B. Lescure, CNRS, Toulouse, France	EF-1, chr.1	63 693
Total		249 137

functional analysis of *Arabidopsis* genes. As a prototype MIPS is currently developing the Yeast Sequence Database (see http://www.mips.biochem.mpg.de). This database will permit restricted access to the owners of the data, to the consortium of ESSA sequencers and to the scientific community accessing published data.

2.2.2.2 Progress made by ESSA

Table 2.1 shows the progress achieved by the ESSA project by the end of 1996. An approximately 1.8 Mb contiguous sequence of the *FCA* region has been distributed by its DNA coordinator, of which 1.7 Mb of a non-redundant contiguous sequence has been determined. Of the *AP2* region 0.3 Mb has been distributed, all of which has been sequenced and submitted to MIPS. In addition, a number of laboratories have submitted 0.26 Mb of several other regions of the genome. An 80 kilobase (kb) region of chromosome 3 around the glyceraldehyde-3-phosphate dehydrogenase (GAPDH) gene and 34 kb of their cognate cDNAs have been determined and thoroughly analysed (Quigley *et al.*, 1996).

Comparison of genomic DNA and corresponding cognate cDNA sequences is regarded as very important in training the DNA sequence analysis programs in recognising plant genes. The detailed analysis of the 80 Kb region showed that 44% of this sequence accounts for 18 genes with an average length of 2300 bp (including intron and 3' non-coding sequence), implying a gene every 4500 bp. Ten of the 18 genes find significant matches with a variety of proteins from *Arabidopsis* or other organisms, while eight do not exhibit any similarity with database entries. No clustering of genes with related function or belonging to the same gene family could be detected, which is different from what is observed for the *FCA* region (see below). Intergenic spaces had a mean length of 2300 bp (from 327 to 5293 bp). Strikingly, genes appeared to be distributed unequally between the Watson and Crick strands, which had 15 and 3 genes respectively. Exons appeared to be less than 67 amino acids in length (from 12 to 991 amino acids), thus relatively small. At least four genes did not contain introns while 12 genes contained 72 introns in total, showing that the number of introns per gene varied from 0 to 26. Cognate cDNA and EST sequence data allowed precise definition of 45 donor and 47 acceptor intron sites giving CAG/GTAAGT... TKCAG/GTT as the consensus sequence in accordance with the consensus sequence determined from database entries. Several types of repeated sequences were present, while microsatellite repeat sequences were also found scattered throughout this region. Although the 80 kb region is relatively small, and may not reflect the *Arabidopsis* genome as a whole, its analysis gave us a first impression of what to expect.

2.2.2.3 Preliminary analysis of the FCA region.

The preliminary analysis of the contiguous DNA sequence in the *FCA* region (ESSA project, unpublished results) showed that, as in the 80 kb region, the information content in all areas sampled is very high, with genes and predicted genes occurring every 5 kb on average. This is similar to data obtained from sequencing 2.2 Mb of *Caenorhabditis elegans* genomic DNA (Wilson *et al.*, 1994; Hodgkin *et al.*, 1995). The intergenic regions varied from 0.1 to 10 kb. These figures indicate that *Arabidopsis* contains about 20 000 protein-coding genes. Repetitive regions comprise retroelements every 100 kb and clustered gene families. In the *FCA* region more than 300 putative or proven genes have been identified, of which 4% had been previously sequenced. Of the putative genes 40% have no known homologues, but 20% of this fraction matched an EST or cDNA sequence present in the databases. The other 60% of the putative genes are similar to genes described earlier in other organisms.

However, these figures have to be used with caution. In many cases identified and putative genes consist of multiple small exons that are less than 50 amino acids in size. To limit the number of ORFs to a manageable number, an arbitrary cut-off value of 100 amino acids was applied in the analysis of the obtained genomic sequence. Therefore there may be many more genes than identified by the computer programs and, as in yeast where a high percentage of ORFs showing no significant homology to genes in other organisms was found, we are clearly confronted with the limits of our knowledge. However, in the absence of homologues, computers can still provide clues about the nature of ORFs. For example, prediction of transmembrane segments in yeast, resulted in the striking conclusion that up to 35–40% of the predicted proteins from chromosome III have transmembrane helices (Goffeau *et al.*, 1993). In addition, 10% of all ORFs are predicted to contain more than four transmembrane helices and might represent transporters. Overall, however, the fraction of yeast proteins showing putative transmembrane segments is not significantly different between ORF encoded proteins and known gene products. Ultimately, the function of ORFs can only be demonstrated by functional analysis (see also below). Construction of a systematic transcript map might shed further light on their function. Such a map of yeast chromosome XI has shown that the distribution of transcript abundance and regulation is essentially the same for ORFs and other genes if highly expressed genes are excluded from the analysis (Richard *et al.*, 1997). It seems therefore that yeast ORFs are as frequently expressed as known genes.

A preliminary functional catalogue of the approximately 400 identified and putative genes in the *FCA* region is shown in Table 2.2. As expected, most of these genes play some role in metabolism, have a regulatory function or are structural proteins. Others are involved in DNA and RNA metabolism, translation, protein modification or (a)biotic stress resistance, and a number of tRNA genes have also been identified. These first results are in good agreement with what is found in yeast. About 50% (i.e. 3000) yeast genes have been classified on the basis of the amino acid sequence homology of the proteins they encode with other proteins of known function (Goffeau *et al.*, 1996). Significant homologies of 11% of the protein-encoding genes were found to genes devoted to metabolism, 3% to energy production and storage, 3% to DNA replication, repair and recombination, 7% to transcription and 6% to translation. About 430 genes code for proteins involved in intracellular trafficking or targeting of proteins to different cell compartments. A total of 250 genes encode proteins that have a structural role and nearly 200 transcription factor genes have been identified as well as 250 primary and secondary transporter genes.

2.2.2.4 Clustering of homologous genes

Surprisingly, copies of a number of putative genes appear to be clustered in the *FCA* region. Six genes encoding cytochrome oxidase P450 homologues are clustered in a region of only 60 kb, while two clusters of three glucosyl transferase genes each are present in less than 20 kb. In addition, heat-shock transcription factors are clustered in groups of 3–5 and four genes encoding CDR1 homologs are clustered in a region of 30 kb. Such multiple copies of a gene might arise because of the different functions of the encoded homologous proteins in the cell. For instance, one gene may encode a protein functional in the cytosol while its homologue might specify the same function in the chloroplasts, mitochondria or another cell compartment. Thus, much of the redundancy in the *FCA* region may be more apparent than real, and detailed experiments are required to determine the function of each putative gene.

Most interestingly, several copies of leucine rich repeats (LRR)-containing disease-resistance gene homologues were found to be clustered and located in the same genomic region where protein kinase encoding genes were present. Such an arrangement has also been found in tomato for the *PRF* gene, an LRR-containing disease-resistance gene and the *PTO* gene that codes for a protein kinase. All these genes are involved in resistance against *Pseudomonas syringae* in tomato (Salmeron *et al.,* 1996). Sequence analysis at CPRO-DLO of a BAC clone located at the *RPP7* locus on chromosome 1 showed that it contained five copies of an LRR-containing disease-resistance gene (Aarts *et al.,* unpublished). Moreover, BAC clone T02O04 located on chromosome 2 and sequenced by The Institute of Genomic Research (TIGR), was shown to contain not less than 10 isologues of a jasmonate inducible protein in a region of only 40 kb. In addition, TIGR detected a cluster of nine tRNA-Pro genes on BAC T01B08 of chromosome 2 (http://www.tigr.org).

Such clustering of homologous genes is not common in yeast. The only example is a cluster that occurs on chromosome 1, where six related but non-identical ORFs specify membrane proteins of unknown function. In addition, in the 20% of the *Caenorhabditis elegans* genome now sequenced, such clustering has not been discovered.

2.2.2.5 Genome sequencing in other laboratories

Besides the progress made by ESSA, other research institutes have started sequencing the *Arabidopsis* genome. TIGR has released six BAC clones to GenBank covering more than 0.6 Mb (http://www.tigr.org). A characteristic

Table 2.2 Preliminary classification and numbers of ORFs identified in the *FCA* region of chromosome 4 of *Arabidopsis thaliana*

Metabolism	ORF	Energy	ORF	Cell growth and division	ORF	Transcription	ORF
Amino acid	5	Glycolysis	2	Cell growth	2	rRNA synthesis	2
Nitrogen and sulphur	1	Gluconeogenesis	8	Meiosis	8	tRNA synthesis	—
Nucleotides	2	Pentose phosphate	—	DNA synthesis	—	mRNA synthesis	2
Phosphate	—	Trichloroacetic acid (TCA) pathway	—	Recombination	—	General transcription factors (TFs)	1
Carbohydrate	16	Respiration	1	Cell cycle	1	Specific TFs	18
Lipid and sterol	15	Fermentation	—	Cytokinesis	—	Chromatin modification	—
Cofactors	1	E—transport	—	Growth regulators	—	mRNA processing	7
		Photosynthesis	—	Other	—	RNA transport	—
Total	40	Total	11	Total	11	Total	30

Protein synthesis	ORF	Protein transport and storage	ORF	Transporters	ORF	Intracellular traffic	ORF
Ribosomal proteins	6	Folding and stability	6	Ions	1	Nuclear	—
Translation factors	—	Targeting	—	Sugars	—	Chloroplast	1
Translation control	—	Modification	—	Amino acids	—	Mitochondrial	—
tRNA synthesis	—	Complex assembly	1	Lipids	1	Vesicular	3
Others	—	Proteolysis	8	Purine/pyrimidines	—	Peroxisomal	—
		Storage proteins	—	Transport ATPases	—	Vacuolar	—
				ATP-binding cassette protein (ABC) type	1	Extracellular	—
				Others	—	Import	—
Total	6	Total	15	Total	3	Total	4

Table 2.2 (continued)

Cell structure	ORF	Signal transduction	ORF	Disease/defence	ORF	Secondary metabolism	ORF
Cell wall	8	Receptors	2	Resistance genes	15	Phenylpropanoids	10
Cytoskeleton	5	Mediators	1	Defence—related	1	Terpenoids	4
Endoplasmic reticulum/Golgi	—	Kinases	11	Cell death	—	Alkaloids	—
Nucleus	—	Phosphatases	1	Cell rescue	—	Non-protein amino acids	—
Chromosomes	1	G-proteins	—	Stress responses	10	Amines	—
Mitochondria	—	Others	2	Detoxification	2	Glucosinolates	1
Peroxisome	—					Others	3
Vacuole	—						
Chloroplast	—						
Total	14	Total	17	Total	28	Total	18

Transposons	ORF	Unclear classification	ORF	Unclassified	ORF
Long terminal repeat (LTR) retroelements	61	Number	10	Number	122
Non-LTR retroelements	—				
Other	—				
Total	61	Total	10	Total	122

feature of the sequence information is the accompanying annotation that makes it a user friendly interface. In total, 139 ORFs were discovered, of which 70% showed homology to genes present in the databases. Of these a remarkable 42% were identified as being likely to have a regulatory function. This figure might be biased by the choice of BACs used for sequencing. Nevertheless, the identification in BAC T06D20 of a region of 50 kb containing 12 genes encoding different kinases and DNA binding proteins is extremely fascinating. Approximately 14% of the identified genes encoded proteins involved in metabolism, 10% in DNA/RNA metabolism and translation, 4% in stress and resistance reactions, 5% in protein modification while 10% are membrane proteins or have some structural function. In addition 15 tRNA genes were identified. Apart from the high number of potential regulatory protein encoding genes this picture is comparable to the preliminary data obtained by ESSA.

Four completed BAC clones have been deposited by SPP, a consortium of Stanford University, the University of Pennsylvania and the Plant Gene Expression Laboratory (Berkeley), in GenBank (http://pgec-genome.pw.usda.gov/spp.html), while the sequence of two BAC clones has been finished by the Cold Spring Harbor Sequencing Consortium (CSHSC) (http://clio.cshl.org/arabweb/). Finally, the Kazusa DNA Research Institute (KI) has already finished the sequence of 27 P1 clones. These, however, have not yet been released to a public database (http://www.kazusa.or.jp/arabi/).

2.2.2.6 *The* Arabidopsis *Genome Initiative*

In 1996 the *Arabidopsis* Genome Initiative (AGI) was established (Bevan *et al.*, 1997). The AGI aims to facilitate the co-operation among *Arabidopsis* genomic sequencing projects in the USA, Japan and the European Union (EU) so that the sequence of the genome can be completed as rapidly and efficiently as possible. Table 2.3 shows the sequencing goals of the AGI participants. The EU *Arabidopsis* Genome Project (EUAGP) which accommodates 18 laboratories and is organised in the same way as the ESSA project, will sequence 5 Mb on chromosome 4 before the end of 1998. Over a three-year period CSHSC has agreed to sequence 6–7 Mb on chromosomes 4 and 5, while KI plans 15 Mb on chromosomes 3 and 5. In the same period, the SPP group will deliver 5 Mb of chromosome 1, TIGR 6–8 Mb of chromosome 2 and CNS (Centre National de Sequenage) 3 Mb per year of chromosome 3. Therefore in the coming years approximately 16 Mb of the *Arabidopsis* genomic sequence will be determined per year, i.e. more than 1 Mb per month. Stanford University (http://genome-www.stanford.edu/Arabidopsis) will further develop the *Arabidopsis* database (AtDB) where all the data generated by the AGI can be searched.

To reach their long-term goals all AGI partners will use the same BAC libraries (Choi *et al.*, 1995; T. Altmann, IGF, Berlin, Germany, unpublished) and P1 library (Liu *et al.*, 1995) for sequencing from now on. BAC and P1 libraries

Table 2.3 Research groups participating in the *Arabidopsis* Genome Initiative

Research group	Genome region	Sequencing scale	Information site
CSHSC[1]	4 and 5	7 Mb in three years	http://clio.cshl.org/arabweb
EUAGP[2]	4	5 Mb in two years	http://www.mips.biochem.mpg.de
KI[3]	3 and 5	15 Mb in three years	htttp://www.kasuza.or.jp
SPP[4]	1	5 Mb in three years	http://sequence-www.stanford.edu
TIGR[5]	2	6–8 Mb in three years	http://www.tigr.org
CNS[6]	3	3 Mb in one year	http://genethon.fr
Stanford University		Development of AtDB	http://genome-www.stanford.edu

[1]Cold Spring Harbor Sequencing Consortium.
[2]EU *Arabidopsis* Genome Project.
[3]Kasuza DNA Research Institute.
[4]Stanford University/University of Pennsylvania/Plant Gene Expression Laboratory consortium.
[5]The Institute of Genome Research; [6]Centre National de Sequenage.

are best suited for sequencing because of their low percentage of rearrangements and chimeras (see chapter 1). To assemble a minimum of overlapping clones for sequencing two complementary approaches are used to identify these tiling paths. TIGR, CNS and SPP use BAC-end sequencing for this purpose to create sequence-tagged connectors (STCs) which allow BAC walking, a new strategy to genome sequencing (Venter *et al.*, 1996). Over 2500 BAC-end sequences from chromosome 1 have been completed by SPP while TIGR determined the end-sequences of 1295 BAC clones covering chromosome 2. EUAGP and CSHSC use the individual YAC clones of the YAC tiling path already established for chromosome 4 and parts of chromosome 5 as hybridisation probes to select a minimal set of clones from the BAC libraries for sequencing. In addition, KI is sequencing P1 clones specifically hybridising with chromosome 5 probes.

2.2.3 cDNA sequencing

Another very productive area of sequence analysis of *Arabidopsis* has been the sequencing of random cDNA clones or expressed sequence tags (EST) (Höfte *et al.*, 1993; Newman *et al.*, 1994; Cooke *et al.*, 1996; Rounsley *et al.*, 1996). Within the framework of the ESSA project 1846 ESTs have been submitted by the French EST sequencing consortium. This was lower than anticipated because of the high redundancy in the cDNA libraries used. For the same reason the US EST sequencing initiatives also ended at the end of 1996.

In total more than 30 000 *Arabidopsis* ESTs have been submitted to dbEST (Bogulski *et al.*, 1993) and 1332 non-redundant *Arabidopsis* transcript sequence entries can be found in GenBank. All these ESTs and non-redundant transcript sequences were analysed by TIGR (www.tigr.org). The EST sequences, after inspection in a quality control process, were reduced to 28 676. Of these, 16 078

ESTs could be grouped into 4153 assemblies or contigs of approximately four-fold redundancy with 12 598 singletons remaining. Among the assemblies, names were assigned to 2031, or approximately 50%, on the basis of sequence similarity studies against protein and nucleotide databases. The matches were categorised in four different classes. About 20% (818) of the assemblies, repre-senting 5390 or 33% of the total redundant ESTs, showed an exact match against a known *Arabidopsis* coding sequence, while 7% (273) showed a non-exact match. A non-exact match to a coding sequence of another plant was detected for 15% (610) assemblies, while 8% (311) showed a non-exact match with non-plant sequences. Only 19 assemblies matched to various forms of contaminating sequences such as *E. coli* or rRNA sequences. No match was found for 2122 (51%) of all assemblies. Of the singletons 5766 (46%) showed a significant hit while 6832 (54%) did not show any homology to sequences present in the database. The analysis of the assemblies and singeltons suggests a maximum of about 16000 different ESTs in the database, representing 80% of the 20000 *Arabidopsis* genes. This is consistent with the proportion (80%) of EST repre-sented sequences among the 1332 *Arabidopsis* transcript sequences included in the assembly process. Considering that the *Arabidopsis* transcript sequences represent a random set of all *Arabidopsis* coding genes, extrapolation of this suggests that 80% of all genes present in *Arabidopsis* are already represented by an EST. It also means that *Arabidopsis* contains approximately 20000 genes, which is in good agreement with the current estimates (Meyerowitz, 1994).

Cooke *et al.* (1997) showed that the ESTs available in dbEST could be used to define unique tags to individual members of multigene families even when the sequences are highly conserved. A total of 106 genes encoding 50 different cytoplasmic ribosomal proteins could be uniquely tagged. Most of these proteins were encoded by two to four genes, confirming earlier findings that many proteins in *Arabidopsis* are encoded by multigene families (Snustad *et al.*, 1992). The existence of multigene families, now also resolved by genomic sequencing (see above), suggests caution in the evaluation of the EST information.

2.2.4 *Perspective of the* Arabidopsis *genomic sequence*

The availability of the complete *Arabidopsis* genome sequence will make it possible to answer scientific questions such as does transplicing exist in *Arabi-dopsis*, are there nested genes in introns of other genes and do operon systems exist in *Arabidopsis* as shown in *C. elegans*? What is the final G+C content and is the sequence biased against CG and TA dinucleotides as has been shown in yeast? Will the whole genome show the same compactness concerning the presence of genes as has been shown in the *FCA* region and the 80 kb genomic sequence analysed thus far or will the *Arabidopsis* genome show alternating gene poor, and gene rich regions, and will the subtelomeric region be gene poor as has been shown for yeast? How about the coding rates of the Watson and Crick strand? Will they be similar or be divergent as found in the 80 kb region? Will there be a similar long-range nucleotide composition with GC peaks in the middle of a chromosome arm as detected in yeast chromosome 3 or will there be more GC peaks per chromosome arm, as occurs in yeast chromosome 11? Further insight will be gained about the length of ORFs, intergenic regions, length of intron and exons, and the boundaries of introns and exons. The distribution of other genetic elements such as transposons, different sorts of repeats, micro-satellites, tRNA and rRNA as well as pseudogenes, will become clear. One of the most intriguing questions is how redundant will the *Arabidopsis* genome be and, if redundant, does this redundancy of genetic information have any biological function? When the complete sequence is known, it will finally become clear how many genes are devoted to the basic construction of a eucaryotic cell and how many are associated with specialisations that are unique to plants. It may be expected that large parts of the genome are performing universal activities so that much of what is learned from *Arabidopsis* will also apply to other organisms.

All these questions will also be studied in yeast and *C. elegans*, so answers should also come from the studies of these organisms. The preliminary analysis of the *FCA* region has already shown that *Arabidopsis* seems to contain unique features such as the clustering of homologous genes. It may be expected that more surprises will appear, making this huge and costly endeavour worthwhile. The cost of sequencing is currently estimated to be US $0.50–1.00 per base pair and may fall to half this amount in a couple of years using present technology. Rough calculations show that sequencing the low-copy regions of the *Arabidopsis* genome will cost US $50–75 million. However, at the Stanford DNA Sequencing and Technology Development Centre instrumentation is being developed to increase the rate of sequencing and reduce its cost. Throughput of shotgun sequencing is aimed at 1 Mb per day with a cost of US $0.01 per base (Marziali *et al.,* 1997). If such a speed was reality now, then the *Arabidopsis* genome could be sequenced before the turn of the century and for less than US$ 10 million. With such a throughput and low cost it is conceivable that not only the rice genome but also low-copy regions of the genomes of other important crop plants can be sequenced.

Notwithstanding the costs, knowledge of the complete *Arabidopsis* genomic sequence will change the lives of plant scientists. It will make it possible to identify orthologous genes in phylogenetically related crops like *Brassica* and also in unrelated crop species. This will shed light on the evolution of DNA sequences and gene function in plants in general and will also have an enormous impact on plant breeding. The genomic sequence of *Arabidopsis* will basically show how a plant is constructed and maintained, in the way that yeast is a model for the eucaryotic cell and *C. elegans* for a multicellular animal.

The design of molecular experiments as we know them will radically change. Laborious experiments aimed at the isolation and sequencing of genes that currently take up much of the time of a plant molecular biologist in the laboratory will become simple and easy to perform by computer desk research. To achieve this a lot of effort must be put into making all the data accessible from easy-to-use databases and integrating the bio-informatic tools in molecular biology.

2.3   Strategies for understanding gene function

The large amount of DNA sequence information from the *Arabidopsis* whole genome and EST sequencing projects will provide the basis for systematic gene function studies. Computer analysis of *Arabidopsis* sequences in combination with integrated databases can lay a foundation to predict gene function and help design appropriate experiments to unravel the biological function of discovered genes. Classic genetic experiments, including mutant phenotypes of single genes and combinations with interacting genes, combined with transgenic approaches to study gene expression and mis-expression, will help provide a complete picture of gene function in plant organisation.

To dissect a developmental, physiological or other biological process, mutants affecting the pathway can be isolated and interaction between them analysed to produce a model of gene interaction. This method has been traditionally approached with classical mutagens and now also by a variety of insertion sequences (Sundaresan, 1996). Insertion sequences that assay the expression patterns of adjacent genes provide a new dimension to *in situ* gene expression analysis. A view of gene activity can also be obtained by measuring the expression of panels of genes or by analysing the cell's protein complement or proteome. The properties of a gene may then be described by its expression pattern, mutant phenotype and interaction with other gene products.

2.3.1   Expression analysis

Genes or ORFs described by genome sequencing can be characterised by expression studies. The most direct method is to use the putative gene sequence as a probe for RNA blot hybridisation with RNA from various tissues or induced conditions. Convenient modifications of this approach are ribonuclease (RNase) protection and reverse transcriptase PCR (RT-PCR) methods that increase sensitivity. This can be supplemented by *in situ* hybridisation experiments to determine the exact location of gene expression. These methods are quite straightforward for most genes with normal expression levels, but become complicated for multigene families and painstaking in a whole genome analysis. In addition, most direct expression descriptions are not complete as all induced conditions are not considered.

To obtain a 'transcriptome' or complete transcription map of the genome, a systematic analysis of gene expression patterns has to be undertaken. One method used in human EST sequencing is single-pass sequencing of random clones from cDNA libraries constructed out of a large number of tissues, cell types or induced conditions so that a catalogue of expressed genes in each tissue or condition can be obtained. This approach has put *Arabidopsis* third on the list for EST sequences but is reaching a state of diminishing return because of saturation or limitations in the diverse cell types available. To obtain a direct quantitative estimate of expressed genes, a technique called serial analysis of gene expression (SAGE) has been described (Velculescu *et al.*, 1995) in which short diagnostic sequence tags are isolated from various cell types, concatenated, cloned and sequenced. This method of analysis reveals cell type specific expression patterns and would make use of the EST information already available.

To measure gene activities simultaneously under a variety of situations, a DNA microarray assay system (Schena *et al.*, 1995) was developed. In this method, cDNA clones are deposited as a microarray on a glass slide, hybridised to fluorescent labelled probes derived from RNA of different tissues, scanned with a laser and the signals analysed by computer. Standard sets of clones, e.g. from a whole genome, can be analysed with different hybridisation probes produced from small amounts of tissue or cell types and from very specific induced conditions. A major advantage of this systematic screening approach is that information on all the transcripts included as cDNAs is obtained, however irrelevant it may seem in some cases. This information can be combined with the results of screening with RNA derived probes from regulatory mutants in specific pathways to provide a gene regulatory network map showing its interaction to other pathways (e.g. metabolic). Improvements in the technology might be possible with Affymetrix chips (Chee *et al.*, 1996) using putative gene/ORF/EST specific oligonucleotides deposited on the DNA chips for hybridisation. The application of DNA chip technology in *Arabidopsis* (Schena, 1996) to study the expression pattern of all the genes simultaneously would be a valuable tool for whole genome analysis.

2.3.2 Mutation analysis

2.3.2.1 Knockout mutants

Mutations have been the classic genetic tool used to dissect a developmental or physiological process and identify the underlying gene structure. In *Arabidopsis* mutations can be conveniently isolated by chemical or physical mutagens, the mutated gene mapped with the help of molecular markers and then isolated by map-based cloning procedures. The availability of dense molecular maps (Schmidt *et al.*, 1995) facilitates chromosome landing, in which the closest linked markers can be directly used to isolate a large YAC/BAC clone containing the gene. With a complete physical map and later the sequence of the *Arabidopsis* genome in hand, precisely mapped mutants will be instrumental for gene identification. To study the interaction between genes, mutants in a particular pathway can be combined and the genetic hierarchy studied.

Mutations caused by known DNA insertion sequences serve the purpose of identifying genes in a specific biological process and also directly lead to gene isolation. The insertion sequence creates a mutant and tags the gene, which can then be isolated with the help of the molecular tag. This method of gene tagging, using well-characterised insertion sequences as molecular tags, does not need prior knowledge of the gene product or expression. Two types of insertion sequences are commonly used in plants: transposable elements and *Agrobacterium tumefaciens* mediated T-DNA (transfer DNA) insertions.

Transposable elements (McClintock, 1948) or transposons were first described in plants such as maize as mobile agents that mutate and rearrange the genome. Since then transposons have been very useful in genetic studies to identify and isolate genes in various organisms, using a technique termed transposon tagging (Bingham *et al.*, 1981). While transposons have without doubt played a role in genome evolution, and are regarded as 'controlling elements' of gene regulation (McClintock, 1956), their exact purpose has not been elucidated. However they are important practical genetic tools for analysing the genome *in situ* (Sundaresan, 1996). Endogenous transposons, for example in maize and snapdragon (Gierl and Saedler, 1992; Walbot, 1992), have been employed in transposon tagging strategies to identify and isolate genes. Usually the phenotypes of knockout mutants are characterised, but quite often transposase-dependent phenotypes of these 'controlling elements' are useful. The introduction of these well-studied transposons into other plant species such as *Arabidopsis* and tomato by transformation has now broadened the perspectives of transposon application.

Foreign insertion sequences, such as T-DNA or introduced transposons, can be engineered *in vitro* to carry selectable marker genes and designed to work as gene detectors with convenient reporter genes to assay the expression of adjacent genes. Initially these mutagenesis systems have been used in a random manner to identify genes, but following the isolation of a large number of inserts that cover the genome, they are being incorporated in reverse genetics strategies to select inserts in or near specific genes of interest by PCR-based screening methods.

2.3.2.2 T-DNA insertions

In plants amenable to transformation *Agrobacterium*-mediated T-DNA insertions in the genome can be utilised in a similar way to transposons to mutate and identify genes. Efficient transformation systems, with minimal tissue culture procedures, developed in *Arabidopsis* have allowed the recovery of numerous T-DNA-tagged genes (Azpiroz-Leehan and Feldmann, 1997). The seed transformation system has been the most effective insertional mutagenesis system (Feldmann, 1991) for the identification and isolation of genes in *Arabidopsis*. Agrobacterium transformation using vacuum infiltration of inflorescences (Bechtold *et al.*, 1993) has been very effective in the recovery of transformants from a wide range of ecotypes. About 14 000 transformants have been generated by seed infection and about 25 000 by vacuum infiltration procedures. On average 1.5 T-DNA inserts are present per transformed line, and the inserts are randomly distributed over the genome. Molecular analyses of characterised tagged mutants (Azpiroz-Leehan and Feldmann, 1997) have shown that insertions can occur all over a gene, even insertions in introns and promoters can reveal mutant phenotypes. Moreover, the tagged genes are distributed randomly over all chromosomal regions. To saturate the genome with T-DNAs, and recover mutations for every gene of interest, about 100 000 independent transformants would be required (Azpiroz-Leehan and Feldmann, 1997). With the continued effort of different research groups to produce transformants efficiently this objective will be met long before the genome is sequenced.

2.3.2.3 Heterologous transposons

The maize transposable element systems *Ac-Ds* (Baker *et al.*, 1986) and *En-I(Spm)* (Pereira and Saedler, 1989; Masson and Fedoroff, 1989) have been shown to function in numerous heterologous hosts (Haring *et al.*, 1991) and have been successfully used for gene tagging purposes. The early experiments employed the autonomous (one element) transposons *Ac* (Schmidt and Willmitzer, 1989) or *En(Spm)* (Cardon *et al.*, 1993) to study heterologous transposition behaviour. To develop effective heterologous transposon tagging strategies, two component systems, comprising a mobile transposon component and an immobilised trans-active transposase source, have subsequently been used (Bancroft *et al.*, 1992; Aarts *et al.*, 1993; Honma *et al.*, 1993). Transposons have been modified *in vitro* to carry various markers or selectable genes and inserted in other assayable genes to monitor excision (Baker *et al.*, 1987; Jones *et al.*, 1989). The transposase source is often immobilised, under the control of heterologous promoters (Swinburne *et al.*, 1992), and by segregating out in progeny stable transposon inserts can be isolated.

The early heterologous tagged mutants in *Arabidopsis* (Aarts *et al.*, 1993; Bancroft *et al.*, 1993; Long *et al.*, 1993) and petunia (Chuck *et al.*, 1993) were obtained by a random tagging strategy in which plants containing transposed elements were selfed and screened for obvious mutant phenotypes. The frequency of random inserts displaying a discernible mutant phenotype is about 1–5%, similar to that observed for T-DNA inserts.

Transposons move preferentially to closely linked locations (Jones *et al.*, 1990; Bancroft and Dean, 1993); a characteristic useful for efficiently isolating mutants for genes near the original position. Directed or targeted tagging is a strategy to tag a specific gene by constructing an appropriate genotype to screen for the target mutation. In heterologous systems, linked transposons can be obtained by mapping a large number of transformed T-DNA inserts containing transposons. The most extensive mutant screening was undertaken in the tomato. With a *Ds* transposon linked at about 3 cM to the target gene, tagged mutants were obtained at a rate of about 1/1000 transpositions (Jones *et al.*, 1994), a much higher frequency than expected for a unlinked transposon. In *Arabidopsis*, a directed tagging strategy (James *et al.*, 1995) was used to isolate the *FAE1* gene with linked *Ac* elements. Because of the difficulty of isolating many independent *Ds* transpositions in *Arabidopsis*, mainly because of early transposition (Scofield *et al.*, 1993), this procedure has not been widely used but many experiments are in progress. Independent transpositions are easier to obtain with the *En-I* system (Aarts *et al.*, 1995a), and a tagged *cer1::I* (Aarts *et al.*, 1995b) mutant was obtained from a line containing a closely linked (<5 cM) T-DNA bearing an *I* element. The targeted tagging strategy has been extended to tag the *CER6* gene using an *I* element from the previously tagged *ap1::I* allele located 4 cM away (Pereira, unpublished). The mapping of more transposon inserts will provide access to target insertions in various chromosomal regions and genes.

2.3.2.4 The insertion spectrum

An analysis of various transposon tagged mutants reveals insertion sites in different parts of a gene, often defining the structure of a gene. These sites are biased to reveal inserts that display a mutant phenotype. Insertions in introns also tend to give a mutant phenotype, often dependent on the presence of the transposase function. In the absence of transposase the insertion allele shows an almost wild-type phenotype, while in presence of the transposase the phenotype is suppressed to display a mutant phenotype. This is characteristic for the *En/Spm* family, which displays this *Suppressor* function, hence the name *Suppressor-Mutator* (*Spm*) given by McClintock (1954). The function is also displayed in *Arabidopsis* for two *ap1::I/dSpm* mutants (Aarts and Pereira, unpublished), suggesting that transposase-dependent conditional mutants can be obtained. This is a promising way to isolate gametophytic or lethal mutants.

In *Arabidopsis*, unselected transposon insertion sites are being characterised in several laboratories for both the *Ac-Ds* (R. Martienssen, personal communication) and *En-I(Spm)* systems. This is done by isolating transposon-flanking DNA by PCR-based methods, followed by sequencing and comparison to available databases. Preliminary evidence (Metz and Pereira, unpublished; J. Jones personal communication) suggests that for the *En-I(Spm)* system insertions are obtained in genes as well as in non-coding regions and a high proportion show homology to genes in databases. There is no indication as yet of insertional specificity, implying that a random distribution of transposon inserts is possible in *Arabidopsis*, unlike transposons in yeast and *Caenorhabiditis*. This approach of producing 'insertion sequence tagged sites' will help identify inserts in genes based on homology, and as contiguous genomic sequence becomes available will provide an extensive insertion map for systematic function analysis.

2.3.3 Gene detector insertions

Most of our present genetic descriptions of developmental and physiological processes come from loss-of-function or knockout mutations. However, in most higher organisms the majority of genes display no obvious knockout phenotype, e.g. in yeast about 60–70% of the genes display no obvious phenotype on disruption (Burns *et al.*, 1994). This suggests that many genes go undetected by a knockout mutational approach and normal methods of screening. Part of the reason could be functional redundancy, where one or more other loci (mostly homologous) can substitute for the same function. Sequential disruption of homologous and redundant genes in an individual genotype might ultimately reveal a mutant phenotype.

Quite often a mutant phenotype reflects only a part of the function of a gene, primarily its first appearance, e.g. in the development of the embryo as embryo lethal mutants, which are the most frequently observed mutations. The phenotypes expressed later in development are obscured owing to their non-recovery. A more thorough screening system should also reveal the more subtle mutants. Plants respond to various cues from the environment, such as pathogens, abiotic stress, etc. These processes need special screening systems to identify conditional mutants.

An enigma of mutations in plants, represented by *Arabidopsis*, is that there are too few lethal mutants known compared to other organisms (Miklos and Rubin, 1996). Considering that many extensive mutant screens have been undertaken, either plants have a different proportion of knockout lethals or a large proportion of gametophytic mutants might be difficult to recover. This suggests the need of other systems for the identification of genes.

An entrapment strategy (Skarnes, 1990) makes use of inserts containing reporter gene constructs, whose expression is dependent on transcriptional regulatory sequences of the adjacent host gene. In this way genes can be identified by their expression pattern, even though they might not directly display an obvious mutant phenotype. The identification of very specific expression patterns, e.g. pathogen induction, will suggest the possible phenotype for which to screen.

Enhancer trap elements are developed to detect enhancers in the genome, which are capable of orientation independent transcriptional activation from a distance. Enhancer trap transposons (Figure 2.1) contain a reporter gene such as β-glucuronidase (GUS) with a weak or minimal promoter, e.g. with a TATA box, situated near the border of the insert. When integrated in the vicinity of an enhancer sequence in the genome, the reporter gene displays the expression pattern of the nearby chromosomal gene. The inserts will be activated from a distance and independent of the orientation of the reporter gene, so the recovery of active enhancer traps is high. One disadvantage of these inserts is that it is difficult to pinpoint the exact location of the trapped gene, which might be located at an unpredictable distance, especially in large complex genomes. In the compact *Arabidopsis* genome, nearby located genes may also influence the expression pattern and complicate the identification of the trapped enhancer or gene. Once contiguous or complete genomic sequence information is available it might be easier to analyse the trapped genes.

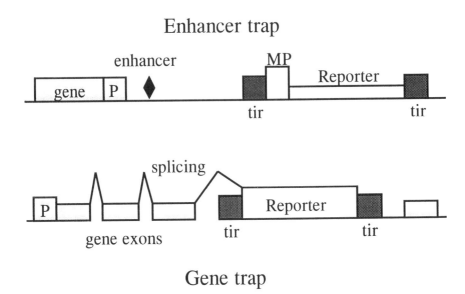

Enhancer trap

Gene trap

Figure 2.1 Structure of entrapment transposon inserts in the plant genome that reveal the regulatory sequences of the adjacent trapped/tagged gene. The chromosomal plant gene promoter sequences are designated P and the plant gene shown lightly shaded. The transposon termini (tir) are shown heavily-shaded and contain a reporter gene such as GUS. The enhancer trap has a minimal promoter, e.g. from the CaMV 35S (−1 to −45 containing the TATA box) and is shown inserted adjacent to the chromosomal enhancer of the plant gene. The gene trap construct has a promoterless reporter gene with splice acceptor sites at the 5' end so that with a transposon insert into the transcription unit of a plant gene splicing to the reporter gene takes place and a fusion transcript or protein is produced.

Gene trap inserts are designed to create fusion transcripts with the target gene (Skarnes, 1990). The prototype was first applied in bacteria, where a promoterless *lacZ* within a transposon is activated by integration into a transcription unit. In plants the reporter genes NPTII and GUS are able to support gene fusions and have been widely used in T-DNA vectors as gene fusion traps (Topping and Lindsey, 1995). Promoter traps consist of a promoterless reporter gene and are expressed when inserted downstream of the chromosomal gene promoter. The exon trap (Figure 2.1) is more versatile as it enables reporter gene fusions to be created at various locations within a gene. The introduction of splice acceptor sites upstream of the reporter creates gene fusions even for insertions in introns and thus increases the recovery of inserts expressing the reporter gene. The trapped gene can then be easily identified as part of a chimeric transcript. Reporter gene fusions can help in localisation of the gene product, e.g. in the nucleus, organelle or membrane. The GUS gene is a useful reporter for most purposes but recent developments in the use of the green fluorescent protein

(GFP) for gene fusions could be extremely useful for studying living cells and protein localisation *in vivo* (Haseloff *et al.*, 1997).

Historically, enhancer traps were first used with the P transposon in *Drosophila* (O'Kane and Gehring, 1987) to identify a large number of genes based on their expression pattern. In plants, T-DNA vectors were first designed as promoter traps (Koncz *et al.*, 1989) with the NPTII reporter gene. Later the GUS reporter gene was employed for enhancer and promoter traps (Kertbundit *et al.*, 1991; Topping and Lindsey, 1995). These studies revealed the high proportion of inserts that displayed reporter gene expression, with about 25% for promoter traps and 50% for the enhancer traps. Most surprisingly, the frequency of expressed inserts was similar in *Arabidopsis* and larger genomes such as tobacco (Koncz *et al.*, 1989), suggesting that T-DNA inserted preferentially in transcriptionally active regions. Although a large collection of T-DNA transformants for enhancer and promoter traps has been produced, analysed and scores of interesting patterns have been identified, it has not been always easy to make the exact correlation between the reporter expression pattern and the resident gene. One reason is the difficulty in determining the position of the expressed trapped gene, especially in enhancer traps and with rearranged or multiple T-DNA inserts. Among the large T-DNA collections now produced, the vacuum-infiltration-derived Versailles T-DNA collection employs a vector for promoter trapping (Bouchez *et al.*, 1993) and should help identify inserts with novel expression patterns.

The first description of a transposon-based promoter and enhancer trap system used the *Ac-Ds* system with a GUS reporter gene (Fedoroff and Smith, 1993). A *Ds* element was engineered to contain a GUS reporter gene with a CaMV minimal promoter (Benfey and Chua, 1990) upstream at the left end of *Ds*. The *Ds* contained a hygromycin resistance marker and was inserted in the chlorsulfuron resistance gene used for the selection of excisions. The *Ac* transposase construct contained the *iaah/tms2* gene, which was used as a negative selection marker for the selection of stable transposed *Ds* elements in progeny. This system led to the isolation of a novel gene, *LRP1*, which was involved in root development (Smith and Fedoroff, 1995). In this insertion no mutant phenotype was visible, probably because it belongs to a small gene family and is redundant, exemplifying the use of enhancer entrapment for function identification. Later the system was also used for the identification of a gene responsible for an embryo-defective lethal mutation (Tsugeki *et al.*, 1996).

An *Ac-Ds*-based gene trap and enhancer trap system has also been developed with a novel selection method for stable transpositions away from the initial T-DNA position (Sundaresan *et al.*, 1995). From starter plants carrying the *Ac-Ds* elements, stable transposed *Ds* elements are selected using the negative selectable *iaaH* marker gene. Independently derived *Ds* inserts were screened for their GUS expression pattern and revealed some GUS reporter gene activity in about 50% of the enhancer trap (*DsE*) and 25% of the gene trap (*DsG*) inserts

(Sundaresan *et al.*, 1995). *PROLIFERA*, a gene involved in megagametophyte and embryo development (Springer *et al.*, 1995), was the first gene identified by transposon gene detection and exemplifies the efficiency of the system. A number of starter *DsG* and *DsE* lines have since been used by different collaborators to produce about 10 000 stable transposed elements, many of which have been screened for their expression pattern. Cumulatively, the recovery of more transposed gene detector inserts will increase the genome coverage to facilitate reverse genetic strategies.

The majority of transposon inserts do not display a mutant phenotype. Even when inserts are in or near genes, the insertion does not result in a mutant phenotype, which is probably because of gene redundancy or the non-essential nature of the organism. In *Arabidopsis* less than 4% of *Ds* insertions yield mutants (Bancroft *et al.*, 1993; Long *et al.*, 1993), thus the entrapment inserts enable the study of the function of genes that do not reveal major mutant phenotypes. They allow for the selection of inserts in specific classes of genes, based on their expression pattern. Subsequent production of double or multiple mutants in a pathway, as indicated by the expression pattern, might finally reveal a mutant phenotype.

2.3.4 Mis-expression mutants

Spatially and temporally targeted mis-expression of individual genes provides an alternative way of perturbing gene regulatory networks. One way this can be achieved is by employing insertion sequences that carry a strong enhancer element near the border, thus activating the adjacent gene ectopically. This method of activation tagging (Walden *et al.*, 1994) has been applied successfully in tobacco, with a T-DNA vector containing multiple strong CaMV enhancer sequences near the border, and selecting transformants for hormone independence. A similar strategy is also being employed in *Arabidopsis*.

A transposon construct, to isolate dominant gain-of-function alleles, bearing a CaMV 35S promoter transcribing outward was used to tag a gene *TINY* (Wilson *et al.* 1996), which was recovered as a semi-dominant overexpression mutant. By this system, dominant mutations might be caused by insertions in the transcription unit driving transcription in the right orientation. As the frequency of such inserts is not expected to be higher than knockout mutations by the same transposon, both types of mutants will be obtained. The recovery of overexpression mutants may be advantageous in cases where positive selection is possible for dominant mutations, or in processes or genes where simple knockout mutants reveal no mutant phenotype.

Functional inactivation of genes by dominant negative mutations was suggested by Herskowitz (1987), using mutant polypeptides that disrupt the activity of the wild-type gene when overexpressed. These may be by mutations in a regulator that change it from an activator to a repressor, or mutations in the DNA-binding domain of multimeric proteins so that non-functional mixed aggregates would compete. Although such mutations in the genome have been obtained, a transgenic approach to introduce them is also possible.

2.3.5 Gene silencing

The transgenic approach to creating silenced genes, and overexpression or ectopic expression mutants, can be used to identify gene function. This method generates mutants without mutation. Typically a complete or partial cDNA clone is placed downstream of a strong promoter, in sense or anti-sense orientation, transformed into the plant and the phenotype examined. In about 10% of the transformants a mutant phenotype is observed for genes that have a visual effect, e.g. developmental regulatory genes, or enzymes in anthocyanin coloration (Meyer and Saedler, 1996). This approach is also applicable for multigene families, where clues to gene function may be obtained, but the interpretation of a phenotype might be complicated by non-specific (homologous) gene suppression. As a general strategy it is too unreliable and time-consuming for making a large number of transformants that have to be tested individually, but the method has particular applications.

2.3.6 Site-specific deletions

Homologous recombination has not been very efficient in plants used to create site-specific mutations and deletions. Heterologous site-specific recombination systems, such as the bacteriophage P1 *Cre-lox* and the yeast *FLP-FRT*, have been shown to work in plants to create deletions, and in combination with transposition can be used for gene identification (van Haaren and Ow, 1993). Recombination occurs between two *lox* sites or *FRT* sites, mediated respectively by the *Cre* or *FLP* recombinase. Constructs have been used (Osborne *et al.*, 1995) that contain the *lox* recombination sites, both within the transposon (*Ds-lox*) and the adjacent T-DNA. After transposition of *Ds-lox* to a closely linked site, the sequence-specific recombination sites are situated close together on the chromosome allowing recombinase mediated small deletions to be recovered (by *Cre*-recombinase line crosses). These specific genomic deletions help to characterise the function of individual genes and of other chromosomal structural elements such as matrix associated regions or general enhancers.

2.3.7 Target-selected insertional mutagenesis

In yeast it is possible to inactivate the genomic copy of a cloned putative ORF or gene by gene replacement, a technique not easily applicable in other eukaryotes. Even in yeast easier, alternative methods to gene replacement-mediated function searches are being sought. A reverse genetics technique called target selected gene inactivation was devised for transposon mutagenesis (Ballinger and Benzer, 1989; Kaiser and Goodwin, 1990) in *Drosophila* and *Caenorhabditis* to inactivate genes that had been sequenced but whose function was unknown. In principle, this strategy involves mutagenizing the entire genome first and then identifying individuals carrying insertions in a gene of interest. The chance of recovering an insert in a target gene is dependent on the population of inserts. In *Arabidopsis* such large populations of inserts, nearing saturation of the genome, can be generated by T-DNA or transposon inserts. About 100 000 inserts would give an insert every kilobase, with about a 99% chance of mutating an average *Arabidopsis* gene.

The selection of defined gene mutations involves PCR using pairs of primers, one of which anneals to the insertion sequence and the other to the specific target gene. If a PCR product specific for the target gene is observed then an insert close to the primer in the gene is present in the population. Specific pooling strategies (Figure 2.2), making use of the sensitivity of the PCR reaction, enable the individual containing the insertion to be pinpointed in a large population. Families containing the homozygous insertion can be produced and the phenotype conferred by the insertion assessed. Currently two strategies for genome saturation with inserts are being pursued: (i) single stable elements, similar to the T-DNA population structure, and (ii) multiple elements from active transposing populations, similar to the endogenous transposon systems.

In maize, snapdragon and petunia (Coen *et al.*, 1989; Gerats *et al.*, 1990; Walbot, 1992) the large number of endogenous transposon copies provides genome saturation. Target-selected insertional mutagenesis has been shown to work in these plants to identify knockout mutations in specific genes (Das and Martienssen, 1995; Koes *et al.*, 1995). In *Arabidopsis* a population of about 8000 T-DNA inserts from 5300 transformants, screened in 53 pools of 100 transformants each, were first used to identify insertions in two members of the actin multi-gene family (McKinney *et al.*, 1995). Later, a hierarchical pooling strategy with DNA from 9100 transformed lines was used to recover inserts in 17 out of the 63 signal transduction genes tested (Krysan *et al.*, 1996). To reduce the numbers of DNA samples for screening, a two-dimensional multiplex pooling system (Azpiroz-Leehan and Feldmann, 1997) was then devised, containing 100 subpools of 10 transformed lines each, arranged in 10 rows and 10 columns to produce 20 DNA pools. Increasing collections of transformants are available through the ABRC. The individual transformants from various collections have to be pooled systematically, DNA has to be prepared from the pools and used for PCR reactions. The sensitivity of the PCR to detect inserts in large pools and the strategy for directly identifying individuals with an insert in the pool are parameters that need attention. The present cumulative T-DNA transformant collection of about 35 000, with about 1.5 inserts/transformant and around 50 000 total inserts, would give a 95% chance of recovering a specific mutant, enough for most purposes.

The transposon populations available at present cumulatively contain about 100 000 inserts and are therefore a valuable complement, especially as they include a variety of insertion types for gene inactivation and gene detection. The

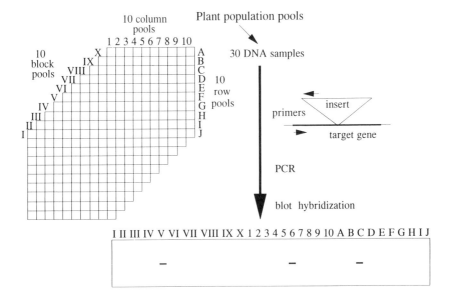

Figure 2.2 Target-selected insertion mutagenesis strategy using three-dimensional pooling. Here 1000 plants/genotypes are arranged in $10 \times 10 \times 10$ trays × columns × rows and pooling of plant material (e.g. leaf) is done to obtain 30 DNA samples for PCR. To identify an insert in the target gene, insert and gene-specific primers are used for PCR reactions, which are transferred to a membrane and hybridised with the target gene-specific probe. Identically sized PCR fragments from the three dimensions hybridise to the probe, indicating the address of the plant containing the insert, in this example V-6-C, whose progeny can be used to identify the insertion mutant and characterise the phenotype.

insertion configurations of T-DNAs are often complex, with tandem-repeat insertion structures and heterogeneity of the left border junctions. In contrast, transposon inserts have a defined terminal sequence rendering them all accessible to PCR selection. Genome saturation with heterologous transposons is not automatically achieved as the transformation process delivers a few copies that yield transposed copies in the genome. One method for saturation is accumulation of high copy number transposons by occasional amplifications or transposon bursts, which have been documented to occur in transgenic plants, for *Ac* in tomato (Yoder, 1990) and *En-I* in *Arabidopsis* (Aarts *et al.*, 1995a). Subsequent inter-crosses between high element-containing lines increase the element copy number additively. The independent transpositions that occur in families made from the high element lines will yield a large population of independent inserts for genome saturation.

Multiple *I/dSpm* element lines (Metz and Pereira, unpublished), containing about 40 elements per plant, are being produced in *Arabidopsis* that are similar to the maize *Mutator* (Das and Martienssen, 1995) and petunia *Tph1* (Koes *et al.*,

1995) mutation machines. With about 2500 independent lines a random population of 100 000 inserts for genome saturation should be achievable. Screening for insertions can be done using a modification of the three-dimensional multiplex pooling strategy (Figure 2.2) (Zwaal *et al.*, 1993; Koes *et al.*, 1995) of about 1000 plants arranged in 10 blocks × 10 rows × 10 columns to yield 30 DNA pools. These pools are screened by PCR and the inserts identified by DNA blot hybridisation to the target gene probe so that the individual plant carrying the target insertion can be directly tested in the progeny. The convenience of this multiple-element system is that inserts in a gene can be obtained by the screening of a relatively small number of plants and DNA pools. A similar strategy using multiple (5–10) autonomous *En* elements (Cardon *et al.*, 1993) is being used in a population of about 5000 plants (E. Wisman, personal communication).

The local transposition behaviour of the maize transposons can also be applied to saturate specific genomic regions with inserts. Thus mapped *Ds* or *I/dSpm* transposons in the genome can serve as donor sites to produce saturated populations of inserts in the linked chromosomal regions. Such a population has been produced with an *I/dSpm* element transposing in a region of chromosome 4, which is being sequenced by ESSA and EUAGP. *I/dSpm* inserts have been recovered for a number of genes present in this sequenced region of chromosome 4 (Speulman and Pereira, unpublished), suggesting that saturation mutagenesis of this region has been achieved. Gene trap *DsG* elements have also been mapped near the top of chromosome 4, at the tagged *PROLIFERA* gene (Springer *et al.*, 1995), and local transpositions are being selected to make a population for PCR screening of mutants in the CSHSC sequenced regions (R. Martienssen, personal communication)

The reverse genetics strategies outlined here can be applied to all the different insertion types. In a systematic approach, the genome sequence information can be used to design primers for genes, and inserts in or near them can be isolated. After undergoing a preliminary screening for mutant phenotype and expression pattern, the insertion lines could be deposited into stock centres for distribution to a network of biologists for specific screening, and all the results entered into public databases.

2.4   Prospects

The complete sequence of the *Arabidopsis* genome will give *Arabidopsis*, and ultimately plant genetics, a new dimension. It will then be possible to elucidate gene functions systematically and efficiently by reverse genetics strategies. The basic techniques for identifying gene function are being developed as outlined above and they will have to be systematically employed to identify the function of all the genes in *Arabidopsis*. In yeast systematic function analysis is already well advanced, with other forthcoming genomes following suit.

In Europe, systematic genome sequencing started with the ESSA program. Gene function analysis network projects have already been initiated in the form of the *Arabidopsis* Insertion Mutagenesis (AIM) project and a Myb gene function project, both aiming at development of reverse genetics strategies. As in yeast, these projects may lead to an international framework to systematically analyse the function of genes and other structural elements discovered by whole genome sequencing. It is very important to link biology to genome analysis so that the results from the model genome can be projected to other crop plant genomes and biological processes of economic interest.

The properties of genes will be described by multi-dimensional information on genome sequence, gene expression and mutant phenotype, individually and in interaction to each other. *Arabidopsis* genome sequence analysis will help describe precisely the synteny between diverse genomes and identify genes by genome comparisons. Finally, complex plant genomes will be understandable because the building blocks from *Arabidopsis* will make sense in the light of evolution.

Note: The sequence of 1.9 Mb of chromosome 4 has been published by the EUAGP (Bevan *et al.* (1998) Analysis of 1.9Mb of contiguous sequence from chromosome 4 of *Arabidopsis thaliana. Nature*, **391** 485-88).

Acknowledgements

We would like to acknowledge funding from the European Union for the ESSA (BIO4-CT93-0075) and AIM (BIO4-CT95-0183) supporting projects. We wish to thank the members of these projects for communicating unpublished information: results of ESSA from Mike Bevan (JIC, Norwich); results from AIM collaborators Jonathan Jones (JIC, Norwich), Ellen Wisman (MPI, Cologne), David Bouchez (INRA, Versailles). We also thank Robert Martienssen (CSHL, New York) for communicating unpublished results. Our colleagues from CPRO-DLO, Mark Aarts, Peter Metz and Elly Speulman, are gratefully acknowledged for sharing unpublished information, and we thank Hans Sandbrink for critically reading the manuscript.

References

Aarts, M.G.M., Dirkse, W., Stiekema, W.J. and Pereira, A. (1993) Transposon tagging of a male sterility gene in *Arabidopsis. Nature*, **363** 715-17.

Aarts, M.G.M., Corzaan, P., Stiekema, W.J. and Pereira, A. (1995a) A two-element *Enhancer-Inhibitor* transposon system in *Arabidopsis thaliana. Mol. Gen. Genet.*, **247** 555-64.

Aarts, M.G.M., Keijzer, C.J., Stiekema, W.J. and Pereira, A. (1995b) Molecular characterization of the *CER1* gene of *Arabidopsis* involved in epicuticular wax biosynthesis and pollen fertility. *Plant Cell*, **7** 2115-27.

Azpiroz-Leehan, R. and Feldmann, K.A. (1997) T-DNA insertion mutagenesis in *Arabidopsis*: going back and forth. *Trends Genet.*, **13** 152-56.

Baker, B., Schell, J., Lörz, H. and Fedoroff, N. (1986) Transposition of the maize controlling element '*Activator*' in tobacco. *Proc. Natl Acad. Sci. USA*, **83** 4844-48.

Baker, B., Coupland, G., Fedoroff, N., Starlinger, P. and Schell, J. (1987) Phenotypic assay for excision of the maize controlling element *Ac* in tobacco. *EMBO J.*, **6** 1547-54.

Ballinger, D.G. and Benzer, S. (1989) Targeted gene mutations in *Drosophila*. *Proc. Natl Acad. Sci. USA*, **86** 9402-9406.

Bancroft, I. and Dean, C. (1993) Transposition pattern of the maize element *Ds* in *Arabidopsis thaliana*. *Genetics*, **134** 1221-29.

Bancroft, I., Bhatt, A.M., Sjodin, C., Scofield, S., Jones, J.D.G. and Dean, C. (1992) Development of an efficient two-element transposon tagging system in *Arabidopsis thaliana*. *Mol. Gen. Genet.*, **233** 449-61.

Bancroft, I., Jones, J.D.G. and Dean, C. (1993) Heterologous transposon tagging of the *DRL1* locus in *Arabidopsis*. *Plant Cell*, **5** 631-38.

Bechtold, N., Ellis, J. and Pelletier, G. (1993) *In planta* Agrobacterium mediated gene transfer by infiltration of adult *Arabidopsis thaliana* plants. *C.R. Acad. Sci.* **316** 1194-99.

Benfey, P.N. and Chua, N.-H. (1990) The cauliflower mosaic virus 35S promoter: combinatorial regulation of transcription in plants. *Science*, **250** 959-66.

Bevan, M., Ecker, J., Theologis, S., Federspiel, N., Davis, R., McCombie, D. *et al.* (1997) Objective: the complete sequence of a plant genome. *Plant Cell*, **9** 476-78.

Bingham, P.M., Levis, R. and Rubin, G.M. (1981) Cloning of DNA sequences from the *white* locus of *D. melanogaster* by a novel and general method. *Cell*, **25** 693-704.

Bogulski, M.S., Lowe, T.M.J. and Tolstoshev, C.M. (1993) dbEST-database for 'expressed sequence tags'. *Nat. Genet.*, **4** 332-33.

Bouchez, D., Camilleri, C. and Caboche, M. (1993) A binary vector based on Basta resistance for *in planta* transformation of *Arabidopsis thaliana*. *C.R. Acad. Sci.*, **316** 1188-93.

Burns, N., Grimwade, B., Ross-Macdonald, P.B., Choi, E.-Y., Finberg, K., Roeder, G.S. *et al.* (1994) Large-scale analysis of gene expression, protein localization, and gene disruption in *Saccharomyces cerevisiae*. *Genes Dev.*, **8** 1087-1105.

Cardon, G.H., Frey, M., Saedler, H. and Gierl, A. (1993) Mobility of the maize transposable element *En/Spm* in *Arabidopsis thaliana*. *Plant J.*, **3** 773-84.

Chee, M., Yang, R., Hubbell, E., Berno, A., Huang, X.C., Stern, D. *et al.* (1996) Accessing genetic information with high-density DNA arrays. *Science*, **274** 610-14.

Choi, S., Creelman, R.A., Mullet, J.E. and Wing, R.A. (1995) Construction and characterization of a bacterial artificial chromosome library of *Arabidopsis thaliana*. *Plant Mol. Biol. Rep.*, **13** 124-28.

Chuck, G., Robbins, T., Nijjar, C., Ralston, E., Courtney-Gutterson, N. and Dooner, H.K. (1993) Tagging and cloning of a petuniaflower color gene with the maize transposable element *Activator*. *Plant Cell*, **5** 371-78.

Coen, E.S., Robbins, T.P., Almeida, J., Hudson, A. and Carpenter, R. (1989) Consequences and mechanisms of transposition in *Antirrhinum majus*, in *Mobile DNA* (eds D.E. Berg and M.M. Howe), American Society of Microbiology, Washington, DC, pp 413-36.

Cooke, R., Raynal, M., Laudié, M., Grellet, F., Delseny, M., Morris, P.-C. *et al.* (1996) Further progress towards a catalogue of all *Arabidopsis* genes: analysis of a set of 5000 non-redundant ESTs. *Plant J.*, **9** 101-24.

Cooke, R., Raynal, M., Laudié, M. and Delseny, M. (1997) Identification of members of gene families in *Arabidopsis thaliana* by contig construction from partial cDNA sequences: 106 genes encoding 50 cytoplasmic ribosomal proteins. *Plant J.*, **11** 1127-40.

Das, L. and Martienssen, R. (1995) Site-selected transposon mutagenesis at the *hcf106* locus in maize. *Plant Cell*, **7** 287-94.

Dujon, B. (1996) The yeast genome project: what did we learn. *Trends Genet.*, **12** 263-70.

Fedoroff, N.V. and Smith, D.L. (1993) A versatile system for detecting transposition in *Arabidopsis*. *Plant J.*, **3** 273-89.

Feldmann, K.A. (1991) T-DNA insertion mutagenesis in *Arabidopsis*-mutational spectrum. *Plant J.*, **1** 71-82.

Gerats, A.G.M., Huits, H., Vrijlandt, E., Maraña, C., Souer, E. and Beld, M. (1990) Molecular characterization of a non-autonomous transposable element (*dTph1*) of petunia. *Plant Cell*, **2** 1121-28.

Gierl, A. and Saedler, H. (1992) Plant transposable elements and gene tagging. *Plant Mol. Biol.*, **19** 39-49.

Goffeau, A., Slonimski, P., Nakai K. and Risler, J.L. (1993) How many yeast genes code for membrane-spanning proteins. *Yeast*, **9** 691-702.

Goffeau, A., Barrell, B.G., Bussey, H., Davis, R.W., Dujon, B., Feldmann, H. *et al.* (1996) Life with 6000 genes. *Science*, **274** 546-67.

Haring, M.A., Rommens, C.M.T., Nijkamp, H.J.J. and Hille, J. (1991) The use of transgenic plants to understand transposition mechanisms and to develop transposon tagging strategies. *Plant Mol. Biol.*, **16** 449-61.

Haseloff, J., Siemering, K.R., Prasher, D.C. and Hodge, S. (1997) Removal of a cryptic intron and subcellular localization of greenfluorescent protein are required to mark transgenic *Arabidopsis* plants brightly. *Proc. Natl Acad. Sci. USA*, **94** 2122-27.

Herskowitz, I. (1987) Functional inactivation of genes by dominant negative mutations. *Nature*, **329** 219-22.

Hodgkin, J., Plasterk, R.H.A. and Waterston, R. (1995) The nematode *Caenorhabditis elegans* and its genome. *Science*, **270** 410-14.

Höfte, H., Desprez, T., Amselem, J., Chiapello, H., Caboche, M., Moisan, A. *et al.* (1993) An inventory of 1152 expressed sequence tags obtained by partial sequencing of cDNAs from *Arabidopsis thaliana*. *Plant J.*, **4** 1051-61.

Honma, M.A., Baker, B.J. and Waddel, C.S. (1993) High-frequency germinal transposition of *Ds^ALS* in *Arabidopsis*. *Proc. Natl Acad. Sci. USA*, **90** 6242-46.

James, D.W.Jr., Lim, E., Keller, J., Plooy, I., Ralston, E. and Dooner, H.K. (1995) Directed tagging of the *Arabidopsis FATTY ACID ELONGATION1* (*FAE1*) gene with the maize transposon *Activator*. *Plant Cell*, **7** 309-19.

Jones, J.D.G., Carland, F.M., Maliga, P. and Dooner, H.K. (1989) Visual detection of transposition of the maize element *Activator* (*Ac*) in tobacco seedlings. *Science*, **244** 204-207.

Jones, J.D.G., Carland, F.M., Lin, E., Ralston, E. and Dooner, H.K. (1990) Preferential transposition of the maize element *Activator* to linked chromosomal locations in tobacco. *Plant Cell*, **2** 701-707.

Jones, D.A., Thomas, C.M., Hammond-Kosack, K.E., Balint-Kurti, P.J. and Jones, J.D.G. (1994) Isolation of the tomato *Cf-9* gene for resistance to *Cladosporium fulvum* by transposon tagging. *Science*, **266** 789-93.

Kaiser, K. and Goodwin, S.F. (1990) 'Site-selected' transposon mutagenesis of *Drosophila*. *Proc. Natl Acad. Sci. USA*, **87** 1686-90.

Kertbundit, S., De Greve, H., De Boeck, F., van Montague, M. and Hernalsteens, J.P. (1991) In vivo random ß-glucuronidase gene fusions in *Arabidopsis thaliana*. *Proc. Natl Acad. Sci. USA*, **88** 5212-16.

Konczg, C., Martini, N., Mayerhofer, R., Koncz-Kalman, Z, Körber, H., Rédei, G.P. and Schell, J. (1989) High frequency T-DNA-mediated gene tagging in plants. *Proc. Natl Acad. Sci. USA*, **86**, 8467-71.

Koes, R., Souer, E., van Houwelingen, A., Mur, L., Spelt, C., Quattrocchio, F. *et al.* (1995) Targeted gene inactivation in petunia by PCR-based selection of transposon insertion mutants. *Proc. Natl Acad. Sci. USA*, **92** 8149-53.

Krysan, P.J., Young, J.C., Tax, F. and Sussman, M.R. (1996) Identification of transferred DNA insertions within *Arabidopsis* genes involved in signal transduction and ion transport. *Proc. Natl Acad. Sci. USA*, **93** 8145-50.

Leutwiler, L.S., Hough-Evans, B.R. and Meyerowitz, E.M. (1984) The DNA of *Arabidopsis thaliana*. *Mol. Gen. Genet.*, **194** 15-23.

Liu, Y.-G, Mitsukawa, N., Vazquez-Tello, A. and Whittier, R.F. (1995) Generation of a high quality P1 library of *Arabidopsis* suitable for chromosome walking. *Plant J.*, **7** 351-58.

Long, D., Martin, M., Sundberg, E., Swinburne, J., Puangsomlee, P. and Coupland, G. (1993) The maize transposable element system *Ac/Ds* as a mutagen in *Arabidopsis*: identification of an *albino* mutation induced by *Ds* insertion. *Proc. Natl Acad. Sci. USA*, **90** 10370-74.

Macknight, R., Bancroft, I., Page, T., Lister, C., Schmidt, R., Love, K., Westphal, L., Murphy, G., Sherson, S., Cobbett, C. and Dean, C. (1997) *FCA*, a gene controlling flowering time in *Arabidopsis*, encodes a protein containing RNA-binding domains. *Cell*, **89** 737-45.

Marziali, A., Federspiel, N. and Davis, R. (1997) Automation for the *Arabidopsis* genome sequencing project. *Trends Plant Sci.*, 2, 71-74.

Masson, P. and Fedoroff, N. (1989) Mobility of the maize *Suppressor-mutator* element in transgenic tobacco cells. *Proc. Natl Acad. Sci. USA*, **86** 2219-23.

McClintock, B. (1948) Mutable loci in maize. *Carnegie Inst. Wash. Yearbook*, **47** 155-69.

McClintock, B. (1954) Mutations in maize and chromosomal aberrations in Neurospora. *Carnegie Inst. Wash. Yearbook*, **54** 254-60.

McClintock, B. (1956) Controlling elements and the gene. *Cold Spring Harbor Symp. Quant. Biol.*, **21** 197-216.

McKinney, E.C., Ali, N., Traut, A., Feldmann, K.A., Belostotsky, D.A., McDowell, J.M. *et al.* (1995) Sequence-based identification of T-DNA insertion mutations in *Arabidopsis*: actin mutants *act2-1* and *act4-1*. *Plant J.*, **8** 613-22.

Meyer, P. and Saedler, H. (1996) Homology dependent gene silencing in plants. *Ann. Rev. Plant Physiol. Plant Mol. Biol.*, **47** 23-48.

Meyerowitz, E.M. (1992) Introduction to the *Arabidopsis* genome, in *Methods in Arabidopsis Research* (eds C. Koncz, N.-H Chua and J. Schell), World Scientific, Singapore, pp 100-18.

Meyerowitz, E.M. (1994) Structure and organization of the *Arabidopsis thaliana* nuclear genome, in *Arabidopsis* (eds E. Meyerowitz and C.R. Somerville), Cold Spring Harbor Laboratory, Cold Spring Harbor, New York, pp 1241-55.

Miklos, G.L.G. and Rubin, G.M. (1996) The role of the genome project in determining gene function: insights from model organisms. *Cell*, **86** 521-29.

Newman, T., Bruijn, F.J. de, Green, P., Keegstra, K., Kende, H., McIntosh, L. *et al.* (1994) Genes galore: a summary of methods for accessing results from large-scale partial sequencing of anonymous *Arabidopsis* cDNA clones. *Plant Physiol.*, **106** 1241-55.

O'Kane, C.J. and Gehring, W.J. (1987) Detection *in situ* of genomic regulatory elements in *Drosophila*. *Proc. Natl Acad. Sci. USA*, **84** 9123-27.

Oliver, S. (1996) A network approach to the systematic analysis of yeast gene function. *Trends Genet.*, **12** 241-42.

Osborne, B.I., Wirtz, U. and Baker, B. (1995) A system for insertional mutagenesis and chromosomal rearrangement using the *Ds* transposon and *Cre-lox*. *Plant J.*, **7** 687-701.

Pereira, A. and Saedler, H. (1989) Transpositional behavior of the maize *En/Spm* element in transgenic tobacco. *EMBO J.*, **8** 1315-21.

Quigley, F., Dao, P., Cottet, A. and Mache, R. (1996) Sequence analysis of an 81 kb contig from *Arabidopsis thaliana* chromosome III. *Nucl. Acids Res.*, **24** 4313-18.

Richard, G-F, Fairhead, C. and Dujon, B. (1997) *J. Mol. Biol.*, **268** 303-21.

Rounsley, S.D., Glodek, A., Sutton, G., Adams, M.D., Somerville, C.R., Venter, J.G. and Kerlavage, A.R. (1996) The construction of *Arabidopsis* expressed sequence tag assemblies. *Plant Physiol.* **112** 1177-83.

Salmeron, J.M., Oldroyd, G.E.D., Rommens, C.M.T., Scofield, S.R., Kim, H-S., Lavelle, D.T. *et al.* (1996) Tomato *Prf* is a member of the leucine-rich repeat class of plant disease resistance genes and lies embedded within the *Pto* kinase gene cluster. *Cell*, **86** 123-33.

Schena, M. (1996) Genome analysis with gene expression microarrays. *BioEssays*, **18** (5) 427-31.

Schena, M., Shalon, D., Davis, R.W. and Brown, P.O. (1995) Quantitative monitoring of gene expression patterns with a complementary DNA microarray. *Science*, **270** 467-70.

Schmidt, R. and Willmitzer, L. (1989) The maize autonomous element *Activator* (*Ac*) shows a minimal germinal excision frequency of 0.2%–0.5% in transgenic *Arabidopsis thaliana* plants. *Mol. Gen. Genet.*, **220** 17-24.

Schmidt, R., West, J., Love, K., Lenehan, Z., Lister, C., Thompson, H. *et al.* (1995) Physical map and organization of *Arabidopsis thaliana* chromosome 4. *Science*, **270** 480-83.

Scofield, S.R., English, J.J. and Jones, J.D.G. (1993) High level expression of the *Activator* transposase gene inhibits the excision of *Dissociation* in tobacco cotyledons. *Cell*, **75** 507-17.

Skarnes, W.C. (1990) Entrapment vectors: a new tool for mammalian genetics. *Biotechnology*, **8** 827-31.

Smith, D.L. and Fedoroff, N.V. (1995) *LRP1*, a gene expressed in lateral and adventitious root primordia of *Arabidopsis*. *Plant Cell*, **7** 735-45.

Snustad, D.P., Haas, N.A., Kopczak, S.D. and Silflow, C.D. (1992) The small genome of *Arabidopsis* contains at least nine expressed beta-tubulin genes. *Plant Cell*, **4** 549-56.

Spradling, A.C., Stern, D.M., Kiss, I., Roote, J., Laverty, T. and Rubin, G.M. (1995) Gene disruptions using *P* transposable elements: an integral component of the *Drosophila* genome project. *Proc. Natl Acad. Sci. USA*, **92** 10824-30.

Springer, P.S., McCombie, W.R., Sundaresan, V. and Martienssen, R.A. (1995) Gene trap tagging of *PROLIFERA*, an essential *MCM2-3-5*-like gene in *Arabidopsis*. *Science*, **268** 877-80.

Stammers, M., Schmidt, R. and Dean, C. (1996) Physical mapping of the *Arabdidopsis thaliana* genome, in *Genomes of Plants and Animals: 21st Stadler Genetics Symposium* (eds J.P. Gustafson and R.B. Flavell), Plenum Press, New York, pp. 73-86.

Sundaresan, V. (1996) Horizontal spread of transposon mutagenesis: new uses for old elements. *Trends Plant Sci.*, **1** 184-90.

Sundaresan, V., Springer, P., Volpe, T., Haward, S., Jones, J.D.G., Dean, C., Ma, H. and Martienssen, R. (1995) Patterns of gene action in plant development revealed by enhancer trap and gene trap transposable elements. *Genes Dev.*, **9** 1797-1810.

Swinburne, J., Balcells, L., Scofield, S.R., Jones, J.D.G. and Coupland, G. (1992) Elevated levels of *Activator* transposase mRNA are associated with high frequencies of *Dissociation* excision in *Arabidopsis*. *Plant Cell*, **4** 583-95.

Topping, J.F. and Lindsey, K. (1995) Insertional mutagenesis and promoter trapping in plants for the isolation of genes and the study of development. *Transgenic Res.*, **4** 291-305.

Tsugeki, R., Kochieva, E.Z. and Fedoroff, N.V. (1996) A transposon insertion in the *Arabidopsis* *SSR16* gene causes an embryo-defective lethal mutation. *Plant J.*, **10** 479-89.

van Haaren, M.J.J. and Ow, D.W. (1993) Prospects of applying a combination of DNA transposition and site-specific recombination in plants: a strategy for gene identification and cloning. *Plant Mol. Biol.*, **23** 525-33.

Velculescu, V.E., Zhang, L., Vogelstein, B. and Kinzler, K.W. (1995) Serial analysis of gene expression. *Science*, **270** 484-87.

Venter, J.C., Smith, H.O. and Hood, L. (1996) A new strategy for genome sequencing. *Nature*, **381** 364-66.

Walbot, V. (1992) Strategies for mutagenesis and gene cloning using transposon tagging and T-DNA insertional mutagenesis. *Ann. Rev. Plant Physiol. & Plant Mol. Biol.*, **43** 49-82.

Walden, R., Fritze, K., Hayashi, H., Miklashevichs, E., Harling, H. and Schell, J. (1994) Activation tagging: a means of isolating genes implicated as playing a role in plant growth and development. *Plant Mol. Biol.*, **26** 1521-28.

Wilson, R., Ainscough, R., Anderson, K., Baynes, C., Berks, M., Bonfield, J. *et al.* (1994) 2.2 Mb of contiguous nucleotide sequence from chromosome III of *C. elegans*. *Nature*, **368** 32-38.

Wilson, K., Long, D., Swinburne, J. and Coupland, G. (1996) A *Dissociation* insertion causes a semidominant mutation that increases expression of *TINY*, an *Arabidopsis* gene related to *APETALA2*. *Plant Cell*, **8** 659-71.

Yoder, J.I. (1990) Rapid proliferation of the maize transposable element *Activator* in transgenic tomato. *Plant Cell*, **2** 723-30.

Zachgo, E.A., Wang, M.L., Dewdney, J., Bouchez, D., Camilleri, C., Belmonte, S. *et al.* (1996) A physical map of chromosome 2 of *Arabidopsis thaliana. Genome Res.*, **6** 19-25.

Zwaal, R.R., Broeks, A., van Meurs, J., Groenin, J.T.M. and Plasterk, R.H.A. (1993) Target-selected gene inactivation in *Caenorhabditis elegans* by using a frozen transposon insertion mutant bank. *Proc. Natl Acad. Sci. USA*, **90** 7431-35.

3 Biochemical genetic analysis of metabolic pathways

Christopher S. Cobbett

3.1 Introduction

In view of the extent of biochemical and physiological studies of aspects of plant metabolism, biochemical genetic studies have lagged behind, particularly with respect to the genetic analysis of other aspects of plant biology. Nevertheless, there have been many notable advances in our understanding of plant metabolic pathways arising from studies of *Arabidopsis*. The isolation and characterisation of mutants deficient in particular metabolic pathways have provided important genetic and, in some cases, molecular confirmation of the biochemistry of those pathways. More importantly, such mutants have provided a valuable resource for studies of the roles of pathways or particular metabolites in various aspects of plant physiology.

The earliest *tour de force* of *Arabidopsis* biochemical genetics was the analysis of photorespiration by Somerville and Ogren. This study is not described in detail here and readers are referred to past reviews (Somerville and Ogren, 1982 a,b; Ogren, 1984). Nonetheless, this study stands as one of the most thorough genetic dissections of a metabolic process in *Arabidopsis*; a standard to which others might aspire. Since that study there has been a number of equally comprehensive studies of particular metabolic pathways in *Arabidopsis*, for example nitrate assimilation, tryptophan biosynthesis and fatty acid biosynthesis, which are reviewed here. In addition, other current, less comprehensive studies are yielding equally exciting new insights into many areas of metabolism not previously subjected to genetic analysis.

While this chapter is broad-ranging it is not exhaustive. There are obvious omissions, not least of which is any real consideration of the pathways of plant hormone biosynthesis. This is best left to other authors in this volume (see chapter 4). A number of interesting mutants have been included in the summary given in Table 3.1 but are not discussed at length in the text. The different pathways discussed here have been grouped in a somewhat arbitrary way in the sections that follow. Each of these pathways deserves to be and, in most cases, has been the subject of entire review articles. In the text the reader is referred to, and encouraged to read, the reviews relating to individual pathways which generally provide integrated accounts of the contributions of biochemical, genetic and physiological studies across the plant kingdom.

Table 3.1 Mutations affecting metabolic pathways in *Arabidopsis*

Locus	Pathway/metabolite affected	Enzyme activity or function affected	Gene isolated	References
Nitrogen and amino acid metabolism				
CHL1	Nitrate uptake	Low-affinity nitrate transporter	NRT1	Tsay et al., 1993; Huang et al., 1996
CHL2	Nitrate assimilation	Nitrate reductase/MoCo		Braaksma and Feenstra, 1982a; Labrie et al., 1992
CHL3/NIA2	Nitrate assimilation	Nitrate reductase	NIA2	Braaksma and Feenstra, 1982a; Wilkinson and Crawford, 1991
CHL4/RGN	Nitrate assimilation	Nitrate reductase/ MoCo		Braaksma and Feenstra, 1982a
CHL5	Nitrate assimilation	Nitrate reductase/MoCo		Braaksma and Feenstra, 1982a
CHL6/CNX	Nitrate assimilation	Nitrate reductase/ MoCo	CNX1	Braaksma and Feenstra, 1982a; Stallmeyer et al., 1995
CHL7	Nitrate assimilation	Nitrate reductase/ MoCo	CHL7	Labrie et al., 1992; Schledzewski et al., 1996
NIA1	Nitrate assimilation	Nitrate reductase	NIA1	Cheng et al., 1988; Wilkinson and Crawford, 1993
NRT2	Nitrate uptake	High-affinity nitrate transporter		Wang and Crawford, 1996
HY? (cr88)	Nitrate assimilation	Regulation of NIA2		Lin and Cheng, 1997
GDH1	NH_4^+ assimilation	Glutamate dehydrogenase	GDH1	Melo-Oliveira et al., 1996
GLS	NH_4^+ assimilation/ photorespiration	Ferridoxin-dependent GOGAT	GLU1	Somerville and Ogren, 1980a
AAT2	Aspartate	Aspartate amino transferase		Lam et al., 1995
AAT3	Aspartate	Aspartate amino transferase		Lam et al., 1995
SPE1	Polyamines	Arginine decarboxylase		R. Malmberg, personal communication
SPE2	Polyamines	Arginine decarboxylase		R. Malmberg, personal communication
TRP1	Tryptophan	Phosphoribosylanthranilate transferase	PAT1	Last and Fink, 1988; Rose et al., 1992, 1997
TRP2	Tryptophan	Tryptophan synthase β-subunit	TSB1	Last et al., 1991; Barczak et al., 1995
TRP3	Tryptophan	Tryptophan synthase α-subunit	TSA1	Radwanski et al., 1995, 1996
TRP4	Tryptophan	Anthranilate synthase β-subunit	ASB1	Niyogi et al., 1993
TRP5/AMT1	Tryptophan	Anthranilate synthase α-subunit	ASA1	Niyogi and Fink, 1992; Li and Last, 1996; Kreps and Town, 1992; Kreps et al., 1996
PAI	Tryptophan	PR-anthranilate isomerase	PAI	Li et al., 1995; Bender and Fink, 1995

Gene		Function		Reference
PAD1,2,3	Phytoalexin	Camalexin biosynthesis?		Glazebrook and Ausubel, 1994
CSR1	Branched-chain amino acids	Acetohydroxyacid synthase	CSR1	Haughn and Somerville, 1986; Haughn *et al.*, 1988
Carbohydrate metabolism				
ADG1	Starch	ADP-glucose pyrophosphorylase		Lin *et al.*, 1988a
ADG2	Starch	ADP-glucose pyrophosphorylase (large subunit)	ADG2	Lin *et al.*, 1988b; Wang *et al.*, 1997
PGM1	Starch	Phosphoglucomutase (plastid)		Caspar *et al.*, 1985
SEX1	Starch	Starch degradation, glucose transport from plastid		Caspar *et al.*, 1991; Trethewey and ap Rees, 1994
IRX1,2,3	Cellulose	Cellulose		Turner and Somerville, 1997
RSW1	Cellulose	Cellulose synthase catalytic subunit	RSW1	Arioli *et al.*, 1996; R. Williamson, personal communication
RSW (other loci)	Cellulose			Baskin *et al.*, 1992; Peng *et al.*, 1996
MUR1	GDP-L-fucose/ cell wall fucose	GDP-D-mannose-4,6-dehydratase	GMD2	Reiter *et al.*, 1993; Bonin *et al.*, 1997
MUR2.3	Cell wall fucose			Reiter *et al.*, 1997
MUR4,5,6,7	Cell wall arabinose			Reiter *et al.*, 1997
MUR8	Cell wall rhamnose			Reiter *et al.*, 1997
MUR9,10,11	Cell wall composition			Reiter *et al.*, 1997
ARA1	L-arabinose salvage pathway	L-arabinose kinase	ARA1	Dolezal and Cobbett, 1991; Cobbett, unpublished
Fatty acids and waxes				
FAB1	Fatty acid biosynthesis	C16 to C18 3-ketoacyl-ACP synthase		Wu *et al.*, 1994
FAB2	Fatty acid biosynthesis	C18:0-ACP desaturase		Lightner *et al.*, 1994

Table 3.1 (continued)

Locus	Pathway/metabolite affected	Enzyme activity or function affected	Gene isolated	References
ACT1	Glycerolipid biosynthesis	C18:1-ACP:glycerol-3-P acyltransferase (chloroplast)		Kunst *et al.*, 1988
FAD2	Fatty acid desaturation	C18:1 desaturase (ER)	FAD2	Miquel and Browse, 1992; Okuley *et al.*, 1994
FAD3	Fatty acid desaturation	C18:2 desaturase (ER)	FAD3	Browse *et al.*, 1993; Arondel *et al.*, 1992
FAD4	Fatty acid desaturation	PG-*sn*-2-C16:0 Δ3-trans-desaturase (chloroplast)		Browse *et al.*, 1985
FAD5	Fatty acid desaturation	MGDG-*sn*-2-C16:0 Δ7-desaturase (chloroplast)		Kunst *et al.*, 1989
FAD6	Fatty acid desaturation	C16:1/C18:1 desaturase (chloroplast)	FAD6	Browse *et al.*, 1989; Hitz *et al.*, 1994; Falcone *et al.*, 1994
FAD7	Fatty acid desaturation	C16:2/C18:2 desaturase (chloroplast)	FAD7	Browse *et al.*, 1986; Iba *et al.*, 1993; Yadav *et al.*, 1993
FAD8	Fatty acid desaturation	C16:2/C18:2 desaturase (chloroplast)	FAD8	McConn *et al.*, 1994; Gibson *et al.*, 1994
ELA1	Seed glycerolipids			Lemieux *et al.*, 1990
TAG1	Seed glycerolipids	Diacylglycerol acyltransferase		Katavic *et al.*, 1995
FAE1	Seed fatty acid elongation	C18 to C22 3-ketoacyl-ACP synthase	FAE1	James and Dooner, 1990; James *et al.*, 1995
DGD1	Digalactosyl diacylglycerol			Dormann *et al.*, 1995
CER1	Waxes	Fatty aldehyde decarbonylase	CER1	Koornneef *et al.*, 1989; Jenks *et al.*, 1995; Aarts *et al.*, 1995
CER2,6	Waxes	C28 to C30 fatty acid elongation	CER2	Koornneef *et al.*, 1989; Jenks *et al.*, 1995; Negruk *et al.*, 1996
CER3	Waxes		CER3	Koornneef *et al.*, 1989; Jenks *et al.*, 1995; Hannoufa *et al.*, 1996
CER4	Waxes	Fatty aldehyde reduction		Koornneef *et al.*, 1989; Jenks *et al.*, 1995

CER8	Waxes	Fatty acid reduction		Koornneef et al., 1989; Jenks et al., 1995
CER (other loci)	Waxes			Koornneef et al., 1989; Jenks et al., 1995
BCF1	Leaf waxes			Jenks et al., 1996
KNB1	Leaf waxes			Jenks et al., 1996
WAX1	Leaf waxes			Jenks et al., 1996

Metabolic pathways responding to environmental stresses

TT3	Anthocyanins	Dihydroflavonol 4-reductase	DFR	Koornneef, 1990; Shirley et al., 1992
TT4	Anthocyanins	Chalcone synthase	CHS	Koornneef, 1990; Shirley et al., 1995
TT5	Anthocyanins	Chalcone isomerase	CHI	Koornneef, 1990; Shirley et al., 1992
TT6	Anthocyanins	Flavonol synthase	Yes	Koornneef, 1990; Shirley et al., 1995; B. Shirley, personal communication
TT7	Anthocyanins	Flavonoid 3'-hydroxylase	F3'H	Koornneef, 1990; Shirley et al., 1995; Cobbett, unpublished
TT8	Anthocyanins	Regulation of dfr		Koornneef, 1990; Shirley et al., 1995
TTG	Anthocyanins	Regulation of dfr		Koornneef, 1990; Shirley et al., 1995
TT (other loci)	Anthocyanins			Koornneef, 1990; Shirley et al., 1995; Peeters et al., 1996
BAN	Anthocyanins in seed coat	Regulation	BAN	Albert et al., 1997
ICX	Anthocyanins	Regulation of chs		Jackson et al., 1995
FAH1	Sinapic acid esters/lignin	Ferulic acid 5-hydroxylase	FAH1	Chapple et al., 1992; Meyer et al., 1996
SNG1	Sinapic acid esters/lignin	Sinapoylglucose:malate sinapoyltransferase		Lorenzen et al., 1996

Table 3.1 (continued)

Locus	Pathway/metabolite affected	Enzyme activity or function affected	Gene isolated	References
PDS1	Plastoquinone/ Phytoene desaturation	4-hydroxyphenylpyruvate dioxygenase	Yes?	Norris et al., 1995
PDS2	Plastoquinone/ Phytoene desaturation	Homogentisic acid prenyl/phytyl transferase		Norris et al., 1995
LUT1	Lutein/xanthophylls	Carotenoid ε-ring hydroxylase		Pogson et al., 1996
LUT2	Lutein/xanthophylls	Carotenoid ε-ring cyclase	Yes?	Pogson et al., 1996
ABA1	Xanthophylls/abscisic acid	Zeaxanthin epoxidase		Rock and Zeevaart, 1991
ABA2,3	Abscisic acid			Schwartz et al., 1997
CAD1	Phytochelatins	Phytochelatin synthase		Howden et al., 1995a
CAD2	Glutathione/phytochelatins	γ-Glutamylcysteine synthetase	GSHA	May and Leaver, 1994; Howden et al., 1995b; Cobbett, unpublished
SOZ1	Ascorbic acid	Biosynthetic		Conklin et al., 1996, 1997
ADH	Hypoxia	Alcohol dehydrogenase		Jacobs et al., 1988; Chang and Meyerowitz, 1986

Other metabolic pathways

Locus	Pathway/metabolite affected	Enzyme activity or function affected	Gene isolated	References
GSM1	Glucosinolates	Transamination of C4–C6 keto acids		Haughn et al., 1991
GSL-ELONG	Glucosinolates	Alkyl side chain elongation		Magrath et al., 1994; Mithen et al., 1995
GSL-OHP/ALK	Glucosinolates	Modification of methylsulphinylpropyl glucosinolate		Magrath et al., 1994; Mithen et al., 1995
GSL-OH	Glucosinolates	Hydroxylation of butenyl glucosinolate		Magrath et al., 1994; Mithen et al., 1995
DCT	Photorespiration	Chloroplast dicarboxylate transport		Somerville and Ogren, 1983
GLD1/2	Photorespiration	Mitochondrial glycine decarboxylase		Somerville and Ogren, 1982c; Artus et al., 1994
PCO	Photorespiration	Phosphoglycolate phosphatase		Somerville and Ogren, 1979

Locus	Pathway	Enzyme/function		References
RCA	Photorespiration	Regulation of carboxylase activation		Somerville et al., 1982
SAT	Photorespiration	Serine-glyoxylate aminotransferase		Somerville and Ogren, 1980b
STM	Photorespiration	Serine transhydroxymethylase		Somerville and Ogren, 1981
APT1	Purines	Adenine phosphoribosyl transferase		Moffatt and Somerville, 1988; Moffatt et al., 1992
BIO1	Biotin	Diaminopelargonic acid aminotransferase		Schneider et al., 1989; Patton et al., 1996a
BIO2	Biotin	Biotin synthase	BIO2	Patton et al., 1996b,c
PY	Thiamine	Pyrimidine biosynthesis		Li and Rédei, 1969
TH1	Thiamine	Thiamine phosphate pyrophosphorylase		Komeda et al., 1988
TH2/3	Thiamine	Thiamine biosynthesis		Li and Rédei, 1969; Koornneef and Hanhart, 1981
TZ	Thiamine	Thiazole biosynthesis		Li and Rédei, 1969
MTO1	Methionine	Biosynthesis		Inaba et al., 1994
CGL	Complex glycans	N-acetyl glucosaminyl transferase		von Schaewen et al., 1993; Gomez and Chrispeels, 1994
STE1	Sterol desaturation	Δ7-sterol-C-5-desaturase		Gachotte et al., 1995
XAN (various)	Chlorophyll			Runge et al., 1995

This table includes genetic loci in *Arabidopsis* associated with metabolic pathways. The loci have been grouped by the pathway affected and the groups have been listed in the order in which they are discussed in the text. In some cases a mutant locus may affect other functions in addition to the metabolic pathway indicated. The references include, where possible, one publication describing the mutant and a second describing the isolation of the gene.

Abbreviations: MoCo, molybdenum co-factor; GOGAT, glutamine-oxoglutarate aminotransferase; ACP, acyl carrier protein; Cn, refers to a chain of n carbon atoms; PG, phosphatidyl-glycerol; ER, endoplasmic reticulum; MGDG, monogalactosyl diacylglycerol.

One aim of this chapter is to provide an account of the mutants of *Arabidopsis* affected in particular metabolic pathways and the metabolites, enzyme activities and genes that have thus far been shown to be affected in these mutants. A second aim is to illustrate the different approaches to the identification of mutants, and a third is to show how these mutants have been or can be used to determine the roles of metabolites or pathways in important aspects of plant physiology. Finally, by encompassing the breadth of biochemical genetic studies in *Arabidopsis*, it is hoped that some of the interest and excitement that has led other workers to pursue this research will be instilled in the reader.

3.2   Nitrogen and amino acids

3.2.1   Assimilation of nitrate

Nitrate is the predominant source of inorganic nitrogen utilised by higher plants. Assimilation of nitrate involves its uptake and successive reduction to nitrite and ammonia (for a review, see Crawford, 1995). Mutants deficient in nitrate assimilation have been isolated in *Arabidopsis* by selecting for resistance to the nitrate analogue, chlorate. When taken up, chlorate is reduced by nitrate reductase to chlorite, which is toxic. Selection for resistance to chlorate would be anticipated to identify mutants affected in either nitrate uptake or nitrate reductase (NR) activity. Traditionally, in *Arabidopsis*, selections using 1 mM chlorate and 2 to 5 mM nitrate have been used to identify mutants affected in nitrate assimilation. This selection was used by Braaksma and Feenstra (1982 a,b) who identified mutants at several different loci that had reduced levels of NR activity.

Two different NR genes have been identified in *Arabidopsis* by using heterologous probes. One of these, NIA2, mapped close to the *CHL3* locus (Cheng *et al.*, 1988; Crawford *et al.*, 1988). Wilkinson and Crawford (1991) demonstrated that NIA2 is *CHL3* by the isolation of a series of chlorate-resistant, gamma-ray-induced mutants at the *CHL3* locus in which the NIA2 gene had been deleted and by complementation of the mutant phenotype with the NIA2 gene. These *chl3* null mutants were viable and retained about 10% of wild-type NR activity in leaves, demonstrating that, while NIA2 accounted for the majority of leaf NR activity, a second gene, presumably NIA1, was also active. Subsequently, by using a modified selection protocol (0.1 mM chlorate with no nitrate) a mutant was isolated that completely lacked NR activity and had mutations in both NIA1 and NIA2 (Wilkinson and Crawford, 1993).

NR activity is dependent on a molybdenum-pterin cofactor (MoCo). Thus, mutants deficient in MoCo would also be expected to be deficient in NR activity. Among the chlorate-resistant mutants of *Arabidopsis*, *chl2, chl4, chl5, chl6* and *chl7* were NR and MoCo deficient (Braaksma and Feenstra, 1982a; Labrie *et al.*, 1992). A gene, *CNX1*, able to complement a MoCo-deficient mutant of *E. coli*, maps to the *chl6* locus and probably corresponds to the *CHL6* gene (Stallmeyer *et al.*, 1995). The *CHL7* gene has recently been isolated by T-DNA tagging and shows homology to gamma adaptin proteins from other organisms that are required for assembling golgi vesicles (Schledzewski *et al.*, 1996).

Because soil nitrate concentrations can vary over a wide range of concentrations, plants have evolved multiphasic nitrate uptake systems in which some components are highly responsive to nitrogen availability. Studies in several species have identified both high-affinity (K_m between 5 and 300 mM) and low-affinity ($K_m > 0.5$ mM) proton/nitrate cotransport systems. Previously, constitutive and nitrate-inducible components of the high-affinity system but only a constitutive low-affinity system had been identified in plants (Glass and Siddiqi, 1995). In the absence of appropriate mutants, or the isolation of the genes involved in nitrate transport, it has not been possible to distinguish between a model where transporters of different kinetic parameters are different forms of the same transporter and one where the transporters are genetically distinct. The highlights of the genetic analysis of nitrate assimilation in *Arabidopsis* have been studies that have resolved some of the complexities of nitrate uptake revealed by earlier physiological and biochemical analyses in other species.

From the original chlorate-resistant mutants, one, *chl1*, had wild-type levels of NR. Early studies by Doddema and Telkamp (1979), confirmed by Huang *et al.* (1996), demonstrated that *chl1* mutants are defective in nitrate/chlorate uptake in the low-affinity range. Cloning *CHL1* by using a T-DNA-tagged chlorate-resistant mutant revealed a gene (NRT1) encoding a member of a characteristic super-family of transporters. Expression of *CHL1* in *Xenopus* oocytes confirmed that it encodes an electrogenic, proton-coupled, nitrate transporter (Tsay *et al.*, 1993). *In situ* hybridisation studies in *Arabidopsis* have shown that the *CHL1* mRNA is expressed predominantly in epidermal cells in young roots but in cortical or endodermal cells in older roots, consistent with a role in nitrate uptake in different cell types during root development. Most significantly, *CHL1* mRNA is highly inducible by nitrate (Tsay *et al.*, 1993), in contrast with previous observations in plants of only a constitutive low-affinity mechanism. This has led to a proposal that the low-affinity system also consists of both constitutive and inducible components. Molecular studies in *Arabidopsis* have identified a candidate gene (NTL1) for the constitutive low-affinity uptake mechanism (referred to in Huang *et al.*, 1996). Using the modified selection for chlorate-resistant mutants, Wang and Crawford (1996) have identified a mutant, *nrt2*, deficient in the high-affinity nitrate uptake mechanism. This provides evidence for genetically distinct transporters controlling the high- and low-affinity transport mechanisms.

No mutants affected in the regulation of NR activity have been identified from the selection regimes described above. Among mutants isolated using a more subtle selection procedure involving exposure of seedlings successively to ammonium-, nitrate- and nitrate/chlorate-supplemented media (Lin and Cheng, 1997) were some that identified new loci. One of these, *cr88*, affects the regulation of nitrate reductase (NIA2) expression and other light-regulated genes. The mutant has an etiolated appearance and defines a new *long hypocotyl (HY)* locus. This and other mutants will be valuable in exploring the complex regulatory mechanisms integrating NR expression with other aspects of plant physiology.

The studies on nitrate assimilation are an example of the systematic genetic dissection of a complex metabolic pathway. A useful selection, systematically applied and modified, has yielded a wealth of mutants, allowing the identification and isolation of many genes in the pathway and an understanding of the roles of those gene products in plant physiology.

3.2.2 Assimilation of ammonium into amide amino acids

In plants ammonium is assimilated into the amino acids glutamine and glutamate by the enzymes glutamine synthetase (GS), glutamine-oxoglutarate aminotransferase (GOGAT) and glutamate dehydrogenase (GDH), and then into aspartate and asparagine by aspartate amino transferase (AspAT) and asparagine synthetase (AS) (Lam *et al.*, 1995, 1996; Oliveira *et al.*, 1997). The main functions of this assimilatory process are the primary assimilation of ammonium derived from nitrate reduction, the reassimilation of ammonium released during the process of photorespiration and the recycling of ammonium released during amino acid turnover, particularly during processes such as senescence. Analysis of this complex, highly regulated process is made especially difficult by the fact that each of these activities is contributed to by multiple isozymes, many of which are expressed in different tissues or cell types or are located in different sub-cellular compartments. In *Arabidopsis* many of the genes encoding these enzymes have been isolated and their sequences have indicated the likely subcellular compartment to which the enzymes are directed.

In view of our currently limited understanding of this complex process it is difficult, in most cases, to predict a likely phenotype, if any, for any particular mutant. A productive approach has been to screen for mutants in *Arabidopsis* by using an electrophoretic enzyme activity assay. GDH and AspAT activities can be detected in extracts electrophoresed on non-denaturing polyacrylamide gels by using different coupled colour reactions (Lam *et al.*, 1995). In this way, mutants at the *aat2*, *aat3* and *gdh1* loci, in which the enzyme activities of AspAT2, AspAT3 and GDH1, respectively, are either absent or have altered mobilities, have been identified (Lam *et al.*, 1996; Melo-Oliveira *et al.*, 1996). In addition, mutants that are deficient in GOGAT activity, one of the enzymes involved in the process of photorespiration, have been identified among photorespiratory mutants (Somerville and Ogren, 1980a). In view of the complexity of this process, what, then, has the isolation of these mutants contributed to our understanding?

In photorespiratory *gls* mutants ferridoxin-dependent (Fd-GOGAT) activity (which constitutes about 95% of total GOGAT activity in leaves) is absent, while

NADH-dependent (NADH-GOGAT) activity is unaffected (Somerville and Ogren, 1980a). Because *gls* mutants exhibit no apparent phenotype under conditions where photorespiration is suppressed, it appears that Fd-GOGAT is required for the reassimilation of ammonium generated by photorespiration but not for the primary assimilation of ammonium. One apparent difficulty is the identification of two genes (GLU1 and GLU2) encoding Fd-GOGAT in *Arabidopsis* whereas a single mutation abolished all leaf Fd-GOGAT activity. However, a plausible explanation for this is that expression of GLU2 is predominant in roots and a mutation in GLU1 alone could effectively abolish leaf Fd-GOGAT activity. Unpublished data have shown that GLU1 is linked to *GLS*, supporting the expectation that *gls* mutations lie in the GLU1 gene (Oliveira *et al.*, 1997). In barley, mutations affecting the chloroplastic form of GS (GS2) also cause a photorespiratory mutant phenotype, indicating that this activity also is primarily involved, in concert with Fd-GOGAT, in the reassimilation of photorespiratory ammonium. That GS1 (cytosolic GS) in barley is unable to compensate for the lack of GS2 can be accounted for by the observation that the two enzymes have different cell-specific expression patterns. Corresponding GS2 mutants were not isolated in extensive photorespiratory mutant screens in *Arabidopsis*. There is a number of possible explanations for this, the most likely of which is either the existence of more than one chloroplastic isozyme or that GS mutations in *Arabidopsis* are lethal.

The role of GDH was once thought to be the primary assimilation of ammonium derived from nitrate reduction. However, the discovery of the GS/GOGAT cycle has called that role into question and this uncertainty is enhanced by the potential of GDH to have either an assimilatory or a catabolic activity. Expression of the GDH1 gene in *Arabidopsis* increases under conditions of either carbon limitation (e.g. dark-adapted plants), suggesting a role in glutamate catabolism, or in light-grown plants supplied with ammonium, suggesting a role in ammonium assimilation (Melo-Oliveira *et al.*, 1996). However, the isolation of photorespiratory GS2 and Fd-GOGAT mutants (in barley and/or *Arabidopsis*) demonstrates that GDH1 is not sufficient for photorespiratory ammonium reassimilation. The *gdh1* mutant provides important clues to the role of GDH. Growth of this mutant is retarded and it has a chlorotic appearance in conditions of high inorganic nitrogen and this phenotype is partly relieved under conditions of low inorganic nitrogen, indicating that GDH1 may have an essential role in ammonium detoxification. Whether GDH1 plays any role in photorespiration has yet to be examined and the effect of the *gdh1* mutation on dark-adapted plants, for example, has not been reported.

AspAT activity results in the formation of aspartate, which is the substrate for many pathways of amino acid biosynthesis. In *Arabidopsis* distinct activities have been identified in the cytosol, mitochondria and plastid/peroxisome, and five different AspAT genes (ASP1-ASP4 and ASP5/AAT1) have been isolated (Schultz and Corruzi, 1995; Wilkie *et al.*, 1995). Because of the complexity of this gene family, mutants such as *aat2* and *aat3,* lacking individual isozymes, will be a valuable resource in assessing the roles of each. Together these studies illustrate the usefulness of an approach to biochemical genetics in *Arabidopsis* in which mutants are identified solely on the basis of loss of enzyme activity. These mutants, individually and in combination, will enhance our understanding of this central process in plant metabolism.

3.2.3 Polyamines

Polyamines are ubiquitous in biology and, in plants, have been implicated in numerous aspects of growth and development (for a review, see Kumar *et al.,* 1997). The predominant polyamines are putrescine, spermidine and spermine, and their biosynthetic pathway has been established through biochemical studies in many organisms. Putrescine can be derived directly from ornithine by the action of ornithine decarboxylase (ODC) and indirectly from arginine via the action of arginine decarboxylase (ADC). Putrescine and decarboxylated S-adenosyl methionine are the substrates for spermidine and, subsequently, spermine biosynthesis. The use of specific inhibitors of ADC and ODC has indicated roles for polyamines in a wide range of developmental processes. However, as is often the case, inhibitor studies are difficult to interpret and, in this pathway, do not allow the roles of individual polyamines or their conjugated forms to be examined. Early mutants isolated in tobacco were refractory to genetic analysis (Malmberg and Rose, 1987), but, more recently, *Arabidopsis* mutants defective in polyamine biosynthesis have been identified by assaying seedlings individually for ADC activity using a sensitive *in vivo* procedure (R. Malmberg, personal communication). Mutants at two loci, *spe1* and *spe2,* which have 20% to 50% of wild-type levels of ADC activity, have been identified. These mutations had obvious effects on root morphology and a double mutant, *spe1,spe2,* exhibited pleiotropic effects on root, leaf and flower development.

3.2.4 Tryptophan

Among the amino acid biosynthetic pathways our level of understanding of the tryptophan pathway in *Arabidopsis* is unique (for a review, see Radwanski and Last, 1995). This is largely owing to the systematic identification of mutant alleles in many of the genes and in most of the steps of the pathway through the careful adaptation to *Arabidopsis* of selection regimes that have proved successful in microorganisms. The pathway for tryptophan biosynthesis is shown in Figure 3.1. Selections for mutants blocked in the pathway have used various analogues of pathway intermediates. For example 5- or 6-methylanthranilate are metabolised by the tryptophan pathway enzymes in wild-type plants to toxic tryptophan analogues. Mutants blocked in the metabolism of an anthranilate analogue would be resistant to its inhibitory effects. Using this selection, *trp1* mutants affecting phosphoribosyl (PR) anthranilate transferase activity (Last and

Fink, 1988; Rose *et al.*, 1992, 1997) and *trp3* and *trp2* mutants affecting the α and β subunits of tryptophan synthase, respectively, have been isolated. Mutants at the *trp2* locus were also isolated by using 5-fluoroindole, which is converted to the toxic tryptophan analogue 5-fluorotryptophan, by the β subunit of tryptophan synthase (Barczak *et al.*, 1995). *trp2* mutants are unable to metabolise the toxic analogue and are resistant to its inhibitory effects. Interestingly, some *trp3* mutants were also isolated using this selection (Radwanski *et al.*, 1996). It is likely that some *trp3* mutations also influence the activity of the β subunit of tryptophan synthase because the α and β subunits form a heterodimeric complex.

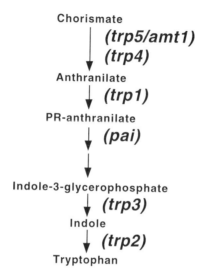

Figure 3.1 Tryptophan biosynthetic pathway. The intermediates of the tryptophan biosynthetic pathway are shown. The steps affected in particular mutants are indicated by the name of the mutant locus. The enzymes and genes affected by each mutation are listed in Table 3.1. PR; phosphoribosyl.

Subsequently, through the ingenious exploitation of a phenotype of *trp1* mutants, *trp4* mutants affected in the first step of the pathway, anthranilate synthase, were isolated. *trp1* mutants that are defective in the conversion of anthranilate to PR-anthranilate accumulate blue-fluorescent anthranilate conjugates. By mutagenising the *trp1-100* mutant and screening for derivatives in which the fluorescent phenotype observed under UV illumination was suppressed, *trp4* mutations that decrease activity of the anthranilate synthase and, consequently, the accumulation of anthranilate were identified (Niyogi *et al.*, 1993). These mutations map to the anthranilate synthase β-subunit, ASB1, gene. The *amt1* and *trp5* mutants selected for resistance to α-methyltryptophan and 6-methylanthranilate, respectively, have anthranilate synthase activity, which is less sensitive than the wild-type enzyme to feedback inhibition by tryptophan. The four independent mutations characterised are identical amino acid substitutions in the anthranilate synthase α subunit, ASA1, gene (Kreps *et al.*, 1996; Li and Last, 1996). Molecular approaches have resulted in the isolation of genes encoding all of the enzymes of the pathway. For many steps multiple genes have been identified. The *trp* loci have been shown by linkage studies, complementation and sequence determination of mutant alleles to correspond to particular genes (see Table 3.1 for relevant references).

The existence of multiple genes encoding most of the tryptophan biosynthetic enzyme activities might have been expected to prevent the isolation of mutants. With the exception of the three PR-anthranilate isomerase genes, loss of function of one member of each gene family has been sufficient to confer resistance to the particular selective analogue or suppression of the blue-fluorescent *trp1* phenotype. In the case of PR-anthranilate isomerase, while mutants in these genes have not been identified in the Columbia and Landsberg ecotypes an interesting unstable blue-fluorescent mutant affecting this activity was discovered in the Wassilewskija (Ws) ecotype. Wild-type Ws has four PR-anthranilate isomerase genes while the Ws *pai* mutant has two tandemly duplicated genes deleted, thus conferring the mutant phenotype (Bender and Fink, 1995). The instability of the mutant phenotype appears to derive from the reactivation, through changes in cytosine methylation of the DNA, of one of the remaining two genes that had been previously silenced in the wild-type.

Mutants affected in tryptophan biosynthesis have provided new approaches to understanding the regulation of the pathway and aspects of the function and metabolism of related indolic compounds. Mutants at the *trp1*, *trp2* and *trp3* loci may be, depending on the severity of the allele, tryptophan auxotrophs. Interestingly, the auxotrophic phenotype is dependent on high light conditions (Last *et al.*, 1991; Rose *et al.*, 1992; Radwanski *et al.*, 1996). One explanation proposed for this observation is that the increased turnover of photosynthetic proteins under high light conditions creates a demand for tryptophan that is exceeded in the tryptophan mutants. It is likely that under low light conditions the expression of other members of each gene family is sufficient to supply the required tryptophan. Metabolites derived from the tryptophan pathway have various roles in plant biology. One such compound is the plant hormone auxin (indole-3-acetic acid, IAA). The most severe auxotrophs exhibit a range of morphological phenotypes, such as reduced apical dominance and reduced growth, which is consistent with a defect in auxin metabolism. The pathway of auxin biosynthesis and turnover in relation to that of tryptophan has been explored using these mutants. Normanly *et al.* (1993,1995) have shown that mutants at the *trp3* and *trp2* loci (encoding α and β subunits of tryptophan synthase, respectively) accumulate IAA esters and amides, indicating that in *Arabidopsis* indole-3-phosphate or indole rather than tryptophan, form the auxin precursor. A related compound, indole-3-acetonitrile (IAN), can be converted to IAA by nitrilase. Although the role of IAN as a precursor for IAA remains a matter of arguement, it too was thought to derive directly from tryptophan. The studies of the *trp2* and *trp3* mutants have also shown this is not the case.

Phytoalexins are low molecular weight, structurally diverse, antimicrobial compounds. In *Arabidopsis* the biosynthesis of the predominant phytoalexin, camalexin, which contains an indole ring, has been explored via the isolation of camalexin-deficient mutants and through the analysis of the *trp* mutants. Three camalexin-deficient mutants, *pad1* to *pad3*, were identified using thin-layer chromatography of extracts of leaves exposed to a bacterial pathogen (Glaze-brook and Ausubel, 1994). Camalexin was reduced to 30% and 10% of wild-type levels in the *pad1* and *pad2* mutants, respectively, and was undetectable in the *pad3* mutant. In view of the antimicrobial activity of camalexin, the *pad* mutants were tested for their sensitivity to virulent and avirulent strains of a bacterial pathogen. The *pad* mutations did not affect the ability of the plants to restrict the growth of the avirulent strain, indicating that camalexin is not required for the response of plants in an incompatible reaction with a pathogen. In contrast, the *pad1* and *pad2* mutants were more susceptible to the virulent strain, suggesting that the role of camalexin is probably to restrict the growth of virulent bacteria. The *pad3* mutant was indistinguishable from the wild-type and it was suggested that while this mutant did not synthesise camalexin it may have synthesised some intermediate of the pathway that was itself a phytoalexin. Since camalexin is an indolic compound the *trp* mutants have been used to explore the connection between the tryptophan and camalexin biosynthetic pathways. A block early in the tryptophan pathway (*trp1*) reduced camalexin biosynthesis in response to the elicitor, silver nitrate, in contrast to the *trp2* and *trp3* mutations, affecting trypto-phan synthase, which had no effect, indicating that camalexin is not derived directly from tryptophan (Tsuji *et al.*, 1993).

3.3   Carbohydrate metabolism

3.3.1   Starch

Starch is a glucose polymer that acts as a short- or long-term store of carbo-hydrate (for reviews, see Beck and Ziegler, 1989 and Martin and Smith, 1995). It is an important intermediate in photosynthetic carbon metabolism where it acts as a short-term store of assimilated carbon, which is subsequently degraded to provide for further metabolism. In addition, starch accumulates as a carbohydrate reserve in storage tissues such as seeds and tubers in which it is subsequently degraded during further development. In photosynthesising chloroplasts starch is formed from fructose-6-phosphate (F-6-P) derived from the Calvin cycle. The presumed pathway of starch synthesis is shown in Figure 3.2. F-6-P is converted through successive hexose phosphates to ADP-glucose, the substrate for starch biosynthesis. In non-photosynthetic plastids, hexose phosphates, imported from

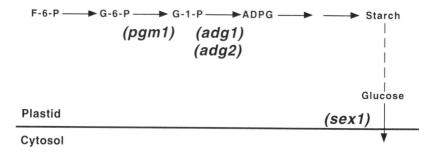

Figure 3.2 Starch biosynthesis and degradation. The presumed pathway of starch biosynthesis in plastids is shown. F-6-P derived from the Calvin cycle is metabolised via the intermediates shown to produce linear and branched forms of starch. The pathway of degradation (indicated by the dashed line) is poorly understood. The steps affected in particular mutants are indicated by the name of the mutant locus. The enzymes and genes affected by each mutation are listed in Table 3.1. F-6-P, fructose-6-phosphate; G-6-P, glucose-6-phosphate; G-1-P, glucose-1-phosphate; ADPG, ADP-glucose.

the cytoplasm, are presumed to be utilised along the same pathway. In *Arabidopsis*, mutants affected in starch metabolism have been identified by extracting leaves of individual plants with ethanol to remove chlorophyll and then staining them with iodine (for a review see Caspar, 1994). After an extended exposure to light, leaves accumulate high levels of starch and can be distinguished from mutants in which starch accumulation is absent or decreased. In contrast, starch is degraded during an extended dark period and under these conditions mutants unable to degrade starch can be distinguished from the wild-type.

Mutants at the *pgm1* locus are deficient in the single plastid, but neither of the two cytosolic phosphoglucomutase (PGM) activities (Caspar *et al.*, 1985). The most severe of these mutants are essentially starchless. Other starch-deficient mutants are affected in the ADP-glucose pyrophosphorylase (ADPG PPase) enzyme, which exists as a heterotetramer (Lin *et al.*, 1988a,b). Mutants at the *adg1* locus lack both subunits of this enzyme and are essentially starchless. In contrast, the *adg2* mutant has reduced enzyme activity and lacks only the larger subunit. Biochemical analysis of the enzymes in these mutants demonstrated that the small subunit contains sites sufficient for allosteric regulation of the enzyme but that both subunits are required for full activity (Li and Preiss, 1992). Recent molecular genetic studies have shown that the *adg2-1* mutation is in the gene encoding the large ADPG PPase subunit (Wang *et al.*, 1997). Additional starch-deficient mutants, including *stf1* (Caspar *et al.*, 1989), are not affected in PGM or ADPG PPase activity and thus appear to identify other essential functions in starch biosynthesis. Among mutants affected in starch degradation identified by Caspar *et al.* (1991), the best characterised is *sex1*. Early studies demonstrated this mutant was unaffected in any of the known activities involved in starch degradation. Subsequent work has demonstrated that this mutant is probably deficient in a hexose transporter in the chloroplast envelope, which is responsible for the transport of the products of starch degradation into the cytosol (Trethewey and ap Rees, 1994).

The *pgm1*, *adg1* and *sex1* mutants have a photoperiod-conditional phenotype such that the mutants are indistinguishable from the wild-type when grown in continuous light but grow more poorly than wild-type plants in a light–dark photoperiod. This indicates that in alternating light–dark photoperiods the storage of starch in chloroplasts and its subsequent degradation is important for optimal growth. The observation that mutants affected in either starch biosynthesis or degradation exhibit similarly decreased growth rates indicates that the flux of carbon through the starch pool, rather than the absolute size of that pool, is the important criterion for normal growth. Together these mutants have been used to study the role of starch metabolism in the processes of photosynthesis and dark respiration and other aspects of carbohydrate metabolism and energy storage, in addition to specific studies of the mechanisms controlling the pathway(s) of starch metabolism (Caspar *et al.*, 1989, 1991; Neuhaus and Stitt, 1990; Sicher and Kremer, 1992). The starchless mutants have also been used in numerous studies to explore the role of starch-filled plastids, called statoliths, in the perception of gravity by plants (Caspar and Pickard, 1989; Kiss *et al.*, 1989; Moore, 1989; Saether and Iverson, 1991). These studies suggest that while starch is not absolutely required for the perception of gravity it may play an important role in that process, particularly in dark conditions.

3.3.2 Cell wall sugar metabolism

Cellulose microfibrils form the structural basis for the plant primary cell wall and control the shape of plant cells through their orientation. In view of the abundance of cellulose and its biological and economic importance, numerous biochemical, molecular and genetic approaches to unravelling the mechanism of its biosynthesis and identifying the genes involved have been used (Brown *et al.*, 1996). A number of genetic studies using *Arabidopsis* have made an important contribution to this endeavour. One approach has been to identify mutants with irregular xylem structures by microscopic examination of stem sections (Turner and Somerville, 1997). Mutants at three *irx* loci have been identified and in each case the mutant has a reduction in the cellulose content of developing stems compared with wild-type. There is no evidence yet indicating the biochemical basis for the observed cellulose deficiency.

Because more severe defects in cellulose biosynthesis were expected to be lethal, an alternative approach to identifying such mutants, rarely used in *Arabidopsis* genetics, has been to identify temperature-conditional mutants. A reduced rate of cellulose deposition in primary cell walls causes a radial swelling of organs such as roots. Thus, mutant seedlings were identified by screening for a root tip radial swelling (*rsw*) phenotype after transfer to the non-permissive temperature (30°C) and then recovered by subsequent growth of the plants at

18°C (Baskin *et al.*, 1992). Mutants with this phenotype identify at least four different loci (Peng *et al.*, 1996) and include a class that is impaired in cellulose biosynthesis. One of these genes, *RSW1*, has been cloned and appears to encode the catalytic subunit of cellulose synthase (Arioli *et al.*, 1996). The *rsw1* mutant at the non-permissive temperature lacks the plasma membrane rosettes that are the sites of the cellulose synthase complexes. This is an exciting advance that illustrates the usefulness of temperature-conditional mutants for the identification of mutants that would otherwise be lethal.

In addition to cellulose, the cell wall contains a wide range of polysaccharides formed from a variety of different sugar components. The composition of the cell walls of *Arabidopsis* has been described by Zablackis *et al.* (1995). The activated sugar donors for polysaccharide biosynthesis are the nucleotide diphosphate (NDP) sugars. In most cases these are derived from UDP-glucose or GDP-mannose via various NDP-sugar interconversions referred to as the *de novo* NDP-sugar biosynthetic pathways. Reiter *et al.* (1993, 1997) argued that alterations to, or deficiencies in, many of the constituents of cell walls may not be lethal and adopted a different approach to identify such mutants. In this approach derivatised, hydrolised leaf cell wall extracts were displayed by gas chromatography and examined for differences in monosaccharide composition. The mutants identified had substantial reductions in the proportions of cell wall fucose, arabinose or rhamnose, grouped into three (*mur1* to *mur3*), four (*mur4* to *mur7*) and one (*mur8*) complementation groups, respectively. In addition, mutants at the *mur9*, *mur10* and *mur11* loci had more complex alterations in monosaccharide composition (Reiter *et al.*, 1997). The *mur1* mutants completely lack fucose in the aerial portions of the plant and exhibit a 60% reduction in roots. Biochemical studies indicate that *mur1* plants are blocked in the L-fucose biosynthetic pathway, which involves the conversion of GDP-D-mannose to GDP-L-fucose in two steps by a GDP-D-mannose-4,6-dehydratase and a GDP-4-keto-6-deoxy-D-mannose-3,5-epimerase-4-reductase. Biochemical, genetic and molecular analyses have shown that the *MUR1* locus encodes a 4,6-dehydratase activity (GMD2) (Bonin *et al.*, 1997).

Only the *mur1*, *mur9* and *mur10* mutants exhibited a mutant phenotype: reduced vigour or dwarfed stature (Reiter *et al.*, 1997). The best characterised mutants to date, *mur1*, have a slightly dwarfed stature and decreased tensile strength in stem segments, suggesting that the almost complete absence of fucose in cell-wall polysaccharides has some, but not a significant, effect on cell wall function. Cell wall xyloglucans can be hydrolysed to form oligosaccharides which have hormone-like effects on growth and have been termed 'oligosaccharins'. One of these, a particular nonasaccharide that contains a single fucose residue, inhibits phytohormone-induced growth of plant cells. In contrast, a related oligosaccharide that lacks the fucose residue does not inhibit growth. Thus, if fucose was an essential component of an oligosaccharin that inhibits growth, it seemed unlikely that a mutant lacking fucose would have a dwarfed phenotype (Fry, 1996). The *mur1* mutant appeared to challenge the idea that fucose is an essential component of biologically active oligosaccharins. To resolve this apparent anomaly Zablackis *et al.* (1996) have determined the structure of xyloglucans and their nonasaccharide derivatives in the *mur1* mutant. They found D-fucose (6-deoxy-L-galactose) was replaced by the related sugar, L-galactose, and that the activity of this oligosaccharin was indistinguishable from that containing fucose. This is a remarkable example of plasticity in plant metabolism that raises further interesting questions about the consequences of this substitution of L-galactose for L-fucose. Fry (1996) has speculated that the oligosaccharin containing L-galactose may be less easily degraded *in vivo* and may have a longer biological half-life and that this may be the cause of the dwarf phenotype.

In view of the fact that few of the *mur* mutations had a visible effect on phenotype, the construction of various double mutant combinations may be an instructive approach to further understanding the contributions of each of the cell wall components to cell wall structure and function. To this end, Reiter *et al.* (1997) have constructed a *mur1,mur4* double mutant which has, in contrast to the slightly dwarfed *mur1* parent, an extreme dwarf phenotype. There are exciting prospects here for understanding both the structural and regulatory contributions of cell wall polysaccharides and their oligosaccharide derivatives.

In addition to the *de novo* pathways of NDP-sugar biosynthesis some free sugars can be metabolised along a sugar salvage pathway to their respective NDP-sugar derivative via a phosphorylated intermediate. For example, free L-arabinose is converted by L-arabinose kinase to L-arabinose-1-phosphate and then to UDP-L-arabinose. The salvage pathways are presumed to recycle sugars released during cell wall turnover through the action of glycosidases, particularly during periods of growth and senescence. The *ara1-1* mutant is sensitive to arabinose and is deficient in arabinose kinase activity. Presumably, in the absence of kinase activity in the mutant, exogenous arabinose accumulates intracellularly and inhibits growth (Dolezal and Cobbett, 1991). The *ARA1* gene has been isolated by positional cloning and has similarities to other known sugar kinase genes (Lecharny and Cobbett, unpublished). Suppressors of the *ara1-1* arabinose-sensitive phenotype have been isolated and these have been predicted to affect the mechanism of arabinose transport (Cobbett *et al.*, 1992). The *ara1-1* mutant and its suppressors and the *mur* mutants altered in arabinose metabolism will together provide a more complete view of cell wall metabolism in plants.

3.4 Glycerolipids and waxes

3.4.1 Membrane glycerolipids

Fatty acids are the most abundant of the different classes of lipids in plant cells, where, in the form of glycerolipids, they constitute the major component of the lipid membranes of the cell and its various subcellular compartments. These different membranes have characteristic proportions of different glycerolipids and distinct fatty acid compositions, which are presumed to be important for the functions of individual membranes. The use of *Arabidopsis* to identify and characterise a, now considerable, number of mutants affected in fatty acid biosynthesis has been a major advance in the last decade. This endeavour has enhanced our understanding of fatty acid biosynthesis and of the contribution of fatty acid composition to membrane function, and stands as an example of the comprehensive genetic dissection of a complex metabolic pathway in a plant (for reviews, see Browse and Somerville, 1991, and Ohlrogge and Browse, 1995).

The major fatty acids of plants have C16 or C18 carbon chains and contain one to three double bonds. In *Arabidopsis* the major fatty acids are palmitic (C16:0), oleic (C18:1), linoleic (C18:2) and linolenic (C18:3). In membranes, most fatty acids are found in the form of glycerolipids consisting of the glycerol three-carbon backbone with fatty acids attached at the *sn*-1 and *sn*-2 positions and a polar headgroup at *sn*-3. There is a variety of glycolipid and phospholipid headgroups which contribute to an extensive range of membrane glycerolipid structures. In plants, in contrast to other eukaryotes that produce fatty acids in the cytosol, fatty acids are synthesised in the plastids (Figure 3.3). The pathway for fatty acid biosynthesis begins with the transfer of a malonyl group from malonyl-CoA to the acyl carrier protein (ACP). This then undergoes a condensation reaction with acetyl-CoA (catalysed by 3-keto-acyl-ACP synthase III, KASIII) followed by successive reduction, dehydration and reduction reactions. Subsequent cycles of condensation (by KASI) with malonyl-ACP, extend the carbon chain by two carbons with each cycle to produce a C16 acyl group. The final condensation (by KASII) elongates the C16 palmitoyl-ACP to C18 stearoyl-ACP. The latter, but not the former, is efficiently desaturated by stearoyl-ACP desaturase to form C18:1-ACP. The elongation cycle is terminated by the removal of the acyl group from ACP.

The incorporation of fatty acids into glycerolipids involves two pathways: the prokaryotic pathway in the chloroplast inner envelope and the eukaryotic pathway in the endoplasmic reticulum (ER). The C16:0 and C18:1 fatty acids synthesised in the chloroplast are used directly in the chloroplast for glycerolipid synthesis or exported as CoA esters to the ER. At these sites fatty acids are transferred to glycerol-3-phosphate to form phosphatidic acid (PA). PA made by the prokaryotic pathway has C16:0 at the *sn*-2 position and, generally, C18:1 at the *sn*-1 position. In contrast, PA formed in the ER generally contains C18:1 at both positions. Polar head groups are added to PA via alternate steps in both pathways: either PA is converted to diacylglycerol (DAG) by a PA-phosphatase or to the nucleotide-activated form, CDP-DAG. DAG and CDP-DAG are the substrates for head group incorporation to form various phospholipids characteristic of extrachloroplast membranes and the galacto- and sulpholipids characteristic of chloroplastic membranes. The DAG derived from phosphatidylcholine

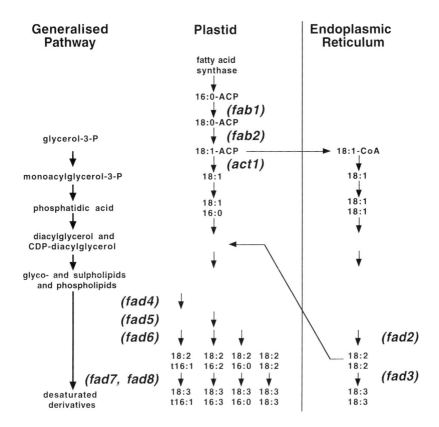

Figure 3.3 A generalised pathway of membrane glycerolipid biosynthesis. A generalised pathway is shown on the left. The particular forms of the classes of compounds that occur in the major steps of glycerolipid biosynthesis in the plastid and endoplasmic reticulum are shown. For example, the major form of phosphatidic acid in the plastid contains two acyl groups of 18 and 16 carbon atoms that contain one (18:1) and nil (16:0) double bonds, respectively. All double bonds are in the *cis*-configuration except for t16:1. Particular mutants are indicated by the name of the mutant locus adjacent to the arrow indicating the metabolic step(s) in the pathway affected in that mutant. The enzymes and genes affected by each mutation are listed in Table 3.1. ACP, acyl carrier protein; CoA, coenzyme A.

formed via the eukaryotic pathway can be returned to the chloroplast and contributes to plastid lipids (see Figure 3.3). The major glycerolipids with their various head groups are first synthesised using only C16:0 and C18:1 acyl groups. These are then desaturated by the membrane-bound desaturases of the chloroplast and ER to produce the various unsaturated forms of glycerolipids found in plant cell membranes.

Mutants affected in many of the steps of these pathways (Figure 3.3) have been identified by gas chromatography of derivatised extracts of leaf lipids, which reveals changes in the profile of lipid components. The *fab1* mutant, identified by increased levels of C16:0, appears to have a defect in KASII affecting the elongation of C16:0 to C18:0 (Wu *et al.*, 1994) while *fab2* mutants probably have decreased 18:0-ACP desaturase activity (Lightner *et al.*, 1994). Both of these mutants are leaky and exhibit relatively small changes in overall membrane composition. The *act1* mutant is deficient in the chloroplast acyl-ACP *sn*-glycerol-3-phosphate acyltransferase activity, the first step of the prokaryotic pathway, and consequently has altered flux of lipid metabolism through the two pathways (Kunst *et al.*, 1988). Mutants at seven other loci (*fad2* to *fad8*) affecting fatty acid desaturation have been identified (Figure 3.3 and Table 3.1). Previous attempts to isolate genes encoding membrane-bound fatty acid desaturases via biochemical approaches proved intractable. However, the identification of these mutants in *Arabidopsis* has led to the isolation of most of the *FAD* genes by gene tagging, chromosome walking and, later, by sequence homology (see Table 3.1 for relevant references). The isolation of these genes will ultimately lead to the manipulation of the lipid composition of different oil-producing crop plants, which are of considerable economic importance.

Glycerolipid biosynthesis must be highly regulated to provide individual lipid structures at levels appropriate for the structural and functional requirements for individual cell membranes. While this is a complex process and little is understood of the signals and points of regulation for glycerolipid biosynthesis, the analysis of various mutants has provided some insights into these regulatory mechanisms. The observation that glycerolipids are desaturated to varying degrees in membranes suggests that desaturase activity may be an important regulatory step. Analysis of the lipid compositions of plants heterozygous for various desaturase mutant alleles demonstrates that, for some loci, heterozygotes have compositions intermediate between the wild-type and mutant (Browse *et al.*, 1985, 1986). This suggests that, in those cases, the level of gene expression is critical in regulating enzyme activity. For other loci the lipid compositions of the heterozygotes were more nearly wild-type, indicating that these enzymes are not limiting in the pathway.

In wild-type plants, chloroplast glycerolipids are derived from both the prokaryotic and eukaryotic biosynthetic pathways (Figure 3.3) in approximately equal amounts. The *act1* mutant is deficient in the activity of the first step of the prokaryotic pathway and, as a result, the greater proportion of the chloroplast lipids are derived via the eukaryotic pathway. Although the levels of some lipid components are altered in this mutant, the eukaryotic pathway largely compensates for the loss of activity of the prokaryotic pathway (Kunst *et al.*, 1988). In addition, the residual metabolic flux through the prokaryotic pathway is largely directed towards phosphatidylglycerol to maintain near-wild-type levels of that component. These observations demonstrate the considerable regulatory flexibility inherent in the mechanisms of glycerolipid biosynthesis in plant cells.

The lipid compositions of membranes change in response to different growth temperatures. The regulation of this response remains unclear. However, at least one of the *FAD* genes of *Arabidopsis* is regulated in response to temperature. Independent *fad7* mutants display a mutant phenotype only above 20°C. This suggests that a second desaturase isozyme may be functioning at the permissive temperature and that, by inference, the expression of this second enzyme is temperature-dependent. The identification of *fad8* mutants confirmed the existence of a second gene and the isolation of the *FAD8* gene confirmed that expression of its mRNA was regulated in response to temperature (Gibson *et al*, 1994). Furthermore, the role of membrane lipid composition in the ability of plants to resist and respond to freezing is an area of study of some agricultural importance. A number of the *Arabidopsis* mutants have provided some insights into this phenomenon. For example, the *fab1* mutant has increased proportions of disaturated phosphatidylglycerol, a membrane component that has been correlated with chilling sensitivity in other species. Although, *fab1* plants are sensitive to extended exposure to cold temperatures, they do not exhibit a classical chilling-sensitive phenotype (Wu and Browse, 1995; Wu *et al*., 1997). This indicates that although increased levels of disaturated PG are deleterious, other factors must be involved in chilling injury. Similar studies exploring the roles of fatty acid desaturation in membrane physiology have been reviewed by Ohlrogge and Browse (1995) and Somerville (1995).

Chloroplast membranes contain high proportions of trienoic, C16:3 and C18:3, fatty acids that derive from the activity of the membrane-bound desaturases of both the chloroplast and ER. By virtue of their abundance it has been assumed that the trienoic acids must have a critical role in chloroplast function. This has been tested by McConn and Browse (1996) in an elegant experiment utilising the panel of *fad* mutants previously isolated. These authors constructed a triple mutant, *fad3, fad7, fad8*, deficient in the single ER and two chloroplastic C16:2/C18:2 desaturases, and which produced no detectable linolenic acid. Growth and photosynthesis of this triple mutant is indistinguishable from the wild-type, demonstrating that linolenic acid is redundant under laboratory conditions. Unexpectedly, the triple mutant was male sterile and applications of linolenic acid or jasmonic acid to pollen restored fertility. Jasmonic acid is a plant growth regulator derived from linolenic acid and the characterisation of this triple mutant demonstrates the importance of this compound in pollen development and may provide clues to the biochemical basis of others of the many male sterile mutants already isolated. Jasmonic acid is also an important signalling molecule in the response of plants to wounding and insect predation. This same triple mutant is highly susceptible to insect predation, demonstrating that jasmonic acid is both necessary and sufficient to protect against insect attack (McConn *et al*, 1997). This mutant will be an important model in further investigations of the role of jasmonic acid in signalling pathways.

In addition to genetic studies of the biosynthesis of the fatty acid side chains, mutants affected in the biosynthesis of the polar headgroups of complex glycerolipids have been identified. In plants the most abundant of the polar lipids are the monogalactosyl and digalactosyl diacylglycerolipids (MGD and DGD, respectively). MGD and DGD are found exclusively in plastid, particularly chloroplast, membranes and it has been assumed that they play an important role in photosynthesis. By separating leaf lipid extracts of individual plants by thin-layer chromatography, Dormann *et al.* (1995) identified a mutant, *dgd1*, with a >90% reduction in DGD. This mutant exhibits stunted growth, a pale green leaf colour and has reduced photosynthetic capability associated with an altered thylakoid membrane structure, confirming the significant role of DGD in photosynthesis. Other studies of the *dgd1* mutant have detected changes in the composition of the photosynthetic apparatus which include effects on the water oxidising complex (C. Benning, personal communication).

3.4.2 Storage lipids

In many plant species lipids constitute a considerable component of seeds, providing a source of energy for germinating seedlings. Oil seeds are of considerable economic importance and their products are used in numerous nutritional, agricultural and industrial applications. The use of molecular approaches to modify the compositions of oil seed crops will improve and extend the usefulness of their products (Ohlrogge, 1994). In *Arabidopsis* C16 and C18 fatty acids form the greater component of seed storage lipids. However, seeds also accumulate very long chain (C20 and C22) fatty acids (VLCFA). The elongation of C18 to C20 and C22 fatty acids occurs via an analogous pathway to that of C18 fatty acid biosynthesis in plastids (see above). To understand better the biosynthesis of VLCFA, mutants affected in the fatty acid composition of seed lipids have been identified (James and Dooner, 1990; Lemieux *et al.*, 1990). The two mutants that have been best characterised are *fae1* and *tag1*. Mutants at the *fae1* locus have reduced levels of seed VLCFA and a deficiency in both the C20 and C22 elongation activities. Because *fae1* mutations have no effect on leaf fatty acid elongation it was predicted that the *FAE1* gene encoded a seed-specific VLCFA condensing enzyme responsible for both the C20 and C22 elongations (Kunst *et al.*, 1992). The *FAE1* gene has been cloned by transposon tagging (James *et al.*, 1995) and, consistent with these expectations, its gene product is similar to other condensing enzymes and is expressed in developing seed but not in leaves. In contrast, the *tag1* mutant has more complex differences in seed lipid composition and these differences have been attributed to a deficiency in diacylglycerol acyltransferase activity (Katavic *et al.*, 1995). The outcome of this analysis prompted these authors to suggest that a more sophisticated analysis of mutants with complex phenotypes may be required.

3.4.3 Waxes

The epicuticular waxes coating the outer surfaces of plants are believed to have numerous roles in plant biology. These include mediating interactions between plants and their insect predators, influencing the water balance of plants and a role in pollen development and possibly in fertilisation. The epicuticular waxes consist largely of VLCFA and their aldehyde, alcohol and ketone derivatives. The biosynthetic pathway is confined to epidermal cells where the primary steps involve the elongation of C18 fatty acids to VLCFA, C20 to C30 in length. This occurs in microsomes in a manner similar to that described for plastid fatty acid biosynthesis above. VLCFA are reduced to primary aldehydes and alcohols and may undergo further metabolism to form alkanes, secondary alcohols and ketones (for reviews see Post-Beittenmiller, 1996, and Lemieux, 1996)

In *Arabidopsis* the *eceriferum* (*cer*) mutants deficient in the production of epicuticular waxes are visibly apparent because of the shiny, bright-green appearance of the surfaces of stems and siliques. More than 21 *cer* complementation groups have been identified thus far (Koornneef *et al.*, 1989). Mutants specifically affected in the wax composition of leaves have also been isolated (Jenks *et al.*, 1996). The genes identified in these mutants are expected to encode biosynthetic enzymes, components for the assembly of biosynthetic complexes, proteins involved in the transport of waxes to the cell surface and regulatory proteins. The analysis of the composition of the waxes of some of these mutants has indicated likely defects in the biosynthetic pathway (Jenks *et al.*, 1995) and in some cases the nucleotide sequence of the gene has provided support for those predictions. The *cer1* mutant, for example, is deficient in the products derived from the C30 aldehyde and was suspected of being deficient in a decarbonylase. Consistent with this, the gene product of *CER1* has similarities to membrane-bound desaturases and alkane hydroxylases (Aarts *et al.*, 1995). *CER2* and *CER3* have been cloned (Negruk *et al.*, 1996; Hannoufa *et al.*, 1996) but their gene products are not clearly related to any known proteins.

There have been few studies of the physiological roles of waxes in *Arabidopsis* using the *cer* mutants. The most apparent role is in pollen/stigma interactions. Some of the *cer* mutants are male sterile (Koornneef *et al.*, 1989) and conversely some mutants isolated as being defective in pollen/stigma interactions turned out to be *cer* mutants (Preuss *et al.*, 1993; Hülskamp *et al.*, 1995). Lipids and proteins in the surface tryphine layer of the pollen grain interact with the stigma surface. Pollen from some *cer* mutants lacking long-chain lipids fails to stimulate the stigma to secrete water and consequently does not hydrate and germinate. Generally the male-sterile phenotype can be suppressed in a high-humidity environment. Thus, long-chain lipids in the tryphine layer are important components for early pollen/stigma interaction. In addition, epicuticular waxes are an important determinant of the feeding habits of insects in some plants. Preliminary studies using *cer* mutants to explore the roles of individual components of waxes in this interaction have been reported (Rashotte and Feldmann, 1996). A thorough analysis of this aspect of wax function using *Arabidopsis* may lead to novel approaches to the management of plant pests.

3.5   Metabolic pathways involved in responses to stress

3.5.1 Phenylpropanoids

In plants there is a wide range of phenylpropanoid metabolic pathways and compounds, many of which have roles in responses to environmental stresses such as wounding, pathogen invasion or UV irradiation. The first step in phenylpropanoid metabolism is the conversion of phenylalanine to cinnamic acid and then to 4-coumaric acid, from which branch the major phenylpropanoid biosynthetic pathways forming, for example, lignins and the sinapic acid esters, and flavonoids such as anthocyanins (Figure 3.4). These are the two pathways that have thus far been subjected to genetic analysis in *Arabidopsis*.

3.5.1.1 Flavonoids
Flavonoid biosynthesis begins with the condensation of 4-coumaryl-CoA and malonyl-CoA by chalcone synthase to form naringenin chalcone. This is converted through a series of steps to the colourless dihydroflavanols followed by their reduction and subsequent conversion to the pelargonidin, cyanidin and delphinidin pigments through various oxidation, dehydration and glycosylation reactions (Shirley *et al.*, 1995; Shirley, 1996). A generalised flavonoid biosynthetic pathway is shown in Figure 3.4.

Anthocyanins, and variations in their expression, are conspicuous in plants by their contribution to flower and fruit colour, in which they play significant roles in pollination and seed dispersal. They also play other important roles in plants, the evidence for some of which has derived from the study of mutants in *Arabidopsis*. Since *Arabidopsis* has white flowers, anthocyanin-deficient mutants have been classically identified in this organism by the transparent testa (*tt*) mutant phenotype that results from reduced anthocyanin expression in the seed coat (Koornneef, 1990). Numerous *tt* loci have been identified, many of which identify the structural genes of the biosynthetic pathway or regulatory genes influencing their expression. The loci for the functions that have been identified are listed in Table 3.1. Many of the genes of the pathway have been isolated and correlated with a specific locus by linkage, complementation or the sequencing of mutant alleles. In addition to chalcone synthase (*tt4*), chalcone isomerase (*tt5*) and dihydroflavanol 4-reductase (*tt3*), genes encoding flavonone hydroxylase, which probably corresponds to *tt6* (B. Shirley, personal communication) and flavonoid-3'-hydroxylase (*tt7*) (Cobbett, unpublished), have been isolated (Table 3.1). A number of genes involved in the regulation of the anthocyanin biosynthetic pathway have also been identified and in some cases isolated.

Most importantly, what do the anthocyanin-deficient mutants tell us about the roles of flavonoids in plants? Evidence demonstrating that flavonoids play an important role in protection from UV radiation has been obtained using *Arabidopsis*. Experiments by Li *et al.* (1993) showed that both *tt4* and *tt5* mutants,

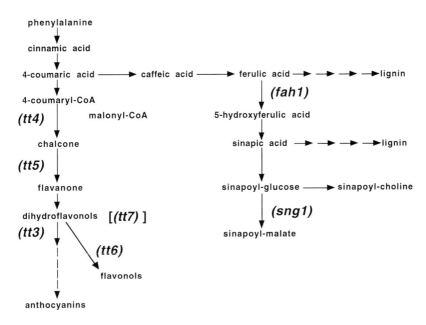

Figure 3.4 A generalised pathway for anthocyanin and sinapic acid ester biosynthesis. The intermediates of the pathways are shown. The steps affected in particular mutants are indicated by the name of the mutant locus. The *tt7* mutation affects the conversion of one dihydroflavanol compound into another. The enzymes and genes affected by each mutation are listed in Table 3.1.

which lack detectable leaf anthocyanin pigmentation, are hypersensitive to UV-B irradiation with the *tt5* mutant exhibiting a more extreme hypersensitivity phenotype. High-performance liquid chromatography (HPLC) separation of UV-absorptive compounds in these mutants showed that in the *tt4* (CHS) mutant biosynthesis of sinapic acid esters, also derived from 4-coumaryl-CoA, is increased, possibly owing to the channelling of the common substrate into a single pathway. In contrast, the *tt5* mutant had decreased levels of sinapic acid esters. These observations led to speculation that sinapic acid esters may play an equally important role in UV-B protection and that there may be a feedback regulatory interaction between flavonoid and sinapic acid ester biosynthesis. This has been confirmed in part through the characterisation of a mutant deficient in sinapic acid esters (see below).

Mutants of maize (Coe *et al*, 1981) or petunia (Taylor and Jorgensen, 1992) that lack flavonoids are infertile owing to the inability of pollen to germinate or grow normally. Fertility can be restored by the direct application of flavonols, but not other flavonoids, to defective pollen. In contrast, mutants of *Arabidopsis*, parsley and snapdragon that lack flavonoids exhibit no loss of fertility suggesting that in these species flavonols are not required for fertility. The possibility that very low levels of flavonols may be sufficient for pollen function in *Arabidopsis* and that the *tt* mutants thus far examined are leaky has been excluded by a thorough demonstration that CHS is encoded by a single gene and that the *tt4* mutation completely disrupts splicing of the mRNA, preventing its expression (Burbulis *et al.*, 1996). These experiments reinforce the observation that some functions of flavonoids vary across different families of plants.

3.5.1.2 Sinapic acid esters

A second major class of phenylpropanoid compounds in plants is the hydroxy-cinnamic acids and their esters. In *Arabidopsis* 4-coumaric acid is converted to sinapic acid via ferulic acid, both of which are converted to substrates for lignin biosynthesis (Figure 3.4). Subsequent steps in the pathway give rise to various sinapic acid esters: sinapoyl glucose, malate and choline. Previous analysis of the *tt* mutants suggested that sinapic acid esters may also be important UV-B protectants (Li *et al.*, 1993). In addition, sinapoyl choline stored in seeds is thought to be a major source of choline for phospholipid biosynthesis during the early development of seedlings (Strack, 1981).

Mutants in two loci involved in this pathway, *fah1* (formerly *sin1*) and *sng1*, were identified by separating methanolic leaf extracts of plants by TLC and visualising the sinapic acid esters by UV illumination. Mutants at the *sng1* locus lack sinapoyl malate but accumulate sinapoyl glucose in leaves, suggesting a block in the conversion of the latter to the former (Lorenzen *et al.*, 1996). This was confirmed by the absence of sinapoyl glucose:malate sinapoyltransferase activity in *in vitro* assays of the mutant. It is likely that the *SNG1* locus encodes this enzyme. Mutants at the *fah1* locus were deficient in all the sinapoyl esters, indicating that the mutation affected a step prior to the terminal enzymes of ester biosynthesis. Assays of enzymatic activities and radio-labelling and precursor feeding experiments suggested a block in ferulate-5-hydroxylase activity although this enzyme activity was not detected in the wild-type (Chapple *et al.*, 1992). The observation that, under UV illumination, the *fah1* mutant plants exhibit red chlorophyll fluorescence compared with the blue–green fluorescence of the wildtype provided a rapid screening procedure for a T-DNA-tagged allele of *fah1* and the subsequent isolation of the *FAH1* gene (Meyer *et al.*, 1996).

The most significant phenotype observed for the *fah1* mutant is, as for the *tt* mutants, sensitivity to UV. The *fah1* mutant is even more sensitive to UV-B irradiation than the *tt5* mutant, confirming the significant role of the sinapic acid esters (Landry *et al.*, 1995). It will be of considerable interest to test the UV-sensitivity of a *tt5,fah1* double mutant. Under normal light conditions the *fah1* mutants had no apparent phenotype and grew at the same rate as the wild-type, in particular during early seedling development, suggesting that a store of sinapoyl choline is not essential for growth (Chapple *et al.*, 1992). However, the mutant accumulates relatively high levels of unesterified choline in seeds and this may serve as an adequate store for phospholipid biosynthesis. Lignin from wild-type and mutant plants was analysed for its relative contents of derivatives of ferulic and sinapic acids. No sinapic acid derivatives were observed in lignin from the mutant, confirming that the chemical of lignin composition can be dramatically influenced by genetic manipulation. The modification of lignin in woody perennials is an important aim of pulp and paper manufacturers. Using *Arabidopsis* to identify and analyse mutants, such as *fah1*, to enhance our understanding of lignin biosynthesis and as a tool for the isolation of important genes in the pathway will be a significant contribution to this endeavour.

The complexity of phenylpropanoid metabolism and the wide range of functions in which they are believed to be involved indicates the genetic dissection of this pathway offers important insights into the contributions of individual compounds and branches of the pathway to particular functions and aspects of plant physiology. Many genes, or candidate genes, involved in this pathway have been identified. Mutants such as those described above will provide a valuable resource for their analysis.

3.5.2   Carotenoids

The carotenoid pigments, carotenes and xanthophylls, are found in all photosynthetic and many non-photosynthetic organisms. In the plastids of higher plants they have essential structural and functional roles in photosynthesis. Carotenoids are synthesised in plastids via a pathway that first requires the condensation of two C20 hydrocarbon molecules by phytoene synthase (Figure 3.5) (for a review see Bartley and Scolnik, 1995). The C40 product, phytoene, undergoes a series of desaturation reactions catalysed by phytoene desaturase and ζ-carotene desaturase to form the the red pigment, lycopene. The pathway then branches with alternative cyclisation reactions to form α-carotene (with a β and an ϵ ring) and β-carotene (with two β rings), respectively. Hydroxylation of the rings of α-carotene, and hydroxylation and epoxidation of the rings of β-carotene, lead to the formation of the xanthophylls, lutein, and violaxanthin and neoxanthin, respectively. The phytohormone abscisic acid is derived from violaxanthin via a series of subsequent reactions.

Carotenoids are found in the multisubunit photosystem complexes in the thylakoid membranes where they have essential structural and functional roles in photosynthesis (Demming-Adams and Adams, 1996; Green and Durnford, 1996; Horton *et al.*, 1996). Lutein is the most abundant carotenoid and is believed to have a role in the assembly and structural integrity of the light harvesting complex. β-carotene, in the reaction centres, and xanthophylls, in the antennae complexes of the photosystems, also have important functional roles in light harvesting and photoprotection by neutralising the reactive oxygen species generated during photosynthesis. The isolation and characterisation of mutants affected in this pathway have yielded exciting and, in some cases surprising, new perspectives on the regulation of this pathway and the roles of its products.

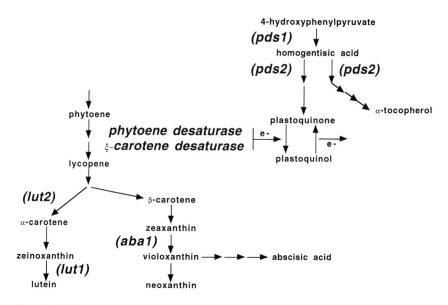

Figure 3.5 Pathway for carotenoid biosynthesis, including a role for plastoquinone. The major intermediates of the carotenoid biosynthetic pathway are shown. The carotenoid desaturases indicated in the figure require plastoquinone as an intermediate electron carrier. The steps affected in particular mutants are indicated by the name of the mutant locus. The enzymes and genes affected by each mutation are listed in Table 3.1.

Mutants defective in the early steps of carotenoid biosynthesis were expected to be lethal owing to lack of photosynthesis resulting from photo-oxidative damage. However, Norris *et al.* (1995) have found that most pigment-deficient mutants can be grown to maturity in an artificial medium supplemented with sucrose where carbon assimilation via photosynthesis is no longer essential. A panel of such mutants grown under these conditions and analysed by HPLC for those which accumulated intermediates of the carotenoid pathway included two, *pds1* and *pds2*, which accumulated phytoene owing to an apparent defect in phytoene desaturation. Unexpectedly, neither locus mapped to the position of the previously isolated structural gene for phytoene desaturase, suggesting the involvement of some other component(s) in this reaction.

Earlier work indicated a role for the quinones in phytoene desaturation and, indeed, both mutants were found to be deficient in plastoquinone and α-tocopherol but accumulated normal levels of ubiquinone. Furthermore, the *pds1* mutant could be rescued by the application of homogentisic acid (HGA), an intermediate in the plastoquinone/α-tocopherol pathway, but not by its immediate precursor, 4-hydroxyphenylpyruvate (OHPP) (Figure 3.5). This suggested a deficency in the enzyme catalysing that reaction, OHPP dioxygenase. This conclusion has been strengthened by recent data showing that a gene encoding this enzyme is linked to the *pds1* locus. In contrast, neither OHPP nor HGA was able to rescue *pds2* mutants, indicating a deficiency in a later step, common to both plastoquinone and α-tocopherol; the most likely candidate is a prenyl/phytyl transferase that acts on HGA.

In contrast to mutations affecting the early steps of the pathway, mutations specifically affecting xanthophyll biosynthesis were not expected to be lethal. The abscisic acid-deficient mutant *aba1*, for example, is deficient in violaxantin and neoxanthin and accumulates zeaxanthin, indicating a deficiency in zeaxanthin β-epoxidase activity (Rock and Zeevaart, 1991). This has been confirmed by the complementation of the *Arabidopsis* mutant with a β-epoxidase gene from *Nicotiana* (Marin *et al.*, 1996). The *aba1* mutation results in altered *in vivo* chlorophyll fluorescence and affects both the ultrastructure and function of chloroplasts. Pogson *et al.* (1996) have identified, by HPLC analysis of the pigment profiles of soil-grown plants, *lut1* and *lut2* mutants that are deficient in the xanthophyll lutein. The *lut1* mutants had low levels of lutein and accumulated its immediate precursor, zeinoxanthin, indicating a defect in the final step of the pathway, the ε ring hydroxylase. The *lut2* mutants had a genetically semidominant lutein-deficiency and had increased levels of violaxanthin and neoxanthin but not the intermediates of the lutein pathway, suggesting a deficiency in an early step, probably the ε ring cyclase enzyme. Consistent with this, a cDNA encoding this enzyme is linked to the *lut2* locus.

The characterisation of these mutants has provided some remarkable insights into the regulation and function of the carotenoid pigments in plants. Most striking is the observation that the *lut* mutations have no apparent effects on growth or chlorophyll content under moderate light conditions. This appears to belie the significance of the structural and functional roles of lutein ascribed to it by previous studies. The interpretation of this observation by Pogson *et al.* (1996) is that other carotenoids can substitute for lutein; a conclusion based on the observation that there is an equimolar increase in the total levels of other xanthophylls to compensate for the deficiency in lutein. This observation also has important implications for the mechanism by which the carotenoid pathway is regulated and implies a critical role for the ε ring cyclase encoded by *lut2*. This is supported by the observation that *lut2* mutants have a semi-dominant lutein-deficiency, suggesting that its gene product is rate limiting in the pathway. Interestingly, a preliminary report that the *lut,aba1* double mutants have no apparent phenotype (Taylor, 1996) suggests that the composition of xanthophylls within the photosynthetic complexes can be radically changed without obvious ill effects on metabolism. Further studies of the mutants described above, similar mutants, such as the abscisic acid-deficient mutants, *aba2* and *aba3* (Schwartz *et al.*, 1997), and various mutant combinations, particularly under various environmental conditions, will extend our understanding of the roles of carotenoids in photosynthesis.

3.5.3 Phytochelatins

In plants, and some yeasts, the predominant heavy-metal-inducible, heavy-metal-binding compounds are the peptides, phytochelatins (PCs). PCs have the structure, $(\gamma\text{-glu-cys})_n\text{-gly}$, and are synthesised from the tripeptide glutathione (GSH, γ-glu-cys-gly) by the enzyme PC synthase, which has been identified in biochemical studies of a number of plant species (for a recent review, see Rauser, 1995). The importance of PCs in heavy metal detoxification has been the subject of some controversy (Rauser, 1995). Cadmium-sensitive mutants identified initially by scoring for inhibition of root growth by cadmium (Howden and Cobbett, 1992) have provided a resolution to this matter. Mutants at the *cad1* locus have wild-type levels of GSH but are deficient in PCs and lack PC synthase activity *in vitro* (Howden *et al.*, 1995a). The *cad1-3* mutant, which has no detectable PCs, has been used to explore the physiological role of PCs in the detoxification of various heavy metals and has shown that PCs are essential for the detoxification of Cd and AsO_3 and, to some extent, Hg and Cu, but are not involved in the detoxification of Zn, Ni or SeO_3 (Howden and Cobbett, 1992; Cobbett, unpublished). PCs have also been ascribed a role in the homeostasis of essential meals such as copper and zinc (Rauser, 1995). This mutant will allow this suggestion to be examined more thoroughly. A second class of heavy-metal-binding proteins, the metallothioneins (MTs), has also been identified in *Arabidopsis* (Zhou and Goldsbrough, 1994). The *cad1-3* mutant will allow the relative contributions of PCs and MTs to heavy metal detoxification and metabolism to be assessed.

3.5.4 Glutathione and ascorbic acid

GSH and ascorbic acid (AsA) are both abundant low molecular weight antioxidant molecules in plant cells. They are presumed to play a significant role in the detoxification of reactive oxygen species, especially in the reduction of hydrogen peroxide via the Asada–Halliwell pathway involving dehydroascorbate reductase and glutathione reductase (Rennenberg 1982; Alscher, 1989). GSH is also the substrate for glutathione-S-transferases, which are involved in the inactivation of xenobiotics in many organisms, particularly herbicides in plants (Timmerman, 1989). GSH is synthesised from its constituent amino acids in two steps by the enzymes γ-glutamylcysteine synthetase (γ-ECS) and glutathione synthetase. A second Cd-sensitive mutant, *cad2-1*, has only 20–30% of wildtype levels of GSH and is consequently deficient in PC biosynthesis (Howden *et al.*, 1995b). Linkage and complementation data (Cobbett, unpublished) have shown that the *CAD2* locus corresponds to the GSHA gene encoding γ-ECS (May and Leaver, 1994).

In contrast, the pathway of AsA biosynthesis in plants is not well characterised. Among the mutants identified as sensitive to ozone (*soz*) are some that are deficient in AsA. The *soz1* mutant has 30% of the wild-type levels of AsA (Conklin *et al.*, 1996). Recent experiments using labelled pathway intermediates indicate that *soz1* is probably deficient in AsA biosynthesis (Conklin *et al.*, 1997). The *soz1* mutant and other, less well characterised, AsA-deficient mutants should allow the steps of the AsA biosynthetic pathway to be identified and the genes involved to be isolated. The *soz1* mutant is hypersensitive to other oxidative stresses, including sulphur dioxide and UV irradiation, confirming the role of AsA as an important antioxidant. In view of the proposed roles of both GSH and AsA as antioxidants it interesting that the *cad2-1* mutant shows no apparent hypersensitivity to oxidative stresses such as ozone and herbicides, for example, paraquat (Cobbett, unpublished), and is not more susceptible to pathogens (May *et al.*, 1996). The *cad2-1,soz1* double mutant will be a combination worth studying to determine the role of GSH under conditions of limiting AsA.

3.6   Other metabolic pathways

3.6.1 Aliphatic glucosinolates

Aliphatic glucosinolates in *Arabidopsis* and related *Brassicas* are believed to be important in mediating plant/predator interactions both as attractants and repellants and in plant/pathogen interactions. They also influence the quality of livestock fodder derived from rape and the flavour of different vegetable crops. Aliphatic glucosinolates consist of a common glucosinolate moiety and a variable aliphatic side chain, which may vary both in length and in the nature of its substituted groups. The primary substrate is presumed to be the amino acid methionine, which is deaminated and then undergoes progressive carbon chain extensions to form a series of keto acids with variable-length aliphatic side chains (with $n = 2$ to 8 carbons). The amino acid moiety of each is then reformed by transamination and further modified to form the common glucosinolate moiety. The aliphatic side chains of the methylthio glucosinolates may then undergo a series of modifications. In *Arabidopsis*, the predominant aliphatic glucosinolates are derivatives of the propyl and butyl side chains. Similar derivatives with longer side chains are minor species.

   In *Arabidopsis* two approaches have been used to identify genes important in glucosinolate biosynthesis. The first by Haughn *et al.* (1991) used HPLC screening of derivatised leaf extracts to identify mutants with variant glucosinolate profiles. The *gsm1* mutants had reduced levels of the $n = 4$, 5 and 6 side chain aliphatic glucosinolates and increased levels of the propyl ($n = 3$) glucosinolates. Levels of the $n = 7$ and 8 glucosinolates were unaffected. These observations coupled with the results of feeding experiments with labelled precursors suggested that *gsm1* mutants were deficient in the transamination of the $n = 4$, 5 and 6 carbon chain keto acid substrates.

   The second approach has been to use intraspecific variation in the glucosinolate profiles of different ecotypes of *Arabidopsis* to identify loci involved in their biosynthesis (Magrath *et al.*, 1994; Mithen *et al.*, 1995). For example, glucosinolates from the Columbia and Landsberg ecotypes differ in their side-chain length and side-chain modifications: Columbia contains predominantly methylsulphinylbutyl glucosinolate while Landsberg contains hydroxypropyl glucosinolate. By examining individuals of F2 and recombinant-inbred populations derived from crosses between these two parental lines, four distinct classes of glucosinolate profiles were observed: a profile of methylsulphinylbutyl together with hydroxypropyl glucosinolates, the two parental profiles and a profile of only methylsulphinylpropyl glucosinolates. For the F2 population the ratio of these classes was 9:3:3:1, respectively, indicating the segregation of two dominant characters: (1) the elongation of propyl to butyl (and longer) chain glucosinolates, which is present in Columbia but deficient in Landsberg, and (2) the modification of methylsulphinylpropyl glucosinolate to hydroxypropyl glucosinolate, which is present in Landsberg but deficient in Columbia. These loci were referred to as GSL-ELONG and GSL-OHP, respectively. Similar studies on crosses between other ecotypes identified a locus GSL-OH controlling the hydroxylation of butenyl glucosinolate and a locus GSL-ALK controlling the conversion of methylsulphinylalkyl to alkenyl glucosinolates. GSL-ALK maps very close to the GSL-OHP locus and may be alleles of the same gene encoding enzymes with different specificities. These studies should lead to the isolation of genes controlling a trait in *Brassicas* of some economic importance. The use of natural variation to identify genes involved in secondary metabolism is an unusual approach that may well be applied to the analysis of other pathways.

3.6.2   Other pathways

Various mutants affected in other metabolic pathways are listed in Table 3.1. These include the photorespiration mutants identified by Somerville and Ogren (1982 a,b), some mutants affected in chlorophyll biosynthesis and others affected in the biosynthesis of purines, pyrimidines, thiamine and biotin.

3.7   Conclusion

This chapter has provided a comprehensive, but not exhaustive, update of progress of the biochemical genetics of metabolic pathways in *Arabidopsis*. The different approaches that have been used to identify mutants, i.e. the intelligent use of selection regimes that have proved useful in microorganisms, the identification of mutants with a physiological phenotype and their subsequent biochemical characterisation, the 'brute-force' screening for mutants with altered enzyme activities or metabolite profiles, and the use of natural variation observed in different ecotypes, have been illustrated. Many of the mutants isolated and characterised thus far have afforded a greater understanding of the metabolic pathways themselves and the roles of these pathways in plant metabolism and physiology. A recurrent theme, if there is one, is the remarkable plasticity of plant metabolism. This can be seen in the different mutants affected in fatty acid, cell wall and carotenoid biosynthesis, for example, where significant changes in the composition of a class of compounds or in the amount of a metabolite, previously believed to be of some essential importance, can be observed without an obvious associated phenotype, at least under laboratory conditions. An implication to be drawn from this is that, ultimately, the 'brute-force' methods of screening for mutants altered in particular metabolite profiles or enzyme activities may be the most informative, since it is these that make no assumptions about phenotype. In time, with the isolation of more mutants affected in related pathways or functions, the creation of multiple-mutant combinations promises to be informative. There is probably an argument, in some cases, for identifying new mutants on the basis of phenotype in the background of an exisiting mutant that lacks a phenotype. The potential for this approach has already been indicated among cell wall function mutants, where significant synergistic effects have been observed between two otherwise relatively innocuous mutations.

Further developments in the identification of genes through the EST database and the creation of extensive pools of insertion element-induced mutations, coupled with PCR techniques, will lead to the routine identification of insertion mutations in any cloned sequence. These will undoubtedly add to our understanding of the physiological roles of those genes. Nevertheless, this approach assumes that we have a clear anticipation of the role of a gene in a pathway or the role of a pathway in plant physiology; in other words, that we think we know what to look for with respect to phenotype. If history is any judge, we will, in many cases, be incorrect. Without selecting or screening among mutagenised populations for a phenotype, or screening for a change in metabolism, many informative mutants are likely to be overlooked. Indeed, even insertion mutants may provide limited information compared with an allelic series of point mutations with variable effects on gene function.

There remains much to be done. Of all the numerous biochemical pathways in *Arabidopsis*, relatively few have been explored and for those that have the explorers have only recently set out on their journey. For example, among the amino acid biosynthetic pathways, with the exception of tryptophan, many genes, but few mutants, have been isolated. It is likely that these pathways are also susceptible to genetic analysis using approaches similar to those described in this chapter. There is much scope here for any scientist with an appreciation of plant biochemistry and physiology and/or microbial biochemical genetics.

Acknowledgements

I wish to thank the many colleagues who have generously provided me with helpful suggestions and information, sometimes prior to publication. I am particularly grateful to Cheryl Grant for her tireless effort in collating the many references. Research in the author's laboratory is supported by the Australia Research Council.

References

Aarts, M.G.M., Keijzer, C.J., Stiekema, W.J. and Pereira, A. (1995) Molecular characterization of the *CER1* gene of *Arabidopsis* involved in epicuticular wax biosynthesis and pollen fertility. *Plant Cell*, **7** 2115-27.

Albert, S., Delseny, M. and Devic, M. (1997) *BANYULS*, a novel negative regulator of flavonoid biosynthesis in the *Arabidopsis* seed coat. *Plant J.*, **11** 289-99.

Alscher, R.G. (1989) Biosynthesis and antioxidant function of glutathione in plants. *Physiol. Plant.*, **77** 457-64.

Arioli, T., Betzner, A., Birch, R., Cork, A., Hofte, H., Burn, J., Plazinski, J., Peng, L., Wittke, W., Herth, W., Redmond, J. and Williamson, R. (1996) Molecular analysis of cellulose biosynthesis, in *40th ASBMB and 36th ASPP Annual Combined Conference*, Australian Society for Biochemistry and Molecular Biology Inc. and Australian Society of Plant Physiologists Inc., Canberra, Abstract SYM-13-06.

Arondel, V., Lemieux, B., Hwang, I., Gibson, S., Goodman, H.M. and Somerville, C.R. (1992) Map-based cloning of a gene controlling omega-3 fatty acid desaturation in *Arabidopsis*. *Science*, **258** 1353-55.

Artus, N.N., Naito, S. and Somerville, C.R. (1994) A mutant of *Arabidopsis thaliana* that defines a new locus for glycine decarboxylation. *Plant Cell Physiol.*, **35** 879-85.

Barczak, A.J., Zhao, J., Pruitt, K.D. and Last, R.L. (1995) 5-Fluoroindole resistance identifies tryptophan synthase β subunit mutants in *Arabidopsis thaliana*. *Genetics*, **140** 303-13.

Bartley, G.E. and Scoinik, P.A. (1995) Plant carotenoids: pigments of photoprotection, visual attraction, and human health. *Plant Cell*, **7** 1027-38.

Baskin, T.I., Betzner, A.S., Hoggart, R., Cork, A. and Williamson, R.E. (1992) Root morphology mutants in *Arabidopsis thaliana*. *Aust. J. Plant Physiol.*, **19** 427-37.

Beck, E. and Ziegler, P. (1989) Biosynthesis and degradation of starch in higher plants. *Ann. Rev. Plant Physiol. Plant Mol. Biol.*, **40** 95-117.

Bender, J. and Fink, G.R. (1995) Epigenetic control of an endogenous gene family is revealed by a novel blue fluorescent mutant of *Arabidopsis*. *Cell*, **83** 725-34.

Bonin, C.P., Potter, I., Vanzin, G.F. and Reiter, W.-D. (1997) The *MUR1* gene of *Arabidopsis thaliana* encodes an isoform of GDP-D-mannose-4,6-dehydratase, catalyzing the first step in the *de novo* synthesis of GDP-L-fucose. *Proc. Natl Acad. Sci. USA*, **94** 2085-90.

Braaksma, F.J. and Feenstra, W.J. (1982a) Isolation and characterization of nitrate reductase-deficient mutants of *Arabidopsis thaliana*. *Theor. Appl. Genet.*, **64** 83-90.

Braaksma, F.J. and Feenstra, W.J. (1982b) Nitrate reduction in the wild type and a nitrate reductase deficient mutant of *Arabidopsis thaliana*. *Physiol. Plant.*, **54** 351-60.

Brown, R.M., Jr., Saxena, I.M. and Kudlicka, K. (1996) Cellulose biosynthesis in higher plants. *Trends Plant Sci.*, **1** 149-56.

Browse, J. and Somerville, C. (1991) Glycerolipid metabolism, biochemistry and regulation. *Ann. Rev. Plant Physiol. Plant Mol. Biol.*, **42** 467-506.

Browse, J.A., McCourt, P.J. and Somerville, C.R. (1985) A mutant of *Arabidopsis* lacking a chloroplast-specific lipid. *Science*, **227** 763-65.

Browse, J.A., McCourt, P.J. and Somerville, C.R. (1986) A mutant of *Arabidopsis* deficient in C18:3 and C16:3 leaf lipids. *Plant Physiol.*, **81** 859-64.

Browse, J., Kunst, L., Anderson, S., Hugly, S. and Somerville, C.R. (1989) A mutant of *Arabidopsis* deficient in the chloroplast 16:1/18:1 desaturase. *Plant Physiol.*, **90** 522-29.

Browse, J., McConn, M., James, D. and Miquel, M. (1993) Mutants of *Arabidopsis* deficient in the synthesis of α-linolenate. Biochemical and genetic characterization of the endoplasmic reticulum linoleoyl desaturase. *J. Biol. Chem.*, **268** 16345-51.

Burbulis, I.E., Iacobucci, M. and Shirley, B.W. (1996) A null mutation in the first enzyme of flavonoid biosynthesis does not affect male fertility in *Arabidopsis*. *Plant Cell*, **8** 1013-25.

Caspar, T. and Pickard, B.G. (1989) Gravitropism in a starchless mutant of *Arabidopsis*: implications for the starch-statolith theory of gravity sensing. *Planta*, **177** 185-97.

Caspar, T., Huber, S.C. and Somerville, C.R. (1985) Alterations in growth, photosynthesis, and respiration in a starchless mutant of *Arabidopsis thaliana* (L.) deficient in chloroplast phosphoglucomutase activity. *Plant Physiol.*, **79** 11-17.

Caspar, T., Lin, T.-P., Monroe, J., Bernhard, W., Spilatro, S., Preiss, J. and Somerville, C.R. (1989) Altered regulation of β-amylase activity in mutants of *Arabidopsis* with lesions in starch metabolism. *Proc. Natl Acad. Sci.*, **86** 5830-33.

Caspar, T., Lin, T.-P., Kakefuda, G., Benbow, L., Preiss, J. and Somerville, C.R. (1991) Mutants of *Arabidopsis* with altered regulation of starch degradation. *Plant Physiol.*, **95** 1181-88.

Caspar, T. (1994) Genetic dissection of the biosynthesis, degradation, and biological functions of starch, in *Arabidopsis* (eds E.M. Meyerowitz and C.R. Somerville), Cold Spring Harbor Laboratory Press, Cold Spring Harbor, New York, pp. 913-36.

Chang, C. and Meyerowitz, E.M. (1986) Molecular cloning and DNA sequence of the *Arabidopsis thaliana* alcohol dehydrogenase gene. *Proc. Natl Acad. Sci.*, **83** 1408-12.

Chapple, C.C., Vogt, T., Ellis, B.E. and Somerville, C.R. (1992) An *Arabidopsis* mutant defective in the general phenylpropanoid pathway. *Plant Cell*, **4** 1413-24.

Cheng, C., Dewdney, J., Nam, H., Den Boer, B.G.W. and Goodman, H.M. (1988) A new locus (*NIA1*) in *Arabidopsis thaliana* encoding nitrate reductase. *EMBO J.*, **7** 3309-14.

Cobbett, C.S., Medd, J.M. and Dolezal, O. (1992) Suppressors of an arabinose-sensitive mutant of *Arabidopsis thaliana*. *Aust. J. Plant Physiol.*, **19** 367-75.

Coe, E.H., McCormick, S.M. and Modena, S.A. (1981) White pollen in maize. *J. Hered.*, **72** 318-20.

Conklin, P.L., Williams, E.H. and Last, R.L. (1996) Environmental stress sensitivity of an ascorbic acid-deficient *Arabidopsis* mutant. *Proc. Natl Acad. Sci. USA*, **93** 9970-74.

Conklin, P.L., Pallanca, J.E., Wheeler, G.L., Last, R.L. and Smirnoff, N. (1997) Ascorbate metabolism in an ascorbate-deficient *Arabidopsis thaliana* mutant. *J. Exper. Bot. Supp.*, **4 8** Abstract P3.10.

Crawford, N.M. (1995) Nitrate: nutrient and signal for plant growth. *Plant Cell*, **7** 859-68.

Crawford, N.M., Smith, M., Bellissimo, D. and Davis, R.W. (1988) Sequence and nitrate regulation of the *Arabidopsis thaliana* mRNA encoding nitrate reductase, a metalloflavoprotein with three functional domains. *Proc. Natl Acad. Sci. USA*, **85** 5006-5010.

Demming-Adams, B. and Adams, W.W., III. (1996) The role of xanthophyll cycle carotenoids in the protection of photosynthesis. *Trends Plant Sci.*, **1** 21-26.

Doddema, H. and Telkamp, G.P. (1979) Uptake of nitrate by mutants of *Arabidopsis thaliana* disturbed in uptake or reduction of nitrate. II. Kinetics. *Physiol. Plant.*, **45** 332-38.

Dolezal, O. and Cobbett, C.S. (1991) An arabinose kinase-deficient mutant of *Arabidopsis thaliana*. *Plant Physiol.*, **96** 1255-60.

Dormann, P., Hoffmann-Benning, S., Balbo, I. and Benning, C. (1995) Isolation and characterization of an *Arabidopsis* mutant deficient in the thylakoid lipid digalactosyl diacylglycerol. *Plant Cell*, **7** 1801-10.

Falcone, D.L., Gibson, S., Lemieux, B. and Somerville, C. (1994) Identification of a gene that complements an *Arabidopsis* mutant deficient in chloroplast omega 6 desaturase activity. *Plant Physiol.*, **106** 1453-59.

Fry, S. (1996) Oligosaccharin mutants. *Trends Plant Sci.*, **1** 326-28.

Gachotte, D., Meens, R. and Benveniste, P. (1995) An *Arabidopsis* mutant deficient in sterol biosynthesis: heterologous complementation by ERG 3 encoding a delta 7-sterol-C-5-desaturase from yeast. *Plant J.*, **8** 407-16.

Gibson, S., Arondel, V., Iba, K. and Somerville, C. (1994) Cloning of a temperature-regulated gene encoding a chloroplast w-3 desaturase from *Arabidopsis thaliana*. *Plant Physiol.*, **106** 1615-21.

Glass, A.D.M. and Siddiqi, M.Y. (1995) Nitrogen absorption by plant roots, in *Nitrogen Nutrition in Higher Plants* (eds H.S. Srivastava and R.P. Singh), Associated Publishing Co, New Delhi, pp. 21-56.

Glazebrook, J. and Ausubel, F.M. (1994) Isolation of phytoalexin-deficient mutants of *Arabidopsis thaliana* and characterization of their interactions with bacterial pathogens. *Proc. Natl Acad. Sci. USA*, **91** 8955-59.

Gomez, L. and Chrispeels, M.J. (1994) Complementation of an *Arabidopsis thaliana* mutant that lacks complex asparagine-linked glycans with the human cDNA encoding N-acetylglucosaminyltransferase I. *Proc. Natl Acad. Sci. USA*, **91** 1829-33.

Green, B.R. and Durnford, D.G. (1996) The chlorophyll-carotenoid proteins of oxygenic photosynthesis. *Ann. Rev. Plant Physiol. Plant Mol. Biol.*, **47** 685-714.

Hannoufa, A., Negruk, V., Eisner, G. and Lemieux, B. (1996) The *CER3* gene of *Arabidopsis thaliana* is expressed in leaves, stems, roots, flowers and apical meristems. *Plant J.*, **10** 459-67.

Haughn, G.W. and Somerville, C.R. (1986) Sulfonylurea resistant mutants of *Arabidopsis thaliana*. *Mol. Gen. Genet.*, **204** 430-34.

Haughn, G.W., Smith, J., Mazur, B. and Somerville, C. (1988) Transformation with a mutant *Arabidopsis* acetolactate synthase gene renders tobacco resistant to sulfonylurea herbicides. *Mol. Gen. Genet.*, **211** 266-71.

Haughn, G.W., Davin, L., Giblin, M. and Underhill, E.W. (1991) Biochemical genetics of plant secondary metabolites in *Arabidopsis thaliana*. The glucosinolates. *Plant Physiol.*, **97** 217-26.

Hitz, W.D., Carlson, T.J., Booth, R., Kinney, A.J., Stecca, K.L. and Yadav, N.S. (1994) Cloning of a higher plant plastid w-6 fatty acid desaturase cDNA and its expression in a cyanobacterium. *Plant Physiol.*, **105** 635-41.

Horton, P., Ruban, A.V. and Walters, R.G. (1996) Regulation of light harvesting in green plants. *Ann. Rev. Plant Physiol. Plant Mol. Biol.*, **47** 655-84.

Howden, R. and Cobbett, C.S. (1992) Cadmium-sensitive mutants of *Arabidopsis thaliana*. *Plant Physiol.*, **100** 100-107.

Howden, R., Goldsbrough, P.B., Andersen, C.R. and Cobbett, C.S. (1995a) Cadmium-sensitive, *cad1* mutants of *Arabidopsis thaliana* are phytochelatin deficient. *Plant Physiol.*, **107** 1059-66.

Howden, R., Andersen, C.R., Goldsbrough, P.B. and Cobbett, C.S. (1995b) A cadmium-sensitive, glutathione-deficient mutant of *Arabidopsis thaliana*. *Plant Physiol.*, **107** 1067-73.

Huang, N.-C., Chiang, C.-S., Crawford, N.M. and Tsay, Y.-F. (1996) *CHL1* encodes a component of the low-affinity nitrate uptake system in *Arabidopsis* and shows cell type-specific expression in roots. *Plant Cell*, **8** 2183-91.

Hülskamp, M., Kopczak, S.D., Horejsi, T.F., Kihl, B.K. and Pruitt, R.E. (1995) Identification of genes required for pollen-stigma recognition in *Arabidopsis thaliana*. *Plant J.*, **8** 703-14.

Iba, K., Gibson, S., Nishiuchi, T., Fuse, T., Nishimura, M., Arondel, V., Hugly, S. and Somerville, C. (1993) A gene encoding a chloroplast omega-3 fatty acid desaturase complements alterations in fatty acid desaturation and chloroplast copy number of the *fad7* mutant of *Arabidopsis thaliana*. *J. Biol. Chem.*, **268** 24099-105.

Inaba, K., Fujiwara, T., Hayashi, H., Chino, M., Komeda, Y. and Naito, S. (1994) Isolation of an *Arabidopsis thaliana* mutant, *mto1*, that overaccumulates soluble methionine. Temporal and spatial patterns of soluble methionine accumulation. *Plant Physiol.*, **104** 881-87.

Jackson, J.A., Fuglevand, G., Brown, B.A., Shaw, M.J. and Jenkins, G.I. (1995) Isolation of *Arabidopsis* mutants altered in the light-regulation of chalcone synthase gene expression using a transgenic screening approach. *Plant J.*, **8** 369-80.

Jacobs, M., Dolferus, R. and van den Bossche, D. (1988) Isolation and biochemical analysis of ethyl methane sulfonate-induced alcohol dehydrogenase null mutants of *Arabidopsis thaliana*. *Biochem. Genet.*, **26** 105-22.

James, D.W. and Dooner, H.K. (1990) Isolation of EMS-induced mutants in *Arabidopsis* altered in seed fatty acid composition. *Theor. Appl. Genet.*, **80** 241-45.

James, D.W., Jr., Lim, E., Keller, J., Plooy, I., Ralston, E. and Dooner, H.K. (1995) Directed tagging of the *Arabidopsis FATTY ACID ELONGATION1 (FAE1)* gene with the maize transposon *Activator*. *Plant Cell*, **7** 309-19.

Jenks, M.A., Tuttle, H.A., Eigenbrode, S.D. and Feldmann, K.A. (1995) Leaf epicuticular waxes of the *Eceriferum* mutants in *Arabidopsis*. *Plant Physiol.*, **108** 369-77.

Jenks, M.A., Rashotte, A.M., Tuttle, H.A. and Feldmann, K.A. (1996) Mutants in *Arabidopsis thaliana* altered in epicuticular wax and leaf morphology. *Plant Physiol.*, **110** 377-85.

Katavic, V., Reed, D.W., Taylor, D.C., Giblin, E.M., Barton, D.L., Zhou, J., Mackenzie, S.L., Covello, P.S. and Kunst, L. (1995) Alteration of seed fatty acid composition by an ethyl methanesulfonate induced mutation in *Arabidopsis thaliana* affecting diacylglycerol acyltransferase activity. *Plant Physiol.*, **108** 399-409.

Kiss, J.Z., Hertel, R. and Sack, F.D. (1989) Amyloplasts are necessary for full gravitropic sensitivity in roots of *Arabidopsis thaliana*. *Planta*, **177** 198-206.

Komeda, Y., Tanaka, M. and Nishimune, T. (1988) A *th-1* mutant of *Arabidopsis thaliana* is defective for a thiamin-phosphate-synthesizing enzyme: thiamin phosphate pyrophosphorylase. *Plant Physiol.*, **88** 248-50.

Koornneef, M. (1990) Mutations affecting the testa color in *Arabidopsis*. *Arabidopsis Inf. Serv.*, **28** 1-4.

Koornneef, M. and Hanhart, C.J. (1981) A new thiamine locus in *Arabidopsis*. *Arabidopsis Inf. Serv.*, **18** 52-58.

Koornneef, M., Hanhart, C.J. and Thiel, F. (1989) A genetic and phenotypic description of *eceriferum* (*cer*) mutants in *Arabidopsis thaliana*. *J. Hered.*, **80** 118-22.

Kreps, J.A. and Town, C.D. (1992) Isolation and characterization of a mutant of *Arabidopsis thaliana* resistant to alpha methyltryptophan. *Plant Physiol.*, **99** 269-75.

Kreps, J.A., Tilak, T., Dong, W. and Town, C.D. (1996) Molecular basis of α-methyltryptophan resistance in *amt-1*, a mutant of *Arabidopsis thaliana* with altered tryptophan metabolism. *Plant Physiol.*, **110** 1159-65.

Kumar, A., Altabella, T., Taylor, M.A. and Tiburcio, A.F. (1997) Recent advances in polyamine research. *Trends Plant Sci.*, **2** 124-30.

Kunst, L., Browse, J. and Somerville, C. (1988) Altered regulation of lipid biosynthesis in a mutant of *Arabidopsis* deficient in chloroplast glycerol phosphate acyltransferase activity. *Proc. Natl Acad. Sci. USA*, **85** 4143-47.

Kunst, L., Browse, J. and Somerville, C. (1989) A mutant of *Arabidopsis* deficient in desaturation of palmitic acid in leaf lipids. *Plant Physiol.*, **90** 943-47.

Kunst, L., Taylor, D.C. and Underhill, E.W. (1992) Fatty acid elongation in developing seeds of *Arabidopsis thaliana*. *Plant Physiol. Biochem.*, **30** 425-34.

Labrie, S.T., Wilkinson, J.Q., Tsay, Y.F., Feldmann, K.A. and Crawford, N.M. (1992) Identification of two tungstate-sensitive molybdenum cofactor mutants, *chl2* and *chl7*, of *Arabidopsis thaliana*. *Mol. Gen. Genet.*, **233** 169-76.

Lam, H.-M., Coschigano, K., Schultz, C., Melo-Oliveira, R., Tjaden, G., Oliveira, I., Ngai, N., Hsieh, M.-H. and Coruzzi, G. (1995) Use of *Arabidopsis* mutants and genes to study amide amino acid biosynthesis. *Plant Cell*, **7** 887-98.

Lam, H.-M., Coschigano, K.T., Oliveira, I.C., Melo-Oliveira, R. and Coruzzi, G.M. (1996) The molecular-genetics of nitrogen assimilation into amino acids in higher plants. *Ann. Rev. Plant Physiol. Plant Mol. Biol.*, **47** 569-93.

Landry, L.G., Chapple, C.C.S. and Last, R.L. (1995) *Arabidopsis* mutants lacking phenolic sunscreens exhibit enhanced ultraviolet-B injury and oxidative damage. *Plant Physiol.*, **109** 1159-66.

Last, R.L. and Fink, G.R. (1988) Tryptophan-requiring mutants of the plant *Arabidopsis thaliana*. *Science*, **240** 305-10.

Last, R.L., Bissinger, P.H., Mahoney, D.J., Radwanski, E.R. and Fink, G.R. (1991) Tryptophan mutants in *Arabidopsis:* The consequences of duplicated tryptophan synthase β genes. *Plant Cell,* **3** 345-58.

Lemieux, B. (1996) Molecular genetics of epicuticular wax biosynthesis. *Trends Plant Sci.,* **1** 312-18.

Lemieux, B., Miquel, M., Somerville, C. and Browse, J. (1990) Mutants of *Arabidopsis* with alterations in seed lipid fatty acid composition. *Theor. Appl. Genet.,* **80** 234-40.

Li, J. and Last, R.L. (1996) The *Arabidopsis thaliana trp5* mutant has a feedback-resistant anthranilate synthase and elevated soluble tryptophan. *Plant Physiol.,* **110** 51-59.

Li, L. and Preiss, J. (1992) Characterization of ADPglucose pyrophosphorylase from a starch-deficient mutant of *Arabidopsis thaliana* (L.). *Carbohydr. Res.,* **227** 227-39.

Li, S.L. and Rédei, G.P. (1969) Thiamine mutants of the crucifer, *Arabidopsis. Biochem. Genet.,* **3** 163-70.

Li, J., Ou-Lee, T.-M., Raba, R., Amundson, R.G. and Last, R.L. (1993) *Arabidopsis* flavonoid mutants are hypersensitive to UV-B irradiation. *Plant Cell,* **5** 171-79.

Li, J., Zhao, J., Rose, A.B., Schmidt, R. and Last, R.L. (1995) *Arabidopsis* phosphoribosylanthranilate isomerase: molecular genetic analysis of triplicate tryptophan pathway genes. *Plant Cell,* **7** 447-61.

Lightner, J., Wu, J. and Browse, J. (1994) A mutant of *Arabidopsis* with increased levels of stearic acid. *Plant Physiol.,* **106** 1443-51.

Lin, Y. and Cheng, C.-L. (1997) A chlorate-resistant mutant defective in the regulation of nitrate reductase gene expression in *Arabidopsis* defines a new *HY* locus. *Plant Cell,* **9** 21-35.

Lin, T.-P., Caspar, T., Somerville, C.R. and Preiss, J. (1988a) Isolation and characterization of starchless mutant of *Arabidopsis thaliana* (L.) Heynh lacking ADPglucose pyrophosphorylase activity. *Plant Physiol.,* **86** 1131-35.

Lin, T.-P., Caspar, T., Somerville, C.R. and Preiss, J. (1988b) A starch deficient mutant of *Arabidopsis thaliana* with low ADPglucose pyrophosphorylase activity lacks one of the two subunits of the enzyme. *Plant Physiol.,* **88** 1175-81.

Lorenzen, M., Racicot, V., Strack, D. and Chapple, C. (1996) Sinapic acid ester metabolism in wild type and a sinapoylglucose-accumulating mutant of *Arabidopsis. Plant Physiol.,* **112** 1625-30.

Magrath, R., Bano, F., Morgner, M., Parkin, I., Sharpe, A., Lister, C., Dean, C., Turner, J., Lydiate, D. and Mithen, R. (1994) Genetics of aliphatic glucosinolates. I. Side chain elongation in *Brassica napus* and *Arabidopsis thaliana. Heredity,* **72** 290-99.

Malmberg, R.L. and Rose, D.G. (1987) Biochemical genetics of resistance to MGBG in tobacco: mutants that alter SAM-decarboxylase or polyamine ratios, and flower morphology. *Mol. Gen. Genet.,* **207** 9-14.

Marin, E., Nussaume, L., Quesada, A., Gonneau, M., Sotta, B., Hugueney, P., Frey, A. and Marion-Poll, A. (1996) Molecular identification of zeaxanthin epoxidase of *Nicotiana plumbaginifolia,* a gene involved in abscisic acid biosynthesis and corresponding to the *ABA* locus of *Arabidopsis thaliana. EMBO J.,* **15** 2331-42.

Martin, C. and Smith, A.M. (1995) Starch biosynthesis. *Plant Cell,* **7** 971-85.

May, M.J. and Leaver, C.J. (1994) *Arabidopsis thaliana* γ-glutamyl-cysteine synthetase is structurally unrelated to mammalian, yeast and *E. coli* homologs. *Proc. Nat. Acad. Sci. USA,* **9** 10059-63.

May, M.J., Parker, J.E., Daniels, M.J., Leaver, C.J. and Cobbett, C.S. (1996) An *Arabidopsis* mutant depleted in glutathione shows unaltered responses to fungal and bacterial pathogens. *Mol. Plant Microbe Interact.,* **9** 349-56.

McConn, M. and Browse, J. (1996) The critical requirement for linolenic acid is pollen development, not photosynthesis, in an *Arabidopsis* mutant. *Plant Cell,* **8** 403-16.

McConn, M., Hugly, S., Somerville, C. and Browse, J. (1994) A mutation at the *fad8* locus of *Arabidopsis* identifies a second chloroplast w-3 desaturase. *Plant Physiol.,* **106** 1609-14.

McConn, M., Creelman, R.A., Bell, E., Mullet, J.E. and Browse, J. (1997) Jasmonate is essential for insect defense in *Arabidopsis*. *Proc. Natl Acad. Sci. USA*, **94** 5473-77.

Melo-Oliveira, R., Oliveira, I.C. and Coruzzi, G.M. (1996) *Arabidopsis* mutant analysis and gene regulation define a nonredundant role for glutamate dehydrogenase in nitrogen assimilation. *Proc. Natl Acad. Sci. USA*, **93** 4718-23.

Meyer, K., Cusumano, J.C., Somerville, C. and Chapple, C.C.S. (1996) Ferulate-5-hydroxylase from *Arabidopsis thaliana* defines a new family of cytochrome P450-dependent monooxygenases. *Proc. Natl Acad. Sci. USA*, **93** 6869-74.

Miquel, M. and Browse, J. (1992) *Arabidopsis* mutants deficient in polyunsaturated fatty acid synthesis. Biochemical and genetic characterization of a plant oleoyl-phosphatidylcholine desaturase. *J. Biol. Chem.*, **267** 1502-1509.

Mithen, R., Clarke, J., Lister, C. and Dean, C. (1995) Genetics of aliphatic glucosinolates. II. Side chain structure of aliphatic glucosinolates in *Arabidopsis thaliana*. *Heredity*, **74** 210-15.

Moffatt, B.A. and Somerville, C. (1988) Positive selection for male-sterile mutants of *Arabidopsis* lacking adenine phosphoribosyltransferase activity. *Plant Physiol.*, **86** 1150-54.

Moffatt, B.A., McWhinnie, E.A., Burkhart, W.E., Pasternak, J.J. and Rothstein, S.J. (1992) A complete cDNA for adenine phosphoribosyltransferase from *Arabidopsis thaliana*. *Plant Mol. Biol.*, **18** 653-62.

Moore, R. (1989) Root graviresponsiveness and cellular differentiation in wild-type and a starchless mutant of *Arabidopsis thaliana*. *Ann. Bot.*, **64** 271-77.

Negruk, V., Yang, P., Subramanian, M., McNevin, J.P. and Lemieux, B. (1996) Molecular cloning and characterization of the *CER2* gene of *Arabidopsis thaliana*. *Plant J.*, **9** 137-45.

Neuhaus, H.E. and Stitt, M. (1990) Control analysis of photosynthate partitioning: Impact of reduced activity of ADP-glucose pyrophosphorylase or plastid phosphoglucomutase on the fluxes to starch and sucrose in *Arabidopsis thaliana* (L.) Heynh. *Planta*, **182** 445-54.

Niyogi, K.K. and Fink, G.R. (1992) Two anthranilate synthase genes in *Arabidopsis*: defense-related regulation of the tryptophan pathway. *Plant Cell*, **4** 721-33.

Niyogi, K.K., Last, R.L., Fink, G.R. and Keith, B. (1993) Suppressors of *trp1* fluorescence identify a new *Arabidopsis* gene, *TRP4*, encoding the anthranilate synthase ß subunit. *Plant Cell*, **5** 1011-27.

Normanly, J., Cohen, J.D. and Fink, G.R. (1993) *Arabidopsis thaliana* auxotrophs reveal a tryptophan-independent biosynthetic pathway for indole-3-acetic acid. *Proc. Natl Acad. Sci. USA*, **90** 10355-59.

Normanly, J., Slovin, J.P. and Cohen, J.D. (1995) Rethinking auxin biosynthesis and metabolism. *Plant Physiol.*, **107** 323-29.

Norris, S.R., Barrette, T.R. and DellaPenna, D. (1995) Genetic dissection of carotenoid synthesis in *Arabidopsis* defines plastoquinone as an essential component of phytoene desaturation. *Plant Cell*, **7** 2139-49.

Ogren, W.L. (1984) Photorespiration: pathways, regulation and modification. *Ann. Rev. Plant Physiol.*, **35** 415-42.

Ohlrogge, J.B. (1994) Design of new plant products: engineering of fatty acid metabolism. *Plant Physiol.*, **104** 821-26.

Ohlrogge, J. and Browse, J. (1995) Lipid biosynthesis. *Plant Cell*, **7** 957-70.

Okuley, J., Lightner, J., Feldmann, K., Yadav, N., Lark, E. and Browse, J. (1994) *Arabidopsis FAD2* gene encodes the enzyme that is essential for polyunsaturated lipid synthesis. *Plant Cell*, **6** 147-58.

Oliveira, I.C., Lam, H.-M., Coschigano, K., Melo-Oliveira, R. and Coruzzi, G. (1997) Molecular-genetic dissection of ammonium assimilation in *Arabidopsis thaliana*. *Plant Physiol. Biochem.*, **35** 185-98.

Patton, D.A., Volrath, S. and Ward, E.R. (1996a) Complementation of an *Arabidopsis thaliana* biotin auxotroph with an *Escherichia coli* biotin biosynthetic gene. *Mol. Gen. Genet.*, **251** 261-66.

Patton, D.A., Johnson, M. and Ward, E.R. (1996b) Biotin synthase from *Arabidopsis thaliana*: cDNA isolation and characterization of gene expression. *Plant Physiol.*, **112** 371-78.

Patton, D., Schetter, A., Franzmann, L., Ward, E. and Meinke, D. (1996c) An embryo-defective mutant (*bio2*) disrupted in the final step of biotin synthesis, in *7th International Conference on Arabidopsis Research* (eds M. Bevan, G. Coupland, C. Dean, R. Flavell and N. Harberd), John Innes Centre, Norwich, Abstract P255.

Peeters, A.J.M., Debeaujon, I., Lon-Kioosterziel, K.M. and Koornneef, M. (1996) The molecular genetic analysis of seed development in *Arabidopsis*, in *7th International Conference on Arabidopsis Research* (eds M. Bevan, G. Coupland, C. Dean, R. Flavell and N. Harberd), John Innes Centre, Norwich, Abstract P235.

Peng, L., Williamson, R., Rolfe, B. and Redmond, J. (1996) Carbohydrate analysis of *Arabidopsis* mutants defective in cellulose synthesis, in *40th ASBMB and 36th ASPP Annual Combined Conference*, Australian Society for Biochemistry and Molecular Biology Inc. and Australian Society of Plant Physiologists Inc., Canberra, Abstract SYM-06-02.

Pogson, B., McDonald, K.A., Truong, M., Britton, G. and DellaPenna, D. (1996) *Arabidopsis* carotenoid mutants demonstrate that lutein is not essential for photosynthesis in higher plants. *Plant Cell*, **8** 1627-39.

Post-Beittenmiller, D. (1996) Biochemistry and molecular biology of wax production in plants. *Ann. Rev. Plant Physiol. Plant Mol. Biol.*, **47** 405-30.

Preuss, D., Lemieux, B., Yen, G. and Davis, R.W. (1993) A conditional sterile mutation eliminates surface components from *Arabidopsis* pollen and disrupts cell signaling during fertilization. *Genes Dev.*, **7** 974-85.

Radwanski, E.R. and Last, R.L. (1995) Tryptophan biosynthesis and metabolism: Biochemical and molecular genetics. *Plant Cell*, **7** 921-34.

Radwanski, E.R., Zhao, J. and Last, R.L. (1995) *Arabidopsis thaliana* tryptophan synthase alpha: gene cloning, expression, and subunit interaction. *Mol. Gen. Genet.*, **248** 657-67.

Radwanski, E.R., Barczak, A.J. and Last, R.L. (1996) Characterization of tryptophan synthase alpha subunit mutants of *Arabidopsis thaliana. Mol. Gen. Genet.*, **253** 353-61.

Rashotte, A. and Feldmann, K. (1996) Epicuticular waxes and aphid resistance in *Arabidopsis cer* mutants and ecotypes. *Plant Physiol. Suppl.*, **111** Abstr. 304.

Rauser, W.E. (1995) Phytochelatins and related peptides. Structure, biosynthesis, and function. *Plant Physiol.*, **109** 1141-49.

Reiter, W.-D., Chapple, C.C.S. and Somerville, C.R. (1993) Altered growth and cell walls in a fucose-deficient mutant of *Arabidopsis*. *Science*, **261** 1032-35.

Reiter, W.-D., Chapple, C. and Somerville, C.R. (1997) Mutants of *Arabidopsis thaliana* with altered cell wall polysaccharide composition. *Plant J.*, **12** 335-45.

Rennenberg, H. (1982) Glutathione metabolism and possible biological roles in higher plants. *Phytochemistry*, **21** 2771-81.

Rock, C.D. and Zeevaart, J.A.D. (1991) The *aba* mutant of *Arabidopsis* is impaired in epoxy-carotenoid biosynthesis. *Proc. Natl Acad. Sci. USA*, **88** 7496-99.

Rose, A.B., Casselman, A.L. and Last, R.L. (1992) A phosphoribosylanthranilate transferase gene is defective in blue fluorescent *Arabidopsis thaliana* tryptophan mutants. *Plant Physiol.*, **100** 582-92.

Rose, A.B., Li, J. and Last, R.L. (1997) An allelic series of blue fluorescent *trp1* mutants of *Arabidopsis thaliana. Genetics*, **145** 197-205.

Runge, S., van Cleve, B., Lebedev, N., Armstrong, G. and Apel, K. (1995) Isolation and classification of chlorophyll-deficient *xantha* mutants of *Arabidopsis thaliana. Planta*, **197** 490-500.

Sæther, N. and Iversen, T.-H. (1991) Gravitropism and starch statoliths in an *Arabidopsis* mutant. *Planta*, **184** 491-97.

Schledzewski, K., Brinkmann, H., LaBrie, S.T., Crawford, N.M. and Mendel, R.R. (1996) Molybdenum cofactor biosynthesis in plants: molecular analysis of the *chl7* locus in *Arabidopsis thaliana*, in *7th International Conference on Arabidopsis Research* (eds M. Bevan, G. Coupland, C. Dean, R. Flavell and N. Harberd), John Innes Centre, Norwich, UK, Abstract P274.

Schneider, R., Dinkins, R., Robinson, K., Shellhammer, J. and Meinke, D.W. (1989) An embryo-lethal mutant of *Arabidopsis thaliana* is a biotin auxotroph. *Dev. Biol.*, **131** 161-67.

Schultz, C.J. and Coruzzi, G.M. (1995) The aspartate aminotransferase gene family of *A. thaliana* encodes isoenzymes localized to three distinct subcellular compartments. *Plant J.*, **7** 61-75.

Schwartz, S.H., Léon-Kloosterziel, K.M., Koornneef, M. and Zeevaart, J.A.D. (1997) Biochemical characterization of the *aba2* and *aba3* mutants in *Arabidopsis thaliana*. *Plant Physiol.*, **114** 161-66.

Shirley, B.W. (1996) Flavonoid biosynthesis: 'new' functions for an 'old' pathway. *Trends Plant Sci.*, **1** 377-82.

Shirley, B.W., Hanley, S. and Goodman, H.M. (1992) Effects of ionizing radiation on a plant genome: analysis of two *Arabidopsis transparent testa* mutations. *Plant Cell*, **4** 333-47.

Shirley, B.W., Kubasek, W.L., Storz, G., Bruggemann, E., Koornneef, M., Ausubel, F.M. and Goodman, H.M. (1995) Analysis of *Arabidopsis* mutants deficient in flavonoid biosynthesis. *Plant J.*, **8** 659-71.

Sicher, R.C. and Kremer, D.F. (1992) Control of carbohydrate metabolism in a starchless mutant of *Arabidopsis thaliana*. *Physiol. Plant.*, **85** 446-52.

Somerville, C. (1995) Direct tests of the role of membrane lipid composition in low-temperature-induced photoinhibition and chilling sensitivity in plants and cyanobacteria. *Proc. Natl Acad. Sci. USA*, **92** 6215-18.

Somerville, C.R. and Ogren, W.L. (1979) A phosphoglycolate phosphatase-deficient mutant in *Arabidopsis*. *Nature*, **280** 833-36.

Somerville, C.R. and Ogren, W.L. (1980a) Inhibition of photosynthesis in mutants of *Arabidopsis* lacking glutamate synthase activity. *Nature*, **286** 257-59.

Somerville, C.R. and Ogren, W.L. (1980b) Photorespiration mutants of *Arabidopsis thaliana* deficient in serine-glyoxylate aminotransferase activity. *Proc. Natl Acad. Sci.*, **77** 2684-87.

Somerville, C.R. and Ogren, W.L. (1981) Photorespiration-deficient mutants of *Arabidopsis thaliana* lacking mitochondrial serine transhydroxymethylase activity. *Plant Physiol.*, **67** 666-71.

Somerville, C.R. and Ogren, W.L. (1982a) Isolation of photorespiration mutants in *Arabidopsis thaliana*, in *Chloroplast Molecular Biology* (eds M. Edelman *et al.*), Elsevier/North-Holland, Amsterdam, pp. 129-38.

Somerville, C.R. and Ogren, W.L. (1982b) Genetic modification of photorespiration. *Trends Biochem. Sci.*, **7** 171-74.

Somerville, C.R. and Ogren, W.L. (1982c) Mutants of the cruciferous plant *Arabidopsis thaliana* lacking glycine decarboxylase activity. *Biochem. J.*, **202** 373-80.

Somerville, C.R. and Ogren, W.L. (1983) An *Arabidopsis thaliana* mutant defective in chloroplast dicarboxylate transport. *Proc. Natl Acad. Sci.*, **80** 1290-94.

Somerville, C.R., Portis, A.R. and Ogren, W.L. (1982) A mutant of *Arabidopsis thaliana* which lacks activation of RuBP carboxylase *in vivo*. *Plant Physiol.*, **70** 381-87.

Stallmeyer, B., Nerlich, A., Schiemann, J., Brinkmann, H. and Mendel, R.R. (1995) Molybdenum co-factor biosynthesis: the *Arabidopsis thaliana* cDNA *cnx1* encodes a multifunctional two-domain protein homologous to a mammalian neuroprotein, the insect protein Cinnamon and three *Escherichia coli* proteins. *Plant J.*, **8** 751-62.

Strack, D. (1981) Sinapine as a supply of choline for the biosynthesis of phosphatidylcholine in *Raphanus sativus* seedlings. *Z. Naturforsch.*, **36c** 215-21.

Taylor, L.P. and Jorgensen, R. (1992) Conditional male fertility in chalcone synthase-deficient petunia. *J. Hered.*, **83** 11-17.

Taylor, C.B. (1996) Control of cyclic carotenoid biosynthesis: No lutein, no problem! *Plant Cell*, **8** 1447-50.

Timmerman, K.P. (1989) Molecular characterization of corn glutathione S-transferase isozymes involved in herbicide detoxification. *Physiol. Plant.*, **77** 465-71.

Trethewey, R.N. and ap Rees, T. (1994) A mutant of *Arabidopsis thaliana* lacking the ability to transport glucose across the chloroplast envelope. *Biochem. J.*, **301** 449-54.

Tsay, Y.F., Schroeder, J.I., Feldmann, K.A. and Crawford, N.M. (1993) The herbicide sensitivity gene CHL1 of *Arabidopsis* encodes a nitrate-inducible nitrate transporter. *Cell*, **72** 705-13.

Tsuji, J., Zook, M., Hammerschmidt, R., Last, R.L. and Somerville, S.C. (1993) Evidence that tryptophan is not a direct biosynthetic intermediate of camalexin in *Arabidopsis thaliana. Physiol. Mol. Plant Pathol.*, **43** 221-29.

Turner, S.R. and Somerville, C.R. (1997) Collapsed xylem phenotype of *Arabidopsis* identifies mutants deficient in cellulose deposition in the secondary cell wall. *Plant Cell*, **9** 689-701.

von Schaewen, A., Sturm, A., O'Neill, J. and Chrispeels, M.J. (1993) Isolation of a mutant *Arabidopsis* plant that lacks N-acetyl glucosaminyl transferase I and is unable to synthesize Golgi-modified complex N-linked glycans. *Plant Physiol.*, **102** 1109-18.

Wang, R. and Crawford, N.M. (1996) Genetic identification of a gene involved in constitutive, high-affinity nitrate transport in higher plants. *Proc. Natl Acad. Sci. USA*, **93** 9297-9301.

Wang, S.-M., Chu, B., Lue, W.-L., Yu, T.-S., Eimert, K. and Chen, J. (1997) *adg2-1* represents a missense mutation in the ADPG pyrophosphorylase large subunit gene of *Arabidopsis thaliana. Plant J.*, **11** 1121-26.

Wilkie, S.E., Roper, J., Smith, A. and Warren, M.J. (1995) Isolation, characterisation and expression of a cDNA clone encoding aspartate aminotransferase from *A. thaliana. Plant Mol. Biol.*, **27** 1227-33.

Wilkinson, J.Q. and Crawford, N.M. (1991) Identification of the *Arabidopsis CHL3* gene as the nitrate reductase structural gene *NIA2. Plant Cell*, **3** 461-71.

Wilkinson, J.Q. and Crawford, N.M. (1993) Identification and characterization of a chlorate-resistant mutant of *Arabidopsis thaliana* with mutations in both nitrate reductase structural genes *NIA1* and *NIA2. Mol. Gen. Genet.*, **239** 289-97.

Wu, J. and Browse, J. (1995) Elevated levels of high-melting-point phosphatidylglycerols do not induce chilling sensitivity in an *Arabidopsis* mutant. *Plant Cell*, **7** 17-27.

Wu, J., James, D.W., Jr., Dooner, H.K. and Browse, J. (1994) A mutant of *Arabidopsis* deficient in the elongation of palmitic acid. *Plant Physiol.*, **106** 143-50.

Wu, J., Lightner, J., Warwick, N. and Browse, J. (1997) Low-temperature damage and subsequent recovery of *fab1* mutant *Arabidopsis* exposed to 2°C. *Plant Physiol.*, **113** 347-56.

Yadav, N.S., Wierzbicki, A., Aegerter, M., Caster, C.S., Pérez-Grau, L., Kinney, A.J., Hitz, W.D., Booth, R., Jr., Schweiger, B., Stecca, K.L., Allen, S.M., Blackwell, M., Reiter, R.S., Carlson, T.J., Russell, S.H., Feldmann, K.A., Pierce, J. and Browse, J. (1993) Cloning of higher plant omega-3 fatty acid desaturases. *Plant Physiol.*, **103** 467-76.

Zablackis, E., Huang, J., Müller, B., Darvill, A.G. and Albersheim, P. (1995) Characterization of the cell-wall polysaccharides of *Arabidopsis thaliana* leaves. *Plant Physiol.*, **107** 1129-38.

Zablackis, E., York, W.S., Pauly, M., Hantus, S., Reiter, W.-D., Chapple, C.C.S., Albersheim, P. and Darvill, A. (1996) Substitution of L-fucose by L-galactose in cell walls of *Arabidopsis mur1. Science*, **272** 1808-10.

Zhou, J. and Goldsbrough, P.B. (1994) Functional homologs of fungal metallothionien genes from *Arabidopsis. Plant Cell*, **6** 875-84.

4   Hormone regulated development

Malcolm Bennett, Joe Kieber, Jérôme Giraudat and Peter Morris

4.1   Introduction

The term 'phytohormone' encompasses a diverse collection of signalling molecules that influence almost every stage of plant development (Davis, 1995). Auxin, ethylene, gibberellins (GA), abscisic acid (ABA), cytokinin and brassino-steroids denote the six classes of phytohormones described to date. Phytohormones are non-peptide-based, low molecular weight organic molecules. Ethylene has the simplest phytohormone structure, whereas GA has the greatest structural diversity, with over 100 variants described to date. Understanding how, when and where phytohormones are synthesised has challenged researchers for over half a century, as has determining the molecular basis of phytohormone action. This chapter highlights the significant contribution *Arabidopsis* molecular genetic research has made to our current understanding of the biosynthesis and action of auxin, ethylene, GA and ABA. Readers interested in the remaining classes of phytohormones are directed to several recent reviews (Binns, 1994; Clouse, 1996; Chory and Li, 1997).

Much of our early knowledge of phytohormone-regulated development comes from so-called 'spray and pray' studies. Plant tissues were either treated with a hormone directly or using compounds that inhibit its biosynthesis, transport or action. As with all pharmacological approaches, questions arise about the dose dependence and specificity of the reagents used in these studies. Mutants, by their very nature, represent the ultimate pharmacological tool. They enable a scientist to study a system, confident that the phenotypic changes observed originate from perturbations in the activity of a single gene product. For example, the developmental importance of GA has been elegantly addressed using mutants that are known to be defective in hormone biosynthesis (see sections 4.3, 4.4 and 4.5 for descriptions of ethylene, GA and ABA biosynthesis, respectively). Many researchers have also adopted a mutational approach in order to gain an insight into the molecular mechanisms regulating hormone biosynthesis and action (Klee and Estelle, 1991). The advent of *Arabidopsis* molecular genetics has facilitated the isolation and characterisation of several important phytohormone genes including *ETR1*, which encodes the first hormone receptor to be described in higher plants (see section 4.4).

Many of these hormone-related gene products are being placed within a framework of interconnecting steps. For hormone biosynthesis, this has been facilitated by the composition of several hormone biosynthetic pathways having being described in detail (see sections 4.3, 4.4 and 4.5). Our knowledge of the intermediates that make up a hormone signal transduction pathway is usually far more limited. These intermediates are likely to consist of protein–protein interactions, covalent modifications and/or secondary messengers. Performing a saturation mutagenesis study of a signalling pathway represents one strategy for identifying the genes that encode all of the transduction components. However, problems such as genetic redundancy and lethality frequently frustrate such efforts. Extragenic suppressor screens represents another approach for identifying additional signalling intermediates within a hormone response pathway. For example, several suppressors of the GA-insensitive mutant, *gai*, have been identified including *gar2* and *gas-1*, which have been proposed to function within the GA signalling pathway (Peng and Harberd, 1993; Wilson and Somerville, 1995). Two hybrid-based studies using a phytohormone signalling component as baits represent a more direct approach to identifying interactive partners (Ulmasov *et al.*, 1997). Reverse genetic approaches in *Arabidopsis* (McKinney *et al.*, 1995) represent a powerful approach to testing the *in planta* role(s) of any interactive partners. Several novel phytohormone response loci are likely to be identified in the near future using this combined strategy.

Once several signalling components have been identified, researchers usually attempt to position them within a transduction pathway. Genetic approaches in *Arabidopsis* allows us to simplify this seemingly complex problem. Epistasis studies using double mutant combinations allow the researcher to predict the relative positions of signalling components within a transduction chain. This has been elegantly demonstrated in the molecular genetic dissection of the ethylene signalling pathway, which appears to mirror the kinase-based pathways described in other eukaryotes (see section 4.3). However, phytohormone signalling should not be considered as an isolated, linear chain of transduction events. Instead, crosstalk will occur between multiple hormone signalling pathways. Several hormone imputs are likely to impinge at key points within a transduction chain. For example, many phytohormones are known to either act agonistically or antagonistically towards one another during plant development. These interactions are likely to reflect the presence of a common signalling component or the modulation of the activity of one element of a hormone transduction chain by another hormone. Understanding how the cell attempts to integrate this signalling information represents one of the greatest challenges facing plant hormone biologists. The isolation of *Arabidopsis* phytohomormone response mutants such as *axr1* (Lincoln *et al.*, 1990), which exhibit multiple hormone resistances, provides a useful starting point for these studies.

Many key questions about phytohormone biology remain to be satisfactorily answered. Where are hormones synthesised? What is the molecular basis for hormone transport? How and where are hormones perceived? This chapter highlights the fact that significant progress is being made towards obtaining answers to these important questions. Judging from the wealth of information obtained to date, *Arabidopsis* is clearly becoming the experimental system of choice with which to conduct such studies in higher plants.

4.2 Auxins

Auxins control plant cell division, elongation and differentiation (Hobbie and Estelle, 1994). As a regulator of such fundamental cellular processes, auxins influence almost every aspect of plant development, ranging from apical dominance and vascular differentiation in shoots to lateral organ initiation and gravitropism in roots. In order to gain greater insight into the molecular basis of auxin biology, many researchers have adopted a mutational approach in *Arabidopsis*. The following four sections describe a selection of *Arabidopsis* mutants that throw new light on auxin biosynthesis, polar transport, intracellular signalling and its downstream responses.

4.2.1 Synthesis, conjugation and degradation

Indole-3-acetic acid (IAA) represents the major form of auxin in higher plants (Figure 4.1). IAA has been proposed to be synthesised using both tryptophan-dependent and tryptophan-independent pathways (Bartel, 1997). In *Arabidopsis*, experimental evidence suggests that IAA is synthesised via the latter route using an indole precursor (Normanly *et al.*, 1993). ^{15}N labelling studies concluded that indole-3-acetonitrile (IAN) is likely to represent the immediate precursor to IAA (Normanly *et al.*, 1993). A series of nitrilase isozymes have been described in *Arabidopsis*, all of which are capable of catalysing the conversion of IAN to IAA (Bartling *et al.*, 1994). To date, four *Arabidopsis* nitrilase genes, termed NIT1-4, have been isolated. These exhibit divergent patterns of tissue-specific expression (Bartel and Fink, 1994). The NIT1 and NIT2 genes encode the nitrilase I and II isozymes, respectively (Bartling *et al.*, 1994). Despite encoding very similar proteins, nitrilase I is soluble, whereas nitrilase II is membrane associated (Bartling *et al.*, 1994). Nitrilase I is more active during vegetative stages of *Arabidopsis* development whereas higher levels of nitrilase II activity are detected during later reproductive stages. This observation has led to suggestions that more than one IAN-related IAA pathway may function during the *Arabidopsis* life cycle.

The *alf3* mutant has been proposed to represent an auxin auxotroph (Calenza *et al.*, 1995; Bartel, 1997). Calenza and co-workers have observed that *alf3* primary and secondary root meristems initially develop but eventually arrest and die unless they are supplemented with either IAA or indole. Laskowski *et al.*

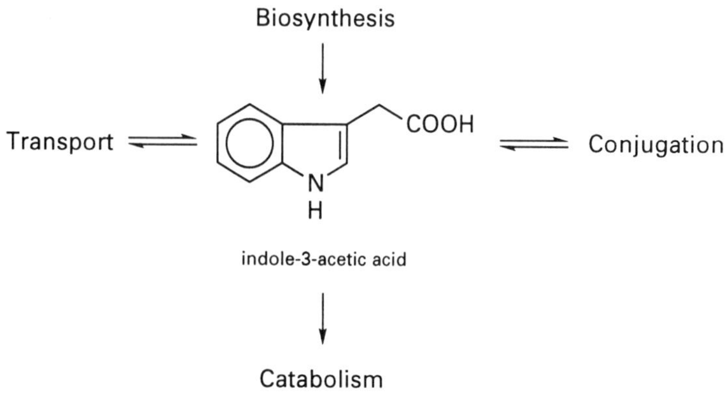

Figure 4.1 Indole-3-acetic acid (IAA) represents the major form of auxin in higher plants. IAA is present as the bioactive, free acid form or conjugated prior to degradation.

(1995) have independently observed that excised lateral root primordia consisting of more than 5 cell layers develop the ability to grow in the absence of exogenous auxin. Hence, ALF3 may facilitate later stages of lateral root meristem development by conferring the ability to synthesise their own auxin.

Conjugative and degradative mechanisms are also used to regulate IAA levels *in planta* (Bartel, 1997). IAA is present within plant tissues either as a free, physiologically active acid or conjugated to a variety of amino acids, peptides and carbohydrates (Figure 4.1). Several mutant alleles of the *rty* gene have been described which overproduce auxin (Boerjan *et al.*, 1995; Calenza *et al.*, 1995; King *et al.*, 1995; Lehman *et al.*, 1996). *rty* mutants accumulate high levels of free and conjugated IAA, resulting in an extreme proliferation of roots that can be phenocopied when wild-type seedlings are grown with auxin. The *RTY* gene has been cloned by T-DNA tagging and encodes a protein that is similar to tyrosine aminotransferase (Gopalraj *et al.*, 1996). As an aminotransferase, RTY must normally act to limit free IAA accumulation. Bartel (1997) has suggested that RTY may perform this task by converting the putative IAA precursor, indole-3-pyruvate, to an alternative product such as tryptophan. Hence in the absence of such an activity within the *rty* mutant background, IAA would be overproduced. The conjugated forms of IAA have been proposed to represent a physiologically inactive store of IAA that can be released by aminohydrolase action. Bartel and Fink (1995) have selected the *Arabidopsis* aminohydrolase mutant *ilr1* on the basis of its inability to cleave the IAA-Leu conjugate. The recombinantly expressed *ILR1* gene product demonstrates hydrolase activity towards selected IAA–amino acid conjugates. Bartel and Fink (1995) have proposed that this apparent substrate specificity may reflect the existance of at least two classes of aminohydrolases. In support of this idea, the authors noted that *ILR1* belongs to a family consisting of at least three homologous genes in *Arabidopsis* and that they had isolated additional aminohydrolases mutants which were resistant to IAA-Ala, but not IAA-Leu.

IAA can be degraded either by decarboxylation or non-decarboxylative oxidation of the indole group. Peroxidase-catalysed decarboxylation has been described in a variety of plants, but the physiological relevence of the IAA oxidase activity by peroxidases is unclear. Instead, Sandberg and co-workers (Tuominen *et al.*, 1994) have proposed that the conjugate, indole-3-acetyl-N-aspartic acid, represents the first step in the irreversible deactivation of IAA, via oxidation and conjugation of glucosides to the indole nitrogen (Figure 4.1).

4.2.2 *Transport*

Many organs cannot synthesise enough IAA to support their continued growth and development. Instead they rely on the continual supply of IAA from the major sites of auxin synthesis in young developing leaves and the shoot apex. Plants employ a specialised delivery system, termed 'polar auxin transport', to convey IAA to its target tissues (see Lomax *et al.*, 1995). IAA moves from cell to cell within phloem parenchyma tissues using an energy-dependent, diffusion-based mechanism. The basipetal direction of auxin movement has been proposed to reside in the asymmetric distribution of auxin transport proteins within these specialised transport cells (Rubery and Sheldrake, 1974; Raven, 1975). The auxin efflux carrier is likely to play an important role in determining the polarity of auxin movement since inhibitors of auxin efflux, such as naphthyl pthalamic acid (NPA), also block polar auxin transport.

Physiological experiments have concluded that the efflux carrier is composed of at least three components: a transmembrane protein, an NPA binding protein and a third, labile component (Morris *et al.*, 1991). Several *Arabidopsis* genes have been proposed to encode components of the efflux carrier complex based on their mutant phenotype. Both the *pin-formed* and *tir3* mutants exhibit significantly reduced rates of polar auxin transport within inflorescence tissues (Okada *et al.*, 1991; Ruegger *et al.*, 1997). Galweiler *et al.* (1996) reported the isolation of the *PIN-FORMED* gene. Interestingly, the *PIN-FORMED* sequence encodes a membrane-localised protein that may represent the transmembrane component of the auxin efflux carrier. Biochemical studies have shown that the *tir3* mutant contains significantly reduced levels of NPA binding, leading Ruegger *et al.* (1997) to suggest that the *TIR3* gene may encode (or regulate the activity of) the NPA-binding protein. The activity of the polar auxin transport machinery may also be regulated by phosphorylation. Garbers *et al.* (1996) observed that auxin efflux in the *rcn1* mutant displays an enhanced sensitivity towards NPA. The RCN1 gene product is similar to regulatory subunit A of the protein phosphatase 2A (PP2A) enzyme. When recombinantly expressed in yeast, the *RCN1* sequence is capable of complementing the PP2A mutant *tpd3-1*.

Polar auxin transport is known to influence a wide range of developmental processes. For example, the polar auxin transport inhibitor NPA is able to disrupt patterning of *in vitro* cultured embryos, leading to suggestions that an IAA-based morphogenic gradient may be important during embryogenesis. In the carrot, NPA disrupts the formation of apical-basal patterning at the globular stage embryogenesis which mimics *Arabidopsis* embryo patterning mutants such as *gnom* and *monopteros* (Schiavone and Racusen, 1997). NPA treatment during later stages of *Brassica* embryogenesis can create fused cotyledon structures that phenocopy the *Arabidopsis pin-formed* mutant (Liu *et al.*, 1993). Polar auxin transport also plays an important role during post-embryonic development. Mutations that disrupt basipetal auxin movement are often associated with morphological changes during vegetative and reproductive development. The *lop1* mutant exhibits alterations in its leaf vascular development, resulting in mid-vein bifurcation and disoriented axial growth (Carland and McHale, 1996), mutations within the *TIR3* gene result in a significant reduction in lateral root numbers and apical dominance (Ruegger *et al.*, 1997), and *pin-formed* mutants develop severely abnormal, unbranched influorescence structures that fail to form floral organs (Okada *et al.*, 1991). Interestingly, Okada *et al.* (1991) observed that the latter phenotype can be recreated in wild-type *Arabidopsis* by subculturing plantlets in the presence of NPA.

4.2.3   Auxin action

A series of *Arabidopsis* mutants has been described that demonstrates altered responses to exogenous auxins (Table 4.1). Characterisation of these response mutants and their wild-type gene products has provided new insight into the molecular basis of auxin action.

Mutations within the *AUX1* gene confer an agravitropic root phenotype (Maher and Martindale, 1980; Figure 4.2A). The *AUX1* gene encodes a hydrophobic polypeptide is similar to a family of plant amino acid uptake carriers (Bennett *et al.*, 1996). AUX1 has been proposed to function as an auxin uptake carrier and several lines of evidence support this hypothesis. For example, mutations within the *AUX1* gene selectively block the action of auxins that require carrier-mediated uptake (Marchant and Bennett, unpublished). Delbarre *et al.* (1996) demonstrated that the auxins IAA and 2,4 dichlorophenoxyacetic acid (2,4-D) require carrier-mediated uptake whereas the lipophilic auxin 1-naphthylacetic acid (1-NAA) is taken up in a carrier-independent fashion. Hence, a defect in auxin uptake should selectively impair responses towards IAA and 2,4-D, but not 1-NAA. We have observed using a root elongation bioassay that *aux1* mutants exhibit a reduced response to IAA and 2,4-D yet retain a wild-type sensitivity towards 1-NAA (Marchant and Bennett, unpublished). Spatially, the *AUX1* gene is expressed within elongating tissue of the root apex (Bennett *et al.*, 1996). AUX1 is likely to regulate root gravitropic curvature by acting in

Table 4.1 *Arabidopsis* auxin mutants

Gene	Map position	Inheritance	Mutant phenotype	Function	Reference
Auxin synthesis, conjugation and degradation mutants					
ALF3	5	Recessive	Lateral root development arrests after emergence		Calenza et al., 1995
RTY	2	Recessive	Auxin overproducer, increased adventitious root formation	Homology with tyrosine amino transferases	King et al., 1995; Boerjan et al., 1995; Calenza et al., 1995; Gopalraj et al., 1996
ILR1	3	Recessive	Insensitive root elongation in the presence of toxic IAA–Leu conjugate	IAA–amino acid conjugate aminohydrolyse	Bartel and Fink, 1995
Auxin transport and signalling mutants					
PIN1	1	Recessive	Altered embryo patterning, aberrant influorescence, meristem development and reduced polar auxin transport	Putative membrane protein encodin auxin efflux carrier	Okada et al., 1991; Galweiler et al., 1996
TIR3	2	Recessive	Reduced lateral root number, apical dominance and polar auxin transport		Ruegger et al., 1997
RCN1	1	Recessive	Excessive root curling in the presence of NPA	Regulatory subunit A protein phosphatase 2A	Garbers et al., 1996
LOP1	5	Recessive	Altered leaf vascular patterning and root development, reduced polar auxin transport		Carland and McHale, 1996
AUX1	2	Recessive	Agravitropic root, reduced lateral root number	Permease-like component of the auxin uptake machinery	Maher and Martindale, 1980; Pickett et al., 1990; Bennett et al., 1996
AXR1	1	Recessive	Reduced stature, lateral root number, defect in cell division	Similarity to the N-terminal domain of ubiquinating enzyme E1	Lincoln et al., 1990; Leyser et al., 1993
SAR1	1	Recessive	Reduced stature, altered leaf morphology and early flowering		Cernac et al., 1997
AXR2	3	Dominant	Agravitropic root, dwarf-like stature, defect in cell elongation		Wilson et al., 1990; Timpte et al., 1994
AXR3	1	Semi-dominant	Severe apical dominance, increased adventitious root formation	IAA/AUX gene family member	Leyser et al., 1996; Leyser, unpublished
AXR4	1	Recessive	Reduced gravitropism, lateral root number		Hobbie and Estelle, 1995
ALF4	5	Recessive	Loss of lateral root development		Calenza et al., 1995
MSG1	5	Recessive	Auxin-insensitive hypocotyl and leaf development		Watahiki and Yamamoto, 1997

A.

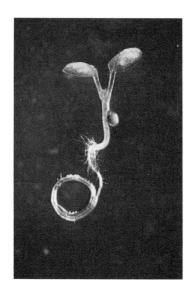

Figure 4.2A The *aux1* auxin response mutation confers an agravitropic root phenotype.

B-E.

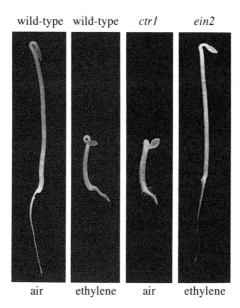

Figure 4.2B-E   B, Wild-type seedlings grown in air or C, 10ppm ethylene demonstrate an etiolated and triple response respectively. D, A *ctr1* seedling grown in air phenocopies wild-type seedlings grown with ethylene. E, An *ein2* seedling grown in the presence of ethylene exhibits an ethylene insensitive phenotype.

F.

G.

Figure 4.2F-G F, The *ga1* mutant (right) is dwarfed owing to gibberellin deficieny but the phenotype can be restored by the exogenous supply of gibberellin (left), G, In contrast to wild-type (left) the ABA insensitive *abi1* mutant (right) demonstrates a wilty phenotype.

conjunction with an efflux carrier to facilitate the movement of auxin between rapidly expanding cells within the elongation zone. *Aux1* mutants also exhibit a significantly reduced rate of root branching (Hobbie and Estelle, 1995). Lateral root initiation within pericycle tissues is an auxin-dependent process (Laskowitz *et al.*, 1995). The permease-like AUX1 protein is likely to stimulate the rate of lateral root initiation by facilitating auxin uptake into dividing pericycle cells. In support of this model for AUX1 action, transgenic lines ectopically expressing an *AUX1* transgene within pericycle tissues exhibit significantly more lateral roots than wild-type *Arabidopsis* (Marchant *et al.*, unpublished).

Axr1 mutants exhibit a pleiotropic phenotype that includes a loss of apical dominance, reduced lateral root number and elevated root length (Estelle and Somerville, 1987; Lincoln *et al.*, 1990). Double mutant studies between the *axr1* and *aux1* mutations have concluded that each mutant alters the *Arabidopsis* auxin response by a different mechanism (Timpte *et al.*, 1995). Several lines of evidence suggest that the *AXR1* gene encodes an important intracellular auxin signalling component. Firstly, the *axr1* mutant is able to block all phenotypic effects associated with the transgenic overexpression of the *iaaM* gene, despite elevated IAA levels (Romano *et al.*, 1993). Secondly, the auxin-mediated induction of the SAUR-AC1 and IAA mRNAs is severely impaired in an *axr1* background (Timpte *et al.*, 1995; Abel *et al.*, 1995). The *AXR1* gene has sequence similarity to the amino terminal domain of the eukaryotic ubiquitin-activating E1 enzymes that regulate protein turnover (Leyser *et al.*, 1993). Ubiquitin-mediated degradation of selected cell cycle components is known to play an important role during cell division of eukaryotic cells. Gene disruption experiments performed in *S. cerevisae* have concluded that a yeast *AXR1* homologue interacts with the cell cycle regulator, the *cdc34* gene (Lammer and Estelle, unpublished). The *Arabidopsis AXR1* gene product is also likely to participate in cell cycle control since morphological studies have concluded that *axr1* mutants have undergone a smaller number of divisions than wild-type (Lincoln *et al.*, 1990).

Estelle and co-workers have recently identified a second auxin-related gene, termed *TIR1*, which further implicates the ubiquitin pathway in auxin action (Ruegger*et al.*, 1998). The TIR1 protein contains a leucine-rich repeat termed an F-box. In humans, the F-box motif has been demonstrated to be important for protein–protein interactions between F-box-containing proteins such as SKP2 and Cyclin A-CDK2, which regulate the G1 to S phase transition (Zhang *et al.*, 1995). In yeast, the F-box containing protein Cdc4p is part of the S-phase promoting complex, which includes the ubiquitin-conjugating enzyme Cdc34p (Bai *et al.*, 1996). Both SKP2 and Cdc4p are proposed to facilitate cell cycle entry by targeting one or more CDK inhibitors for ubiquitin-dependent degradation. The synergistic phenotype of the *axr1,tir1* double mutant suggests that TIR1 is involved in AXR1-mediated processes. Several extragenic suppressors of *axr1* have been identified that are likely to encode additional components within the *AXR1* signalling pathway (Cernac *et al.*, 1997). The *sar1, 2* and *3* mutations are capable of suppressing many of the pleiotropic effects associated with the *axr1* phenotype. Collectively, the *AXR1*, *TIR1* and *SAR* genes represent exciting tools for the elucidation of the relationship between auxin and plant cell division.

The dominant *axr2* mutation confers a severe dwarf-like aerial phenotype (Wilson *et al.*, 1990). Histological studies have observed that *axr2* contains smaller cells than wild-type, suggesting that the dwarf phenotype results from a defect in cell elongation, rather than division. The *AXR2* gene product is likely to mediate an early step in the auxin response pathway since the induction of several auxin responsive genes is severely disrupted within the *axr2* mutant (Timpte *et al.*, 1994). However, *axr2*, like *axr1* and *aux1*, exhibits an altered response towards phytohormones other than auxin (Wilson *et al.*, 1990; Lincoln *et al.*, 1990; Pickett *et al.*, 1990). The *axr4* mutant is unusual because it is specifically resistant to auxin (Hobbie and Estelle, 1995). Mutations within the *AXR4* gene reduce the rates of gravitropic curvature and lateral initiation in roots. Hobbie and Estelle (1995) investigated the *AXR4* signalling pathway by constructing *axr4,axr1* and *axr4,aux1* double mutant combinations. These studies conclude that the *AXR1* and *AXR4* normally act in separate pathways. The genetic interaction between *AUX1* and *AXR4* appears to be more complex. *AUX1* is epistatic towards *AXR4* during root elongation, whereas it regulates lateral root initiation in an additive fashion. Further approaches are clearly necessary in order to comprehend the nature of the interaction between *AXR4* and *AUX1*.

All previously described auxin response mutants were selected on the basis of the reduced sensitivity of their root elongation towards inhibitory concentrations of exogenous auxin. Watahiki and Yamamoto (1997) have employed an alternative strategy leading to the isolation of the massugu (*msg1*) mutant. The *msg1* mutant was selected by the inability of its hypocotyl to bend in response to an unilateral application of lanolin paste containing IAA. The *MSG1* locus defines a new auxin-response gene based on its map position and physiological characteristics. For example, the *msg1* mutant defects are limited to the hypocotyl and leaf tissues and appear to be auxin specific, in contrast to other auxin-response mutants. The identification of *msg1* mutation highlights the need for a battery of novel screens in order to gain a comprehensive understanding of the auxin signalling pathways that operate within every *Arabidopsis* tissue.

4.2.4 Downstream auxin responses

Auxin elicits two types of rapid responses within elongating tissues: an increase in the pH of the apoplastic space, resulting in acid-induced cell growth (for a review see Rayle and Cleland, 1992), and the induction of gene transcription, which is necessary to underpin the biosynthetic expenditure associated with prolonged cell expansion (Abel and Theologis, 1996). Auxin-induced transcripts are classified as either early or late response mRNAs (Abel and Theologis,

1996). In *Arabidopsis* early response genes include SAUR-AC1, *ACS4* and members of the IAA gene family. *ACS4* encodes 1-aminocyclopropane-1-carboxylic acid (ACC) synthase, which represents a key enzyme within the ethylene biosynthetic pathway (Abel *et al.,* 1995b). The *ACS4* gene is the only member of the ACC synthase family in *Arabidopsis* that is specifically induced by IAA within etiolated *Arabidopsis* seedlings, suggesting that the up-regulation of this gene may contribute to auxin-stimulated ethylene synthesis.

The IAA genes encode a family of short lived, nuclear localized proteins which are proposed to function as transcription factors (Abel and Theologis, 1996). Abel *et al.* (1995) have isolated 14 IAA genes from *Arabidopsis*. IAA polypeptides share several features: four conserved domains (termed boxes I to IV), a nuclear targeting signal and an amphipathic $\beta\alpha\alpha$ fold found in several prokaryotic repressor proteins. The recent isolation of the *AXR3* gene has served to throw new light on the functional importance of the IAA gene products. *AXR3* encodes a member of the IAA gene family (H.M.O. Leyser, personal communication). The *AXR3* gene was originally defined by two semi-dominant mutants which affect several auxin-regulated developmental processes (Leyser *et al.,* 1996). The *SAUR-AC1* promoter-GUS transgene is ectopically expressed within vascular tissues of *axr3* plants, suggesting alterations in the regulation of the mutant's transcriptional machinery. The *AXR3* gene product, as a member of the IAA family, could therefore represent a direct link between auxin and gene transcription.

Several lines of evidence suggest that the early response genes are primary targets for the auxin signalling pathway(s) (Abel *et al.,* 1995). Firstly, gene induction is very rapid (between 4 and 30 minutes) following auxin treatment. Secondly, gene induction cannot be blocked by cycloheximide treatment, indicating that their transcription does not require the synthesis of any additional factors. Thirdly, the relative abundance of these early response mRNAs are reduced within the auxin mutants, *aux1, axr1* and *axr2* (Timpte *et al.,* 1994, 1995; Abel *et al.,* 1995).

The promoters of the early response genes represent targets for the auxin transduction pathway. Several elements within these promoters have been demonstrated to regulate auxin-inducible expression (Ulmasov *et al.,* 1995). Ulmasov *et al.* (1997) have identified an *Arabidopsis* transcription factor, termed *ARF1*, whose amino terminal domain binds an auxin-responsive promoter element. The carboxyl terminus of the *ARF1* protein contains the box III and IV motifs found in IAA polypeptides. This domain appears to facilitate protein–protein interactions since two hybrid cloning experiments have identified an interactive partner that contains box III- and IV-like motifs. This observation raises the intriguing possibility that *ARF1* activity may be regulated by dimerisation with other box III- and IV-containing polypeptides, including IAA family members.

4.3 Ethylene

4.3.1 Ethylene biosynthesis

The ethylene biosynthetic pathway has been elucidated in a series of elegant studies (Figure 4.3). The two key enzymes are ACC synthase and ACC oxidase, which catalyze the conversion of S-adenosyl-L-methionine (SAM) to 1-amino-cyclopropane-1-carboxylic acid (ACC) and ACC to ethylene, respectively. ACC synthase is usually the rate-limiting step; its activity is highly regulated and closely parallels the level of ethylene biosynthesis. ACC synthase is generally encoded by a small gene family. There are at least five ACC synthase genes in *Arabidopsis*, three of which are expressed and encode active ACC synthase enzymes (Van der Straeten 1990; Liang *et al.* 1992).

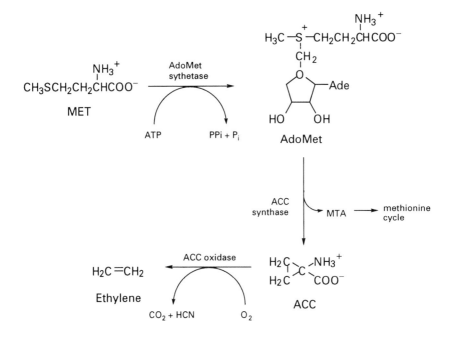

Figure 4.3 Ethylene biosynthetic pathway. The enzymes catalyzing each step are shown above the arrows. AdoMet S-adenyl-methionine, MET methionine, ACC 1-aminocyclopropane-1-carboxylic acid, MTA methylthioadenine.

The final step of ethylene biosynthesis, the conversion of ACC to ethylene, is catalyzed by the enzyme ACC oxidase. Studies suggest that ACC oxidase may also play a role in regulating ethylene biosynthesis, especially during conditions of high ethylene production. ACC oxidase, like ACC synthase, also appears to be encoded by a gene family. Examination of the EST database reveals that there are at least five genes that display homology to ACC oxidases in *Arabidopsis*, although only one has been characterized in any detail (Gomez-Lim *et al.*, 1993).

Almost all plant tissues have the capacity to make ethylene, although in most cases the amount produced is very low. Ethylene production increases dramatically during a number of developmental events such as germination, fruit ripening and leaf and flower senescence and abscission (Abeles *et al.*, 1992; Yang and Hoffman, 1984). There is a diverse group of factors that increase the level of ethylene biosynthesis in *Arabidopsis* and many other plant species, including auxin, cytokinin, inorganic ions such as Li^+ and Cu^{2+}, and various stresses. The emerging picture is that different ACC synthase genes are expressed in response to various developmental, environmental and hormonal factors. In *Arabidopsis*, the three ACS genes show distinct patterns of expression. Inhibition of protein synthesis by cycloheximide treatment induces expression of all three genes, suggesting that they are under negative control (Liang *et al.*, 1992). The steady-state level of ACS2 displayed a slower but more sustained rise in response to cycloheximide than *ACS5* (Liang *et al.*, 1996). Auxin specifically and rapidly induces expression of *ACS4* (Abel *et al.*, 1995). ACS5 has been implicated in the response to cytokinin in etiolated seedlings (see below) and is also induced by Li^+ (Liang *et al.*, 1996). The pattern of expression of *ACS2* was examined by fusions to a GUS reporter gene (Rodrigues-Pousada *et al.*, 1993). This study suggested that *ACS2* expression is high in young tissues and is switched off as the tissue matures. In addition, expression of *ACS2* was correlated with lateral root formation. The *ACS3* gene is most likely a pseudogene and *ACS1* encodes a non-functional ACC synthase (Liang *et al.*, 1995).

A number of *Arabidopsis* mutants that are affected in the production of ethylene using the triple response screen have been isolated (see below and Table 4.2). One class, the Eto mutants, produce from ten- to 50-fold more ethylene as etiolated seedlings, but do not produce highly elevated levels of ethylene in the light or as adults. This suggests that perhaps ethylene biosynthesis is regulated independently in seedlings and adults or in light- and dark-grown plants. Cytokinin induces ethylene biosynthesis in etiolated *Arabidopsis* seedlings, resulting in a triple response morphology. The *cin5* mutant was isolated in a screen for etiolated seedlings that failed to display the triple response in the presence of cytokinin (Vogel *et al.*, 1998). *cin5* was found to disrupt the *ACS5* gene. Furthermore, activation of ACS5 by cytokinin was found to be primarily post-transcriptional. The *cin5* mutation mapped very close to the dominant mutation *eto2* and molecular analysis revealed that the *eto2* mutation is a single base pair insertion in the 3' end of *ACS5* that is predicted to disrupt the carboxyl-terminal 11 amino acids. Thus, *eto2* is the result of an increase in the activity of the ACS5 isoform and *cin5* is the result of a loss-of-function *ACS5* mutation.

Table 4.2 *Arabidopsis* ethylene mutants

Mutant	Inheritance	Comments	Reference
Insensitive mutations			
etr1/ein4/etr2	Dominant	Ethylene receptor; two-component histidine kinase	Bleecker et al., 1988; Roman et al., 1995
ers/ers2	Dominant	Made by transformation with a mutated homolog of ETR1	Hua et al., 1995
ein2	Recessive	12-pass trans-membrane protein	Guzman and Ecker, 1990
ein3	Recessive	Weak phenotype; gene family; transcription factor?	Roman et al., 1995
ein5/ain1	Recessive	Weak phenotype	Roman et al., 1995; Van der Straeten et al., 1993
ein6	Recessive	Weak phenotype; increased sensitivity to taxol	Roman and Ecker, 1996; Roman et al., 1995
aux1; axr1, 2, 3		Isolated as auxin-resistant mutants; display resistance to various hormones, including ethylene	Hobbie and Estelle, 1994
Constitutive mutations			
eto1	Recessive	Overproduce ethylene as etiolated seedling	Guzman and Ecker, 1990
eto2	Dominant	Overproduce ethylene; *ACS5* allele	Kieber et al., 1993
eto3	Dominant	Overproduce ethylene	Kieber et al., 1993
ctr1	Recessive	Similar to Raf family of protein kinases	Kieber et al., 1993
ctr2	Recessive	Rosette-lethal phenotype	Woeste and Kieber, unpublished

4.3.2   Ethylene action

When seedlings from many different dicotyledonous species are grown in the dark in the presence of ethylene they adopt a striking morphology referred to as the triple response (Figure 4.2B,C). In *Arabidopsis*, the triple response consists of an inhibition of root and hypocotyl elongation, radial swelling of the hypocotyl and an exaggeration of the curvature of the apical hook. This response of *Arabidopsis* seedlings provides a facile screen for the isolation of ethylene response mutants and has been the key in unravelling the molecular basis of ethylene signalling. Mutants have been identified that fail to display the triple response in the presence of exogenous ethylene (insensitive) as have mutations that constitutively display this response (Table 4.2; Figures 4.2D, E). These mutations have an impact on virtually all ethylene responses in both seedlings and adult plants, suggesting that they affect central components in ethylene signalling. An additional class of mutants alter ethylene responses in only a subset of tissues, such as the root *(eir1)* or the apical hook *(hls1)* and may act late in the ethylene response pathway.

4.3.3   Ethylene response mutants

Ethylene-insensitive mutants fail to display the triple response in the presence of saturating levels of exogenous ethylene and are readily identifiable as tall seedlings protruding above the 'lawn' of short, wild-type seedlings. The first such mutant identified was *etr1* (<u>et</u>hylene <u>r</u>esistant), which is inherited as a single gene, dominant mutation (Bleecker *et al.*, 1988). *etr1* leaves bind only 20% as much exogenous ethylene as wild-type leaves, which is consistent with data suggesting that *ETR1* encodes an ethylene receptor (see below). The *ein4* and *etr2* mutations, like *etr1*, are also dominant and act upstream of *ctr1* (Chang, 1996; Roman *et al.*, 1995). The absence of recessive, loss-of-function alleles at these loci is probably owing to the fact that these genes encode homologous proteins and may be genetically redundant. The *ein2* mutation is recessive and displays a relatively strong ethylene-insensitive phenotype (Guzman and Ecker, 1990). On infection with various strains of virulent bacteria, *ein2* mutants, unlike the other ethylene-insensitive mutants, fail to display typical disease symptoms, although the growth of the pathogen is unaffected.

Mutations that result in a weaker ethylene-insensitive phenotype have been identified at four other genetic loci. The *ain1(ein5)*, *ein3* and *ein6* mutations are recessive whereas the *ein7* mutation is semi-dominant (Roman *et al.*, 1995; Van der Straeten *et al.*, 1993). The weak phenotype of these mutations could be the result of partial redundancy of their gene products or of mutations that are only partial loss of function. *ein6* displays increased sensitivity to taxol, which suggests that it may affect a protein that plays a role in microtubule function.

ctr1 mutants constitutively display the triple response and are recessive, suggesting that the wild-type function of CTR1 is to negatively regulate ethylene signalling (Figure 4.2D). The *ctr1* mutation has dramatic effects on the morphology and development of both seedlings and adult plants. Etiolated *ctr1* seedlings take longer to open the apical hook and expand their cotyledons when shifted to light than wild-type seedlings. *ctr1* leaves and roots are smaller, the plants flower later and the inflorescence is much more compact than that of wild-type plants. These phenotypes can be copied by growing wild-type plants in ethylene, suggesting that the various *ctr1* phenotypes reflect constitutive ethylene responses. These phenotypes correlate with a failure of *ctr1* mutant cells to expand, consistent with the effect of ethylene in other plant species. An additional phenotype of *ctr1* is the formation of ectopic root hairs (Dolan *et al.*, 1994). *Arabidopsis* root hairs are almost always located in epidermal cells that overlie a junction between adjacent cortical cells (Dolan *et al.*, 1993). In *ctr1* mutant roots, extra root hairs form in ectopic locations in the epidermis, suggesting that *ctr1*, and by inference ethylene, may act to negatively regulate the decision to adopt a hair cell fate.

The screen for ethylene response mutations is not yet saturated as only single recessive alleles at several loci have been identified. Furthermore, additional genes involved in ethylene signalling may not have been detected because they are functionally redundant or their loss of function may result in lethality or infertility. Thus, it is likely that additional elements will be identified by a combination of further genetic screens and biochemical and molecular methods.

4.3.4 Molecular analysis of mutants

The first gene identified in ethylene signalling was *CTR1* (Kieber *et al.*, 1993). The carboxyl terminus of CTR1 has all the hallmark features of a serine/threonine protein kinase and expression of CTR1 in insect cells using baculoviral vectors confirms that the protein does have intrinsic Ser/Thr protein kinase activity (Huang *et al.*, unpublished). The amino acid sequence of CTR1 is most similar to the Raf family of protein kinases. Raf-1 encodes a Ser/Thr protein kinase that is part of a cascade of conserved protein kinases involved in the transduction of a number of external regulatory signals (reviewed in Marshall, 1994; Moodie and Wolfman, 1994; Morrison, 1990 and Rapp, 1991). The receptors for these signals generally activate Raf indirectly, through the small GTP binding protein Ras, which interacts with the amino terminal domain of Raf (Avruch *et al.*, 1994; Daum *et al.*, 1994). The lack of homology of CTR1 to the amino-terminal portion of Raf suggests that CTR1 may be regulated by different upstream factors and/or by a distinct mechanism. Consistent with this, recent results suggest that ETR1 and CTR1 interact, suggesting that ETR1 may directly regulate CTR1.

The major physiological target of Raf is the dual-specificity protein kinase MEK, which in turn activates MAP kinase via phosphorylation. The activated MAPK then phosphorylates numerous downstream targets, including a number of transcription factors such as c-Myc and c-Jun. The region of Raf responsible for the recognition of MEK is completely contained within the protein kinase catalytic domain (Van Aelst *et al.*, 1993). Since CTR1 displays high similarity to the kinase domain of Raf, it seems plausible that CTR1 may phosphorylate a similar target.

Most of the mutations in the CTR1 gene that have been analysed are predicted to disrupt its kinase activity, including three single amino acid changes in residues that are extremely conserved in protein kinases (Kieber *et al.*, 1993). Coupled with the recessive nature of *ctr1* mutations these results indicate that the kinase activity of CTR1 is required to negatively regulate the ethylene response pathway.

The amino-terminal region of the predicted *ETR1* protein contains three transmembrane domains and the carboxyl-terminus displays similarity to both the histidine kinase and response regulator domains of bacterial two-component sensing systems (Chang *et al.*, 1993; reviewed in Parkinson, 1993 and Stock *et al.*, 1990). Two-component regulators are the major route by which bacteria sense and respond to various environmental cues. The two components consist of a sensor and an associated response regulator. The sensor is responsible for perceiving the signal (the input domain), which induces autophosphorylation of the histidine kinase domain on a conserved histidine residue. This phosphate is then transferred to a conserved aspartate residue on the receiver domain of the cognate response regulator, which in turn regulates the activity of the output domain. Based on its primary amino acid sequence and binding to ethylene, the ETR1 protein is predicted to contain a sensor component fused to a receiver domain of a response regulator.

ETR1 is found as a membrane-associated, disulphide-linked dimer in extracts of *Arabidopsis* and when expressed in yeast (Schaller *et al.*, 1995). The disulphide linkage in ETR1 was localized by expression of truncated forms and by *in vitro* mutagenesis to two cystines at the amino terminus (Cys4 and Cys6) in a presumed extracellular domain of the protein. ETR1 is also associated with membranes when extracted from *Arabidopsis* (Schaller *et al.*, 1995), although it is unclear which membrane system.

Yeast cells that express wild-type ETR1 protein were shown to bind ethylene with a high affinity and this binding was saturable, whereas yeast expressing a mutant version of the ETR1 protein (*etr1-1*) did not display detectable saturable binding of ethylene (Schaller and Bleecker, 1995). The K_d of binding was 0.04 µl/l, which is close to the dose required for the half-maximal response in the *Arabidopsis* triple response (Chen and Bleecker, 1995). This binding was reduced by *trans*-cyclooctene and 2,5-norbornadiene, two competitive inhibitors of ethylene binding in many plant species, including *Arabidopsis* (Sisler and Yang, 1984; Guzman and Ecker, 1990; Sisler *et al.*, 1990). These results, together with the observation that *etr1* mutant seedlings display reduced ethylene binding (Bleecker *et al.*, 1988) and genetic epistasis analysis placing ETR1 as the earliest acting of the ethylene mutations (Kieber *et al.*, 1993), provide compelling evidence that ETR1 encodes an ethylene receptor. An important experiment to confirm this is to demonstrate that ethylene binding alters the functionality of the ETR1 protein.

The binding of ethylene was localized to the amino-terminal, hydrophobic domain of ETR1 (Schaller and Bleecker, 1995). The four *etr1* mutations all occur in this amino-terminal hydrophobic domain within the three trans-membrane segments and one of these mutations, *etr1-1*, has been shown to disrupt ethylene binding both in *Arabidopsis* and when expressed in yeast. Binding in a membrane environment is not unexpected as ethylene is 12 times more soluble in lipids than in water under physiological conditions (Wilhelm *et al.*, 1977).

ETR1 is present as a small gene family in *Arabidopsis*. One homolog, ERS was cloned by low stringency hybridization to ETR1 (Hua *et al.*, 1995). ERS is 67% identical to ETR1 at the amino acid level but lacks the receiver domain of the response regulator. ERS does not map to any identified ethylene-insensitive mutation. A mutation analogous to the *etr1*-4 mutation (a Ile to Phe change in the second trans-membrane segment) was introduced into ERS and found to result in dominant ethylene insensitivity when expressed in *Arabidopsis*. Three other homologs of ETR1 have been implicated in ethylene signalling (*ETR2, ERS2* and *EIN4*; Hua *et al.*, 1997). An intriguing possibility is that heterodimers may form between the various ETR1 proteins, allowing for modulation of the sensitivity of a tissue to ethylene.

Two other genes downstream of CTR1 have been cloned (Ecker *et al.*, 1996; Chao *et al.*, 1997). *EIN3* encodes a nuclear-localized protein that is present as a small gene family in *Arabidopsis*. The predicted EIN2 protein contains a series of trans-membrane domains in the amino-terminal domain and the carboxyl-terminus is composed of a highly charged, nuclear-localized domain.

4.3.5 Downstream components

One of the primary effects of ethylene is to alter the expression of various target genes. Ethylene treatment results in an increase in the level of mRNA of numerous plant genes, including cellulase, chitinase, peroxidase, chalcone synthase, a basic-type PR gene and β1-3 glucanase, as well as ripening related genes and ethylene biosynthetic genes (for examples see: Ecker and Davis, 1987; Felix and Meins, 1987; Broglie *et al.*, 1989; Theologis, 1992; Gray and Grierson, 1993; reviewed in Broglie, 1991). DNA sequences (ethylene response elements, ERE) that confer ethylene responsiveness to a minimal promoter have been identified from the ethylene-regulated pathogenesis-related (PR) genes and a GCCGCC repeat motif was identified as both necessary and sufficient for this regulation (Ohme-Takagi and Shinshi, 1995). Four proteins that bind to ERE sequences were identified in tobacco and the steady-state level of RNA for these genes increases dramatically and rapidly in response to ethylene. These ERE binding proteins (EREBPs) may be primary targets of the ethylene response pathway and may regulate the expression of other secondary response genes such as the PR genes.

Ethylene regulates several plant responses that require differential cell elongation, such as epinasty, tendril coiling and apical hook formation (Abeles *et al.*, 1992). Growth in the presence of ethylene results in a marked exaggeration in the curvature of the apical hook in etiolated *Arabidopsis* seedlings. The apical hook is the result of differential cell expansion of the hypocotyl caused by the influence of ethylene and auxin, as well as other factors. The *hookless1* (*hls*) mutation disrupts apical hook formation in both air- and ethylene-grown *Arabidopsis* seedlings (Guzman and Ecker, 1990). The *HLS1* gene has recently been cloned and found to show similarity to a diverse group of N-acetyltransferases from bacteria to mammalian cells (Lehman *et al.*, 1996). *HLS1* expression is uniform transversely across the apical hook, suggesting that it does not influence hook formation by differential expression in the adaxial and abaxial tissues of the hook. However, the steady-state level of *HLS1* mRNA is induced by ethylene and overexpression of HLS1 resulted in seedlings that displayed a constitutively exaggerated apical hook. This suggests that increased HLS1 expression is sufficient to induce hook formation and that ethylene may influence hook curvature by regulating the expression of *HLS1*. The pattern of expression of two primary auxin-upregulated genes, *AtAux2-11* and *SAUR-AC1*, were found to be altered in *hls1* mutants in the tissue that normally forms the apical hook but not in other parts of the seedling, suggesting that HLS1 may play a role in auxin metabolism or transport in this tissue.

4.3.6 *Model for ethylene signalling*

Ethylene signalling begins with binding to a receptor(s), which almost certainly includes members of the ETR1 gene family (Figure 4.4). This binding has been hypothesized to be mediated through a transition metal, probably copper, coordinated within the hydrophobic region of ETR1, which then induces a conformational change of ETR1 that may alter the rate of *trans*-phosphorylation between subunits (Bleecker and Schaller, 1996). In the absence of ethylene, ETR1 acts to induce the kinase activity of CTR1, and binding of ethylene to ETR1 is postulated to block this activation. The function of CTR1 is to negatively regulate ethylene responses, and genetic and molecular data suggest that the kinase activity of CTR1 is required for this repression. Presumably, the activity of CTR1 is regulated through its amino-terminal domain, as is the case with Raf and other protein kinases with amino-terminal extensions, such as STE11 and protein kinase C. CTR1 then regulates the activity of the *EIN2* and *EIN3* gene products, perhaps via a MAP kinase cascade. While this model is consistent with all the data obtained, it is almost certainly incomplete and the details of these putative interactions are only just beginning to be elucidated.

4.4 Gibberellins

The gibberellins (GAs) form a complex family of structurally related compounds, not all of which are directly biologically active, based on a 19 or 20 carbon diterpene-derived gibbane carbon skeleton. Gibberellin biosynthesis starts from acetyl CoA, which is converted through mevalonic acid to *ent*-kaurenoic acid and from this to the first true GA, GA12. From GA12, many different

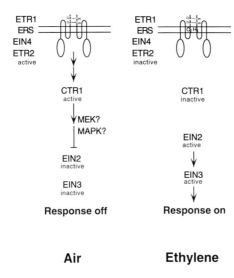

Figure 4.4 Model for ethylene signal transduction in *Arabidopsis*. Arrows represent positive regulatory steps and the flat symbol represents a negative regulatory step. Each arrow may represent several steps in the transduction pathway as no direct interactions have yet been definitively demonstrated. See text for details.

forms of GA may be produced by reactions taking place in three parallel but interlinked pathways (Figure 4.5) (reviewed by Finkelstein and Zeevart, 1994).

The active GAs have many physiological roles, as determined by exogenous application of GAs or inhibitors of GA biosynthesis and by the analysis of GA biosynthetic and response mutants. Important processes that GAs are involved in include:

- stem elongation (cell division and elongation)
- flower formation
- floral organ development
- fruit development
- seed germination.

4.4.1 GA biosynthesis

Since GAs are involved in seed germination and stem elongation, these characteristics were used to screen for GA biosynthetic mutants in *Arabidopsis*; candidate GA mutants were those that failed to germinate on basal medium but did subsequently germinate when transferred to medium containing GA, or dwarf plants that produced short stems which elongated in response to GA application. Five different GA-responsive loci were isolated (*ga1*, *ga2*, *ga3*, *ga4* and *ga5*) (Koornneef and van Veen, 1980; Table 4.3, Fig. 4.2F). Alleles of the *ga1*, *ga2* and *ga3* mutants generally require GA in order to germinate and in the absence of GA grow as sterile, dark green dwarves. The mutants *ga4* and *ga5* do not need GA in order to germinate, but in the absence of GA have a semi-dwarf phenotype.

The *ga1*, *ga2* and *ga3* mutants have been analysed for their endogenous GA content by gas chromatography selected ion monitoring, for their response to exogenous GA and GA precursors and for their response to GA biosynthesis inhibitors (reviewed by Finkelstein and Zeevart, 1994). The *ga1*, *ga2* and *ga3* mutants contain very low GA levels, indicating a lesion early on in the biosynthetic pathway. Both *ga1* and *ga2* respond with normal growth if *ent*-kaurene or later metabolites are applied exogenously. The inhibitor tetcyclacis interferes with GA biosynthesis by preventing the steps leading to the oxidation of *ent*-kaurene to *ent*-kaurenoic acid, and so leads to *ent*-kaurene accumulation. Neither *ga1* nor *ga2* accumulate *ent*-kaurene after tetcyclacis treatment. The conclusion drawn was that these mutants are impaired in the steps between geranylgeranyl pyrophosphate (which also leads to carotenoids and chlorophyll, both of which are normal in *ga1* and *ga2*) and *ent*-kaurene. The *ga3* mutant does not normally respond to *ent*-kaurene, but does if *ent*-kaurenal or *ent*-kaurenoic acid are applied. This implies that the *ga3* mutation is in a gene encoding an enzyme responsible for oxidation of *ent*-kaurenol to *ent*-kaurenal, possibly a cytochrome P450 monooxygenase.

The *ga4* and *ga5* mutants have also been analysed for their GA content (Talón *et al.*, 1990a). These mutants contain lower amounts of multiple specific GAs, suggesting lesions in single enzymes responsible for the interconversion of GA. The *ga4* mutation has reduced levels of 3-hydroxy-GA, and hence seems to block 3β-hydroxylation, and the *ga5* mutation, which has reduced levels of C-19 GA, blocks the oxidation and elimination of C-20. Application of GA downstream from the deduced lesion results in phenotypic reversion. For example, application of the 3-hydroxy GA_4 to the *ga4* mutant gave a wild-type phenotype, but GA_9 (converted to GA_4 by 3β-hydroxylation) had no affect on *ga4*. The *ga4* mutation can be phenocopied by application of the 3β-hydroxylation inhibitor BX-112 to wild-type plants (resulting in reduced levels of 3-hydroxy-GAs) and the effect of BX-112 on wild-type plants can be overcome by application of 3-hydroxy-GA such as GA_4.

The biochemically deduced role of the proteins encoded by some of the genes mutated in the *ga* series has been confirmed by the analysis of the cloned genes. The *GA1* locus was cloned by a genomic subtraction approach, made possible because the *ga1-3* allele results from a 5 kb deletion that removes the 5' half of the gene (Sun *et al.*, 1992). Transformation of the *ga1-3* mutant with the *GA1* cDNA was able to complement the mutation. Expression of *GA1* in an *E. coli* strain engineered to produce geranylgeranyl pyrophosphate (the presumed substrate for GA_1) resulted in the expected accumulation of the copalyl pyrophosphate product, confirming that *GA1* encodes the enzyme *ent*-kaurene synthetase A (Sun and Kamiya, 1994). The *ga1-3* allele is frequently used as an experimental tool in GA research; however, *ga1-3* still contains a small amount of GA, suggesting that it retains residual activity or perhaps another locus can to some extent supply the missing function of GA_1 (Sun and Kamiya, 1994).

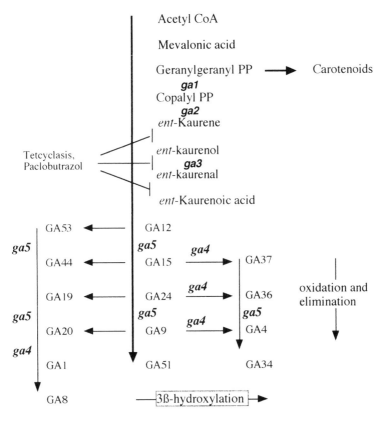

Figure 4.5 An abbreviated scheme of gibberellin biosynthesis. Only intermediates (prior to GA12) that are mentioned in the text are illustrated. The proposed steps blocked by the mutants *ga1, ga2, ga3, ga4* and *ga5* are indicated, as is the inhibition by the monooxygenase inhibitors tetcyclasis and paclobutrazol. Adapted from Finkelstein and Zeevart (1994).

Table 4.3 *Arabidopsis* GA mutants

Mutant	Symbol	Location (Meinke)	Reference	Function?	Cloned?
Gibberellin requiring 1	*ga1*	4–5	Koornneef and Van der Veen, 1980	*ent*-kaurene synthetase A	Sun *et al.*, 1992
Gibberellin requiring 2	*ga2*	1–119	Koornneef and Van der Veen, 1980	*ent*-kaurene synthetase B?	
Gibberellin requiring 3	*ga3*	5–33	Koornneef and Van der Veen, 1980	P450 mono-oxygenase?	Chiang *et al.*, 1995
Gibberellin requiring 4	*ga4*	1–22	Koornneef and Van der Veen, 1980	Hydroxylase	
Gibberellin requiring 5	*ga5*	4–53	Koornneef and Van der Veen, 1980	GA 20 oxidase	Xu *et al.*, 1995
Gibberellin insensitive	*gai*	1–22	Koornneef *et al.*, 1985		
Suppressors of *gai*	*gai-d1* to *gai-d4*	1–22	Peng and Harberd, 1993		
Suppressor of *gai*	*gas1*	? (extragenic)	Carol *et al.*, 1995		
Suppressor of *gai*	*gar2*	? (extragenic)	Wilson and Somerville, 1995		
Spindly	*spy*	3–12	Jacobsen and Olszewski, 1993	Tetratricopeptide repeat protein	Jacobsen *et al.*, 1996

The *GA4* locus was characterised by identifying a T-DNA insertion mutation allele, *ga4-2*. Analysis of the flanking sequences from the mutant and wild-type genomic and cDNA sequences showed that the gene encodes, as expected from the previous biochemical studies, a hydroxylase enzyme. Curiously, the ethyl methanesulfonate-induced allele, *ga4-1*, has a higher level of *GA4* transcript accumulation than the wild-type (Chiang *et al.*, 1995)

The *GA5* gene has been identified by using a heterologous pumpkin GA_{20}-oxidase clone to isolate the *Arabidopsis* homologue. The *Arabidopsis* clone showed tight linkage to the *GA5* locus, and the mutant *ga5* allele contained a point mutation leading to a premature stop codon in the deduced protein coding sequence. The *Arabidopsis* GA 20-oxidase cDNA, when overexpressed in *E.coli*, showed GA oxidase activity (i.e. it was able to convert GA_{53} to GA_{44} and GA_{19} to GA_{20}). Expression of the gene in *Arabidopsis* was reduced by exogenous GA_4 treatment, suggesting end-product repression (Xu *et al.*, 1995). At least three different GA_{20}-oxidase genes are described for *Arabidopsis*, with different patterns of expression in the plant (Phillips *et al.*, 1995). Recent results show that overexpression of GA_{20}-oxidase results in *Arabidopsis* plants with long hypocotyls and accelerated bolting, whereas antisense repression (using a constitutive promoter) shows for some constructs shortened hypocotyls and delayed bolting (P. Hedden, personal communication).

4.4.2 *GA response mutations*

Many mutants of *Arabidopsis* that show altered GA responses have been described. The first of these to be isolated was the semi-dominant *gibberellin insensitive* (*gai*) mutation, which phenotypically resembles the *ga* mutant series being a dark green dwarf with reduced germination and reduced apical dominance. However, *gai* is GA-insensitive, showing only a residual response to applied GAs, which is best seen in a *ga1* background (Koornneef *et al.*, 1985). The *gai* mutant shows greatly increased levels of C-19 GAs, such as (the bioactive) GA_1 and GA_4 (Talón *et al.*, 1990b), indicating that the mutant phenotype is not simply the result of inactivation of GA, and also implying a feedback mechanism regulating GA synthesis.

In a search for suppressors of *gai*, a series of intragenic or deletion mutants (*gai-d1* to *gai-d4*) were isolated that reverted the *gai* phenotype to that of the wild-type. Since the original *gai* is semi-dominant, these results imply that the *gai* phenotype results from to a gain of function of the *GAI* protein, and the *gai-d* mutants are loss-of-function alleles. Since *gai* may be a gain-of-function mutant, it is not absolutely certain that wild-type *GAI* is actually involved in GA perception. In addition, these results imply that *GAI* may be dispensable since a homozygous *GAI* null allele is viable (Carol *et al.*, 1995). In further work (Carol *et al.*, 1995), an unlinked partial suppressor of *gai* was found (*gas-1*), which appears to reduce the GA dependency of plant growth since GA responsiveness is not restored to *gas-1,gai* double mutants. Wilson and Somerville (1995) also looked for *gai* suppressors and found a number of probable null alleles of *GAI*. However, the dominant mutation *gar2* acted as an independent dominant suppressor and thus may also represent a new gene involved in GA signalling. The relationship between *gas-1* and *gar2* is not yet known.

Another GA response mutant that has been described has the opposite phenotype to *gai*. An allelic series of recessive *spindly* (*spy*) mutants was isolated by their ability to germinate and grow in the presence of paclobutrazol, an inhibitor of monooxygenase enzyme activity and thus of GA biosynthesis (*ent*-kaurene to *ent*-kaurenoic acid). The *spy* mutant behaves as if it is constitutively responding to GA; it has long hypocotyls, pale leaves, increased stem length, is early flowering and shows parthenocarpy. The wild-type function of *SPY* must be to act as a repressor of GA signalling. *spy* is partially epistatic over *ga1*, since the double-mutant combination *ga1-2,spy* does not require GA for germination and some of the normal vegetative and floral features of wild-type plants are regained. When GA is applied to *spy* plants, the effect on hypocotyl elongation is additive. This suggests that in *spy* mutants a basal level of GA signal tranduction is activated but this can be augmented by further GA treatment. This may mean that the GA signal transduction pathway is functionally redundant or that *spy* is modulating cross-talk between the GA signal transduction pathway and other GA independent pathways (Jacobsen and Olszewski, 1993).

Screening of T-DNA lines uncovered a tagged spy allele *spy-4*, which is thought to be a null allele and, unlike the previous alleles, semi-dominant for some phenotypic characteristics such as flowering time under long days. *spy-4* is also partially epistatic over *ga1* but completely epistatic over *gai*, supporting the fact that GAI acts upstream of *SPY*. The tagged allele allowed *SPY* to be cloned, and the protein predicted to be encoded by the *SPY* gene is of the tetratricopeptide repeat (TPR) class. The TPR motif is found in a number of prokaryotic and eukaryotic proteins of diverse regulatory function, such as transcriptional repression, cell cycle regulation and protein kinase inhibition. The TPR motif is thought to be involved in protein–protein interactions (Jacobsen *et al.*, 1996).

4.4.3   GA and phytochrome

It is striking that many of the physiological processes controlled by GA are also events that are regulated by the phytochrome family of light receptors, and the evidence suggests that in some cases at least phytochrome may exert its effects through GA. For example, germination of mature *Arabidopsis* seeds is GA and light dependent: GA biosynthetic mutants such as *ga1* and the insensitive *gai* mutant exhibit poor germination (Koornneef and van der Veen, 1980; Koornneef *et al.*, 1985), as do *Arabidopsis* phytochrome mutants such as *hy1* (a chromophore mutant) and *phyB* (*hy3*) (Spruit *et al.*, 1980). *Arabidopsis* germination is red/far-red reversible within the first 14 hours of germination, through PHYB (Cone, 1982; Shinomura *et al.*, 1996), although mature seeds are thought to contain small amounts of active Pfr phytochrome, allowing some lines to germinate in darkness (Cone, 1982; Cone and Kendrick, 1985).

The poor germination of *gai* can be partially overcome by high levels of exogenous GA in the light, but not in the dark (however, this is seed-lot dependent) (Koorneef *et al.*, 1985). Wild-type seeds that do not germinate in the dark can be induced to do so by GA treatment or by after-ripening at 2°C in the light. The seeds are then more sensitive to exogenous GA. The GA biosynthesis inhibitor tetcyclasis, if applied during preincubation in the light, will negate the effect of this on subsequent germination. This suggests that the effect of light, through phytochrome, is to induce GA biosynthesis and increase GA sensitivity of the seed (Derkx and Karssen, 1993). As the double-mutant combination *ga1,aba1* is fully capable of germination (Koorneef *et al.*, 1982), the probable role of GA during *Arabidopsis* germination is to counteract the effect of the germination inhibitor abscisic acid (Koorneef and Karssen, 1994).

Other workers have studied the link between phytochrome and GAs during vegetative growth and flowering. The elongated hypocotyl phenotype of *Arabidopsis phyB* null mutants was found to be GA-dependent, since elongation was inhibited by paclobutrazol. The *ga1-3,phyB* double mutant was three times as responsive to GA3 (in terms of hypocotyl elongation) as *ga1-3* alone. The *phyB* mutant did not have a significantly different GA content to the wild-type, suggesting that in contrast to the situation in seeds, the function of phytochrome in hypocotyls is to desensitise the tissue to GA (Reed *et al.*, 1996). This fits in with the observation that over-expression of *PHYB* in *Arabidopsis* led to plants with shortened hypocotyls, and treatment of these plants with exogenous GA3 led to an increase in hypocotyl length, but the plants were not fully restored to the wild-type phenotype, even with 10 μm GA3 (Wagner *et al.*, 1991). Peng and Harberd (1997) have also thoroughly investigated the phenotype of phytochrome/GA double mutants and found hypocotyl length and stem elongation of *gai,hy1* and *gai,phyB* to be intermediate between that of the parental types, again consistent with the hypothesis that phytochrome desensitises the stem and hypocotyl to GA. The *ga1,hy1* and *ga1,phyB* double mutants were also intermediate to the parental type for hypocotyl elongation but *ga1* was epistatic to *hy1* and *phyB* for stem length. This may mean that stems are less sensitive to GA than hypocotyls or alternatively this is an artifact resulting from residual exogenous GA (required for *ga1* germination) influencing hypocotyl elongation. The cotyledons of *hy1*, *phyB*, *ga1* and *gai* mutants were found to be smaller than those of the wild-type, and those of the double mutants *ga1,hy1*; *ga1,phyB*; *gai,hy1* and *gai,phyB* were smaller still than the parent controls, suggesting here an additive effect on cotyledon expansion for GA and phytochrome. Seedling rosette leaves of *ga1* and *gai* were found to be smaller than wild-type, *hy1* or *phyB*. The double mutants *ga1,hy1*; *ga1,phyB*; *gai,hy1* and *gai,phyB* had smaller early rosette leaves than the *ga1* or *gai* controls, but mature rosette leaves were comparable to these controls. Chlorophyll content of *gai,hy1* or *ga1,phyB* mutants was reduced, like that of the *hy1* parent and *gai,phyB* and *ga1,phyB* showed an intermediate phenotype to the parental types, again suggesting that phytochrome may be desensitising the plant to the effects of GA with respect to chlorophyll accumulation.

Since the positive interactions between phytochrome and GA are seen in organs and processes which are properties of the seed (germination, cotyledon growth and the first leaf primordia which are laid down in the seed) and negative interactions are seen in organs and processes which occur post-germination (chlorophyll accumulation, hypocotyl growth, further vegetative growth and stem elongation), it appears that the relationship between GA and phytochrome reverses post-germination.

Chien and Sussex (1996) have demonstrated that trichome formation in *Arabidopsis* is under the control of photoperiod and GA. In wild-type plants, early leaves bear only adaxial trichomes but the distribution changes with the age of the plant so that the newer leaves carry both adaxial and abaxial trichomes and on cauline leaves the number of adaxial trichomes is reduced. Plants under long days (LD) produce earlier and more abaxial trichomes than plants under short days (SD). Plant under SD treated with exogenous GA produced precocious abaxial trichomes, whereas the *ga1-3* mutant produced no abaxial and only a few adaxial trichomes. Application of GA3 to *ga1-3* restored adaxial and abaxial trichome formation.

Arabidopsis shows early flowering under LD and delayed flowering under SD. Such photoperiodic responses are associated with phytochrome and indeed the phytochrome mutants *hy1, hy2* (both chromophore mutants) and *phyB* are early flowering under SD (Goto *et al.*, 1991). Blue and far-red light accelerate flowering whereas red light inhibits flowering. This suggests that the Pfr form of phytochrome B represses the floral transition under LD (Eskins, 1992; Martinez-Zapater *et al.*, 1994). The *phyA* mutant (*hy8*), a mutation in the photoreceptor responsible for the low fluence response, is in contrast late flowering and thus PHYA may act under certain environmental circumstances as a counter to PHYB (Johnson *et al.*, 1994).

The *ga1-3* mutant will not flower at all under SD, and flowering is somewhat delayed under continuous light. The *gai* mutant is also delayed in flowering under SD (Wilson *et al.*, 1992); the *spy* mutant is in contrast late flowering (Jacobsen and Olszewski, 1993). This suggests that GA is important for regulating flowering, primarily under SD. In studies undertaken on some late flowering *Arabidopsis* mutants it was found that exogenous GA significantly reduced time to flowering; this was also found to be the case under LD (Bagnall, 1992; Chandler and Dean, 1994).

The *CONSTANS* (*CO*) gene codes for a zinc finger-containing transcription factor that, when mutated, causes *Arabidopsis* to be late flowering under LD. The transcript is more abundant under LD than SD; overexpression of *CO* results in early flowering under SD. The double mutants *ga1-3,co* and *gai,co* both flower later than any of the parent mutants under LD, showing that both GA and CO are needed for flowering under LD. However, the LD *co* phenotype can only be partially relieved by exogenous GA. The double mutants *hy-1,co* and *phyB,co* both flower slightly earlier than *co* under LD and SD, but they do not flower as early as wild-type or *PhyB*, hence it appears that CO is required for early flowering of *hy1* and *phB* mutants under both LD and SD. These data suggest that CO plays a primary role in promoting flowering in LD and GA plays a primary role in SD, but the functions of each overlap and are complementary (Putterill *et al.*, 1995).

Although the evidence is not fully conclusive, it would not be surprising if phytochrome mediates its effects on flowering through GA, as seems to be the case for germination, hypocotyl elongation and trichome formation. One might predict that under SD, Pfr-form PHYB accumulation results in reduced endogenous GA content and reduced GA sensitivity, as in hypocotyls, and this in turn causes flowering to be delayed. Under LD, Pfr-PHYB is reduced, GA levels and sensitivity increase and the CO protein is more abundant, and perhaps this further sensitises the plant to GA.

Recent research has also implicated GAs and phytochrome in the control of floral meristem identity. The *LFY* gene determines meristem identity and, if mutated, results in plants with inflorescence meristems in place of floral meristems; heterozygous *lfy* plants are phenotypically wild-type. *AG* is a floral homeotic gene and when disrupted the flowers of homozygotes show a stamen to petal and carpel to sepal transformation. Okamuro *et al.* (1996) have demonstrated that under SD but not LD both heterozygous *lfy* and homozygous *ag* mutants show floral meristem reversion, that is in place of flowers, shoots are produced. When double mutants were made with *hy1* (*ag,hy1* and *lfy,hy1*), floral meristem reversion was abolished. Application of GA abolished SD floral meristem reversion and the double mutant *lfy,spy* also showed no SD floral meristem reversion.

The *ap1* and *ap2* mutants are also defective in flower meristem identity and produce axillary secondary flowers (*ap2* under SD only). Again, a *hy1* background, a *spy* background or exogenous GA abolished SD axillary flowers from *ap2*. Further, the *ap2,spy* double mutant showed under LD-enhanced transformation of first whorl organs from sepals to carpels and the suppression of second whorl organs. The double mutant *ap1,spy* showed much reduced axillary flowering under both SD and LD, and the flowers are less inflorescence-like. Exogenous GA also suppressed axillary branching in *ap1* and promoted petal production. Considering these data, it may be suprising that GA biosynthetic mutants or response mutants do not show changes in floral meristem or primordium identity, but one must suppose that the developmental regulatory network can compensate for changes in GA signalling unless that network itself is perturbed, in which case flower development becomes very sensitive to other disturbances (Okamuro *et al.*, 1997).

4.5 Abscisic acid

4.5.1 ABA biosynthesis

ABA-deficient mutants of *Arabidopsis* have been isolated on the basis of their reduced requirement for GA during seed germination. Recessive mutations at the *ABA1* (designated as *ABA* until recently), *ABA2* and *ABA3* loci were identified by screening either for revertants of the non-germinating GA-deficient *ga1* mutant (Koornneef *et al.*, 1982) or for germinating seeds on the GA biosynthesis inhibitor paclobutrazol (Léon-Kloosterziel *et al.*, 1996).

The biosynthetic defects characterized in the ABA-deficient mutants from *Arabidopsis* and other species provide conclusive evidence that higher plants synthesize ABA by the indirect pathway shown in Figure 4.6 (Taylor, 1991; Schwartz *et al.*, 1997). The *Arabidopsis aba1* mutant is affected in the epoxidation of zeaxanthin and antheraxanthin to all-*trans*-violaxanthin, the direct C_{40} precursors of ABA (Duckham *et al.*, 1991; Rock and Zeevaart, 1991). The *Nicotiana plumbaginifolia ABA2* gene, which is the orthologue of the *Arabidopsis ABA1*, has been cloned and shown to encode a protein with *in vitro* zeaxanthin epoxidase activity (Marin *et al.*, 1996). The *Arabidopsis aba2* mutant is blocked in the conversion of xanthoxin to ABA-aldehyde, whereas *aba3* is impaired in the production of a molybdenum cofactor that is required for the oxidation of ABA-aldehyde to ABA (Schwartz *et al.*, 1997).

The *aba* mutants contain markedly reduced levels of endogenous ABA in seeds and in leaves, and display only a weak accumulation of ABA on water stress (Koornneef *et al.*, 1982; Karssen *et al.*, 1983; Léon-Kloosterziel *et al.*, 1996). These single mutants, however, retain significant residual amounts of ABA that can be further reduced by combining non-allelic *aba* mutations in double mutants (Léon-Kloosterziel *et al.*, 1996). Although *ABA2* appears to be a single-copy gene in *N. plumbaginifolia*, leaves of a presumably null *aba2* mutant allele also retain substantial amounts of ABA (Marin *et al.*, 1996). These intriguing observations suggest the possible existence of some secondary biosynthetic pathways, and cloned biosynthetic genes should provide novel insights into the regulation and the sites of ABA biosynthesis.

The phenotypes of the *aba* mutants of *Arabidopsis* demonstrate the involvement of endogenous ABA in several important physiological processes. Mutant *aba* plants are characterized by reduced growth, wilting and increased transpiration, and the wild-type phenotype can be restored by spraying the mutants with ABA or its analogue LAB 173.711 (Koornneef *et al.*, 1982; Léon-Kloosterziel *et al.*, 1996). These observations indicate that wild-type levels of endogenous ABA are needed to limit water loss through transpiration by decreasing stomatal aperture, and biophysical studies support the hypothesis that ABA promotes stomatal closure by modifying the activity of a number of ion channels in stomatal guard

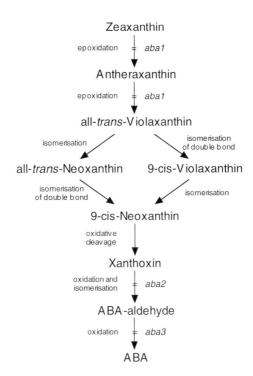

Figure 4.6 Pathway of ABA biosynthesis in higher plants. Higher plants synthesize the sesquiterpenoid ABA ($C_{15}H_{20}O_4$) by an indirect pathway from xanthophyll carotenoids (C_{40}), with the cleavage product xanthoxin as C_{15} intermediate. The metabolic blocks in the *Arabidopsis aba* mutants are indicated.

cells (Armstrong *et al.*, 1995; Pei *et al.*, 1997). The *aba* mutants are also affected in more long-term molecular and physiological responses associated with adaptation to environmental stresses such as drought, salinity and cold, indicating that ABA is one of the endogenous regulators of stress tolerance (Giraudat *et al.*, 1994; Shinozaki and Yamaguchi-Shinozaki, 1996).

During seed development, induction of dormancy is impaired by the *aba* mutations (Koornneef *et al.*, 1982; Léon-Kloosterziel *et al.*, 1996). Reciprocal crosses between wild-type and the *aba1* mutant showed that the onset of dormancy is correlated with an ABA peak of zygotic origin (Karssen *et al.*, 1983). The *aba* mutants also display a marked reduction in their GA requirement for germination (Koornneef *et al.*, 1982; Léon-Kloosterziel *et al.*, 1996), suggesting that the impact of GA is dependent on the degree of dormancy, which is in turn determined by the amount of ABA present during seed development (Karssen and Laçka, 1986). Finally, the characterisation of double-mutant seeds that combine a biosynthetic *aba* with the leaky ABA-insensitive *abi3-1* mutation indicates that endogenous ABA at least contributes to the accumulation of storage proteins and the acquisition of desiccation tolerance (Koornneef et al., 1989; Meurs et al., 1992; Léon-Kloosterziel et al., 1996).

4.5.2 ABA action

Mutations at six loci affecting ABA responses in Arabidopsis have been characterised to a substantial extent. The ABA-insensitive ABI1 to ABI5 loci were all identified by selecting for seeds capable of germinating in the presence of inhibitory concentrations of exogenous ABA (Koornneef et al., 1984; Finkelstein, 1994a). However, as will be discussed below, the various abi mutations lead to additional distinctive phenotypes in seeds or vegetative tissues (Table 4.4, Fig. 4.2G). Mutations at the enhanced-response-to-ABA ERA1 locus confer to germinating seeds a hypersensitivity to applied ABA, and markedly increase seed dormancy (Cutler et al., 1996). The ERA1 gene encodes the β subunit of a farnesyl transferase that may possibly function as a negative regulator of ABA signalling by modifying signal transduction proteins for membrane localisation (Cutler et al., 1996). Further phenotypic and genetic analysis is needed to determine the exact relationships between the ERA1 and ABI loci.

Table 4.4 ABA-related mutants

Gene	Chromosome	Mutant phenotype	Function	References
ABA1	5	ABA deficiency	Zeaxanthin epoxidase	Koornneef et al., 1982; Marin et al., 1996
ABA2	1	ABA deficiency	Conversion of xanthoxin to ABA aldehyde[a]	Léon-Kloosterziel et al., 1996; Schwartz et al., 1997
ABA3	1	ABA deficiency	Addition of sulphur to molybdenum cofactor of ABA oxidase[a]	Léon-Kloosterziel et al. 1996; Schwartz et al. 1997
ABI1	4	ABA insensitivity	Protein phosphatase 2C	Koornneef et al., 1984; Leung et al., 1994; Meyer et al., 1994
ABI2	5	ABA insensitivity	Protein phosphatase 2C	Koornneef et al., 1984; Leung et al., 1997
ABI3	3	ABA insensitivity in seeds	Seed-specific transcription factor	Koornneef et al., 1984; Giraudat et al., 1992
ABI4	2	ABA insensitivity in seeds	?	Finkelstein, 1994a
ABI5	2	ABA insensitivity in seeds	?	Finkelstein, 1994a
ERA1	?	ABA hypersensitivity	β subunit of farnesyl transferase	Cutler et al., 1996

[a] Based on biochemical characterisation of corresponding mutants.

4.5.2.1 Vegetative tissues

The *abi1* and *abi2* mutations affect ABA sensitivity in vegetative tissues. In particular, both mutants display an increased rate of transpiration (Koornneef *et al.*, 1984), which has been traced to the inability of their stomata to close in response to ABA (Armstrong *et al.*, 1995; Roelfsema and Prins, 1995; Pei *et al.*, 1997). The *abi1* and *abi2* mutants are also affected in various ABA-mediated responses to stress (Finkelstein and Somerville, 1990; Giraudat *et al.*, 1994). The *ABI1* and *ABI2* genes have been cloned and shown to encode homologous proteins (Leung *et al.*, 1994, 1997; Meyer *et al.*, 1994). These proteins are composed of a novel N-terminal domain, and a C-terminal domain with serine/threonine phosphatase 2C (PP2C) activity (Bertauche *et al.*, 1996; Leung *et al.*, 1997). The N-terminal domain of ABI1 contains a local sequence motif that matches one of the consensus patterns proposed for EF-hand Ca^{2+}-binding sites (Leung *et al.*, 1994; Meyer *et al.*, 1994). Subsequent analyses, however, revealed that ABI1 is unlikely to contain a functional EF-hand and that, like classical PP2Cs, the *in vitro* phosphatase activity of ABI1 is not regulated by physiological concentrations of Ca^{2+} (Bertauche *et al.*, 1996).

Remarkably, both *abi1-1* and *abi2-1* mutants contain identical Gly to Asp substitutions at equivalent positions in the ABI1 and ABI2 phosphatase domains, respectively (Leung *et al.*, 1994, 1997; Meyer *et al.*, 1994). This single amino acid change maps in close vicinity to invariant residues of the PP2C catalytic site, and markedly reduces the PP2C activity of the ABI1 and ABI2 proteins (Bertauche *et al.*, 1996; Leung *et al.*, 1997). The degree of dominance of the *abi1-1* and *abi2-1* mutations over their respective wild-type alleles has been a matter of controversy (Koornneef *et al.*, 1984; Finkelstein, 1994b). Results indicate that, as would be anticipated for identical mutations in homologous proteins, the *abi1-1* and *abi2-1* mutations are both zygotic and have similarly dominant or semi-dominant effects depending on the particular phenotype considered (Leung *et al.*, 1997).

In the absence of recessive mutant alleles, the possibility that the dominant *abi1-1* and *abi2-1* might be neomorphic gain-of-function mutations cannot be formally excluded. Alternatively, *abi1-1* and *abi2-1* could be dominant negative mutations, and the mutant proteins would have dominant effects possibly because they form poison complexes with the substrates of the wild-type ABI1 and ABI2 phosphatases. The *abi1-1* and *abi2-1* mutations have not only qualitatively but also quantitatively very similar effects on several responses, including the ABA sensitivity of seed germination (Leung *et al.*, 1997). This suggests that the homologous ABI1 and ABI2 proteins have overlapping functions in ABA signalling and might, for instance, recognise common substrates involved in the ABA responses delineated by the shared mutant phenotypes. This scenario is reminiscent of the proposed functional redundancy between ETR1 and its homolog ERS in the ethylene response pathway (Hua *et al.*, 1995) and would explain why recessive *abi1* and *abi2* mutant alleles have not been identified on the basis of seed germination on ABA. The roles of the ABI1 and ABI2 proteins are, however, unlikely to be completely redundant for all ABA responses. Some of the ABA-dependent responses triggered by water stress are indeed significantly more affected in *abi1-1* than in the *abi2-1* mutant, or conversely more affected in *abi2-1* than in *abi1-1* (see references in Leung *et al.*, 1997). Furthermore, whereas the inhibitory effect of *abi1-1* on stomatal regulation by ABA could be rescued by applying kinase antagonists (Armstrong *et al.*, 1995; Pei *et al.*, 1997), intriguingly this was not the case for *abi2-1* (Pei *et al.*, 1997).

Identifying the regulatory proteins as well as the physiological substrates of ABI1 and ABI2 should provide further insights into the biological functions of these two proteins. In addition, combining genetics with pharmacological and microinjection approaches should help to establish the exact relationships between the ABI proteins and other components of the ABA signalling cascades at the single-cell level (Armstrong *et al.*, 1995; Sheen, 1996; Pei *et al.*, 1997).

4.5.2.2 Seeds

Unlike *abi1* and *abi2*, the recessive *abi3*, *abi4* and *abi5* mutants do not display phenotypic alterations in vegetative tissues (Koornneef *et al.*, 1984; Finkelstein, 1994a; Giraudat *et al.*, 1994). In the case of *ABI3*, this is directly correlated with the fact that this gene is expressed exclusively in seeds and in young germinating seedlings (Giraudat *et al.*, 1992; Parcy *et al.*, 1994).

The *Arabidopsis ABI3* is the orthologue of the seed-specific *VIVIPAROUS-1* (*VP1*) gene from maize (Giraudat *et al.*, 1992). By analogy to VP1 (McCarty *et al.*, 1991; Suzuki *et al.*, 1997), the ABI3 protein is a putative transcription factor. Like the *aba* biosynthesis mutants, *abi3* mutants lack seed dormancy (Koornneef *et al.*, 1984; Ooms *et al.*, 1993). *Abi3* mutant seeds contain decreased levels of various storage protein and late embryogenesis abundant mRNAs (Parcy *et al.*, 1994; Nambara *et al.*, 1995), and these normally seed-specific mRNAs can be induced by ABA in transgenic leaves that ectopically express the wild-type ABI3 protein (Parcy *et al.*, 1994; Parcy and Giraudat, 1997). Intriguingly, the marked reductions in storage protein accumulation, chlorophyll breakdown and desiccation tolerance exhibited by seeds of severe *abi3* mutants are not observed in *aba* single mutants (Ooms *et al.*, 1993; Parcy *et al.*, 1994; Nambara *et al.*, 1995). The phenotypes of *aba,abi3-1* double mutants nevertheless support the fact that ABA contributes to these developmental processes (Koornneef *et al.*, 1989; Meurs *et al.*, 1992; Léon-Kloosterziel *et al.*, 1996). A first explanation could be that these ABI3-dependent responses are strictly controlled by ABA and that the residual ABA content of *aba* mutants is sufficient to ensure near wild-type responses (provided that signal transduction is not simultaneously reduced by the *abi3-1* mutation). An alternative possibility is that the above-mentioned processes are jointly regulated by ABA and additional developmental factors, and that ABI3 mediates both types of signals. Severe *abi3* mutations would then inhibit both pathways and hence have stronger effects than *aba* mutations inhibiting only the

ABA pathway. The latter scenario is supported by evidence that VP1 can achieve, via distinct *cis*-acting elements within a given promoter, gene activation both in synergy with ABA and independently from ABA (Hattori *et al.*, 1992; Vasil *et al.*, 1995). The ABI3 and VP1 proteins may thus conceivably integrate ABA and additional developmental signals, for instance by interacting with multiple other transcription factors.

Putative partners of ABI3 include FUSCA 3 (FUS3) and LEAFY COTY-LEDON 1 (LEC1), two other known key regulators of seed maturation in *Arabidopsis* (Meinke, 1995). Double-mutant analysis indicates that *ABI3* acts synergistically with *FUS3* and with *LEC1* to control multiple elementary processes during seed maturation, including sensitivity to ABA (Parcy *et al.*, 1997). In contrast, several uncertainties remain as to the relationships between *ABI3* and other *ABI* loci in regulating seed development. The *ABI1* and *ABI2* genes are expressed in seeds (Leung *et al.*, 1997), and the *abi1* and *abi2* mutations markedly reduce seed dormancy (Koornneef *et al.*, 1984). Nevertheless, like *aba* mutants, the *abi1* and *abi2* mutants do not exhibit the additional characteristic defects observed in seeds of severe *abi3* mutants. As discussed above for the *aba* mutants, this phenotypic difference may be explained by hypothesising that ABI1 and ABI2 act in the same ABA cascade as ABI3, but that ABI3 mediates in addition some other regulatory signals (Parcy and Giraudat, 1997). Alternatively, ABI1 and ABI2 may belong to an ABA signalling cascade different from that of ABI3, and both cascades would have overlapping effects only on seed germination and dormancy (Finkelstein and Somerville, 1990). The *abi4* and *abi5* mutants display certain seed phenotypes in common with *abi3* mutants, suggesting that all three loci may act in a common regulatory pathway in developing seed (Finkelstein, 1994a). Intriguingly, however, unlike *abi3*, available *abi4* and *abi5* mutations do not seem to reduce seed dormancy.

Seed maturation thus appears to be coordinated by a complex regulatory network in which ABA most likely acts in concert with additional putative developmental signals. Furthermore, the genetic interactions between loci may vary depending on the particular physiological process considered. The identification of additional mutants should help to resolve the remaining gaps in our understanding of the ABA signalling cascades in seed.

4.6 Concluding remarks

Within the last five years there have been revolutionary changes in our understanding of plant hormone biology. This has been largely due to the 'adoption' of mutational approaches in *Arabidopsis* and other genetically amenable plants such as maize and tomato. Our knowledge of hormone production has advanced considerably following the isolation of several ABA, GA and ethylene biosynthetic mutations and their corresponding genes. Genetic approaches are proving

particularly effective for isolating and characterising membrane-bound biosynthetic enzymes, for example. Frustratingly, auxin biosynthesis remains largely
recalcitrant to study. Several pathways have been proposed which may operate at
different stages of development. A more comprehensive knowledge of the relative importance of these individual pathways could facilitate the design of
screens for auxin biosynthetic mutants. One of the most important challenges
facing hormone biosynthesis involves dissecting the molecular mechanisms that
regulate their production. Probes derived from biosynthetic genes are proving
useful tools for addressing the spatial and temporal control of hormone production. However, the identification of hormone overproducing mutants such as *eto1*
are likely to provide a more direct approach to obtain a handle on these
regulatory proteins.

Several signalling components have recently been described for auxin,
ethylene, ABA and GA. The isolation of the receptor kinase *ETR1* gene product
has provided a new understanding of ethylene sensing. However, the mechanism
of perception for the three other hormones remains elusive. The continued isolation and characterisation of hormone-related gene products is likely to provide
further insight. Alternative strategies, including activation tagging and expression cloning approaches, should also be pursued in parallel. We must also remain
aware that plant hormones could be perceived at a variety of subcellular locations. For example, symplastic connections between plant cells render exclusively plasma-membrane-based hormone sensing mechanisms unlikely. Furthermore, the isolation of the permease-like *AUX1* gene product serves to highlight
that uptake into root cells may be necessary prior to intracellular auxin perception. Understanding how hormone transduction chains integrate their signalling
information with other cellular components that regulate elongation and division
presents a long-term challenge. However, impressive advances are being made
by researchers who adopt parallel approaches in yeast and *Arabidopsis*, reaping
the benefits of both experimental systems. For example, Estelle and co-workers
have made significant progress towards understanding how the orthologous
AXR1 and *TIR1* gene products control the yeast cell cycle, hence providing fresh
insight into auxin-regulated cellular processes in *Arabidopsis*. In summary, we
are at last getting to grips with the molecular components that mediate hormone
biosynthesis, transport, perception and transduction. It is clear that *Arabidopsis*
has, does and will continue to play an important role in studying plant hormone
biology into the next century.

Acknowledgements

The authors would like to thank colleagues for providing unpublished information and to acknowledge support from the BBSRC (UK), EU, INRA (France)
and NSF (USA) funding bodies. This work was supported in part by the Centre

National de la Recherche Scientifique (France), the European Community BIO-TECH program (BI04-CT96-0062), the International Human Frontier Science Program (RG-303/95) and a NASA/NSF grant # IBN-9416017 and a USDA grant # 95-37304-2294 to J.J.K.

References

Abel, S. and Theologis, A. (1996) Early genes and auxin action. *Plant Physiol.*, **111** 9-17.

Abel, S., Nguyen, M.D. and Theologis, A. (1995a) The PS-IAA4/5-like family of early auxin-inducible mRNAs in *Arabidopsis thaliana*. *J. Mol. Biol.*, **251** 533-49.

Abel, S., Nguyen, M., Chow, W. and Theologis, A. (1995b) *ACS4*, a primary indole acid-responsive gene encoding 1-aminocyclopropane-1-carboxylate synthase in *Arabidopsis thaliana*. *J. Biol. Chem.*, **270** 19093-99.

Abeles, F.B., Morgan, P.W. and Saltveit, M.E., Jr. (1992) *Ethylene in Plant Biology*, Academic Press, San Diego, CA.

Armstrong, F., Leung, J., Grabov, A., Brearley, J., Giraudat, J. and Blatt, M.R. (1995) Sensitivity to abscisic acid of guard cell K⁺ channels is suppressed by *abi1-1*, a mutant *Arabidopsis* gene encoding a putative protein phosphatase. *Proc. Natl Acad. Sci. USA*, **92** 9520-24.

Avruch, J., Zhang, X.-F. and Kyriakis, J. (1994) Raf meets Ras: completing the framework of a signal transduction pathway. *Trends Biochem. Sci.*, **19** 279-83.

Bagnall, D.J. (1992) Control of flowering in *Arabidopsis thaliana* by light, vernalisation and gibberellins. *Aust. J. Plant Physiol.*, **19** 401-409.

Bai, C., Partha, S., Hofmann, K., Ma, L., Goebl, M., Harper, J.W. and Elledge, S.J. (1996) SKP1 connects the cell cycle regulators to the ubiquitin proteolysis machinery through a novel motif, the F-box. *Cell*, **86** 263-74.

Bartel, B. (1997) Auxin biosynthesis. *Ann. Rev. Plant Physiol. Plant Mol. Biol.*, **48** 49-64.

Bartel, B. and Fink, G.R. (1994) Differential regulation of auxin producing nitrilase gene family in *Arabidopsis thaliana*. *Proc. Natl Acad. Sci. USA*, **91** 6649-53.

Bartel, B. and Fink, G.R. (1995) ILR1, an amidohydrolase that releases active indole-3-acetic acid from conjugates. *Science*, **268** 1745-48.

Bartling, D., Seedorf, M., Schmidt, R.C. and Weiler, E.W. (1994) Molecular characterisation of two cloned nitrilases from *Arabidopsis thaliana*: key enzymes in the biosynthesis of the plant hormone indole-3-acetic acid. *Proc. Natl Acad. Sci. USA*, **91** 6021-25.

Bennett, M.J., Marchant, A., Green, H.G., May, S.T., Ward, S.P., Millner, P.A., Walker, A.R., Schulz, B. and Feldmann, K.A. (1996) *Arabidopsis* AUX1 gene: a permease-like regulator of root gravitropism. *Science*, **273** 949-50.

Bertauche, N., Leung, J. and Giraudat, J. (1996) Protein phosphatase activity of abscisic acid insensitive 1 (ABI1) protein from *Arabidopsis thaliana*. *Eur. J. Biochem.*, **241** 193-200.

Binns, A.N. (1994) Cytokinin accumulation and action: biochemical, genetic and molecular approaches. *Ann. Rev. Plant Physiol. Plant Mol. Biol.*, **45** 173-96.

Bleecker, A. and Schaller, G. (1996) The mechanism of ethylene perception. *Plant Physiol.*, **111** 653-60.

Bleecker, A., Estelle, M., Somerville, C. and Kende, H. (1988) Insensitivity to ethylene conferred by a dominant mutation in *Arabidopsis thaliana*. *Science*, **241** 1086-89.

Boerjan, W., Cervera, M.T., Delarue, M., Breekman, T., Dewitte, W., Bellini, C., Caboche, M., Van Onckelen, H., Van Montagu, M. and Inze, D. (1995) Superroot, a recessive mutation in *Arabidopsis*, confers auxin overproduction. *Plant Cell*, **7** 1405-19.

Broglie, K. and Brogue, R.B. (1991). Ethylene and gene expression. In *The Plant Hormone Ethylene*, (eds A. Mattoo and J. Suttle), CRC Press, Boca Raton, FL, pp. 101-14.

Broglie, K., Biddle, P., Cressman, R. and Broglie, R. (1989) Functional analysis of DNA sequences responsible for ethylene regulation of a bean chitinase gene in transgenic tobacco. *Plant Cell,* **1** 599-607.

Calenza, J., Grisafi, P.L. and Fink, G.R. (1995) A pathway for lateral root formation in *Arabidopsis thaliana. Genes Dev.,* **9** 2131-42.

Carland, F.M. and McHale, N.A. (1996) LOP1: a gene involved in auxin transport and patterning in *Arabidopsis. Development,* **122** 1811-19.

Cernac, A., Lincoln, C. and Estelle, M. (1997) The SAR1 gene of *Arabidopsis* acts downstream of the AXR1 gene in auxin response. *Development,* **124** 1583-91.

Chandler, J. and Dean, C. (1994) Factors influencing the vernalization response and flowering time of late flowering mutants of *Arabidopsis thaliana* (L.) Heynh. *J. Exper. Bot.,* **278** 1279-88.

Chang, C. (1996) The ethylene signal transduction pathway in *Arabidopsis*: an emerging paradigm? *Trends Biochem. Sci.,* **21** 129-33.

Chang, C., Kwok, S., Bleecker, A. and Meyerowitz, E. (1993) *Arabidopsis* ethylene-response gene *ETR1*: similarity of product to two-component regulators. *Science,* **262** 539-44.

Chao, Q., Rothenberg, M., Solano, R., Roman, G., Terzaghi, W. and Ecker, J.R. (1997) Activation of the ethylene gas response pathway in *Arabidopsis* by the nuclear protein ETHYLENE-INSENSITIVE3 and related proteins. *Cell,* **89** 1133-44.

Chen, Q. and Bleecker, A. (1995) Analysis of ethylene signal transduction kinetics associated with seedling-growth responses and chitinase induction in wild-type and mutant *Arabidopsis. Plant Physiol.,* **108** 596-607.

Chiang, H.-H., Hwang, I. and Goodman, H.M. (1995) Isolation of the *Arabidopsis* GA4 locus. *Plant Cell,* **7** 195-201.

Chien, J.C. and Sussex, I.M. (1996) Differential regulation of trichome formation on the adaxial and abaxial leaf surfaces by gibberellins and photoperiod in *Arabidopsis thaliana* (L.) Heynh. *Plant Physiol.,* **111** 1321-28.

Chory, J. and Li, J. (1997) Gibberellins, brassinosteroids and light-regulated development. *Plant Cell Environ.,* **20** 801-806.

Clouse, S. (1996) Molecular genetic studies confirm the role of brassinosteroids in plant growth and development. *Plant J.,* **10** 1-8.

Cone, W.J. (1982) The escape from photocontrol of seed germination of *Arabidopsis thaliana. Arabidopsis Inf. Ser.,* **19** 35-38

Cone, W.J. and Kendrick, R.E. (1985) Fluence-response curves and action spectra for promotion and inhibition of seed germination in wildtype and long-hypocotyl mutants of *Arabidopsis thaliana* L. *Planta,* **163** 43-54.

Cutler, S., Ghassemian, M., Bonetta, D., Cooney, S. and McCourt, P. (1996) A protein farnesyl transferase involved in abscisic acid signal transduction in *Arabidopsis. Science,* **273** 1239-41.

Daum, G., Eisenmann-Tappe, I., Fries, H.-W., Troppmair, J. and Rapp, U. (1994) The ins and outs of Raf kinases. *Trends Biochem. Sci.,* **19** 474-79.

Davis, P.J. (ed.) (1995) *Plant Hormones,* Kluwer, Dordreck.

Delbarre, A., Muller, P., Imhoff, V. and Guern, J. (1996) Comparisons of mechanisms controlling uptake and accumulation of 2,4-dichlorophenoxy acetic acid, naphthalene-1-acetic acid, and indole-3-acetic acid in suspension-culutred tobacco cells. *Planta,* **198** 532-41.

Derkx, M.P.M. and Karssen, C.M. (1993) Effects of light and temperature on seed dormancy and gibberellin-stimulated germination in *Arabidopsis thaliana*: studies with gibberellin-deficient and -insensitive mutants. *Physiol. Plant.,* **89** 360-68.

Dolan, L., Janmaat, K., Willemsen, V., Linstead, P., Poethig, S., Roberts, K. and Scheres, B. (1993) Cellular organization of the *Arabidopsis thaliana* root. *Development,* **119** 71-84.

Dolan, L., Duckett, C., Grierson, C., Linstead, P., Schneider, K., Lawson, E., Dean, C. and Roberts, K. (1994) Clonal relationships and cell patterning in the root epidermis of *Arabidopsis. Development,* **120** 2465-74.

Duckham, S.C., Linforth, R.S.T. and Taylor, I.B. (1991) Abscisic acid deficient mutants at the *aba* gene locus of *Arabidopsis thaliana* are impaired in the epoxidation of zeaxanthin. *Plant Cell Environ.*, **14** 601-606.

Ecker, J. and Davis, R. (1987) Plant defense genes are regulated by ethylene. *Proc. Natl. Acad. Sci. USA*, **84** 5202-5206.

Ecker, J., Roman, G., Rothenberg, M., Lehman, A., Lubarsky, B., Cho, Q., Raz, V., Alonso, J.M., Nourizadeh, S.D. and Solano, R. (1996) Genes and gene interactions controlling ethylene signal transduction. *7th International Conference on Arabidopsis Research*, Norwich.

Eskins, K. (1992) Light-quality effects on *Arabidopsis* development. Red, blue and far-red regulation of flowering and morphology. *Physiol. Plant.*, **86** 439-44.

Estelle, M. and Somerville, C.R. (1987) Auxin-resistant mutants of *Arabidopsis* with an altered morphology. *Mol. Gen. Genet.*, **206** 200-206.

Felix, G. and Meins, F.J. (1987) Ethylene regulation of β-1,3-glucanase in tobacco. *Planta*, **172** 386-92.

Finkelstein, R.R. (1994a) Mutations at two new *Arabidopsis* ABA response loci are similar to the *abi3* mutations. *Plant J.*, **5** 765-71.

Finkelstein, R.R. (1994b) Maternal effects govern variable dominance of two abscisic acid response mutations in *Arabidopsis thaliana*. *Plant Physiol.*, **105** 1203-1208.

Finkelstein, R.R. and Somerville, C.R. (1990) Three classes of abscisic acid (ABA)-insensitive mutations of *Arabidopsis* define genes that control overlapping subsets of ABA responses. *Plant Physiol.*, **94** 1172-79.

Finkelstein, F.F. and Zeevart, J.A.D. (1994) Gibberellin and abscisic acid biosynthesis and response, in *Arabidopsis* (eds E.M. Meyerowitz and C.R. Somerville), Cold Spring Harbor Laboratory Press, Cold Spring Harbour, New York, pp. 523-54.

Galweiler, L., Wismann, E., Yephremov, A. and Palme, K. (1996) Using a transposon tagged *pin1* mutant of *Arabidopsis thaliana* to elucidate the role and molecular mechanisms of auxin signalling in plants, *7th International Conference on Arabidopsis Research*, Norwich, Abstract P307.

Garbers, C., Delong, A., Deruere, J., Bernasconi, P. and Soll, D. (1996) A mutation in protein phosphatase-2a regulatory subunit-A affects auxin transport, in *Arabidopsis*. *EMBO J.*, **15** 2115-24.

Giraudat, J., Hauge, B.M., Valon, C., Smalle, J., Parcy, F. and Goodman, H.M. (1992) Isolation of the *Arabidopsis ABI3* gene by positional cloning. *Plant Cell*, **4** 1251-61.

Giraudat, J., Parcy, F., Bertauche, N., Gosti, F., Leung, J., Morris, P.-C., Bouvier-Durand, M. and Vartanian, N. (1994) Current advances in abscisic acid action and signaling. *Plant Mol. Biol.*, **26** 1557-77.

Gomez-Lim, M.A., Valdes-Lopez, V., Cruz-Hernandez, A., Saucedo-Arias and L.J. (1993) Isolation and characterization of a gene involved in ethylene biosynthesis from *Arabidopsis thaliana*. *Gene*, **134** 217-21.

Gopalraj, M., Tseng, T.-S. and Olszewski, N. (1996) The ROOTY gene of *Arabidopsis* encodes a protein with high similarity to aminotransferases. *Plant Physiol.*, **111S** 114.

Goto, N., Kumagai, T. and Koornneef, M. (1991) Flowering responses to light-breaks in photomorphogenic mutants of *Arabidopsis thaliana*, a long day plant. *Physiol. Plant.*, **83** 209-15.

Gray, R. and Grierson, D. (1993) Molecular genetics of tomato fruit ripening. *Trends Genet.*, **9** 438-43.

Guzman, P. and Ecker, J.R. (1990) Exploiting the triple response of *Arabidopsis* to identify ethylene-related mutants. *Plant Cell*, **2** 513-23.

Hattori, T., Vasil, V., Rosenkrans, L., Hannah, L.C., McCarty, D.R. and Vasil, I.K. (1992) The *Viviparous-1* gene and abscisic acid activate the *C1* regulatory gene for anthocyanin biosynthesis during seed maturation in maize. *Genes Dev.*, **6** 609-18.

Hobbie, L. and Estelle, M. (1995) The axr4 auxin-resistant mutants of *Arabidopsis thaliana* define a gene important for root gravitropism and lateral root initiation. *Plant J.*, **7** 211-20.

Hua, J., Chang, C., Sun, Q. and Meyerowitz, E.M. (1995) Ethylene insensitivity conferred by *Arabidopsis ERS* gene. *Science*, **269** 1712-14.

Jacobsen, S.E. and Olszewski, N.E. (1993) Mutations at the SPINDLY locus of *Arabidopsis* alter gibberellin signal transduction. *Plant Cell*, **5** 887-96.

Jacobsen, S.E., Binkowski, K.A. and Olszewski, N.E. (1996) SPINDLY, a tetratricopeptide repeat protein involved in gibberellin signal transduction in *Arabidopsis*. *Proc. Natl Acad. Sci. USA*, **93** 9292-96.

Johnson, E., Bradley, M., Harberd, N.P. and Whitelam, G.C. (1994) Photoresponses of light-grown phyA mutants of *Arabidopsis*: phytochrome A is required for the perception of daylight extensions. *Plant Physiol.*, **105** 141-49.

Karssen, C.M. and Laçka, E. (1986) A revision of the hormone balance theory of seed dormancy: studies on gibberellin and/or abscisic acid deficient mutants in *Arabidopsis thaliana*, in *Plant Growth Substances* (ed. M. Bopp), Springer-Verlag, Heidelberg, pp. 315-23.

Karssen, C.M., Brinkhorst-van der Swan, D.L.C., Breekland, A.E. and Koornneef, M. (1983) Induction of dormancy during seed development by endogenous abscisic acid: studies on abscisic acid deficient genotypes of *Arabidopsis thaliana* (L.) Heynh. *Planta*, **157** 158-65.

Kieber, J.J., Rothenburg, M., Roman, G., Feldmann, K.A. and Ecker, J.R. (1993) CTR1, a negative regulator of the ethylene response pathway in *Arabidopsis*, encodes a member of the Raf family of protein kinases. *Cell*, **72** 427-41.

King, J.J., Stimart, D.P., Fisher, R.H. and Bleecker, A.B. (1995) A mutation altering auxin homeostasis and plant morphology in *Arabidopsis*. *Plant Cell*, **7** 2023-37.

Klee, H. and Estelle, M. (1991) Molecular genetic approaches to plant hormone biology. *Ann. Rev. Plant Physiology Plant Mol. Biol.*, **42** 529-51.

Koornneef, M. and van der Veen, J.H. (1980) Induction and analysis of gibberellin sensitive mutants in *Arabidopsis thaliana* (L.) Heynh. *Theor. Appl. Genet.*, **61** 385-93.

Koornneef, M. and Karssen, C.M. (1994) Seed dormancy and germination, in *Arabidopsis* (eds. E.M. Meyerowitz and C.R. Somerville), Cold Spring Harbor Laboratory Press, Cold Spring Harbour, New York, pp. 313-34.

Koornneef, M., Jorna, M.L., Brinkhorst-van der Swan, D.L.C. and Karssen, C.M. (1982) The isolation of abscisic acid (ABA) deficient mutants by selection of induced revertants in nongerminating gibberellin sensitive lines of *Arabidopsis thaliana* (L.) Heynh. *Theor. Appl. Genet.*, **61** 385-93.

Koornneef, M., Reuling, G. and Karssen, C.M. (1984) The isolation and characterization of abscisic acid-insensitive mutants of *Arabidopsis thaliana*. *Physiol. Plant.*, **61** 377-83.

Koornneef, M., Elgersma, A., Hanhart, C.J., van Loenen-Martinet, E.P., van Rijn, L. and Zeevaart, J.A.D. (1985) A gibberellin insensitive mutant of *Arabidopsis thaliana*. *Physiol. Plant.*, **65** 33-39.

Koornneef, M., Hanhart, C.J., Hilhorst, H.W.M. and Karssen, C.M. (1989) *In vivo* inhibition of seed development and reserve protein accumulation in recombinants of abscisic acid biosynthesis and responsiveness mutants in *Arabidopsis thaliana*. *Plant Physiol.*, **90** 463-69.

Laskowski, M.J., Williams, M.E., Nusbaum, H.C. and Sussex, I.M. (1995) Formation of lateral root meristems is a two-stage process. *Development*, **121** 3303-10.

Lehman, A., Black, R. and Ecker, J. (1996) *Hookless1*, an ethylene response gene, is required for differential cell elongation in the *Arabidopsis* hook. *Cell*, **85** 183-94.

Léon-Kloosterziel, K.M., Alvarez Gil, M., Ruijs, G.J., Jacobsen, S.E., Olszewski, N.E., Schwartz, S.H., Zeevaart, J.A.D. and Koornneef, M. (1996) Isolation and characterization of abscisic acid-deficient *Arabidopsis* mutants at two new loci. *Plant J.*, **10** 655-61.

Leung, J., Bouvier-Durand, M., Morris, P.-C., Guerrier, D., Chefdor, F. and Giraudat, J. (1994) *Arabidopsis* ABA-response gene *ABI1*: features of a calcium-modulated protein phosphatase. *Science*, **264** 1448-52.

Leung, J., Merlot, S. and Giraudat, J. (1997) The *Arabidopsis ABSCISIC ACID-INSENSITIVE 2* (*ABI2*) and *ABI1* genes encode redundant protein phosphatases 2C involved in abscisic acid signal transduction. *Plant Cell*, **9** 759-71.

Leyser, H.M.O., Lincoln, C.A., Timpte, C., Lammer, D., Turner, J. and Estelle, M. (1993) *Arabidopsis* auxin-resistance gene *AXR1* encodes a protein related to the ubiquitin-activating enzyme E1. *Nature*, **364** 161-64.

Leyser, H.M.O., Pickett, F.B., Dharmasiri, S. and Estelle, M. (1996) Mutations within the AXR3 gene of *Arabidopsis* result in altered auxin response including ectopic expression from the SAUR-AC1 promoter. *Plant J.*, **10** 403-13.

Liang, X., Abel, S., Keller, J., Shen, N. and Theologis, A. (1992). The 1-aminocyclopropane-1-carboxylate synthase gene family of *Arabidopsis thaliana*. *Proc. Natl Acad. Sci. USA*, **89** 11046-50.

Liang, X., Oono, Y., Shen, N.F., Köhler, C., Li, K., Scolnik, P.A. and Theologis, A. (1995). Characterization of two members (ACS1 and ACS3) of the 1-aminocyclopropane-1-carboxylate synthase gene family of *Arabidopsis thaliana*. *Gene*, **167** 17-24.

Liang, X., Shen, N.F. and Theologis, A. (1996). Li+-regulated 1-aminocyclopropane-1-carboxylate synthase gene expression regulation in *Arabidopsis thaliana*. *Plant J.*, **10** 1027-36.

Lincoln, C., Britton, J.H. and Estelle, M. (1990) Growth and development of the axr1 mutants of *Arabidopsis*. *Plant Cell*, **2** 1071-80.

Liu, C.-M., Xu, Z.-H and Chua, N.-H. (1993) Auxin polar transport is essential for the establishment of bilateral symmetry during early plant embryogenesis. *Plant Cell*, **5** 621-30.

Lomax, T.L., Muday, G.K. and Rubery, P.H. (1995) Auxin transport in *Plant Hormones* (ed. P.J. Davies), Kluwer, Dordrecht, pp. 509-530.

Maher, E.P. and Martindale, S.J.B. (1980) Mutants of *Arabidopsis* with altered responses to auxins and gravity. *Biochem. Genet.*, **18** 1041-53.

Marin, E., Nussaume, L., Quesada, A., Gonneau, M., Sotta, B., Hugueney, P., Frey, A. and Marion-Poll, A. (1996) Molecular identification of zeaxanthin epoxidase of *Nicotiana plumbaginifolia*, a gene involved in abscisic acid biosynthesis and corresponding to *ABA* locus of *Arabidopsis thaliana*. *EMBO J.*, **15** 2331-42.

Marshall, C. (1994) MAP kinase kinase kinase, MAP kinase kinase, and MAP kinase. *Curr. Opin. Genet. Dev.*, **4** 82-89.

Martinez-Zapater, J.M., Coupland, G., Dean, C. and Koornneef, M. (1994) The transition to flowering in *Arabidopsis*, in *Arabidopsis* (eds. E.M. Meyerowitz and C.R. Somerville), Cold Spring Harbor Laboratory Press, Cold Spring Harbour, New York, pp. 403-34.

McCarty, D.R., Hattori, T., Carson, C.B., Vasil, V., Lazar, M. and Vasil, I.K. (1991) The *viviparous-1* developmental gene of maize encodes a novel transcriptional activator. *Cell*, **66** 895-905.

McKinney, E.C., Ali, N., Traut, A., Feldmann, K.A., Belostotsky, D.A., McDowell, J.M. and Meagher, R.B. (1995) Sequence-based identification of T-DNA insertion mutations in *Arabidopsis* actin mutants *act2-1* and *act4-1*. *Plant J.*, **8** 613-22.

Meinke, D.W. (1995) Molecular genetics of plant embryogenesis. *Ann. Rev. Plant Physiol. Plant Mol. Biol.*, **46** 369-94.

Meurs, C., Basra, A.S., Karssen, C.M. and van Loon, L.C. (1992) Role of abscisic acid in the induction of desiccation tolerance in developing seeds of *Arabidopsis thaliana*. *Plant Physiol.*, **98** 1484-93.

Meyer, K., Leube, M.P. and Grill, E. (1994) A protein phosphatase 2C involved in ABA signal transduction in *Arabidopsis thaliana*. *Science*, **264** 1452-55.

Moodie, S. and Wolfman, A. (1994) The 3Rs of life: Ras, Raf and growth regulation. *Trends Genet.*, **10** 44-48.

Morris, D.A., Rubery, P.H., Jarman, J. and Sabater, M. (1991) Effects of inhibitors of protein synthesis on transmembrane auxin transport in Cucurbita pepo L hypocotyl segments. *J. Exper. Bot.*, **42** 773-83.

Morrison, D. (1990) The Raf-1 kinase as a transducer of mitogenic signals. *Cancer Cells*, **2** 377-382.

Nambara, E., Keith, K., McCourt, P. and Naito, S. (1995) A regulatory role for the *ABI3* gene in the establishment of embryo maturation in *Arabidopsis thaliana*. *Development*, **121** 629-36.

Normanly, J., Cohen, J.D. and Fink, G.R. (1993) *Arabidopsis thaliana* auxotrophs reveal a tryptophan-independent biosynthetic pathway for indole-3-acetic acid. *Proc. Natl Acad. Sci. USA,* **90** 10355-59.

Ohme-Takagi, M. and Shinshi, H. (1995) Ethylene-inducible DNA binding proteins that interact with an ethylene-responsive element. *Plant Cell,* **7** 173-82.

Okada, K., Ueda, J., Komaki, M.K., Bell, C.J. and Shimura, Y. (1991) Requirement of the polar auxin transport system in early stages of *Arabidopsis* floral bud formation. *Plant Cell,* **3** 677-84.

Okamuro, J.K., den Boer, B.G.W., Lotys-Prass, C., Szeto, W. and Jofuku, K.D. (1996) Flowers into shoots: photo and hormonal control of a meristem identity switch in *Arabidopsis. Proc. Natl Acad. Sci. USA,* **93** 13831-36.

Okamuro, J.K., Szeto, W., Lotys-Prass, C. and Jofuku, K.D. (1997) Photo and hormonal control of meristem identity in the *Arabidopsis* flower mutants apetala2 and apetala1. *Plant Cell,* **9** 37-47.

Ooms, J.J.J., Léon-Kloosterziel, K.M., Bartels, D., Koornneef, M. and Karssen, C.M. (1993) Acquisition of desiccation tolerance and longevity in seeds of *Arabidopsis thaliana.* A comparative study using abscisic acid-insensitive *abi3* mutants. *Plant Physiol.,* **102** 1185-91.

Parcy, F. and Giraudat, J. (1997) Interactions between the *ABI1* and the ectopically expressed *ABI3* genes in controlling abscisic acid responses in *Arabidopsis* vegetative tissues. *Plant J.,* **11** 693-702.

Parcy, F., Valon, C., Raynal, M., Gaubier-Comella, P., Delseny, M. and Giraudat, J. (1994) Regulation of gene expression programs during *Arabidopsis* seed development: roles of the *ABI3* locus and of endogenous abscisic acid. *Plant Cell,* **6** 1567-82.

Parcy, F., Valon, C., Kohara, A., Miséra, S. and Giraudat, J. (1997) The *ABSCISIC ACID-INSENSITIVE 3 (ABI3), FUSCA 3 (FUS3)* and *LEAFY COTYLEDON 1 (LEC1)* loci act in concert to control multiple aspects of *Arabidopsis* seed development. *Plant Cell,* **9** 1265-77.

Parkinson, J. (1993) Signal transduction schemes of bacteria. *Cell* **73** 857-71.

Pei, Z.-M., Kuchitsu, K., Ward, J.M., Schwarz, M. and Schroeder, J.I. (1997) Differential abscisic acid regulation of guard cell slow anion channels in *Arabidopsis* wild-type and *abi1* and *abi2* mutants. *Plant Cell,* **9** 409-23.

Peng, J. and Harberd, N.P. (1993) Derivative alleles of the *Arabidopsis* gibberellin-insensitive (*gai*) mutation confer a wild-type phenotype. *Plant Cell,* **5** 351-60.

Phillips, A.L., Ward, D.A., Uknes, S., Appleford, N.E.J., Lange, T., Huttley, A.K., Gaskin, P., Graebe, J.E. and Hedden, P. (1995) Isolation and expression of three gibberellin 20-oxidase cDNA clones from *Arabidopsis. Plant Physiol.,* **108** 1049-57.

Pickett, F.B., Wilson, A.K. and Estelle, M. (1990) The *aux1* mutation of *Arabidopsis* confers both auxin and ethylene resistance. *Plant Physiol.,* **94** 1462-66.

Putterill, J., Robson, F., Lee, K., Simon, R. and Coupland, G. (1995) The CONSTANS gene of *Arabidopsis* promotes flowering and encodes a protein showing similarities to zinc finger transcription factors. *Cell,* **80** 847-57.

Rapp, U.R. (1991) Role of Raf-1 serine/threonine protein kinase in growth factor signal transduction. *Oncogene,* **6** 495-500.

Raven, J.A. (1975) Transport of indoleacetic acid in plant cells in relation to pH and electrical potential gradiets, and its significance for polar IAA transport. *New Phytol.,* **74** 163-72.

Rayle, D.L. and Cleland, R.E. (1992) The acid growth theory of auxin-induced cell elongation is alive and well. *Plant Physiol.,* **99** 1271-74.

Reed, J.W., Foster, K.R., Morgan, P.W. and Chory, J. (1996) Phytochrome B affects responsiveness to gibberellins in *Arabidopsis. Plant Physiol.,* **112** 337-42.

Rock, C.D. and Zeevaart, J.A. (1991) The *aba* mutant of *Arabidopsis thaliana* is impaired in epoxy-carotenoid biosynthesis. *Proc. Natl Acad. Sci. USA,* **88** 7496-99.

Rodrigues-Pousada, R., De Rycke, R., Dedonder, A., van Caeneghem, W., Engler, G., van Montagu, M. and Van der Straeten, D. (1993) The *Arabidopsis* 1-aminocyclopropane-1-carboxylate synthase gene 1 is expressed during early development. *Plant Cell,* **5** 897-911.

Roelfsema, M.R.G. and Prins, H.B.A. (1995) Effect of abscisic acid on stomatal opening in isolated epidermal strips of *abi* mutants of *Arabidopsis thaliana. Physiol. Plant.*, **95** 373-78.

Roman, G. and Ecker, J. (1996) Genetic analysis of a seedling stress response to ethylene in *Arabidopsis. Philos. Trans. Roy. Soc. Lond. Ser.* B 75-81.

Roman, G., Lubarsky, B., Kieber, J., Rothenberg, M. and Ecker, J. (1995) Genetic analysis of ethylene signal transduction in *Arabidopsis thaliana*: five novel mutant loci integrated into a stress response pathway. *Genetics*, **139** 1393-1409.

Romano, C.P., Robson, P.R.H., Smith, H., Estelle, M. and Klee, H. (1993) Transgene-mediated auxin overproduction in *Arabidopsis*—hypocotyl elongation phenotype and interactions with the *hy6-1* hypocotyl elongation and *axr1* auxin-resistant mutants. *Plant Mol. Biol.*, **27** 1071-83.

Rubery, P.H. and Sheldrake, A.R. (1974) Carrier-mediated auxin transport. *Planta*, **188** 101-21.

Ruegger, M., Dewey, E. and Estelle, M. (1997) Reduced naphthylphthalamic acid binding in the tir3 mutant of *Arabidopsis* is associated with a reduction in polar auxin transport and diverse morphological defects. *Plant Cell,* **9** 745-57.

Ruegger, M., Dewey, E., Gray, W.M., Hobbie, L., Turner, J. and Estelle, M. (1998) The TIR1 protein of *Arabidopsis* functions in auxin reponse and is related to human SKP2, and yeart Grrlp. *Genes Dev.*, **12** 198-207.

Schaller, G. and Bleecker, A. (1995) Ethylene-binding sites generated in yeast expressing the *Arabidopsis ETR1* gene. *Science*, **270** 1809-11.

Schaller, G., Ladd, A., Lanahan, M., Spanbauer, J. and Bleecker, A. (1995) The ethylene response mediator ETR1 from *Arabidopsis* forms a disulfide linked dimer. *J. Biol. Chem.*, **270** 12526-30.

Schiavone, F.M. and Racusen, R.H. (1997) Unusual patterns of somatic embryogenesis in domesticated carrot: developmental effects of endogenous auxins and auxin transport inhibitors. *Cellular Differentiation*, **21** 53-62.

Schwartz, S.H., Léon-Kloosterziel, K.L., Koornneef, M. and Zeevaart, J.A.D. (1997) Biochemical characterization of the *aba2* and *aba3* mutants in *Arabidopsis thaliana. Plant Physiol.*, **114** 161-66.

Sheen, J. (1996) Ca^{2+}-dependent protein kinases and stress signal transduction in plants. *Science*, **274** 1900-1902.

Shinomura, T., Nagatani, A., Hanzawa, H., Kubota, M., Watanabe, M. and Furuya, M. (1996) Action spectra for phytochrome A- and B-specific photoinduction of seed germination in *Arabidopsis thaliana. Proc. Natl Acad. Sci. USA*, **93** 8129-33.

Shinozaki, K. and Yamaguchi-Shinozaki, K. (1996) Molecular responses to drought and cold stress. *Curr. Opin. Biotechnol.*, **7** 161-67.

Sisler, E. and Yang, S. (1984) Anti-ethylene effects of cis-2-butene and cyclic olefins. *Phytochemistry,* **23** 2765-68.

Sisler, E., Blankenship, S. and Guest, M. (1990) Competition of cyclooctenes and cyclooctadienes for ethylene binding and activity in plants. *Plant Growth Reg.*, **9** 157-64.

Spruit, C.J.P., van den Boom, A. and Koornneef, M. (1980) Light induced germination and photochrome content of seeds of some mutants of *Arabidopsis. Arabidopsis Inf. Serv.*, **17** 137-41

Stock, J., Stock, A. and Mottonen, J. (1990) Signal transduction in bacteria. *Nature*, **344** 395-400.

Sun, T.-P. and Kamiya, Y. (1994) The *Arabidopsis* GA1 locus encodes the cyclase ent-kaurene synthetase A of gibberellin biosynthesis. *Plant Cell,* **6** 1509-18.

Sun, T-P., Goodman, H.M. and Ausubel, F.M. (1992) Cloning the *Arabidopsis* GA1 locus by genomic substraction. *Plant Cell,* **4** 119-28.

Suzuki, M., Kao, C.Y. and McCarty, D.R. (1997) The conserved B3 domain of VIVIPAROUS 1 has a cooperative DNA binding activity. *Plant Cell,* **9** 799-807.

Talón, M., Koornneef, M. and Zeevaart, J.A.D. (1990a) Endogenous gibberellins in *Arabidopsis thaliana* and possible steps blocked in the biosynthetic pathways of the semidwarf ga4 and ga5 mutants. *Proc. Natl Acad. Sci. USA*, **87** 7983-87.

Talón, M., Koornneef, M. and Zeevaart, J.A.D (1990b) Accumulation of C19-gibberellins in the gibberellin-insensitive dwarf mutant gai of *Arabidopsis thaliana* (L.) Heyhn. *Planta*, **182** 501-505.

Taylor, I.B. (1991) Genetics of ABA synthesis, in *Abscisic Acid Physiology and Biochemistry* (eds W.J. Davies and H.G. Jones), BIOS Scientific Publishers, Oxford, pp. 23-37.

Theologis, A. (1992) One rotten apple spoils the whole bushel: the role of ethylene in fruit ripening. *Cell,* **70** 181-84.

Timpte, C., Wilson, A.K. and Estelle, M. (1994) The *axr2-1* mutation of *Arabidopsis thaliana* is a gain-of-function mutation that disrupts an early step in auxin response. *Genetics,* **138** 1239-49.

Timpte, C., Lincoln, C., Pickett, F.B., Turner, J. and Estelle, M. (1995) The *Axr1* and *Aux1* genes of *Arabidopsis* function in separate auxin-response pathways. *Plant J.,* **8** 561-69.

Tuominen, H., Ostin, A., Sanberg, G. and Sundberg, B. (1994) A novel metabolic pathway for indole-3-acetic acid in apical shoots of *Populus tremula* (L.)× *Populus tremuloides* (Michx.). *Plant Physiol.,* **106** 1511-20.

Ulmasov, T., Liu, Z.-B., Hagen, G. and Guilfoyle, T.J. (1995) Composite structure of auxin response elements. *Plant Cell,* **7** 1611-23.

Ulmasov, T., Hagen, G. and Guilfoyle, T.J. (1997) ARF1, a transcription factor that binds to auxin response elements. *Science,* **276** 1865-68.

Van Aelst, L., Barr, M., Marcus, S., Polverino, A. and Wigler, M. (1993). Complex formation between RAS and RAF and other protein kinases. *Proc. Natl Acad. Sci. USA,* **90** 6213-17.

Van der Straeten, D., Van Wiemeersch, L., Goodman, H. and Van Montagu, M. (1990). Cloning and sequence of two different cDNAs encoding 1-aminocyclopropane-1-carboxylate synthase in tomato. *Proc. Natl. Acad. Sci. USA,* **1987** 4859-63.

Van der Straeten, D., Djudzman, A., Van Caeneghem, W., Smalle, J. and van Montagu, M. (1993) Genetic and physiological analysis of a new locus in *Arabidopsis* that confers resistance to 1-aminocyclopropane-1-carboxylic acid and ethylene and specifically affects the ethylene signal transduction pathway. *Plant Physiol.,* **102** 401-408.

Vasil, V., Marcotte, W.R., Jr, Rosenkrans, L., Cocciolone, S.M., Vasil, I.K., Quatrano, R.S. and McCarty, D.R. (1995) Overlap of Viviparous1 (VP1) and abscisic acid response elements in the *Em* promoter: G-box elements are sufficient but not necessary for VP1 transactivation. *Plant Cell,* **7** 1511-18.

Vogel, J.P., Woeste, K.E., Theologis, A. and Kieber, J.J. (1998) Recessive and dominant mutations in the ethylene biosynthetic gene *ACS5* of *Arabidopsis* confer cytokinin insensitivity and ethylene overproduction respectively. *Proc. Natl Acad. Sci. USA,* **95** 4766-71.

Wagner, D., Tepperman, J.M. and Quail, P.H. (1991) Overexpression of phytochrome B induces a short hypocotyl phenotype in transgenic *Arabidopsis. Plant Cell,* **3** 1275-88.

Watahiki, M.K. and Yamamoto, K.T. (1997) The massugu1 mutation of *Arabidopsis* identified with failure of auxin-induced growth curvature of hypocotyl confers auxin insensitivity to hypocotyl and leaf. *Plant Physiol.,* **115** 419-26.

Wilhelm, E., Battino, R. and Wilcock, R. (1977) Low pressure solubility of gases in liquid water. *Chem. Rev.,* **77** 219-62.

Wilson, R.N. and Somerville C.R. (1995) Phenotypic suppression of the gibberellin-insensitive mutant (gai) of *Arabidopsis. Plant Physiol.,* **108** 495-502.

Wilson, A.K., Pickett, F.B., Turner, J.C. and Estelle, M. (1990) A dominant mutation in *Arabidopsis* confers resistance to auxin, ethylene and abscisic acid. *Mol. Gen. Genet.,* **222** 377-83.

Wilson, R.N., Heckman, J.W. and Somerville C.R. (1992) Gibberellin is required for flowering in *Arabidopsis thaliana* under short days. *Plant Physiol.,* **100** 403-408.

Xu Y.-L., Li, L., Wu, K., Peeters, A.J.M., Gage, D.A. and Zeevaart, J.A.D. (1995) The GA5 locus of *Arabidopsis thaliana* encodes a multifunctional gibberellin 20-oxidase: molecular cloning and functional expression. *Proc. Natl Acad. Sci. USA,* **92** 6640-44.

Yang, S. and Hoffman, N. (1984) Ethylene biosynthesis and its regulation in higher plants. *Ann. Rev. Plant Physiol.,* **35** 155-89.

Zhang, H., Kobayashi, R., Galaktionov and Beach, D. (1995) p19skp1 and p45skp2 are essential elements of the cyclin A-CDK2 S phase kinase. *Cell,* **82** 915-25.

5 The secretory system and machinery for protein targeting

Susannah Gal

5.1 Introduction

All eukaryotic cells contain a number of organelles, each performing a particular function. These distinct functions are possible only through the presence of distinct classes of proteins in these compartments. The mechanism for delivery of these proteins to the particular organelles is therefore of intense interest; in particular because it may allow these systems to be harnessed for the targeting of heterologous proteins expressed in plants. A group of these organelles in eukaryotic cells is interconnected by membrane-bound vesicles. The secretory system, also called the endomembrane pathway, is comprised of the endoplasmic reticulum (ER), the Golgi apparatus, the *trans*-Golgi network, the vacuole (similar to the mammalian lysosome), the plasma membrane and secreted proteins as well as the membrane-bound vesicles that deliver proteins between these compartments (Figure 5.1). All eukaryotic cells appear to have a similar relationship between these organelles. Some aspects of the protein targeting machinery are similar in plant, mammalian and yeast cells but a few are distinct. The similarities and differences will be pointed out in this chapter. This chapter will focus on some general principles and how the secretory machinery has been studied, then on what has been found in terms of targeting signals and the receptors recognising those signals, and finally on the means by which the proteins are shuttled around the system. What we know in *Arabidopsis* is expanding rapidly because of the availability of a few mutants and the expressed sequence tag (EST) databases, but some other plant systems have been better characterised so will be given as examples of the kinds of experiments that could be done in this more genetically tractable system. Several recent reviews of the system have appeared (Gal and Raikhel, 1993; Bar-Peled *et al.*, 1996; Bassham and Raikhel, 1996).

5.1.1 General principles

Proteins generally enter the endomembrane system via the ER during a co-translational translocation event. This is mediated by the signal peptide generally located at the amino terminus of most soluble secretory proteins. Most proteins are then transferred from the ER to the Golgi and then either are secreted or

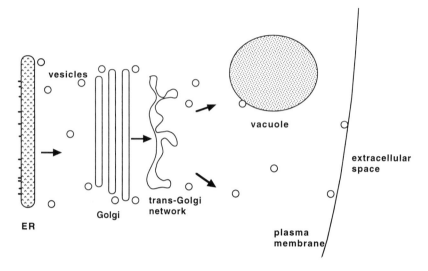

Figure 5.1 The secretory system. This system is comprised of the endoplasmic reticulum (ER), the Golgi apparatus, the *trans*-Golgi network, the vacuole, the plasma membrane and the extracellular space, as well as the vesicles that deliver components between the different organelles and spaces. The general flow of proteins synthesised on the rough ER is indicated by the arrows.

targeted to vacuoles. In the seeds of some plants, there is evidence for direct delivery of proteins from the ER to the vacuole or protein storage body, by-passing the Golgi. If a soluble protein contains only a signal peptide and no other retention sequence, it is secreted by the cells. This is the so-called default pathway for soluble proteins and appears to be a common pathway in all eukaryotic cells. Proteins retained in organelles of this system require a second targeting motif to remain in any of the cellular compartments linked to the ER. The specific signals used for retention are discussed below. The default location for membrane proteins is not as clearly defined nor is it the same in all systems.

5.1.2 Methods for study

To study the secretory machinery of cells a number of standard approaches has been followed with *Arabidopsis thaliana*. The first and most conventional is the isolation of the different compartments of the system and analysis of the components. Organelle fractionation protocols for a number of cell types have been published and recently adopted for the use of *Arabidopsis* suspension culture cells as well as whole leaves (Bar-Peled and Raikhel, 1997; da Silva Conceição *et al.*, 1997). These protocols take advantage of the different buoyant densities of the organelles, separating them via centrifugation. Marker enzymes are used to monitor the purity of the organelles in the different fractions. There is also a number of protein markers (monitored with antibodies) used to determine organelle purity in different fractions. These protocols have been instrumental in defining the compartments and correlating immunocytochemical localisation data with biochemical fractionation results.

A second very useful method that has been used to examine *Arabidopsis* components of the secretory system is the identification of components similar to known proteins from the secretory system in other cells. This has been done using antibodies to those components or gene homologies through the numerous sequences available. ESTs have been used to isolate the *Arabidopsis* homologues for a number of secretory components. With the continual appearance of comparable sequences in the databases, information on secretory components will be constantly updated. A third very useful method for identification of secretory components is via activity assays. Several assays for enzymes within the endomembrane system have been used to isolate *Arabidopsis* homologues to these components first identified in another system. One very powerful example of an activity-based isolation method is complementation. A number of yeast and mammalian cell lines exist that have mutations in various secretory machinery components. Transformation of these mutant cells with an *Arabidopsis* expression library and identification of revertant clones has been useful in isolating a number of genes for proteins involved in the secretory pathway. Finally, once a gene has been cloned and a function proposed mutants lacking this protein provide confirmation of the hypothesised role and give information about the effect of its loss on plant physiology. This approach is particularly important for *Arabidopsis* as transgenic plants expressing an antisense of the gene or ones with a T-DNA tag within the gene (McKinney *et al.*, 1995; Krysan *et al.*, 1996) are relatively easily generated. Isolation of mutants with altered levels or regulation of specific secretory genes has been successful in a number of cases. However, mutants with altered targeting of specific secretory proteins have not been isolated. Therefore, there are a number of means for analysing the secretory system in *Arabidopsis*. Specific examples of how these approaches have been used are discussed in detail below.

5.2   Endoplasmic reticulum

5.2.1   Entry into the ER

The mechanism of protein entry into the secretory system appears to be common to all eukaryotic cells (for a review see Rapoport, 1992). For most soluble proteins, an amino-terminal peptide sequence called the signal sequence is recognised during translation on free ribosomes and bound by a complex called the signal recognition particle (SRP). The SRP is comprised of several proteins and one RNA component. This SRP-nascent polypeptide chain–ribosome complex is arrested in further translation until an interaction at the ER membrane occurs via a ribosome receptor and a translocation apparatus. Translation continues on the rough ER membrane after hydrolysis of GTP and release of the SRP protein. The rest of the polypeptide chain is co-translationally inserted into the lumen of the ER.

There are at least two pieces of evidence that the signal peptide mediated targeting to the ER is common to all eukaryotic cells. First, common components of the SRP have been isolated from a number of organisms. From *Arabidopsis*, SRP54 has been isolated and has a high degree of sequence similarity with the similar protein from mammals and yeast (Lindstrom *et al.*, 1993). This protein is involved in the recognition of the signal sequence and contains a GTP-binding domain. Interestingly, in *Arabidopsis* three divergent SRP54 genes have been isolated which appear to have tissue-specific differences in expression (Chu *et al.*, 1994). Tomato appears to contain two genes for SRP54. Yeast appears to have only one gene while the copy number for SRP54 in mammalian systems has not yet been described. The RNA component of the SRP has also been isolated from a number of plant systems, including *Arabidopsis* (Shimomura *et al.*, 1993). Gene sequences of proteins in the translocation apparatus from *Arabidopsis* and mammals show some common components (Hartmann and Prehn, 1994).

The second piece of evidence that entry into the ER is common to all eukaryotes is that signal peptide sequences on translocated proteins appear similar. Although these sequences do not share common amino acids, they do show certain similar properties (von Heijne, 1985). In addition, several signal peptides have been tested in heterologous systems and have been shown to correctly deliver the protein to the ER. These include the heterologous expression of the signal sequence from one plant to another plant (barley lectin in tobacco, Bednarek *et al.*, 1990; *Arabidopsis* 2S albumin in tobacco, De Clercq *et al.*, 1990; patatin signal sequence in tobacco, Iturriaga *et al.*, 1989; bean phytohemagglutinin in tobacco and *Arabidopsis*, von Schaewen *et al.*, 1993) as well as expression of yeast or mammalian secretory proteins in plants (Hiatt *et al.*, 1989; Sjimons *et al.*, 1990; Chrispeels, 1991; Galbraith *et al.*, 1992; Ma *et al.*, 1995).

5.2.2   ER soluble and membrane proteins and their retention

The ER compartment serves as the entry point for all the secretory proteins of the cell. It is the site of initial protein folding and glycosylation as well as lipid synthesis. Thus the proteins contained within this compartment perform these functions. These proteins include the chaperones immunoglobulin-binding protein (BiP), heat-shock proteins, calnexin and calreticulin as well as refolding proteins such as protein disulphide isomerase (PDI) and the oligosaccharyl transferase. In several cases, the *Arabidopsis* homologue of these proteins has been isolated

(Bar-Peled *et al.*, 1995; Helm *et al.*, 1995; Nelson *et al.*, 1997; X. Li and H. Sze, personal communication). Calreticulin expression was found in all tissues but was particularly abundant in distinct floral tissues (Nelson *et al.*, 1997). Calnexin was identified in *Arabidopsis* using monoclonal antibodies to the homologue from oat. A 64–67 kDa protein was visualised in oat and co-fractionated with BiP on sucrose gradients. Sze and co-workers have used reciprocal immunoprecipitation to show an association of oat calnexin with various subunits of the vacuolar ATPase, suggesting that this protein has a role in protein folding and possibly assembly of the multimeric proton pump (X. Li and H. Sze, personal communication).

These ER proteins are retained in this organelle by binding to a specific receptor, Erd2, via a carboxyl-terminal sequence, most commonly Lys–Asp–Glu–Leu (KDEL) or related sequences. This receptor system was first identified in yeast using mutants that secreted ER-localised proteins. The *Arabidopsis* isoform was identified from the EST database by sequence comparison with the yeast and mammalian isoforms (Lee *et al.*, 1993). All of the Erd2 protein homologues have seven membrane-spanning domains and are membrane embedded. The plant protein was able to complement the yeast *erd2* mutant while the mammalian homologue was unable to do so (Lee *et al.*, 1993). The mammalian Erd2 has been localised to the *cis*-Golgi membranes and on ligand binding relocates to the ER membrane. This has been used to develop a model for ER protein retention that involves retrieval of the KDEL-containing proteins from the Golgi after they have 'escaped' from the ER. This is consistent with evidence from yeast that the ER proteins obtain sugar modifications from enzymes present in the *cis*-Golgi cisternae (reviewed in Bar-Peled *et al.*, 1995).

The *Arabidopsis* homologue for *ERD2*, *aERD2*, appears to be a single-copy gene and is expressed in several tissues, although it is notably higher in roots, inflorescent stems and flowers (Lee *et al.*, 1993). These same tissues have higher transcript levels for *aPDI* and two other genes involved in vesicle traffic between the ER and Golgi, *aSAR1* and *aSEC12* (Bar-Peled *et al.*, 1995) (see below for more discussion of Sar1p and Sec12p). It is not entirely clear why roots would have apparently more secretory protein activity than leaves. Two agents which block secretory protein transport, tunicamycin (blocks protein glycosylation) and cold shock (slows down membrane flow), lead to marked increase in aERD2 gene expression in leaves. Both *aSAR1* and *aPDI* were induced in leaves after tunicamycin treatment (Bar-Peled *et al.*, 1995). Interestingly, heat shock caused no change in the level of expression of this gene, but does induce a specific heat shock protein found in the ER of *Arabidopsis* (K. Helm, personal communication).

5.2.3   *ER membrane proteins and their retention*

An ER membrane protein with two membrane-spanning regions has been isolated from *Arabidopsis* (Campos and Boronat, 1995). This protein, 3-hydroxy-

3-methylglutaryl coenzyme A reductase (HMGR), is involved in the synthesis of mevalonate, a precursor of the isoprenoid biosynthetic pathway. Campos and Boronat (1995), using an *in vitro* system, have shown that this protein contains two hydrophobic sequences recognised by the SRP to mediate targeting to the ER. These workers believe the protein is retained in the ER, although movement to other organelles is still a possibility.

Another specific membrane protein in the ER from *Arabidopsis* has been isolated and partially characterised. Sze and co-workers have isolated a Ca^{2+} ATPase called *ECA1* which encodes a 116 kDa polypeptide with more significant homology to sarcoplasmic Ca^{2+} pumps than ones located in the plasma membrane (Liang *et al.*, 1997). Antibodies to the carboxyl-terminal region of Eca1 recognised a protein associated with ER membranes separated on a sucrose gradient. Interestingly, this protein complements yeast mutants defective in a Golgi Ca^{2+} pump or both a Golgi and vacuolar Ca^{2+} pumps by allowing these cells to grow on a Ca^{2+} chelator, EGTA (Liang *et al.*, 1997). Cloning of an *Arabidopsis* homologue of the Sac1 protein, which is an integral membrane protein of the yeast ER and Golgi complex, has been reported (H.-J. Wu, *et al.*, personal communication).

Targeting of Type I membrane proteins to the ER in mammalian and yeast systems seems to involve a di-lysine (K–K or K–X–K) motif located at the carboxyl terminus of the cytoplasmic region of the protein (Jackson *et al.*, 1990). Work that mutates that region in a component of the yeast oligosaccharyl transferase complex indicates that the ER retention and binding to a protein complex called coatamer are linked (Cosson and Letourneur, 1994). Coatamer is an important component of budding vesicles from the ER en route to the *cis*-Golgi (see below). The sequence of the *Arabidopsis ECA1* contains the sequence K–X–K–X–X at its carboxyl terminal (Liang *et al.*, 1997) although the importance of this sequence for ER-membrane targeting has yet to be demonstrated in plants. It is also not clear whether other *Arabidopsis* ER resident membrane proteins have a similar motif.

5.3   Golgi apparatus

The Golgi apparatus is one of the more mysterious organelles in plants. It is comprised of a series of flattened cisternae held together by proteinaceous components between the stacks (Cluett and Brown, 1992). The cisternae are arranged as *cis*-, medial and *trans*-Golgi with the orientation based on the forward flow of proteins in this apparatus from the ER. The organelle serves to modify carbohydrate moeities on proteins by the removal and addition of specific sugars (Moore *et al.*, 1991; Zhang and Staehelin 1992; reviewed by Faye *et al.*, 1989, and Driouich *et al.*, 1993). In all eukaryotic cells there is a conversion of a specific subset of Asn-linked glycans from a high mannose form to a complex glycan through the action of a variety of enzymes. In plants, the complex glycans are characterised by the presence of xylose and fucose, not found in yeast or mammalian cells (reviewed by Faye *et al.*, 1989, and Driouich *et al.*, 1993). The modification enzymes have been localised to specific compartments using antibodies against xylose or fucose in a conjugated form, the former being first found in the medial-Golgi while fucose-containing glycoconjugates are found primarily in the *trans*-Golgi (Fitchette-Lainé *et al.*, 1994).

One of the first enzymes acting in the conversion of glycoconjugates from high mannose to complex glycan type is N-acetyl glucosaminyltransferase I. Using an antibody specific for complex glycans, Chrispeels and co-workers have isolated an *Arabidopsis thaliana* mutant that lacks complex sugars on all Asn-linked glycans (von Schaewen *et al.*, 1993). The glycans on proteins in these plants are maintained as high mannose chains. Characterisation of this mutant revealed that it lacked N-acetyl glucosaminyltransferase I. The plant lacking complex carbohydrates is able to complete its life cycle, suggesting that the form of these glycans does not play an essential role in plant development. The function missing in these mutants appears to have been complemented by the transformation of these plants with the human gene for N-acetyl glucosaminyl-transferase I (Gomez and Chrispeels, 1994). The glycoproteins from these trans-formants contain a near-normal level of complex glycans. The human transferase enzyme co-localises on sucrose gradients with the N-acetyl glucosaminyl-transferase I activity in wild-type plants, suggesting that the targeting of proteins to the Golgi apparatus of plants and animals may be similar (Gomez and Chrispeels, 1994). One may be able to use these mutants that lack the enzyme as recipients for different modified forms of the mammalian homologue so that Golgi targeting signals can be identified.

The Golgi apparatus is the major site of the synthesis of cell wall components using some of the same enzymes that convert glycans on proteins to their mature form (Driouich *et al.*, 1993). A few cell wall synthesis enzymatic activities have been characterised from pea, including a UDPase from stems (Orellana *et al.*, 1997), a xyloglucan fucosyltransferase from epicotyl membranes (Maclachlan *et al.*, 1992) and xyloglucan galactosyl- and fucosyltransferase activities from epicotyl microsomes (Faïk *et al.*, 1997). *Arabidopsis* mutants in cell wall components have been isolated using a few different screens. In one case, mutants were screened by analysing acid hydrolysates of cell walls in a gas chromatograph and looking for ones with altered sugar content (Reiter *et al.*, 1993). One line identified was completely devoid of fucose and showed altered growth and more fragile cell walls (Reiter *et al.*, 1993). Dhugga and Ray (1994) obtained sequence from a 40 kDa protein that is reversibly glycosylated by UDP-glucose, UDP-xylose and UDP-galactose. *Arabidopsis* ESTs have been isolated with a similar sequence (Z. Wang *et al.*, personal communication).

The *trans*-Golgi network in mammalian cells is a complicated tubular and vesicular grouping at the far *trans*-face of this apparatus. In animal cells it appears to be the site of protein sorting between proteins destined for the lysosome with a targeting signal and those destined to be secreted either constitutively or in a regulated manner (reviewed in Chrispeels, 1991, and Bar-Peled *et al.*, 1996). Although a potential vacuolar targeting receptor been isolated from plants, it does appear to localise to the *trans*-Golgi network and in clathrin-coated vesicles that emanate from this organelle. The specific properties of this putative receptor are discussed in the next section.

5.4 Vacuoles

5.4.1 Soluble vacuolar components

The large central vacuole of plant cells performs several functions (reviewed by Wink, 1993). It acts as a storage organelle in seeds and some vegetative tissues, and it gives the plant cell turgor and shape against the firm plant cell wall. It also is a site of component degradation as it contains a variety of hydrolases of proteins, nucleic acids and other compounds. There are a number of proteases that have been localised here, including the aspartic proteinase from *Arabidopsis* seeds (Mutlu and Gal, unpublished) and three *Arabidopsis* homologues of the vacuolar processing enzymes (Kinoshita *et al.*, 1995 a,b) shown to be in protein storage vacuoles of castor bean and soybean seeds. The expression pattern of these three forms has been analysed using β-glucuronidase promoter fusions in transgenic tobacco. It was found that one form of the vacuolar processing enzymes was expressed in seeds, while the others were expressed in vegetative tissues (I. Hara-Nishimura *et al.*, personal communication). The major seed storage protein in *Arabidopsis* is 2S albumin found in the protein storage organelles, which are functionally similar to the vacuole (De Clercq *et al.*, 1990). As in other plants, the vacuolar form of chitinase is likely to be localised here as well (Samac *et al.*, 1990). The role of this enzyme has been proposed to be in plant defence as its synthesis is increased during pathogen attack.

5.4.2 Soluble vacuolar protein targeting

There appears not to be a common means for targeting proteins to the eukaryotic vacuole/lysosome (reviewed in Bar-Peled *et al.*, 1996). There are both mannose 6-phosphate-dependent and independent systems for targeting proteins to the mammalian lysosome. In yeast, amino-terminal propeptides on proteins contain the vacuolar-targeting motif. However, in plants there appear to be at least three different signals targeting to the plant vacuole, all involving peptide components.

The three different peptide components used for protein targeting to the plant vacuole were distinguished initially by their location on the targeted protein (reviewed in Gal and Raikhel, 1993, and Bar-Peled *et al.*, 1996). Barley aleurain and sweet potato sporamin are targeted via amino-terminal propeptides, barley lectin and vacuolar chitinase and glucanase from *Nicotiana* species by carboxyl-terminal propeptides, and bean phytohemagglutinin and legumin via an internal peptide. These signals have been identified primarily in transgenic tobacco cells and plants where the studied proteins have been manipulated and then the location of the mutated protein determined. With the two propeptides, if they are removed, the expressed protein is secreted by the transgenic cells, suggesting that this region is necessary for targeting. These sequences are sufficient for targeting as their addition to a normally secreted protein redirects the fusion protein to the plant vacuole. This has been shown using the amino-terminal propeptide of aleurain added to the secreted cysteine proteinase EP-B, and with the carboxyl-terminal propeptides of barley lectin and tobacco vacuolar chitinase fused to the secreted cucumber chitinase. When a 30 amino acid internal sequence of bean phytohemagglutinin is added to the secreted yeast protein invertase, it redirects approximately 50% of this protein to the vacuoles of tobacco protoplasts.

As with the many other examples given in this chapter, several groups had hoped to use yeast as a system to study the vacuolar targeting systems of plants. Yeast uses peptide sequences to target proteins to the vacuole, and a variety of mutants in the secretory pathway exist that could be used to explore the targeting of the heterologous plant proteins. Unfortunately, this has not proved to be feasible (reviewed in Gal and Raikhel, 1993). The first group to test this analysed the targeting of phytohemagglutinin fused with the secreted yeast invertase and narrowed down the vacuolar targeting signal recognised in yeast to a short stretch of amino acids. However, when this fusion protein was tested in plants it was found not to redirect the yeast invertase to the vacuole. Sporamin and sporamin lacking the amino-terminal propeptide were expressed in yeast and both were found in yeast vacuoles. Finally, fusion proteins of full-length barley lectin to invertase were sequestered by yeast but this was not dependent on the presence of the carboxyl-terminal propeptide known to contain the plant vacuolar targeting signal (Gal and Raikhel, 1994). Thus in all cases these vacuolar proteins were retained by the yeast cells but only because of other signals on the proteins, not those that are recognised by the plant to target proteins to vacuoles.

Most of the targeting experiments have been done in tobacco cells or plants although similar results are expected in *Arabidopsis*. The invertase fusion proteins with phytohemagglutinin peptides were tested in *Arabidopsis* protoplasts and found to cause vacuolar retention of this secreted yeast protein (von Schaewen and Chrispeels, 1993). The barley lectin carboxyl-terminal propeptide functions in this plant, as the full-length protein is in the vacuoles of transgenic *Arabidopsis* plants while the protein lacking the carboxyl-terminal propeptide is secreted (Bednarek and Raikhel unpublished). Full-length sweet potato sporamin is also found in the vacuoles of *Arabidopsis* plants (K. Nakamura, personal communication).

The targeting of one of the native *Arabidopsis* seed proteins, 2S albumin, was studied in transgenic tobacco seeds. This protein contains both amino-terminal and carboxyl-terminal propeptides as well as an intervening propeptide between the large and small subunits of the mature protein. Removal of any of these individual regions did not disrupt the targeting (D'Hondt *et al.*, 1993), suggesting that targeting information is contained within the mature polypeptides. This contrasts sharply with the work of Saalbach and colleagues (1996) using the 2S albumin from Brazil nut. These workers found that removal of the four amino acid carboxyl-terminal propeptide caused secretion of the modified protein in tobacco. One possibly significant difference between these experiments is that the Brazil nut protein was expressed using a strong constutive promoter in all tissues while the *Arabidopsis* protein was expressed using the native seed specific promoter. In addition, the carboxyl-terminal propeptides in these proteins are significantly different. The Brazil nut protein contains four amino acids with a sequence of Ile-Ala-Gly-Phe while the protein from *Arabidopsis* has a carboxyl-terminal propeptide of only two amino acids, Phe-Tyr.

5.4.3 The search for a vacuolar targeting receptor

Because the vacuolar targeting signals were so varied it was not clear whether there would be one or multiple classes of targeting receptors involved in protein targeting to the plant vacuole. Targeting receptors for the mammalian mannose 6-phosphate-containing proteins and for the amino-terminal targeting peptides of yeast vacuolar proteins have been identified (reviewed in Bar-Peled *et al.*, 1996). The sorting of proteins destined for the vacuole from those destined for secretion appears to take place in the *trans*-Golgi network, where those proteins bound by membrane-embedded receptors are segregated from proteins in the lumen of this organelle. Packaged in clathrin-coated vesicles after budding, these vesicles then undergo acidification releasing the bound proteins. The membranes with the receptor recycle back to the Golgi while the contents are carried further to be deposited in the vacuole. Acidification of the transport vesicles is important for vacuolar targeting in tobacco as treatments that disrupt the ATPase are likely to produce the change in pH that causes misdirection of the vacuolar proteins (Matsuoka *et al.*, 1997). A protein enriched in clathrin-coated vesicles from developing pea cotyledons has been isolated that binds the amino-terminal targeting determinant of barley aleurain (Kirsch *et al.*, 1994). This protein has a molecular weight of 80 kDa and has been shown to be embedded in the membrane by various biochemical means. This protein bound the amino terminus of proaleurain at neutral pH, but released it at pH 4.0. When the binding of other propeptides to this protein were tested, it was found that the secreted protein EP-B did not bind, but the amino-terminal targeting peptide of sporamin did, as did the carboxyl-terminal region of the Brazil nut 2S albumin (Kirsch *et al.*, 1996). Mutations in sporamin that cause secretion also interfered with binding to the 80 kDa peptide. Interestingly, this potential receptor did not bind to the carboxyl-terminal propeptide of barley lectin suggesting that there may be another class of targeting receptors (Kirsch *et al.*, 1996). Alternatively, the binding of this protein to barley lectin, may not occur under the experimental conditions used.

There are indications that there may be functionally distinct vacuolar compartments as well as distinct means of targeting. Although sporamin and barley lectin are found in the same compartment in transgenic tobacco leaf and root cells (Schroeder *et al.*, 1993), barley lectin appeared to be in a separate compartment from the proteolytic enzyme aleurain in barley roots (Paris *et al.*, 1996). The barley aspartic proteinase appeared to co-localize with both proteins. Wortmannin, an inhibitor of phosphatidyl inositol 3-kinase, differentially disrupts the targeting of sporamin, which contains an amino-terminal propeptide, and barley lectin, which contains a carboxyl-terminal propeptide in tobacco cells (Matsuoka *et al.*, 1995). Low concentrations of this drug totally disrupted the targeting of barley lectin but did not disturb the targeting of sporamin. In a dose-dependent analysis, disruption of barley lectin targeting and phospholipid synthesis correlated, suggesting a link between these two processes.

5.5   Secreted proteins from *Arabidopsis*

A few secreted proteins have been isolated from *Arabidopsis*. These are usually identified as being similar to secreted proteins from other systems. A number of secreted chitinases and glucanases, whose synthesis is enhanced on pathogen attack as in other systems, have been identified. Clones for *Arabidopsis* isoforms of chitinase with and without a carboxyl-terminal propeptide (shown to be the vacuolar targeting peptide in tobacco) have been isolated (Samac *et al.*, 1990). One member of the chitinase gene family is found in developing seeds so a role in endosperm development or embryogenesis has been proposed (P.A. Passarinho *et al.*, personal communication). The mRNA is also present throughout the adult plant, suggesting other roles for this secreted protein. Several *Arabidopsis* glucanase clones have been isolated because of their induction after pathogen attack (Dong *et al.*, 1991). These proteins are presumed to be secreted as they lack a carboxyl-terminal propeptide found in the vacuolar forms of these enzymes (Melchers *et al.*, 1993). The gene for a secreted purple acid phosphatase has been identified from *Arabidopsis* (K. Patel *et al.*, personal communication). This single gene encodes a 45 kDa signal-sequence-containing protein whose presumed expression in the roots in response to low phosphate allows cells to scavenge phosphate from various sources to minimise the effect of phosphate starvation.

5.6   Membrane proteins

5.6.1   Protein of the tonoplast

The tonoplast is an important part of the vacuole as it regulates pH, turgor and ion flow across the membrane (reviewed by Wink, 1993 and Rausch et al., 1996). A number of tonoplast proteins have been identified which perform these functions. The best-characterised protein is the vacuolar ATPase, which comprises a large peripheral subunit (V_1) and an integral membrane subunit (V_0). Each of these subunits comprises multiple polypeptides and is analogous to the well studied F_1-ATPase of chloroplasts and mitochondria (reviewed by Rausch et al., 1996, and Lüttge and Ratajczak, 1997). Homologues of this pump are found in mammalian and yeast systems. The V_1 subunit is composed of six subunits, including the site of ATP hydrolysis and a regulatory subunit. The regulatory subunit has been cloned from Arabidopsis and contains a putative nucleotide-binding sequence (Manolson et al., 1988). The V_0 subunit is composed of four subunits, one of which is the 16 kDa polypeptide that is responsible for forming the proton conductance pathway. In both Arabidopsis and oat, the 16 kDa protein is encoded by a multigene family, with four genes in Arabidopsis (Perera et al., 1995). The four different polypeptides are almost identical and three of the genes are expressed in all plant tissues.

A plant-specific tonoplast protein is the proton pyrophosphatase, which uses pyrophosphatase as the energy source to translocate a proton across the membrane. With the vacuolar ATPase, this protein contributes strongly to the proton electrochemical potential gradient across the membrane. The pyrophosphatase from Arabidopsis was cloned (Sarafian et al., 1992) and, using yeast, Kim and colleagues (1994) were able to demonstrate that this membrane-intrinsic peptide is sufficient for pyrophosphate-dependent proton transport. The tonoplast also contains a number of ion channels, including a malate-regulated anion channel (Cerana et al., 1995), a potassium channel (Colombo et al., 1994) and two proton–calcium antiporters (Hirschi et al., 1996), all isolated from Arabidopsis.

Some of the most intriguing tonoplast proteins are the tonoplast intrinsic proteins (TIPs) related to the membrane intrinsic proteins (MIPs) found in many other organisms, including bacteria, yeast, mammalian and Drosophila. These proteins have in common a six transmembrane domain structure containing several conserved amino acids (reviewed in Chrispeels and Maurel, 1994). Some members of the MIP family are found in the tonoplast while others are found in the plasma membrane. These proteins form pores for water, ions or small metabolites. In Arabidopsis, several different gene sequences have been obtained and the expression of two of these proteins is differentially regulated (Höfte et al., 1992). For instance, γ-TIP is found in vegetative tissues while α-TIP is found specifically in developing embryos. Isoforms are expressed in bean seeds, in tobacco roots and are induced by water stress in pea (reviewed in Chrispeels and Maurel, 1994). Höfte and colleagues (1992) proposed that proteins with similar expression patterns from different organisms (e.g. in seeds or vegetative tissues) are more closely related to one another than those isoforms in different organs from the same organism, suggesting an evolutionary link between them. The bean-seed-specific α-TIP was shown to be modified by phosphorylation in

Xenopus oocytes, which regulated its water channel activity (Maurel *et al.*, 1995). However, the serines, which are phosphorylated, are not conserved in other TIP isoforms (Höfte *et al.*, 1992; Chrispeels and Maurel, 1994)

5.6.2 Plasma-membrane proteins

The plasma membrane is considered to be part of the secretory machinery as it is the deposition site of vesicles derived from the *trans*-Golgi network. There is a number of plasma membrane proteins including H^+-ATPases (AHAs), K^+ channels, chlorate and peptide transporters. Their structure and tissue specific localisation have led researchers to propose physiological functions for these membrane proteins in the plant (reviewed in Sussman, 1994).

The plasma membrane H^+-ATPases transport protons out of the cell at the expense of hydrolysing ATP, producing an electrical potential as well as a pH gradient. These were the first plant plasma-membrane proteins purified and so, much information has been accumulated about them. Through the identification of a number of tryptic peptides several genomic and cDNA clones for these enzymes have been isolated from a variety of plants. A large family of H^+-ATPases (AHAs) has been found in *Arabidopsis*, as many as 10 different forms, and the possibility of more is suggested by degenerative polymerase chain reaction experiments (Harper *et al.*, 1994). The amino acid structures have homology with the plasma membrane H^+-ATPases from other organisms. The plasma-membrane localisation of one of the members of this class of proteins was shown using epitope tagged AHA3 and is presumed for the other members of this gene family (DeWitt and Sussman, 1995). These different forms seem to be expressed in tissue specific patterns shown using promoter β-glucuronidase gene fusions and epitope tagging of the AHA proteins. For instance, AHA3 is found in phloem companion cells (DeWitt and Sussman, 1995), AHA9 is restricted to anther tissue (Houlné and Boutry, 1994) and AHA10 confined to seeds (Harper *et al.*, 1994). These differences may reflect differences in the requirement of the cells for a driving force in solute transport, namely the H^+ gradient. Isolation of plant lines with T-DNA lodged in specific AHA genes (Krysan *et al.*, 1996) will allow a more thorough investigation of the role of the individual members of the plasma membrane H^+-ATPase gene family in the plant.

Incubation of plant cells with the phytotoxin fusicoccin in the presence of K^+ causes an activation of the plasma-membrane H^+-ATPase, resulting in proton extrusion and K^+ uptake. The mechanism of this effect is believed to be mediated by binding of fusicoccin to its receptor, a protein identified to be a member of the 14-3-3 family (De Michelis *et al.*, 1996). Binding of the toxin is strictly correlated with the activation of the H^+-ATPase in *Arabidopsis* and radish seedlings (De Michelis *et al.*, 1996). A direct interaction of one H^+-ATPase (AHA2) and three proteins of the 14-3-3 family has been demonstrated using the yeast two-hybrid system (A.T. Fuglsang and M. Palmgren, personal communication). An

Arabidopsis mutant (called 5-2) that is less sensitive to the toxic effects of fusicoccin shows a decreased capacity of the H^+-ATPase to respond to the toxin, an effect that is independent of toxin binding or the K^+ uptake system (Marré *et al.*, 1995). The actual mutation in this plant awaits further characterisation but would be consistent with an alteration in the H^+-ATPase itself or a factor controlling its activity.

The transport of K^+ across membranes is extremely important to plant cells for growth and osmoregulation. From electrophysiological studies, there appear to be two transport systems for K^+, one high affinity and one lower affinity. Using the patch clamp technique, one group reported Ca^{2+} permeable K^+ channels in the plasma membrane of *Arabidopsis* leaf mesophyll cells (L.A. Romano *et al.*, personal communication). Two *Arabidopsis* genes were isolated by functional complementation of yeast cells lacking K^+ transporters (reviewed in Sussman, 1994). These cDNAs (KAT1 and AKT1) allowed the yeast to grow on micromolar K^+ while the mutant must be kept on millimolar K^+. Two other clones were isolated by homology to the first clones (Cao *et al.*, 1995; Ketchum and Slayman, 1996). Interestingly, all of these genes have little homology with the yeast genes but much more to a locus called *shaker* from *Drosophila* known to encode K^+ channels. All of the *Arabidopsis* forms have a similar structure, and mutations in the pore-like region alter the cation selectivity of the channel when it is expressed in *Xenopus* (Uozumi *et al.*, 1995). As in the case of the AHA-isoforms, the different K^+ channel genes are expressed differentially in plants; AKT1 is expressed primarily in the roots and AKT2 highly in leaves (Cao *et al.*, 1995; Lagarde *et al.*, 1996). As the four genes are not identical, they may also have distinct functions in the K^+ metabolism in the plants. Interestingly, the KAT1 channel is regulated not only by voltage but also by pH, ATP and cGMP (Hoshi, 1995). There is also some evidence that these channels may be regulated by phosphorylation (Armstrong *et al.*, 1995). A novel K^+ transporter, which can also mediate the uptake of rubidium, has been isolated from wheat and *Arabidopsis* cDNA libraries (E.J. Kim *et al.*, personal communication).

Another approach to studying K^+ channels was undertaken using cesium. *Arabidopsis* plants are sensitive to this cation, whose uptake is blocked by K^+. The high affinity potassium channel in roots is relatively insensitive to cesium, while the low affinity site is more sensitive (Maathuis and Sanders, 1996). *Arabidopsis* seedlings were selected on high concentrations of cesium and those surviving were characterised with regard to K^+ transport. Maathuis and Sanders (1996) have shown that the K^+ inward current is low in one such mutant. Interestingly, both pathways for K^+ uptake are present in the mutant, but in the presence of low levels of K^+ the mutant increases the level of the high affinity channel (which is less sensitive to cesium) more than the wild-type plant does. This suggests that this mutation does not affect a particular K^+ channel but instead shifts the total uptake system towards the one less sensitive to cesium (Maathuis and Sanders, 1996). Thus the mutation that segregates as a single recessive gene lies within a gene that regulates the co-ordination of the several K^+ transport systems that affect total uptake of this ion.

Several receptor-like protein kinases from *Arabidopsis* have been characterised. In animal cells, this class of membrane proteins is important for intracellular signalling in response to a number of hormones. The enzymes were detected in the membranes of *Arabidopsis* biochemically and have properties similar to the mammalian serine/threonine class of protein kinases (Schaller and Bleecker, 1993). These membrane proteins were proposed to have molecular weights between 115 and 135 kDa determined by renaturation after protein blotting. Kohorn and colleagues have isolated another receptor-like protein kinase with a molecular weight of 68 kDa that they have recently shown is localised in the plasma membrane and associated with the cell wall (He *et al.*, 1996). This protein is proposed to have an extracellular domain with epidermal growth factor-like repeats and is found in all vegetative tissues of *Arabidopsis*. Hervé and colleagues (1996) identified a putative plant receptor kinase with an extracellular lectin-like domain from the EST database. Another group has identified a receptor-like protein kinase with leucine-rich repeats in its extracellular domain that is associated with plasma membranes (T.-L. Jinn *et al.*, personal communication). All of these proteins appear to be serine/threonine kinases. Interestingly, one of the ethylene receptors, ETR1 associated with the plasma membrane has been shown to have histidine kinase activity. This might indicate a means by which ethylene signalling could be modulated (G.E. Schaller and M.L. Coonfield, personal communication).

The MIP family (see section 5.6.1) has members not only in the tonoplast (TIP) but also in the plasma membrane (Chrispeels and Maurel, 1994). These proteins have in common six membrane-spanning domains with both amino and carboxyl termini facing the cytosol. Several members of this family have been shown to be water channels or aquaporins using functional assays in *Xenopus* oocytes or by the reduction in water transport capacity in plant cells expressing an antisense construct (Kaldenhoff *et al.*, 1995). One group subclassified this family of genes encoding aquaporins into two groups, PIP1 and PIP2, with the former having at least five expressed members and the latter at least six in *Arabidopsis* (Robinson *et al.*, 1996). One member of the *Arabidopsis* plasma-membrane localised MIP family is blue-light responsive and expressed in expanding and differentiating cells (Kaldenhoff *et al.*, 1995). Another is turgor responsive and present in leaves and roots of well-watered plants, but not in seeds (Daniels *et al.*, 1994).

During seed germination and organ senescence, rapid transport of amino acids and peptides is required to provide or remove compounds from cells (reviewed in Frommer *et al.*, 1994). Several different amino acid transporters with overlapping specificity for different classes of amino acids have been identified from *Arabidopsis* using yeast complementation (Frommer *et al.*, 1995; Fischer *et al.*, 1995). The specificity and kinetics of these transporters can be analysed in yeast and in *Xenopus* oocytes (Fischer *et al.*, 1995; Boorer *et al.*, 1996). Analysis of some of these transporters in *Arabidopsis* plants shows that each protein has a different pattern of expression: one in seeds and vascular tissues, another exclusively in roots, while another is found in all tissues (Fischer *et al.*, 1995; Frommer *et al.*, 1995). One of these transporters (NTR1), originally isolated for histidine permease activity, has structural similarity to the peptide and nitrate transporter superfamily and was found to also transport oligopeptides (Rentsch *et al.*, 1995). NTR1 was found exclusively in embryos, suggesting its role in seed development. Another peptide transporter from *Arabidopsis* was isolated using yeast lacking this system. AtPTR2 is specific for the transport of di- and tri-peptides, has homology to NTR1 and is expressed in roots (Steiner *et al.*, 1994). Using antibodies, AtPTR2 has been shown to localise to the plasma membrane (C. Zhang *et al.*, personal communication). These peptide transporters have homology to the nitrate transporter CHL1, which was isolated from an *Arabidopsis* mutant sensitive to chlorate (Tsay *et al.*, 1993), a herbicide, which enters the system via the nitrate transporter and is converted to the toxic chlorite in the cells. Analysis of nitrate fluxes in this mutant suggest that there may be two systems for low-affinity transport of nitrate, only one of which is altered in the mutant (Touraine and Glass, 1997)

Anion channels that transport Cl^- ions across the plasma membrane have been identified in *Arabidopsis*. Using patch clamp techniques with hypocotyl cells of etiolated seedlings Cho and Spalding (1996) found that blue light activated an anion channel, suggesting a role for the transport of Cl^- ions in the perception process. This channel is activated by calcium ions although calcium does not appear to play a role in the response of this channel to blue light (Lewis *et al.*, 1997). Cloning of four Cl^- channels from *Arabidopsis* was achieved using ESTs homologous to a tobacco protein (Hechenberger *et al.*, 1996). Transcripts for the four genes are found in nearly all tissues, but one of them is also highly expressed in siliques. Only that particular homologue was able to rescue a yeast mutant lacking its internal membrane chloride channel, suggesting that this *Arabidopsis* protein has an intracellular function in the plant.

An integrin-like protein has been identified in membranes of *Arabidopsis* using antibodies to chicken β-1-integrin (Katembe *et al.*, 1997). This protein acts as a cell surface receptor and links the extracellular matrix to the cytoskeleton in animals. The protein was found in root cap cells. Preliminary results indicate that one member of the family of calcium-dependent protein kinases is associated with the plasma membrane, potentially via a putative myristoylation site at its amino terminus (E.M. Hrabak, personal communication).

5.6.3 Membrane protein targeting

The understanding of membrane protein targeting mechanisms is in its infancy in plants (reviewed in Bassham and Raikhel, 1996). We do not as yet know for certain which site is the default localisation for membrane proteins. In yeast when the signal for retention is deleted from Golgi membrane proteins they appear in the vacuolar membrane while in animal cells, default for membrane proteins is the plasma membrane. The situation in plant cells is not known. The approach used for analysis of soluble protein amino- and carboxyl-terminal propeptides, namely to cut and paste domains onto a marker protein, may not be feasible for probing the targeting of membrane proteins as the targeting sequences are likely to be within the internal portion of the proteins. These changes may destabilise the protein structure. There is evidence for a separate pathway for targeting tonoplast and soluble vacuolar proteins. Gomez and Chrispeels (1993) showed that brefeldin A and monensin both block delivery of phytohemagglutinin to the vacuole of transgenic tobacco plants although at different stages. The targeting of a second transgene, α-TIP, to the tonoplast was not disrupted by either treatment.

Höfte and Chrispeels (1992) studied vacuolar membrane targeting by fusing the sixth transmembrane domain and the 18 amino acid cytoplasmic tail of α-TIP to the marker protein phosphinotricine acetyltransferase. When this fusion protein was expressed in transgenic tobacco cells, it was retained in the vacuoles, suggesting that this region contains the tonoplast-targeting domain. Next, an attempt was made to remove the targeting information from α-TIP to misdirect the protein. The removal of the membrane-spanning domain caused the α-TIP protein to be unstable, but removal of 15 amino acids from the cytoplasmic tail did not disrupt targeting. Although not conclusive this work suggests that the membrane-spanning domain may confer tonoplast targeting, but definitive experiments await determination of the default location of membrane proteins. The availability of members of the MIP family found in both plasma membrane and tonoplast may allow one to swap the domains of these proteins to identify the parts necessary for targeting without dramatically altering the protein structure. Interestingly, the evidence for two separate vacuoles based on differential localisation of soluble components (see above) has been extended using TIP proteins where two different tonoplast membrane proteins were shown by confocal microscopy in peas, barley and tobacco to have distinct distributions within the tissues (Paris et al., 1996).

Targeting of K^+ channels in plants could be analysed using some of the mutant channels mentioned above. Presumably more drastic mutations could be produced to maintain the main structure of the protein, but these might modulate its location within the cell. The recent isolation of a mutant plant with a T-DNA lodged in the AKT1 gene will prove interesting to characterise (Krysan et al., 1996). Initial analysis using patch clamp techniques shows no inward K^+ currents in roots from these plants and they show growth defects at low K^+ concentrations (R.E. Hirsch et al., personal communication). Root cells from these plants could be used as recipients of the genes for mutated channels. One could monitor membrane localisation of different AKT1 mutants and identify regions necessary for plasma membrane accumulation in plants.

When *AHA*1, 2 and 3 are expressed in yeast, these proteins are correctly folded but are retained in the ER of these cells (Palmgren and Christensen, 1994). Interestingly, a mutant of AHA2 lacking a portion of the carboxyl-termi-nus is subsequently found in the plasma membrane of yeast and is able to rescue cells containing a defective proton pump (Sussman, 1994). The availability of mutant lines with T-DNA lodged in the some of the *AHA* genes (Krysan *et al.*,1996) may be a good source of cells to receive various forms of the AHA genes to monitor which region is necessary and sufficient for plasma-membrane targeting.

In one case, a tomato Ca-ATPase appears to localise to two different sites, the tonoplast and plasma-membrane fractions (Ferrol and Bennett 1996). The proteins in the two locations have slightly different molecular weights. This size difference may indicate that proteolytic processing reveals different targeting domains or may simply be a result of proteases in the different locations. It is possible that the plasma membrane form is simply on its way to the tonoplast, arriving there after endocytosis as is the case with the mammalian lysosomal acid phosphatase (Peters *et al.*, 1990).

5.7   The mechanism of membrane trafficking in cells

Thus far, this chapter has focused on the components within the organelles of the secretory system, the targeting signals (when known) and the receptors believed to recognise those signals (if identified), but how is the physical movement of these components within the system accomplished? The budding of vesicles from one organelle, the movement of vesicles from one site to another and the specific fusion of a vesicle with the appropriate organelle next in the progression is another layer of targeting within this system. How is that accomplished and where does the specificity lie to distinguish what should be within a budding vesicle and with which membrane should the vesicle fuse in its journey? The specificity of this system is essential for the maintenance of the components of the organelles and to avoid mis-delivery of components to inappropriate com-partments. Our understanding of the general system has been made possible by work in animal and yeast systems, but there is a variety of evidence to suggest that the same mechanisms are at work in all eukaryotic cells, including plants.

General proteins involved in vesicle budding and fusion have been identified using an *in vitro* system for monitoring the movement of Golgi proteins between the cisternae of this organelle (for reviews see Balch, 1990 and Rothman, 1994). Vesicle budding involves the interaction of a small GTP-binding protein (like ARF) with the donor membrane and the recruitment of vesicle coat proteins (either coatomer for ER and *cis*-, medial- and *trans*-Golgi membranes, or clathrin for *trans*-Golgi network and plasma membranes) which induce pinching off from the donor membrane (Figure 5.2). The direct interaction of the ER membrane protein retention motif K–K with coatamer has been demonstrated (Cosson and Letourneur, 1994). The GTP-binding protein associates with the membrane after bound GDP is exchanged for GTP, a step that is enhanced by various nucleotide exchange proteins. After vesicle formation, the GTP-binding protein and coat proteins are removed from the vesicle, fuelled by hydrolysis of the GTP, to allow docking of the vesicle to the target membrane (Figure 5.2). An N-ethyl-malemide-sensitive factor (called NSF) and a soluble NSF attachment protein (SNAP) are recruited to the docked vesicle and cause fusion fuelled by hydrolysis of ATP.

The SNARE (SNAP receptor) hypothesis was proposed to explain the specificity of fusion of vesicles or donor membranes with specific organelles or target membranes (for reviews see Rothman, 1994, and Bassham and Raikhel, 1996). This hypothesis states that each donor membrane has a specific protein termed a v-SNARE that interacts with a protein on the target membrane called a t-SNARE (Figure 5.2). A whole variety of v- and t-SNAREs exist in cells to allow the many specific vesicle–organelle fusion reactions to occur, such as ER-derived vesicles with the cis-Golgi and trans-Golgi network-derived vesicles either with the plasma membrane or the vacuole. The interaction of v-SNARE and t-SNARE proteins is only possible on a vesicle stripped of its protein coat after vesicle budding. This promotes fusion of vesicles with target organelles and not the two organelles directly. The vesicle docks on the target membrane via this interaction, NSF and SNAP are recruited and then membrane fusion occurs.

The evidence for this hypothesis comes from a variety of experiments, originally in mammalian cells and mutants available in yeast, and the universality of this system in eukaryotic cells comes from the observation that a number of similar components of this system have been identified. The fusion system for intra-Golgi transport in mammalian cells can use proteins derived from wheat germ or the slime mold Dictyostelium (Pâquet et al., 1986). Once NSF and a-SNAP were isolated, the homology with the secretion mutants from yeast sec18 and sec17 was clear, as was the role of these two proteins in vesicle transport throughout the secretory pathway. Yeast complementation of a number of mutants in protein transport using Arabidopsis cDNA libraries demonstrates the extension of this hypothesis to the plant kingdom. Some components have been identified using antibodies to the specific yeast or mammalian component. Still more proteins have been cloned that are presumably involved in this pathway via homology to known animal and yeast proteins of this system. The Arabidopsis EST database has proved invaluable in this analysis.

5.7.1 Proteins in Arabidopsis involved in the SNARE hypothesis

The budding of yeast vesicles from the ER requires a number of proteins, including Sar1p, Sec23/24p and Sec13/31p, which are part of the coated vesicle intermediate, and Sec12p, which is part of the ER membrane (for a review see

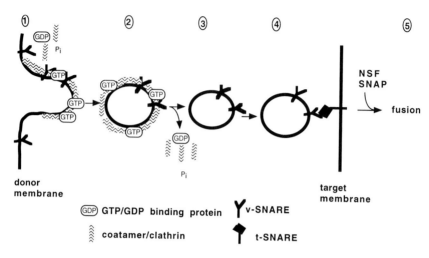

Figure 5.2 The mechanism of membrane trafficking in cells. There are several steps in the process. They include vesicle budding from the donor membrane (step 1), which is mediated by the recruitment of a GTP-binding protein and coat proteins, either coatamer or clathrin to produce the coated vesicle (step 2). The uncoating of the vesicle (step 3) is mediated by the hydrolysis of GTP to release the GTP-binding protein and the vesicle coat proteins. Docking of the vesicle to the target membrane is mediated by the specific interaction of the v- and t-SNAREs (step 4). The final step of vesicle fusion with the target membrane is facilitated by NSF and SNAP (step 5). For more details see the text and references such as Rothman (1994) and Bassham and Raikhel (1996).

Bar-Peled *et al.*, 1996). Sar1p is a GTP-binding protein whose exchange of GDP for GTP, promoting membrane binding, is enhanced by Sec12p. This binding recruits the other proteins to the ER membrane, one of which, Sec23p, is a GTPase activating protein whose action uncoats the vesicle. The fusion of these uncoated ER-derived vesicles to the *cis*-Golgi is mediated by a small GTP-binding protein, Ypt1p, and the t-SNARE, Sed5p. Several of the homologues of these components have been isolated from *Arabidopsis* either by functional complementation of the yeast mutants (d'Enfert *et al.*, 1992) or by sequence similarity from the EST database (Bar-Peled *et al.*, 1995). A coatamer subunit homologue in *Arabidopsis* has been found in the EST database (Guo *et al.*, 1994) as undoubtedly many other components will be in the future as more sequences are deposited. Using yeast complementation, the t-SNARE homologue for vesicle fusion to the vacuole *aPEP12* was isolated (Bassham *et al.*, 1995) and a t-SNARE-related product was identified as the complementing clone for the embryo mutant *KNOLLE* (Lukowitz *et al.*, 1996). Another *Arabidopsis* mutant, *EMB30* (also known as *GNOM*), which shows a wide variety of embryo mutant phenotypes, is complemented by a *SEC7* homologue, a protein important for trafficking in yeast (Shevell *et al.*, 1994; Busch *et al.*, 1996).

In addition to the aSar1p, numerous examples of other GTP-binding proteins from *Arabidopsis* homologous to yeast and/or mammalian proteins have been isolated. These include the 5 *ARA* genes (Matsui *et al.*, 1989; Anai *et al.*, 1991), *ARF* homologues (Regad *et al.*, 1993; Bar-Peled *et al.*, 1995), a *RAB6* homologue (Bednarek *et al.*, 1994), a cDNA related to the mammalian high molecular weight GTP-binding protein dynamin (Dombrowski and Raikhel, 1995) and a family of GTPases related to *rho* (Z. Yang, personal communication). Several of these genes were isolated by scanning the EST databases for clones containing characteristic sequences while others were isolated using clones homologous to ones previously isolated. Many of them have been transformed into yeast mutants lacking the corresponding gene and the *Arabidopsis* clones in some cases can complement the mutations.

Expression of several of the *Arabidopsis* clones of the components of this pathway has been analysed and found to vary in different tissues. Dombrowski and Raikhel (1995), analysing the dynamin homologue, found the message present in all tissues but lower in mature leaves, consistent with there being similar relative levels of other endomembrane mRNAs in these tissues (*aERD2*, Lee *et al.*, 1993; *aSEC12*, Bar-Peled *et al.*, 1995; *aPEP12*, Bassham *et al.*, 1995). However, Anai and colleagues (1995) found the *ARA4* message lower in roots but present in all of the other tissues, and the distribution of the *aRAB6* message is highest in roots and flowers (Bednarek *et al.*, 1994). The mRNA for the *ARF* homologue was present during all stages of cell growth in suspension culture cells (Regad *et al.*, 1993), while one member of the *rho* family is specifically expressed in pollen (Z. Yang, personal communication). The *KNOLLE* gene is expressed highly in flowers and siliques, lower in leaves and seedlings, and lowest in roots (Lukowitz *et al.*, 1996). The *EMB30* mRNA is present in all tissues, although it is slightly lower in leaves and seedlings than in flowers, stems and roots (Shevell *et al.*, 1994). Bar-Peled and Raikhel (1997), studying the *Arabidopsis* Sar1p homologue, found the protein in all tissues but slightly lower amounts in mature leaves. The levels of this protein decline after cold shock of tissue culture cells, suggesting this treatment may adversely affect transport between the ER and Golgi. However, the levels of the *Arabidopsis* Sec12p did not change during these conditions (Bar-Peled and Raikhel, 1997) thus a consistent distribution of the expression of all of the secretory machinery genes in tissues is not apparent, suggesting that each may function in different cell types or conditions or be required for different types of protein trafficking in these tissues.

Within the cell, each GTP-binding protein presumably has a given membrane interaction, as seen in yeast and mammalian cell experiments. Some of this can be inferred from the complementation experiments of *Arabidopsis* clones into yeast mutants, but in a few cases the workers have identified the specific membrane association for one or more GTP-binding proteins. Using transgenic tobacco plants expressing a heat shock inducible ARA4 gene, Ueda and colleagues

(1996a) found the protein to be primarily localised within Golgi cisternae, Golgi-derived vesicles and the *trans*-Golgi network. AtSar1p was found localised both in the cytosol and in ER-derived membranes, while the AtSec12p was only found in the ER-derived membranes (Bar-Peled and Raikhel, 1997). Consistent with its proposed role in vesicle to vacuole fusion, aPep12p was localised to a late post-Golgi compartment in plants (da Silva Conceição *et al.*, 1997).

The small GTP-binding proteins undergo a continual shift in GTP–GDP exchange and GTP hydrolysis during the numerous cycles of membrane budding and fusion reactions (reviewed by Balch, 1990). In yeast and mammalian cells, several classes of proteins have been identified that promote or inhibit these reactions. A gene homologue to the GAP protein *Sec23* from *Arabidopsis* has been isolated as well (mentioned in Bar-Peled and Raikhel, 1997). Anai and colleagues (1994) have found evidence for GTP-ase activating proteins (GAP) in plants using *ARA2* and *ARA4*. The same research group has also isolated a GDP-dissociation inhibitor (GDI) protein using suppression of the deleterious effect of heterologous *ARA4* expression in the yeast *ypt1* mutant (Ueda *et al.*, 1996b). Other suppressors of this phenotype have been isolated (T. Ueda *et al.*, personal communication). Future sequences available from the EST and genomic sequencing projects will probably uncover more homologues of yeast and mammalian proteins whose function in membrane trafficking is clear. Proteins isolated using a two-hybrid screen (Fields and Sternglanz, 1994; Chien-Ting *et al.*, 1996) with proteins within the endomembrane will probably uncover other interacting proteins in this system.

5.8   Future directions

Clearly our understanding of the secretory machinery has advanced dramatically in the last few years. The use of molecular tools in cloning various components of this pathway will allow researchers to analyse the system better. However, there are still several gaps in our understanding which provide areas ripe for future research endeavours.

As stated above, there are a few Golgi markers available but a complete understanding of the inter-workings of this compartment is lacking. Purification, cloning and preparation of specific antibodies to Golgi proteins will facilitate the next phase in the analysis. The mechanism of general membrane protein targeting in plant cells is still a mystery, but the availability of membrane proteins naturally in the tonoplast and plasma membrane as well as in the ER and Golgi provide the tools necessary for advances on this important topic. As there are still only a few publications on membrane protein targeting in yeast and mammalian cells and these systems do not at present provide a consistent model for the mechanism, information provided from plant systems may greatly advance our understanding of this pathway in all eukaryotic cells.

We are approaching a significant point in studies of vacuolar protein targeting as one potential targeting receptor has been isolated which recognises some of the targeting motifs (Kirsch *et al.*, 1994, 1996). Future work will be needed to determine if this receptor or another recognises those vacuolar proteins not yet found to bind to the isolated receptor *in vitro*. A report has been published describing a protein whose localisation and structure are similar to other targeting receptors (Ahmed *et al.*, 1997), but more work is required before the definitive role of this protein in targeting can be established. Several lines of evidence have accumulated, suggesting two different targeting mechanisms for soluble proteins (see above). Continued work will be needed to determine if that is the general observation in plant cells or if this is specific to certain cell types or developmental stages.

Tissue-specific regulation of a number of genes involved in the sorting or trafficking of proteins throughout the secretory system suggests that there may be differences in the secretory competence of different cell types. The effect of various conditions known to alter the secretory system, such as pathogen attack (causing secretion of a number of pathogenesis-related proteins), heat shock (denaturing proteins) and cold-shock (slowing down membrane trafficking), will be very interesting to analyse for all of the components known. An overall compilation of these results will allow us to manipulate the system better, for our benefit.

The next major questions pertain to the role of each of the identified components in the secretory machinery of specific cells and in plant physiology. This work demonstrates the power of the genetic system of *Arabidopsis* and is fast becoming a common way to analyse a gene. Altering the expression of specific genes has been in many cases via antisense approaches. However, with the numerous gene families of some of the components described here, this approach is limited as the expression of many members may be altered by the transformation of one antisense clone. The technique of gene replacement useful in bacteria, yeast and mammalian cells needs to be improved in plants, requiring researchers to use other strategies to knockout genes. The approach (McKinney *et al.*, 1995; Krysan *et al.*, 1996) using the T-DNA transformed plant collections and a PCR strategy to identify plants with T-DNA lodged within a gene, will provide an important means of generating plant mutants in specific genes. This will allow the effect of the loss of one gene within a family and the effect on the expression or activity of other genes to be analysed. Another approach using transposon tagging may also prove to be useful, but this requires the gene of interest to be mapped to a particular chromosomal region in order to choose an appropriate line for the analysis (Long *et al.*, 1997). No doubt future techniques will improve our ability to obtain plants lacking specific genes within this pathway to probe function within the cell and in specific secretion competent organs.

The availability of yeast mutants in various steps within the secretory pathway has advanced our understanding of this system substantially. In many cases these mutants can be complemented with an *Arabidopsis* library to identify a plant cDNA with a role potentially similar to the yeast product. This has been used in a number of the cases mentioned above. There also exist several mammalian cell mutants with alterations in the secretory pathway that could be used as recipients in a similar manner. However, the next advance in our understanding will come as means of isolating plant-specific mutants in the secretory pathway are found. These approaches include trying to isolate mutants secreting a particular retained component or retaining of a normally secreted one. A number of different elegant strategies are being developed and the next few years will undoubtedly uncover promising mutants from several of these approaches.

The other major advance will come as we identify the specific interacting components of the trafficking system and begin to piece together this complicated puzzle. Identification of interacting components is one of the uses proposed for the yeast two-hybrid screen, which requires two components to interact for expression of a selectable or screenable marker (Fields and Sternglanz, 1994; Chien-Ting *et al.*, 1996). Other types of screens using suppression of mutant phenotypes in yeast, mammalian or plant cells will also identify interacting components of the system. Recipients for these screens may come from plant-targeting mutants that are identified in the above screens or from plants transformed with dominant negative forms of proteins. Modifications of these approaches or new strategies will undoubtedly appear to improve the efficiency of detecting interacting components.

A compilation of these approaches will allow us to put together the parts of a very complicated system in ways that yeast and mammalian cell experiments have yet to do. The plant work will provide more tissue- and cell-type-specific information as well as a means for understanding environmental effects on the system. Our better understanding of the secretory system in plants will allow us to manipulate this system in the future to express compounds of agricultural or medicinal use and therefore expand the use of plants in applied areas.

References

Ahmed, S.U., Bar-Peled, M. and Raikhel, N.V. (1997) Cloning and subcellular location of an *Arabidopsis* receptor-like protein that shares common features with protein-sorting receptors of eukaryotic cells. *Plant Physiol.*, **114** 325-36.

Anai, T., Hasegawa, K., Watanabe, Y., Uchimiya, H., Ishizaki, R. and Matsui, M. (1991) Isolation and analysis of cDNAs encoding small GTP-binding proteins of *Arabidopsis thaliana*. *Gene*, **108** 259-64.

Anai, T., Matsui, M., Nomura, N., Ishizaki, R. and Uchimiya, H. (1994) *In vitro* mutation of *Arabidopsis thaliana* small GTP-binding proteins and detection of GAP-like activities in plant cells. *FEBS Lett.*, **346** 175-80.

Anai, T., Aspuria, E.T., Fujii, N., Ueda, T., Matsui, M., Hasegawa, K. and Uchimiya, H. (1995) Immunological analysis of a small GTP-binding protein in higher plant cells. *J. Plant Physiol*, **147** 48-52.

Armstrong, F., Leung, J., Grabov, A., Brearley, J., Giraudat, J. and Blatt, M.R. (1995) Sensitivity to abscisic acid of guard-cell K⁺ channels is suppressed by abi1-1, a mutant *Arabidopsis* gene encoding a putative protein phosphatase. *Proc. Natl Acad. Sci. USA*, **92** 9520-24.

Balch, W.E. (1990) Small GTP-binding proteins in vesicular transport. *Trends Biol. Sci.*, **15** 473-77

Bar-Peled, M. and Raikhel, N.V. (1997) Characterization of AtSEC12 and AtSAR1. *Plant Physiol.*, **114** 315-24.

Bar-Peled, M., da Silva Conceição, A., Frigerio, L. and Raikhel, N.V. (1995) Expression and regulation of a ERD2, a gene encoding the KDEL receptor homolog in plants, and other genes encoding proteins involved in ER-Golgi vesicular trafficking. *Plant Cell*, **7** 667-76.

Bar-Peled, M., Bassham, D.C. and Raikhel, N.V. (1996) Transport of proteins in eucaryotic cells: more questions ahead. *Plant Mol. Biol.*, **32** 223-49.

Bassham, D.C. and Raikhel, N.V. (1996) Transport proteins in the plasma membrane and the secretory system. *Trends Plant Sci.*, **1** 15-20.

Bassham, D.C., Gal, S., da Silva Conceição, A. and Raikhel, N.V. (1995) An *Arabidopsis* syntaxin homologue isolated by functional complementation of a yeast *pep*12 mutant. *Proc. Natl Acad. Sci. USA*, **92** 7262-66.

Bednarek, S.Y., Wilkins, T.A., Dombrowski, J.E. and Raikhel, N.V. (1990) A carboxy-terminal propeptide is necessary for proper sorting of barley lectin to vacuoles of tobacco. *Plant Cell*, **2** 1145-55.

Bednarek, S.Y., Reynolds, T.L., Schroeder, M., Grabowski, R., Hengst, l., Gallwitz, D. and Raikhel, N.V. (1994) A small GTP-binding protein from *Arabidopsis thaliana* functionally complements the yeast YPT6 null mutant. *Plant Physiol.*, **104** 591-96.

Boorer, K.J., Frommer, W.B., Bush, D.R., Kreman, M., Loo, D.D.F. and Wright, E.M. (1996) Kinetics and specificity of a H⁺/amino acid transporter from *Arabidopsis thaliana*. *J. Biol. Chem.*, **271** 2213-2220.

Busch, M., Mayer, U. and Jürgens, G. (1996) Molecular analysis of the *Arabidopsis* pattern formation gene GNOM: gene structure and intragenic complementation. *Mol. Gen. Genet.*, **250** 681-91.

Campos, N. and Boronat, A. (1995) Targeting and topology in the membrane of plant 3-hydroxy-3-methylglutaryl coenzyme A reductase. *Plant Cell*, **7** 2163-74.

Cao, Y., Ward, J.M., Kelly, W.B., Ichida, A.M., Gaber, R.F., Anderson, J.A., Uozumi, N., Schroeder, J.I. and Crawford, N.M. (1995) Multiple genes, tissue specificity, and expression-dependent modulation contribute to the functional diversity of potassium channels in *Arabidopsis thaliana*. *Plant Physiol.*, **109** 1093-1106.

Cerana, R., Giromini, L. and Colombo, R. (1995) Malate-regulated channels permeable to anions in vacuoles of *Arabidopsis thaliana*. *Aust. J. Plant Physiol.*, **22** 115-21.

Chien-Ting, C., Bartel, P.L., Sternglanz, R. and Fields, S. (1996) The two-hybrid system: a method to identify and clone genes for proteins that interact with a protein of interest. *Proc. Natl Acad. Sci. USA*, **88** 9578-82.

Cho, M.H. and Spalding, E.P., (1996) An anion channel in *Arabidopsis* hypocotyls activated by blue light. *Proc. Natl Acad. Sci. USA*, **93** 8134-38.

Chrispeels, M.J. (1991) Sorting of proteins in the secretory system. *Ann. Rev. Plant Physiol. Plant Mol. Biol.*, **42** 21-53.

Chrispeels, M.J. and Maurel, C. (1994) Aquaporins: the molecular basis of facilitated water movement through living plant cells? *Plant Physiol.*, **105** 9-13.

Chu, B., Lindstrom, J.T. and Belanger, F.C. (1994) *Arabidopsis thaliana* expresses three divergent Srp54 genes. *Plant Physiol.*, **106** 1157-62.

Cluett, E.B. and Brown W.J. (1992) Adhesions of Golgi cisternae by proteinaceous interactions: intercisternal bridges as putative adhesive structures. *J. Cell Sci.*, **103** 773-84.

Colombo, R., Cerana, R. and Giromini, L. (1994) Tonoplast K⁺ channels in *Arabidopsis thaliana* are activated by cytoplasmic ATP. *Biochem. Biophys. Res. Com.*, **200** 1150-54.

Cosson, P. and Letourneur, F. (1994) Coatomer interaction with di-lysine endoplasmic reticulum retention motifs. *Science*, **263** 1629-31.

Daniels, M.J., Mirkov, T.E. and Chrispeels, M.J. (1994) The plasma membrane of *Arabidopsis thaliana* contains a mercury-insensitive aquaporin that is a homolog of the tonoplast water channel protein TIP. *Plant Physiol.*, **106** 1325-33.

da Silva Conceição, A., Marty-Mazars, D., Bassham, D.C., Sanderfoot, A.A., Marty, F. and Raikhel, N.V. (1997) The syntaxin homolog AtPEP12p resides on a late post-Golgi compartment in plants. *Plant Cell*, **9** 571-82.

De Clercq, A., Vandewiele, M., De Rycke, R., Van Damme, J., Van Montagu, M., Krebbers, E. and Vandekerckhove, J. (1990) Expression and processing of an *Arabidopsis* 2S albumin in transgenic tobacco. *Plant Physiol.*, **92** 899-907.

d'Enfert, C., Gensse, M. and Gaillardin, C. (1992) Fission yeast and a plant have functional homologues of the Sar1 and Sec12 proteins involved in ER to Golgi traffic in budding yeast. *EMBO J.*, **11** 4205-11.

De Michelis, M.I, Rasi-Caldogno, F., Pugliarello, M.C. and Olivari, C. (1996) Fusicoccin binding to its plasma membrane receptor and the activation of the plasma mambrane H$^+$-ATPase. *Plant Physiol.*, **110** 957-64.

DeWitt, N.D. and Sussman, M.R. (1995) Immunocytological localization of an epitope-tagged plasma membrane proton pump (H$^+$-ATPase) in phloem companion cells. *Plant Cell*, **7** 2053-67.

D'Hondt, K., Van Damme, J., Van Den Bossche, C., Leejeerajumnean, S., De Rycke, R., Derksen, J., Vandekerckhove, J. and Krebbers, E. (1993) Studies of the role of the propeptides of the *Arabidopsis thaliana* 2S albumin. *Plant Physiol.*, **102** 425-33.

Dhugga, K.S. and Ray, P.M. (1994) Purification of the reversibly glycosylated polypeptides from pea: purified polypeptides exhibit the same properties as the Golgi-bound form. *Plant Physiol. Suppl.*, **105** 126.

Dombrowski, J.E. and Raikhel, N.V. (1995) Isolation of a cDNA encoding a novel GTP-binding protein of *Arabidopsis thaliana*. *Plant Mol. Biol.*, **28** 1121-26.

Dong, X., Mindrinos, M., Davis, K.R. and Ausubel, F.M. (1991) Induction of *Arabidopsis* defense genes by virulent and avirulent *Pseudomonas syringae* strains and by a cloned avirulence gene. *Plant Cell*, **3** 61-72.

Driouich, A., Faye L. and Staehelin, L.A. (1993) The plant Golgi apparatus: a factory for complex polysaccharides and glycoproteins. *Trends Biochem. Sci.*, **18** 210-14.

Faïk, A., Chileshe, C., Sterling, J. and Maclachlan, G. (1997) Xyloglucan galactosyl- and fucosyltransferase activities from pea epicotyl microsomes. *Plant Physiol.* **114** 245-54.

Faye, L., Johnson, K.D., Sturm, A. and Chrispeels, M.J. (1989) Structure, biosynthesis, and function of asparagine-linked glycans on plant glycoproteins. *Physiol. Plant.*, **75** 309-14.

Ferrol, N. and Bennett, A.B. (1996) A single gene may encode differentially localized Ca^{2+}-ATPases in tomato. *Plant Cell*, **8** 1159-69.

Fields, S. and Sternglanz, R. (1994) The two-hybrid system: an assay for protein–protein interactions. *Trends Genet.*, **10** 286-91.

Fischer, W.-N., Kwart, M., Hummel, S. and Frommer, W.B. (1995) Substrate specificity and expression profile of amino acid transporters (AAPs) in *Arabidopsis*. *J. Biol. Chem.*, **270** 16315-20.

Fitchette-Lainé, A.-C., Gomord, V., Chekkafi, A. and Faye, L. (1994) Distribution of xylosylation and fucosylation in the plant Golgi apparatus. *Plant J.*, **5** 673-82.

Frommer, W.B., Kwart, M., Hirner, B., Fischer, W.N., Hummel, S. and Ninnenmann, O. (1994) Transporters for nitrogeneous compounds in plants. *Plant Mol. Biol.*, **26** 1651-70.

Frommer, W.B., Hummel, S., Unseld, M., and Ninnenman, O. (1995) Seed and vascular expression of a high-affinity transporter for cationic amino acids in *Arabidopsis*. *Proc. Natl Acad. Sci. USA*, **92** 12036-40.

Gal, S. and Raikhel, N.V. (1993) Protein sorting in the endomembrane system of plant cells. *Curr. Opin. Cell Biol.*, **5** 636-40.

Gal, S. and Raikhel, N.V. (1994) A carboxy-terminal plant vacuolar targeting signal is not recognized by yeast. *Plant J.*, **6** 235-40.

Galbraith, D.W., Zeither, C.A., Harkins, K.R., and Afonso, C.L. (1992) Biosynthesis, processing and targeting of the G-protein of vesicular stomatitis virus in tobacco protoplasts. *Planta*, **186** 324-36.

Gomez, L. and Chrispeels, M.J. (1993) Tonoplast and soluble vacuolar proteins are targeted by different mechanisms. *Plant Cell*, **5** 1113-24.

Gomez, L. and Chrispeels, M.J. (1994) Complementation of an *Arabidopsis thaliana* mutant that lacks complex asparagine-linked glycans with the human cDNA encoding N-acetylglycosaminyl-transferase 1. *Proc. Natl Acad. Sci. USA*, **91** 1829-33.

Guo, Q., Vasile, E. and Krieger, M. (1994) Disruptions in Golgi structure and membrane traffic in a conditional lethal mammalian cell mutant are corrected by e-COP. *J. Cell Biol.*, **125** 1213-24.

Harper, J.F., Manney, L. and Susman, M.R. (1994) The plasma membrane H⁺-ATPase gene family in *Arabidopsis:* genomic sequence of AHA10 which is expressed primarily in developing seeds. *Mol. Gen. Genet.*, **244** 572-87.

Hartmann, E., and Prehn, S., (1994) The N-terminal of the α-subunit of the TRAP complex has a conserved cluster of negative charges. *FEBS Lett.*, **349** 324-26.

He, Z.-H., Fujiki, M. and Kohorn, B.D. (1996) A cell wall-associated, receptor-like protein kinase. *J. Biol. Chem.*, **271** 19789-93.

Hechenberger, M., Schwappach, B., Fischer, W.N., Frommer, W.B., Jentsch, T.J. and Steinmeyer, K. (1996) A family of putative chloride channes from *Arabidopsis* and functional complementation of a yeast strain with a CLC gene disruption. *J. Biol. Chem.*, **271** 33632-38.

Helm, K.W., Schmeits, J. and Vierling, E. (1995) An endomembrane-localized small heat-shock protein from *Arabidopsis thaliana*. *Plant Physiol.*, **107** 287-88.

Hervé, C., Dabos, P., Galaud, J.-P., Rouge, P.K, and Lescure, B. (1996) Characterization of an *Arabidopsis thaliana* gene that defines a new class of putative plant receptor kinases with an extracellular lectin-like domain. *J. Mol. Biol.*, **58** 778-88.

Hiatt, A., Cafferkey, R. and Bowdish, K. (1989) Production of antibodies in transgenic plants. *Nature*, **342** 76-78.

Hirschi, K.D., Zhen, R., Cunningham, K.W., Rea, P.A. and Fink, G.R. (1996) CAX1, an H⁺/Ca²⁺ antiporter from *Arabidopsis*. *Proc. Natl Acad. Sci. USA*, **93** 8782-86.

Höfte, H. and Chrispeels, M. (1992) Protein sorting to the vacuolar membrane. *Plant Cell*, **4** 995-1004.

Höfte, H., Hubbard, L., Reizer, J., Ludevid, D., Herman, E.M. and Chrispeels, M.J. (1992) Vegetative and seed-specific forms of tonoplast intrinsic protein in the vacuolar membrane of *Arabidopsis thaliana*. *Plant Physiol.*, **99** 561-70.

Hoshi, T. (1995) Regulation of voltage dependence of the KAT1 channel by intracellular factors. *J. Gen. Physiol.*, **105** 309-28.

Houlné, G. and Boutry, M. (1994) Identification of an *Arabidopsis thaliana* gene encoding a plasma membrane H⁺-ATPase whose expression is restricted to anther tissues. *Plant J.*, **5** 311-17.

Iturriaga, G., Jefferson, R.A. and Bevan, M.W. (1989) Endoplasmic reticulum targeting and glycosylation of hybrid proteins in transgenic tobacco. *Plant Cell*, **1** 381-90.

Jackson, M.R., Nilsson, T. and Peterson, P.A. (1990) Identification of a consensus motif for retention of transmembrane proteins in the endoplasmic reticulum. *EMBO J.*, **9** 3153-62.

Kaldenhoff, R., Kolling, A., Meyers, J., Karmann, U., Ruppel, G. and Richter, G. (1995) The blue light-responsive AthH2 gene of *Arabidopsis thaliana* is primarily expressed in expanding as well as in differentiating cells and encodes a putative channel protein of the plasmalemma. *Plant J.*, **7** 87-95.

Katembe, W.J., Swatzell, L.J., Makaroff, C.A. and Kiss, J.Z. (1997) Immunolocalization of integrin-like proteins in *Arabidopsis* and *Chara. Physiol. Plant.*, **99** 7-14.

Ketchum, K.A. and Slayman, C.W. (1996) Isolation of an ion channel gene from *Arabidopsis thaliana* using the H5 signature sequence from voltage-dependent K⁺ channels. *FEBS Lett.*, **378** 19-26.

Kim, E.J., Zhen, R.-G. and Rea, P.A. (1994) Heterologous expression of plant vacuolar pyro-phosphase in yeast demonstrates sufficiency of the substrate-binding subunit for proton transport. *Proc. Natl Acad. Sci. USA*, **91** 6128-32.

Kinoshita, T., Nishimura, M. and Hara-Nishimura, I. (1995a) Homologues of a vacuolar processing enzyme that are expressed in different organs in *Arabidopsis thaliana. Plant Mol. Biol.*, **29** 81-89.

Kinoshita, T., Nishimura, M. and Hara-Nishimura, I. (1995b) The sequence and expression of the g-VPE gene, one member of a family of three genes for vacuolar processing enzymes in *Arabidopsis thaliana. Plant Cell Physiol.*, **36** 1555-62.

Kirsch, T., Paris, N., Butler, J.M., Beevers, L. and Rogers, J.C. (1994) Purification and initial characterization of a potential plant vacuolar targeting receptor. *Proc. Natl Acad. Sci. USA*, **91** 3403-3407.

Kirsch, T., Saalbach, G., Raikhel, N.V. and Beevers, L. (1996) Interaction of a potential vacuolar targeting receptor with amino-and carboxyl-terminal targeting determinants. *Plant Physiol.*, **11** 469-74.

Krysan, P.J., Young, J.C., Tax, F. and Sussman, M.R. (1996) Identification of transferred DNA insertions within *Arabidopsis* genes involved in signal transduction and ion transport. *Proc. Natl Acad. Sci. USA*, **93** 8145-50.

Lagarde, D., Basset, M., Lepetit, M., Conejero, G., Gaymard, F., Astruc, S. and Grignon, C. (1996) Tissue-specific expression of *Arabidopsis* AKT1 gene is consistent with a role in K$^+$ nutrition. *Plant J.*, **9** 195-203.

Lee, H., Gal, S., Newman, T.C., and Raikhel, N.V. (1993) The *Arabidopsis* endoplasmic reticulum retention receptor functions in yeast. *Proc. Natl Acad. Sci. USA*, **90** 11433-37.

Lewis, B.D., Karlin-Neumann, G., Davis, R.W. and Spalding, E.P. (1997) Ca^{2+}-activated anion channels and membrane depolarizations induced by blue light and cold in *Arabidopsis* seedlings. *Plant Physiol.*, **114** 1327-34.

Liang, F., Cunningham, K.W., Harper, J.F., and Sze, H. (1997) ECA1 complements yeast mutants defective in Ca^{2+} pumps and encodes an ER-type Ca^{2+}-ATP-ase in *Arabidopsis thaliana. Proc. Natl Acad. Sci. USA*, **94** 8579-84.

Lindstrom, J., Chu, B. and Belanger, F.C. (1993) Isolation and characterization of an *Arabidopsis thaliana* gene for the 54 kDa subunit of the signal recognition particle. *Plant Mol. Biol.*, **23** 1265-72.

Long, D., Goodrich, J., Wilson, K., Sundberg, E., Martin, M., Puangsomlee, P. and Coupland, G. (1997) Ds elements on all five *Arabidopsis* chromosomes and assessment of their utility for transposon tagging. *Plant J.*, **11** 145-48.

Lukowitz, W., Mayer, U. and Jürgens, G. (1996) Cytokinesis in the *Arabidopsis* embryo involves the syntaxin-related KNOLLE gene product. *Cell*, **84** 61-71.

Lüttge, U. and Ratajczak, R. (1997) The physiology, biochemistry and molecular biology of the plant vacuolar ATPase. *Adv. Bot. Res.*, **25** 253-96.

Ma, J., Hiatt, A., Hein, M., Vine, N.D., Wang, F., Stabila, P., Van Dolleweerd, C., Mostov, K. and Lehner, T. (1995) Generation and assembly of secretory antibodies in plants. *Science*, **268** 716-19.

Maathius, F.J.M. and Sanders, D. (1996) Characterization of csi52, a Cs$^+$ resistant mutant of *Arabidopsis thaliana* altered in K$^+$ transport. *Plant J.*, **10** 579-89.

Maclachlan, G., Levy, B. and Farkas, V. (1992) Acceptor requirements for GDP-fucose: xyloglucan 1,2-α-L-fucosyltransferase activity solubilized from pea epicotyl membranes. *Arch. Biochem. Biophys.*, **294** 200-205.

Manolson, M.F., Quellette, B.F.F., Fillion, M. and Poole, R.J. (1988) cDNA sequence and homologies of the '57 kDa' nucleotide-binding subunit of the vacuolar ATPase from *Arabidopsis*. *J. Biol. Chem.*, **263** 17987-94.

Marré, M. T., Venegoni, A., Talarico, A., Soave, C. and Marré, E. (1995) Evidence that the partial resistance of the *Arabidopsis* 5-2 mutant to fusicoccin depends on a decreased capacity of the H$^+$ pump to respond to activating factors. *Plant Cell Environ.*, **18** 651-59.

Matsui, M., Sasamoto, S., Kunieda, T., Nomura, N. and Ishizaki, R. (1989) Cloning of *ara*, a putative *Arabidopsis thaliana* gene homologue to the ras-related gene family. *Gene*, **76** 313-19.

Matsuoka, K., Bassham, D.C., Raikhel, N.V. and Nakamura, K. (1995) Different sensitivity to wortmannin of two vacuolar sorting signals indicates the presence of distinct sorting machineries in tobacco cells. *J. Cell Biol.*, **130** 1307-18.

Matsuoka, K., Higuchi, T., Maeshima, M. and Nakamura, K. (1997) A vacuolar-type H+ATPase in a nonvacuolar organelle is required for the sorting of soluble vacuolar protein precursors in tobacco cells. *Plant Cell*, **9** 533-46.

Maurel, C., Kado, R.T., Guern, J. and Chrispeels, M.J. (1995) Phosphorylation regulates the water channel activity of the seed-specific aquaporin α-TIP. *EMBO J.*, **14** 3028-35.

McKinney, E.C., Ali, N., Traut, A., Feldmann, K.A., Belostotsky, D.A., McDowell, J.M. and Meagher, R.B. (1995) Sequence-based identification of T-DNA insertion mutations in *Arabidopsis*: actin mutants *act2-1* and *act4-1*. *Plant J.*, **8** 613-22.

Melchers, L.S., Sela-Buurlage, M.B., Vloemans, S.A., Woloshuk, C.P., Van Roekel, J.S.C., Pen, J., van den Elzen, P.J.M. and Cornelissen, B.J.C. (1993) Extracellular targeting of the vacuolar to-bacco proteins AP24, chitinase and β-glucanase in transgenic plants. *Plant Mol. Biol.*, **21** 583-93.

Moore, P.J., Swords, K.M.M., Lynch, M.A. and Staehelin, L.A. (1991) Spatial organization of the assembly pathways of glycoproteins and complex polysaccharides in the Golgi apparatus of plants. *J. Cell Biol.*, **112** 589-602.

Nelson, D.E., Glaunsinger, B. and Bohnert, H.J. (1997) Abundant accumulation of the calcium-binding molecular chaperone calreticulin in specific floral tissues of *Arabidopsis thaliana*. *Plant Physiol.*, **114** 29-37.

Orellana, A., Neckelmann, G. and Norambuena, L. (1997) Topography and function of Golgi uridine-5'-diphosphatase from pea stems. *Plant Physiol.*, **114** 99-107.

Palmgren, M.G. and Christenen, G. (1994) Functional comparisons between plant plasma membrane H+-ATPase isoforms expressed in yeast. *J. Biol. Chem.*, **269** 3027-33.

Pâquet, M.R., Pfeffer, S.R., Burczak, J.D., Click, B.S. and Rothman, J.E. (1986) Components responsible for transport between successive Golgi cisternae are highly conserved in evolution. *J. Biol. Chem.*, **261** 4367-70.

Paris, N., Stanley, C.M., Jones, R.L. and Rogers, J.C. (1996) plant cells contain two functionally distinct vacuolar compartments. *Cell*, **85** 563-72.

Perera, I.Y., Li, X. and Sze, H. (1995) Several distinct genes encode nearly identical 16kDa proteolipids of the vacuolar H+-ATPase from *Arabidopsis thaliana*. *Plant Mol. Biol.*, **29** 227-44.

Peters, C., Braun, M., Weber, B., Wendland, M., Schmidt, B., Pohlmann, R., Waheed, A. and von Figura, K. (1990) Targeting of a lysosomal membrane protein: a tyrosine-containing endocytosis signal in the cytoplasmic tail of lysosomal acid phosphatase is necessary and sufficient for targeting to lysosomes. *EMBO J.*, **9** 3497-3506.

Rapoport, T.A. (1992) Transport of proteins across the endoplasmic reticulum membrane. *Science*, **258** 931-36.

Rausch, T., Kirsch, M., Löw, R., Lehr, A., Viereck, R. and Zhigang, A. (1996) Salt stress responses of higher plants: the role of proton pumps and Na+/H+- antiporters. *J. Plant Physiol.*, **148** 425-33.

Regad, F., Bardet, C., Tremousaygue, D., Moisan, A., Lescure, B. and Axelos, M. (1993) cDNA cloning and expression of an *Arabidopsis* GTP-binding protein of the ARF family. *FEBS Lett.*, **316** 133-36.

Reiter, W., Chapple, C.C.S. and Somerville, C.R. (1993) Altered growth and cell walls in a fucose-deficient mutant of *Arabidopsis*. *Science*, **261** 1032-35.

Rentsch, D., Laoi, M., Rouhana, I., Schmelzer, E., Delrot, S., and Frommer, W.B. (1995) NTR1 encodes a high affinity oligopeptide transporter in *Arabidopsis*. *FEBS Lett.*, **370** 264-68.

Robinson, D.G., Sieber, H., Kammerloher, W., and Schäffner, A.R. (1996) PIP1 aquaporins are concentrated in plasmalemmasomes of *Arabidopsis thaliana* mesophyll. *Plant Physiol.*, **111** 645-49.

Rothman, J.E. (1994) Mechanisms of intracellular protein transport. *Nature*, **372** 55-63.

Saalbach, G., Rosso, M. and Schumann, U. (1996). The vacuolar targeting signal of the 2S albumin from Brazil nut resides at the C terminus and involves the C-terminal propeptide as an essential element. *Plant Physiol.*, **112** 975-85.

Samac, D.A., Hironaka, C.M., Vallaly, P.E. and Shah, D.M. (1990) Isolation and characterization of the genes encoding basic and acidic chitinase in *Arabidopsis thaliana. Plant Physiol.*, **93** 907-14.

Sarafian, V., Kim, Y., Poole, R.J. and Rea, P.A. (1992) Molecular cloning and sequence of cDNA encoding the pyrophosphate-energized vacuolar membrane proton pump of *Arabidopsis thaliana. Proc. Natl Acad. Sci. USA*, **89** 1775-79.

Schaller, G.E. and Bleecker, A.B. (1993) Receptor-like kinase activity in membranes of *Arabidopsis thaliana. FEBS Lett.*, **333** 306-10.

Schroeder., M.R., Borkhesenious, O.N., Matsuoka, K., Nakamura, K. and Raikhel, N.V. (1993) Colonization of barley lectin and sporamin in vacuoles of transgenic tobacco plants. *Plant Physiol.*, **101** 451-58.

Shevell, D.E., Leu, W.-M., Gillmore, C.S., Xia, G., Feldmann, K.A. and Chua, N.-H. (1994) EMB30 is essential for normal cell division, cell expansion and cell adhesion in *Arabidopsis* and encodes a protein that has similarity to Sec7. *Cell*, **77** 1051-62.

Shimomura, S., Liu, W., Inohara, N., Watanabe, S. and Futai, M. (1993) Structure of the gene for an auxin-binding protein and a gene for 7SL RNA from *Arabidopsis thaliana. Plant Cell Physiol.*, **34** 633-37.

Sijmons, P.C., Dekker, B.M.M., Schrammeijer, B., Verwoerd, T.C., van den Elzen, P.J.M. and Hoekema, A. (1990) Production of correctly processed human serum albumin in transgenic plants. *Biotechnology*, **8** 217-21.

Steiner, H.-Y., Song, W., Zhang, L., Nalder, F., Becker, J. and Stacey, G. (1994) An *Arabidopsis* peptide transporter is a member of a new class of membrane transport proteins. *Plant Cell*, **6** 1289-99.

Sussman, M.R. (1994) Molecular analysis of proteins in the plant plasma membrane. *Annu. Rev. Plant Physiol. Plant Mol. Biol.*, **45** 211-34.

Touraine, B. and Glass, A.D.M. (1997) NO_3^- and ClO_3^- fluxes in the *chl1-5* mutant of *Arabidopsis thaliana. Plant Physiol.*, **114** 137-44.

Tsay, Y.-F., Schroeder, J.I., Feldmann, K.A. and Crawford, N.M. (1993) The herbicide sensitivity gene CHL1 of *Arabidopsis* encodes a nitrate-inducible nitrate transporter. *Cell*, **72** 705-13.

Ueda, T., Anai, T., Tsukaya, H., Hirata, A. and Uchimiya, H. (1996a) Characterization and subcellular localization of a small GTP-binding protein (Ara-4) from *Arabidopsis*: conditional expression under control of the promoter of the gene for heat-shock protein Hsp81-1. *Mol. Gen. Genet.*, **250** 533-39.

Ueda, T., Matsuda, N., Anai, T., Tsukaya, H., Uchimiya, H. and Nakano, A. (1996b) An *Arabidopsis* gene isolated by a novel method for detecting genetic interaction in yeast encodes the GDP dissociation inhibitor of Ara4 GTPase. *Plant Cell*, **8** 2079-91.

Uozumi, N., Gassmann, W., Cao, Y. and Schroeder, J.I. (1995) Identification of strong modifications in cation selectivity in an *Arabidopsis* inward rectifying potassium channel by mutant selection in yeast. *J. Biol. Chem.*, **270** 24276-81.

von Heijne, G. (1985) Signal sequences: the limits of variation. *J. Mol. Biol.*, **184** 99-105.

von Schaewen, A. and Chrispeels, M.J. (1993) Identification of vacuolar sorting information in phytohemagglutinin, an unprocessed vacuolar protein. *J. Exper. Bot.*, **44** 339-42.

von Schaewen, A., Sturm, A., O'Neill, J. and Chrispeels, M.J. (1993) Isolation of a mutant *Arabidopsis* plant that lacks N-acetyl glucosaminyl transferase 1 and is unable to synthesize Golgi-modified complex N-linked glycans. *Plant Physiol.*, **102** 1109-18.

Wink, M. (1993) The plant vacuole: a multifunctional compartment. *J. Exp. Biol.*, **44** 231-46.

Zhang, G.F. and Staehelin, L.A. (1992) Functional compartmentation of the Golgi apparatus of plant cells. *Plant Physiol.*, **99** 1070-83.

6 Sexual reproduction: from sexual differentiation to fertilization

Hen-Ming Wu and Alice Y. Cheung

6.1 Introduction

Plants enter the reproductive phase in response to either internal (e.g. hormones) or external (e.g. day length and temperature) stimuli. This change is characterized by the conversion of the vegetative shoot apical meristem into an inflorescence meristem from which the floral meristems emerge. The key questions in plant reproduction include how sexual organs are determined, how the male and female gametophytes (pollen and embryo sacs, respectively) interact to effect gamete fusion and how fertilized ovules develop into mature seeds.

Plants have evolved diverse strategies for sexual reproduction. Most angiosperms are hermaphrodites, producing bisexual flowers that are either self-fertile or self-incompatible. A low percentage of angiosperms produce unisexual flowers, bearing flowers of a single sex on one plant (dioecious) or producing flowers of either the male or female sex on the same plant (monoecious). A number of different mechanisms are known to underlie sex determination in unisexual plants to allow the expression of one of the two sexual organ developmental programs in an individual flower. In bisexual flowers, several key regulatory genes have been identified to control the production of male and female reproductive organs (see Table 6.1).

Bisexual flowers (also referred to as perfect flowers) have four major structural parts: two vegetative organs (sepal and petal) and the male (stamen) and female (carpel or pistil) reproductive organs. They are arranged in concentric whorls from the outside to the center of the flower and are often referred to as whorls 1 to 4, respectively (Figure 6.1, and see chapter 9). Primordia for these floral organs emerge sequentially from the flanks of the floral meristems, with the carpel emerging last and occupying the central dome of the meristem. Successful fertilization relies on a series of intimate interactions between pollen (male gametophytes, which are produced in the anthers of the stamen) and a number of cell types in various pistil parts (the stigma, style and ovary) and the embryo sacs (female gametophytes, which develop inside the ovary). Mechanisms to allow compatible and to preclude incompatible male–female interactions operate in concert to accomplish productive fertilization. Success in producing the second generation relies on continued development of the fertilized ovule and involves the participation of both maternal and fertilized tissues to produce viable seeds.

Table 6.1 A list of the genes related to sex differentiation in the bisexual and unisexual flowers discussed in the text

Bisexual					Unisexual					
Arabidopsis	Function[a]	*Silene*	Expression in young flowers[a]		*Rumex acetosa*	Expression in young flowers[a] (later development)		*Zea mays*	Expression in young flowers[a]	
			Male	Female		Male	Female		Male	Female
AP3	2, 3	SLM3	2, 3, 4	2, 3	RAD1, 2	3 (3)	3 (–)			
PI	2, 3	SLM2	2, 3, 4	2, 3						
AG	3, 4				RAP1	3, 4 (3)	3, 4 (4)	ZAG1	3, 4 <<	3, 4
SUP	3, 4							ZMM2	3, 4 >>	3, 4

[a] Whorls of the flower: 2, sepal; 3, stamen; 4, carpel.

A.

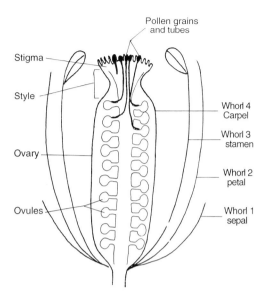

B.

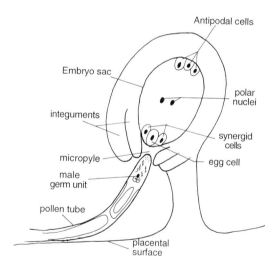

Figure 6.1 Schematic representations of (A) the structure of a flower showing the four whorls and elongating pollen tubes in the ovary; (B) the entrance of a pollen tube into an ovule.

In this chapter, we focus on understanding the mechanisms underlying the various stages of sexual reproduction in self-compatible plants, emphasizing the differentiation of the sexual organs, pollination, fertilization and post-fertilization developments. Examples in *Arabidopsis* will be used as often as practical. However, a large amount of research on sexual reproduction has been conducted in other plant systems and these will be discussed whenever they are more informative.

6.2   Sexual organ development

Whether plants produce bisexual or unisexual flowers, and whether primordia for stamens or carpels are to be organized at the floral meristems of bisexual flowers, decisions have to be made by the plants as to which sexual differentiation pathway is to be expressed by particular cell clusters at the apical domes. During early development, the flowers of most unisexual plants are not distinguishable from hermaphrodite flowers. Primordia for both stamens and carpels are produced at their floral meristems. This suggests that the exclusive development of one sex or another in unisexual flowers is likely to be a consequence of regulatory programs superimposed on the basic mechanisms that underlie sexual organogenesis to suppress further differentiation of the unwanted sex. In between dioecy and monoecy exists a spectrum of intermediate species: variations of dioecy in which plants produce hermaphrodite and female flowers or hermaphrodite and male flowers, and variations of monoecy in which plants produce bisexual and unisexual flowers. The spectrum of plants with properties intermediate of monoecy and dioecy probably represents modulations of the basic regulatory programs for unisexuality by endogenous or exogenous factors. A few unisexual plants appear not to have a hermaphrodite phase, producing primordia only for a unique sexual organ. Although this seems to suggest a sex determination mechanism bypassing a bisexual phase and not requiring the abolition of an unwanted sex, it remains to be ruled out that the hermaphrodite phase in these plants may be too transient and terminates too early in floral development to be readily observable.

Genetics plays predominant roles in sexual organogenesis in bisexual plants and in sex determination in several unisexual plants. These genetic programs inevitably operate within the context of endogenous physiological and biochemical conditions, which are conceivably more labile than genetic programming and are easily influenced by environmental conditions. Manipulations which alter these endogenous conditions, such as exposure to different day lengths or temperatures, can induce sex conversion in some unisexual plants. In bisexual plants, environmental factors can also alter the characteristics of reproductive development. These various aspects of sexual differentiation are discussed here. Several reviews have also provided thorough overviews of the genetic and molecular mechanisms known to operate in the sexual development of unisexual (Irish and Nelson, 1989; Dellaporta and Calderon-Urrea, 1993, 1994; Grant *et al.*, 1994; Lebel-Hardenack and Grant, 1997) and bisexual plants (see chapter 9; Coen and Meyerowitz, 1991; Ma, 1994; Weigel and Meyerowitz, 1994).

6.2.1 *Genetic control of sex determination in unisexual plants*

Both sex-chromosome-based and sex-determining-gene(s)-based mechanisms are known to operate in unisexual plants. In a number of dioecious plants, heteromorphic chromosomes, like the XY chromosomes in animals, operate to control sex determination. Males are usually the heterogametic sex (XY) and females are homogametic (XX). Some plants, e.g. *Silene*, utilize an active Y-chromosome for sex determination (Westergaard, 1958). Presence of a single Y chromosome is determinant for maleness and can override even a triple dose of X chromosomes. *Asparagus* also has an active Y system for male determination, but the extent of stylar degeneracy may be modulated, sometimes yielding perfect flowers on a male plant (Pierce and Currence, 1962; Franken, 1970; Bracale *et al.*, 1991). In other plants, e.g. *Rumex acetosa*, sex is determined by the ratio of X to autosomes. An X:autosome ratio of 1 and higher determines female, a ratio of 0.5 and lower determines male, while plants with ratios between 0.5 and 1 produce hermaphrodite, partially male or partially female flowers (Parker and Clark, 1991).

Efforts to isolate DNA fragments from the Y chromosomes of plants with a chromosome-based sex determination mechanism have been initiated (Ye *et al.*, 1991; Donnison *et al.*, 1996; Lebel-Hardenack and Grant, 1997). Deletion mutations on the Y chromosome of *Silene* are associated with asexuality (lacking sexual development) and with bisexuality in male plants. Therefore male-determination, genes for male organogenesis, and female suppression are all located on the *Silene* Y chromosome. The combined use of these deletion mutations and DNA fragments isolated from the Y chromosome should provide useful molecular landmarks to identify and isolate genes related to sexual differentiation in *Silene*.

The monoecious species maize probably has the best characterized sex determination system based on either a single gene, or several unlinked genes working in combination. Briefly, mutations in several *DWARF* genes confer dwarfness and result in andromonecious individuals which develop normal male flowers in the tassel but hermaphrodite flowers in the ears because stamen development is not aborted in the female florets. A mutation in another gene, *ANTHER EAR 1* (*AN1*), also causes stamens to develop in ears and induces semi-dwarfism. Mutations in another set of genes, *TASSELSEED*, have the opposite effect and allow pistil differentiation to be exhibited to various extents in normally staminate florets. *AN1* and one of the *TASSELSEED* genes, *TS2*, have been cloned (see below) (DeLong *et al.*, 1993; Bensen *et al.*, 1995). Molecular analyses of these and other sex determination genes in maize, together with physiological and biochemical information already existing for these mutants, should allow the understanding of how various regulatory aspects interplay to control sex determination in a major crop plant.

6.2.2   *Physiological and biochemical regulation of sex determination in unisexual plants*

Genetics may play a critical role in governing whether a plant is unisexual or bisexual and which sexual developmental pathway is undertaken by unisexual plants. These genetic parameters ultimately either provide for, or interact with, the biochemical and physiological conditions present at the floral meristem for one or the other sex to be manifested.

In some unisexual plants, e.g. *Silene*, sex determination is unaffected by exogenous factors such as the application of gibberellins (GAs), auxins or cytokinins. On the other hand, the genetic programs for sex determination in a number of plants appear to be subjected to modulation by endogenous or exogenous factors (Chailakhyan, 1979), sometimes producing a continuum of different sexual habits within one plant. For example, in monoecious cucumber lower nodes produce male flowers, intermediate nodes produce both male and female flowers, while the uppermost nodes produce female flowers exclusively (Shifriss, 1961; Durand and Durand, 1984). This mixed sexual pattern in cucumber already suggests that the program for sex determination, which is controlled by three major genetic loci (Galun, 1961; Pierce and Wehner, 1990), is modulated during development. Conditions at the apical meristem must fluctuate as the plant grows, leading to the expression of different sexual characteristics as the plant ages. Mutations also exist to shift towards feminization or to enhance the hermaphrodite habit. Long days, high temperatures and GAs masculinize cucumber plants by delaying the emergence of female flowers, thus limiting them to the most terminal nodes (Yin and Quinn, 1995). Short days, low temperatures and auxins cause feminization by shortening the male and mixed flower phases, so female flowers emerge at earlier nodes. The feminization of cucumber by auxin is likely to be mediated by ethylene, since both ethylene and its precursor aminocyclopropane-1-carboxylic acid (ACC) also induce feminization. An ACC synthase gene is closely linked to the single dominant female locus in cucumber (Trebitsh *et al.*, 1997). Monoecious cucumber has a single copy of this gene, while gynoecious (female only) plants have additional copies.

The interchangeability between sexes in response to endogenous and exogenous conditions is consistent with a floral meristem programmed to adopt a single sex differentiation pathway but still maintain the potential for bisexual development. The observations that growth regulators may cause sex conversion in unisexual plants suggest that sex determining genes may directly or indirectly regulate the endogenous hormonal conditions, such as the levels or the availability of biologically active forms of these hormones or the sensitivity of plants to these hormones.

Although hormones clearly play an important role in the expression of sex, their effects are not identical for every plant that has been examined. While GAs masculinize cucumber and hemp (Atal, 1959), they feminize maize (Phinney, 1956; Rood et al., 1980). In wild-type maize, GA_1 is 100-fold higher in the lower shoots, which produce ears, than in the upper tassel shoots. Tassels are feminized by direct application of GAs, or by exposure to conditions, such as short day length and low light, that induce increased endogenous GA levels (Nickerson, 1957; Heslop-Harrison, 1960; Krishnamoorthy and Talukdar 1976; Hansen et al., 1976). In dwarf mutants, application of GAs reverses dwarfism and prevents stamen development in ears. Chemical analysis showed that dwarf mutants have lower GA_1 levels than the wild-type, and different GA biosynthetic steps are defective in the various dwarf and the an1 mutants (Phinney, 1961, 1984). Characterization of AN1 showed that it encodes a protein in the GA biosynthetic pathway (Bensen et al., 1995). The effect of GAs on tasselseed mutants is not as defined as that on the dwarf mutants. Analysis of TS2 showed that it encodes a protein with homology to steriod-specific dehydrogenases from bacteria (DeLong et al., 1993). TS2 is expressed in the carpel primordia of the male floral meristem prior to the abortion of the female tissues, consistent with a function to ensure the arrest of female development in a hermaphrodite meristem by causing, directly or indirectly, cell death in the carpel primordia. A steroid hormone, brassinosteroid (Clouse, 1996; Yokota, 1997), is important for pollen fertility in a number of plants. Whether the TS2 gene product functions via its probable ability to modify a steroid-like substance to arrest pistil development remains to be determined.

Cytokinins also affect sex expression in a number of plants. A synthetic cytokinin can induce male flowers of indigenous forms of vinifera grapes to become hermaphrodite, producing mature fruits (Negri and Olmo, 1966). In Mercurialis annua, three major genetic loci control the sex of an individual plant (Louis and Durand, 1978; Louis, 1989). However, the manifestation of the two sexes appears to be correlated with extreme levels of endogenous indoleacetic acid (IAA) and cytokinins. Female floral apices have almost exclusively trans-zeatin and low levels of IAA. Male floral apices have no detectable trans-zeatin, but have zeatin nucleotide and high levels of the auxin (Dauphin et al., 1979; Dauphin-Guerin et al., 1980; Champault et al., 1985; Hamdi et al., 1987). Genetic males are feminized by the application of cytokinins or by grafting of genotypes with different levels of endogenous cytokinins (Durand, 1967). Nodes from genetic females may be masculinized in in vitro culture by auxins to develop male flowers (Champault, 1973). Interestingly, callus tissues derived from either male or female apices retained the same hormonal constitutions as those in the original explants, despite a lack of morphological differentiation, suggesting that maintaining endogenous hormone levels could be a function of the sex determining genes either parallel to, or independent of, the morphological developmental pathways that they regulate (Champault *et al.*, 1985). These studies together suggest that while genetic programming may play a fundamental role in sex determination, the expression of these programs is still subject to modulations by endogenous and exogenous factors that affect the physiology and the biochemistry of plant cells.

6.2.3 *Genetic control of sexual organogenesis in bisexual plants*

Floral organogenesis has been most extensively examined at the genetic and molecular levels in plants producing perfect flowers, such as *Arabidopsis*, *Antirrhinum* and *Petunia* (see chapter 9; Coen and Meyerowitz, 1991; Ma, 1994; Weigel and Meyerowitz, 1994). These studies have led to the identification of a collection of genes that regulate the production of different floral parts. Mutant analyses suggest that simple genetic rules underlie floral organogenesis. An ABC model was proposed to predict how three classes of regulatory genes (A, B and C) interact to produce the four whorls of a flower (Figure 6.1). This model predicts that class A genes act in whorls 1 and 2, class B genes act in whorls 2 and 3, and class C genes act in whorls 3 and 4. Sepal and carpel development requires only the actions of class A and C genes, respectively. Petal development requires the actions of class A and B genes, and stamen development requires the actions of B and C genes. Mutations in some of these genes result in conversion of organ identity within the flower, including changes involving the sexual organs. The model also stipulates that class A and C genes have to mutually inhibit each other. The key aspects of the ABC model have proved to be correct, although it is constantly being fine-tuned as more information is becoming available rapidly.

In *Arabidopsis*, the genes that are most relevant for sexual organ development are the class B genes *APETELA 3* (*AP3*) and *PISTILATA* (*PI*) and the class C gene *AGAMOUS* (*AG*) (Yanofsky *et al.*, 1990; Jack *et al.*, 1992; Goto and Meyerowitz, 1994). Analogous genes for these MADS box homeotic transcription factors have been identified and characterized in other plants and the key aspects of these genes are similar to those described for the *Arabidopsis* genes. *AP3* and *PI* are both required for proper petal and stamen development, and are expressed specifically in these two whorls. *ap3* and *pi* mutants have conversion of petals to sepals, and stamens to carpels, essentially producing a female flower. *AG* is required for stamen and carpel organogenesis and *ag* mutants produce asexual flowers with a reiterating (sepal, petal, petal)$_n$ phenotype. This phenotype suggests that in the absence of *AG* functions, petals replace stamens, carpels do not develop and the apical dome is no longer determinant, thus resulting in the continuous emergence of floral organs, giving rise to the reiterating floral structure. Ectopic expression of *AP3* results in carpels being replaced by stamens (Jack *et al.*, 1994). Ectopic expression of *AG* in the sepals and petals in transgenic plants resulted in the appearance of carpelloid sepals and stamenoid petals (Mizukami and Ma, 1992; Kempin *et al.*, 1993). The carpelloid sepals in these transgenic plants also possess the biochemical and functional capacities normally expressed only in the carpels (Cheung *et al.*, 1996). These major regulatory genes apparently are capable of inducing the differentiation of particular floral organs when a certain threshold level of their gene products is present, even ectopically in some non-reproductive tissues. Therefore, to produce a normal flower, these gene systems must be able to be modulated or fine-tuned in order for them to function optimally together with all the other developmental events that take place in flower development. Many additional genes have been identified which interact with each other in floral organogenesis. The interplay of these genes is beyond the scope of this chapter and has been reviewed in many articles (see, for example, Ma, 1994; Weigel and Meyerowitz, 1994).

An *Arabidopsis* gene that plays a key role in sexual organogenesis is *SUPERMAN* (*SUP*), which encodes a putative transcription factor (Schultz *et al.*, 1991; Bowman *et al.*, 1992; Sakai *et al.*, 1995). Not only do *sup* mutants either lack or have defective carpels, but they also produce extra numbers of stamens, which emerge from their normal position in the apical meristem and also from the central domain, which is normally occupied by the carpel primordia (Figure 6.2B). These mutant phenotypes suggest that *SUP* is needed for normal carpel production and to spatially confine stamen development to its proper domain in the floral meristem. The *sup* mutant phenotype is similar to flowers in transgenic plants that ectopically express *AP3*. In wild-type plants, *SUP* mRNA is detected in a narrow zone of cells in the floral meristem bordering the stamen and the carpel primordia. In *sup* mutants, the expression of *AP3,* which is normally restricted to the petal and stamen primordia, extends into the carpel primordium. Although the exact functional mechanism of *SUP* remains to be determined, at least one possible aspect of its function may be to set up a barrier for whorl 3 function genes to extend into the whorl 4 domain and function there also.

Analyses of ovary mutants have led to the suggestion that the ovary be considered as the fifth whorl of a flower. *Arabidopsis* genes important for ovule development include *AINTEGUMENTA, BELL1* (*BEL1*), *INNER NO OUTER*, and *SHORT INTEGUMENTS 1* (*SIN1*) (Angenent and Colombo, 1996; Baker *et al.*, 1997). Each of these genes affects different aspects of ovule development, although mutations in some also affect the embryo sac and have pleiotropic effects on other aspects of plant development, suggesting additional functional roles for these genes. An important aspect of ovule development in most angiosperms is the asymmetric growth of the integuments leading to their amphitropous shape, which brings the micropyle into close proximity with the placenta (Figure 6.1B), so enhancing the probability of pollen tube penetration. *SUP* appears to also play a role in the asymmetric growth of the ovule outer coats because mutations in this gene also result in some ovules assuming an unbent, elongated shape (Gaiser *et al.*, 1995).

Figure 6.2 Scanning micrographs of (A) wild-type *Arabidopsis* flower with some of the sepals and petals removed to expose the bicarpelloid pistil and the six stamens; (B) a *sup* flower with the outer two whorls removed, showing no carpel development and excess stamens; (C) an abnormal carpel from a *sup* flower; (D) a wild-type *Arabidopsis* flower treated with BAP, arrowheads indicate the stigmatoid tips of appendages which developed from the suture between the two carpels of the gynoecium

In *Petunia*, two related MADS box regulatory genes, *FBP7* and *11*, are adequate for ovule initiation (Angenent *et al.*, 1995; Colombo *et al.*, 1995). Ectopic expression of *FBP11* led to ectopic ovule development on sepals and petals. *FBP7* and *11*, like the *BEL1* gene in *Arabidopsis*, also appear to function in suppressing the carpel development pathway in places where ovules normally initiate because the absence or reduction of these gene products results in the appearance of carpel structures within the ovary chamber.

6.2.4   *Physiological regulation of floral development in bisexual flowers*

Environmental factors, such as day length and temperature, and hormonal factors, such as GAs and cytokinins, are known to be closely associated with the change from the vegetative to the reproductive phase of plants (Bernier *et al.*, 1993). Several studies have examined how environmental and physiological factors affect the relationship between inflorescence and floral meristems, and have demonstrated that phytochrome, GAs and cytokinins regulate the interconversion between them (Okamuro *et al.*, 1996, 1997; Venglat and Sawhney, 1996). Mutants deficient in brassinosteroids from a range of plant species, including *Arabidopsis*, and GA-deficient *Arabidopsis* mutants are male sterile (Sun *et al.*, 1992; Clouse, 1996; Yokota, 1997). These studies are indicative of the important roles physiological factors play in reproductive development.

Analysis of a regulatory gene for maize leaf development, *KNOTTED 1* and its analogue *KNAT*, from *Arabidopsis* showed that constitutive expression of these genes in transgenic plants produces phenotypes suggesting there are elevated cytokinin levels in these plants (Sinha *et al.*, 1993; Lincoln *et al.*, 1994). This implies that the developmental pathways defined by these regulatory genes, at least in part, involve the control of endogenous hormone levels. It also raises the possibility that genes regulating floral development also function via physiological conditions directly controlled or conferred by the developmental programs that they regulate. We have observed that when the *Arabidopsis SUP* is constitutively expressed in transgenic tobacco, it feminizes the transgenic flowers by inducing increased carpel numbers in the gynoecium, stigmatoid stamens and ectopic ovules on the sepals and petals (Wu and Cheung, unpublished). These transgenic tobacco plants also show vegetative phenotypes similar to plants harbouring elevated cytokinin levels. At a low frequency, tobacco plants sprayed with benzylamino purine (BAP) display a phenotype that mimicks some of the *SUP*-induced phenotypes, including tricarpelloid gynoecium. Spraying of wild-type *Arabidopsis* with BAP induces profuse outgrowth of appendages with stigmatoid apices from the carpel (Figure 6.2D). *sup* mutants respond to BAP application by producing some stigmatoid appendages among the ectopic stamens. These results suggest that process involving cytokinin are at least part of the program under the control of *SUP*, which must also induce pathways necessary for other aspects of stamen and carpel organogenesis. They also support a role for cytokinins in conferring femininity in *Arabidopsis* and tobacco (Wu and Cheung, unpublished).

6.2.5 Shared molecular factors in unisexual and bisexual plants

As most unisexual plants have a hermaphroditic phase, it seems plausible that they utilize similar themes in sexual organ primordia determination as bisexual plants in the early phase of floral development. Genes analogous to the *Arabidopsis AG*, *PI* and *AP3*, and the *Antirrhinum DEFECIENS* (*AP3* homologue) and *PLENA* (*AG* homologue) have been isolated and characterized from *Silene*, *Rumex acetosa* and maize (Schmidt *et al.*, 1993; Hardenack *et al.*, 1994; Ainsworth *et al.*, 1995). They have been used as markers to follow male and female organ developments in these plants.

During the hermaphroditic phase in *Silene*, the gynoecium primordium in male flowers is significantly smaller than that in female flowers and its development is arrested as soon as it emerges. On the other hand, the abortion of male development in female flowers is relatively late and takes place when tapetum development begins. *Silene SLM2* and *SLM3* (homologues of *PI* and *AP3*) mRNAs are confined to petal and stamen primordia (as in the bisexual *Arabidopsis* flowers) in both emerging male and female flowers. In the male flower, their expression domains extend further into the central region of the meristem, which normally defines the carpel domain. This suggests that invasion of the *SLM2* and *SLM3* gene products into the carpel domain of a male flower may play a role in inhibiting carpel development in *Silene*, reminiscent of the invasion of *AP3* mRNA into the carpel domain of the floral apices of *sup* mutant flowers and in transgenic flowers ectopically expressing *AP3*. Unlike these *Silene* genes, the *Rumex acetosa* class B *RAD1* and *RAD2* (*DEFICIEN* homologues) mRNAs do not invade the central dome of the meristem of the male flowers. This suggests demarcation of the spatial extent for the stamen and carpel primordia at the floral apex may not be a mechanism utilized by *Rumex* in sex determination. *RAD1* and *RAD2* mRNAs are detected in the stamen whorls of very young buds of male and female flowers. Their levels decline in the female flowers and are barely detectable when the gynoecium approaches maturity. In male *Rumex* flowers, *RAD1* and *RAD2* mRNAs are present in the stamen whorl throughout development.

In *Rumex acetosa*, morphological analyses show that there is no gynoecial development in male flowers and only vestigial stamens develop in the female flowers, suggesting that the hermaphroditic phase in these flowers is absent or extremely short. *In situ* hybridization showed that the *RAP1* (*PL* analogue) mRNAs accumulate very briefly in the stamen and carpel domains in the meristem of both male and female flowers, but they are depleted from the primordia, that do not develop further, consistent with the presence of a short hermaphroditic phase. Therefore, *in situ* examination using molecular probes may provide more refined assessment of the early reproductive development in unisexual plants.

Analogues of *AG*, *ZAG1* and *ZMM2* have been isolated from maize (Schmidt *et al.*, 1993; Mena *et al.*, 1996). Like *Arabidopsis*, both *ZAG1* and *ZMM2* are expressed in the carpels and stamens of male and female florets. There appears to be a division of function for *ZAG1* and *ZMM2*. *ZAG1* mRNAs accumulate to high levels in female florets while the *ZMM2* mRNA level is higher in male florets. Consistent with this expression pattern, a loss-of-function mutation in *ZAG1* results in defects in the gynoecium of female flowers only.

These studies represent emerging efforts to take advantage of the information available for the reproductive developmental pathways of both unisexual and bisexual flowers. Further analyses should reveal the fundamental similarities and differences in the mechansims utilized by plants with two different reproductive habits.

6.3 Pollination and fertilization

6.3.1 Maturation of pollen and pistil

After the initial organogenesis of stamens and pistil, many biochemical and cellular processes occur to produce the competent male gametophytes and the receptive female reproductive organ (Knox, 1984; Bedinger, 1992; Goldberg *et al.*, 1993; Gasser and Robinson-Beers, 1993). Microsporogenesis occurs in the anther and produces haploid microspores, which are encased in a callose walled tetrad. Callose dissolution releases individual microspores. The microspores and mature pollen grains secrete the proteinaceous and polysaccharide-containing inner pollen wall (intine) and some of the materials in the outer wall (exine), which is largely composed of materials synthesized and secreted by the tapetum of the anther. Many male sterile mutants in different plant species have defects in tapetal development (Echlin, 1971; Goldberg *et al.*, 1993). Genetically-engineered transgenic plants in which the tapetum is specifically ablated by the expression of a tapetum-specific cytotoxin transgene are also male sterile (Mariani *et al.*, 1990). These observations confirm the essential role the anther tapetum plays in pollen development and provide a highly specific method for introducing nuclear male sterility, an important agronomic trait, into crop plants.

The predominant molecules on the pollen exine surface are carotenoid molecules known as sporopollenin. Lipids, proteins, polysaccharide-containing molecules and other pigment molecules are also spread along the exine in a surface layer known as the tryphine. The highly sculptured pollen grain exine not only confers the surface structure characteristics of pollen grains from different species, which presumably contributes to the the initial cell–cell recognition process between pollen and stigmas, but also contains molecules essential for the initial pollination events in some plants (see below).

The gynoecium, or the pistil, is made up of either single or multiple carpels. The pistil is composed of three major structural parts: the stigma, the style and the ovary (Figure 6.1). The stigma has a pollen receptive surface and several underlying secretory cell layers. The style connects the stigma to the basally located ovary, which contains the ovules. Carpels emerge from the centre of the floral meristem. In multi-carpelloid flowers, these carpels may remain individual organs, or fuse on their adnating faces as they emerge from the meristem and grow into contact with one another (Walker, 1978; Hill and Lord, 1989; Verbeke, 1992). The top of the carpels differentiates into the stigma. The epidermal and subepidermal layers differentiate into the papillar surface and the secretory zone of the stigma, respectively. In multi-carpelloid pistils with fused carpels, cells along the adaxial surface layers of the carpels undergo a process known as progenital fusion once the carpels grow into contact with one another. These epidermal cells de-differentiate and re-differentiate into highly secretory cells, which constitute the transmitting tract of the style. These cells secrete copious amounts of materials to their extracellular space (Rosen and Thomas, 1970; Bell and Hicks, 1976). Pollen tubes elongate exclusively within this mucilaginous extracellular medium of the transmitting tracts. In multi-carpelloid pistils, the most basal parts of the carpels may also fuse and the partitioning cell layers between the carpels develop into walls separating individual locules of the ovary.

6.3.2 *Pollen–pistil interactions: an overview*

Intimate interactions between pollen and pistils are the requisites for successful sexual reproduction (Figure 6.1) (Mascarenhas, 1975, 1993; Knox, 1984; Herrero, 1992; Bedinger, 1992; Bedinger *et al.*, 1994; Cheung, 1995, 1996a,b). A mature pollen grain consists of a vegetative cell and, depending on the plant species, either a generative cell or two sperm cells embedded within the vegetative cell cytoplasm. Fertilization depends on further cellular activities in the pollen on encountering the pistil. Initial pollen–stigma interactions elicit the polarized outgrowth of a pollen tube from each grain. Continued interactions between the pollen tube and the style are associated with the mitotic division of the generative cell to two sperm cells for bicellular pollen grains. Pollen tubes penetrate the cuticle of the stigmatic papillar cells and invade the extracellular matrix of the underlying secretory cells. On entering the style, pollen tubes are confined to the centrally located transmitting tract. Pollen tubes either elongate in the extracellular matrix of the transmitting cells, which fill the solid styles of some plants, or along the surface and in the mucilage of the transmitting canal in the hollow styles of other plants. They emerge from the bottom of the style to enter the ovary to penetrate the ovules.

The female gametophyte develops within the ovule. Pollen tubes gain access to the embryo sacs via the micropyle, an ovule aperture oriented towards the placental surface (Figure 6.1B). The embryo sac comprises a haploid egg cell surrounded by two haploid synergid cells at the micropylar end of the ovule, three haploid antipodal cells distally located from the micropyle, and a central cell with two haploid nuclei. Once inside the ovary, each pollen tube turns from its basally oriented growth direction to enter the ovule via the micropyle and penetrate the embryo sac. The pollen tube tip bursts in the embryo sac to release two male germ cells, one to fuse with the egg cell and the other with the binuclear central cell to produce a diploid zygote and a triploid endosperm, accomplishing a double fertilization process unique to plants.

Successful fertilization depends on highly specific interactions between pollen and pistil since incompatible pollen may be arrested at the stigma or along the pathway of pollen tube elongation (Nasrallah *et al.*, 1994; Kao and McCubbin, 1996). Prior to embryo sac penetration, extracellular interactions between pollen and various pistil tissues and cells are probably the most significant in eliciting the necessary pollen and pistil responses to support compatible pollen tube growth and to preclude incompatible pollination. Interactions between the male gametes and factors within the embryo sac probably determine which of the two sperm cells fuse with the egg cell or the central cell to accomplish double fertilization. Pollen has a high capacity to support its activity during germination and tube growth, but pistil tissues also provide physical and chemical supports and directional guidance to the pollen tube growth process. Transduction of these interactions into the pollen cytoplasm to elicit cytoplasmic activities is essential for the pollen tube elongation process. Therefore, successful plant sexual reproduction involves intricate cell–cell interactions and signal transduction mechanisms.

6.3.3 The cytoskeletal basis of pollen tube elongation

Pollen germination involves a polarized extrusion of the pollen cytoplasmic contents into the pollen tube. Pollen tubes extend by a tip-growth process, sharing many of the characteristics of other tip-growth processes in plants, fungi and animals (Heslop-Harrison, 1987; Steer and Steer, 1989; Pierson and Cresti, 1992; Cai et al., 1997). It involves the continuous deposition at the tube apex of cell membrane and wall materials, which are synthesized in the slightly distally located organelles and transported to the apex via secretory vesicles. The vegetative cell cytoplasm and nucleus, and the two sperm cells migrate with, and are confined to, the pollen tube tip by the periodic deposition of a callose [(1-3)-β-glucan] plug behind the cytosolic region of the tube as it extends. The cytosolic regions of pollen tubes may function independently of the spent tubes, since they are competent for further elongation and fertilization even when the distal regions have been excised (Jauh and Lord, 1995). Therefore, pollen tube elongation has been described as a unique form of plant cell migration in which the cytoplasm moves forward, leaving behind a trail of extracellular materials (Lord and Sanders, 1992; Sanders and Lord, 1992).

Pollen tubes exhibit active cytoplasmic streaming, which transports secretory vesicles from distally located Golgi and endoplasmic reticulum to the tube apex to deposit new membrane and wall materials at the tip to support its extension. Cytoskeletal proteins, including actins, myosins, profilins, kinesins, spectrin, and dynein-like molecules, have been detected in pollen tubes. Both microfilaments and microtubules are abundant in pollen tube cytoplasm. Microfilament bundles are found throughout the pollen cytoplasm and are predominantly oriented parallel to the long axis of the pollen tube, except at the extreme tip region where fewer and shorter actin filaments are found in a random organization. The microtubules also extend along the length of the pollen tube, but are usually more concentrated along the cortical region along the plasma membrane of the tube. Cytochalasin B, which interferes with actin polymerization, inhibits cytoplasmic streaming in pollen tubes, whereas colchicine, which causes microtubules to depolymerize, has little effect on pollen tube growth. Pollen organelles have been shown to be translocated along Characeae actin filament bundles in an

ATP- and Ca^{2+}-dependent manner (Khono and Shimmen, 1988). Pollen-tube-expressed myosin genes and myosins have all been identified (Tang *et al.*, 1989; Yokota and Shimmen, 1994; Miller *et al.*, 1995). These observations provide strong evidence that an actomyosin-based motor is the primary force for the cytoplasmic streaming that occurs in pollen tube growth. The microtubule system is apparently not strictly required for pollen tube extension and may play subtle roles in regulating pollen tube growth and in the translocation of the male germ unit within the pollen tube.

Ca^{2+} is essential for pollen tube growth and is believed to control this process by regulating the activity of the pollen cytoskeleton (Steer and Steer, 1989; Hepler, 1997). A tip-focused Ca^{2+} concentration gradient exists in elongating pollen tubes (Pierson *et al.*, 1994, 1996; Malho *et al.*, 1992, 1994, 1995; Malho and Trewavas, 1996). *In vitro* grown pollen tubes have a steep, tip-focused intracellular Ca^{2+} gradient (averaging from about 3 μM at the apex to 0.2 μM at about 20 μm from the tip for lily pollen tubes) and a tip-directed influx of extracellular Ca^{2+} occurs across the membrane at the pollen tube tip surface. The intracellular Ca^{2+} gradient is dissipated by ionophoretic application of Ca^{2+}-chelating buffers, accompanied by the abolition of Ca^{2+} influx and arrest of pollen tube growth. The normal cytoplasmic streaming pattern is perturbed and organelles in the subapical zone of the pollen tube invade the apex. These observations suggest that the Ca^{2+} gradient and Ca^{2+} influx are closely coupled and that Ca^{2+} is important to regulate cytoskeletal activities in order to maintain the proper cytoplasmic streaming to sustain pollen tube growth. Furthermore, the Ca^{2+} influx is stimulated by membrane stretching and is dissipated by a loss in turgor pressure at the tip, suggesting that a stretch-activated channel may be involved in its regulation. The tip-focused intracellular Ca^{2+} gradient fluctuates up to four-fold during normal tube elongation, which exhibits periodic oscillations in pollen tube growth rates. The magnitudes of the Ca^{2+} gradient correlate positively with and slightly precede the pollen tube growth rate changes. Pollen tube growth cessation is accompanied by the dissipation of the Ca^{2+} gradient. The recovery of the Ca^{2+} gradient also precedes pollen tube growth recovery and predicts the new growth orientation. These results suggest a causal role for the Ca^{2+} gradient in affecting pollen tube growth activity. Local intracellular increase in free Ca^{2+} concentration accomplished by manipulating the release of caged Ca^{2+} or a caged Ca^{2+} chelator also precedes reorientation and regrowth. Manipulations of Ca^{2+} influx at the tip by increasing extracellular Ca^{2+} concentration or blocking Ca^{2+} channel activities, which presumably also alter the intracellular Ca^{2+} concentration, result in pollen tube growth oriented towards the region of a higher Ca^{2+} concentration region. This suggests that elevated Ca^{2+} concentration leads the way for pollen tube growth. Calcium-binding proteins, such as calmodulin, calcium-dependent calmodulin-independent protein kinase and annexin, are present in the pollen tubes. These proteins are candidate molecules that could interact with Ca^{2+} to regulate pollen tube growth activities.

Although self-incompatible pollination (Nasrallah *et al.*, 1994; Kao and McCubbin, 1996) is not included in our discussion here, it should nonetheless be indicated that Ca^{2+} appears to be an important factor in the self-incompatible response in *Papaver rhoseas* (Franklin-Tong *et al.*, 1993, 1995). Incompatible response induced by the self-incompatible protein from an incompatible origin in an *in vitro* pollen tube growth assay is followed by a transient increase in the internal Ca^{2+} concentration in the pollen tubes, and it is associated with the increased phosphorylation of a 26 kD protein via a Ca^{2+} and calmodulin-dependent phosphorylation pathway. Photo-activation of microinjected caged Ca^{2+} within the pollen tube also results in inhibition of pollen tube growth, mimicking the self-incompatible response. In *Nicotiana alata*, a pollen tube expressed protein kinase with features similar to Ca^{2+}-dependent protein kinases phosphorylates the self-incompatible RNases (S-RNases) from itself but not those from *Lycopersicon peruvianum*. (Kunz *et al.*, 1996). It is plausible that this Ca^{2+}-dependent phosphorylation of S-RNases, which are taken up by pollen tubes (Gray *et al.*, 1991), is an aspect of the self-incompatibility response in *Nicotiana alata*.

6.3.4   *Pollen grain–stigma interactions: the pollen aspect*

Pollination begins with the deposition of pollen on the stigmatic surface (Knox, 1984; Dickinson, 1995). The stigmatic papilla and underlying cell layers deposit secretory materials and provide the environment for the initial pollen–pistil interactions necessary to allow pollen germination and tube growth to proceed. Water absorption by pollen grains is an essential aspect of pollen germination. On wet stigmas, compatible pollen grains often hydrate and germinate readily. On dry stigmas, the papillar cells are stimulated on pollination to secrete water and exudate onto the stigmatic surface. Interactions between pollen grains and the dry stigmatic surface have been most vividly demonstrated in *Brassica* and *Arabidopsis* (Dickinson, 1995; Kandasamy *et al.*, 1994). In *Brassica*, pollen adheres tightly to the stigmatic surface by its exudates and water is secreted by the papillar cells to hydrate the pollen grain. It appears that direct contact between pollen and the stigmatic surface is critical for this hydration process because pollen grains removed from the stigma, but attached to another pollen grain that is already hydrated, fail to hydrate. Similarly, in *Arabidopsis,* the lipoidal and proteinaceous tryphine materials from mature pollen grains are mobilized within five minutes after pollination to form an adhesion zone for the pollen to attach to the papillar cell surface. Intracellular activities in preparation for pollen tube emergence are detectable 10 minutes after this adhesion. Only microspores from the most mature bud behave in a similar way to mature pollen grains, while those from even a slightly younger bud stage (6–8 hours younger) do not have the competence for the various steps in pollination. Surface molecules on the pollen and the stigma both actively participate in this process since mutations affecting the pollen coat, washing the stigmatic surface and genetic ablation of stigmatic tissues (Elleman *et al.*, 1992; Kandasamy *et al.*, 1993; Preuss *et al*, 1993; Hulskamp *et al.*, 1995 a; Dickinson, 1995) obliterate pollen grain hydration.

Surface wax-deficient mutants and flavonoid-deficient mutants in several plant species are male sterile, implying a role for long-chain lipids and flavonoids in pollen function (Coe *et al.*, 1981; Mo *et al.*, 1992; Taylor and Jorgensen, 1992; Preuss *et al.*, 1993; Hulskamp *et al.*, 1995 b; McConn and Browse, 1996). Mature pollen grains from the male-sterile waxless *Arabidopsis* mutants lack a tryphine. The analysis of wax composition in the pollen from one of these mutants, *pop-1(cer 6-2)*, showed that it lacks long-chain lipids (C-29 and C-30). These mutant pollen grains fail to germinate on mutant and wild-type stigmas and they do not stimulate water secretion by the stigmas. However, they can germinate and grow tubes normally *in vitro*. High humidity during pollination can restore a normal hydration process for these mutant pollen and restore their fertility. Mixing mutant with wild-type pollen, or with wild-type coat extracts can rescue the hydration of the mutant grains on the stigma and restore fertility. These results suggest that long-chain lipids on the tryphine are important pollen coat components for signalling the early pollen–stigma interactions necessary to initiate pollen germination.

A triple *Arabidopsis* mutant defective in linoleic acid production and containing a negligible level of trienoic acid is male sterile. The tricellular pollen grains in this mutant have low viability and fail to germinate. Application of linolenate (18:3) or jasmonic acid, which is synthesized from 18:3, restored fertility. A quantitative analysis of these treatments, however, showed that very low amounts of these compounds were taken up by the pollen grains. It was suggested that trienoic acids function as precursors for oxylipin, which may signal and regulate the final maturation process and the release of pollen grains.

Flavonoid-deficient plants in maize and *Petunia* are sterile and their flavonoid-deficient pollen fail to germinate. A pollen surface flavonoid, kaempherol, can restore germination and tube growth *in vitro*. Co-inoculation of flavonoid-deficient mutant stigmas with flavonoid-deficient pollen and kaempherol also restores germination and normal fertility. Kaempherol is also found among the stigmatic exudate of wild-type pistils at concentrations capable of supporting the germination of flavonoid-deficient pollen grains. Therefore, both pollen and pistils have evolved the capability to produce this essential flavonoid molecule to ensure successful pollen germination. However, the *Arabidopsis* mutant *tt4*, which has a mutated chalcone synthase gene and is completely devoid of flavonoids, including in the reproductive organs, is fully fertile and no pollen tube growth aberrations have been observed in these plants (Burbulis *et al.*, 1996; Ylstra *et al.,* 1996). Similarly, *Antirrhinum* and parsley mutants with defective chalcone synthase genes have not been reported to have sterility problems. Despite high levels of flavonoids in the pollen of most plants and an apparent role for the germination of pollen, these pigment molecules are not universally required for male fertility.

Polypeptides are also present on the pollen grain surface. Among these are hydroxyproline-rich glycoproteins and proteins collectively called 'allergens'. These proteins most probably contribute to the surface architecture of pollen and are thus likely candidates for recognition molecules between pollen and stigmas. These pollen proteins also often continue to be present in the pollen tubes and their functions may extend to the pollen tube elongation process. A pollen-specific extensin-like extracellular matrix protein, *PEX1*, has been characterized in maize. The *PEX1* protein is most predominantly expressed in mature pollen and has a deduced structure similar to the sexual agglutinins of *Chlamydomonas*, which mediate the initial recognition of opposite mating types. This has led to the proposal that *PEX1* protein may mediate recognition between pollen and the pistil. Alternatively, these pollen surface proteins may play roles in providing wall structures compatible with the dynamics of pollen tube growth (see below) (Rubenstein *et al.*, 1995a,b).

Two low molecular weight pollen coat proteins, PCP7 (pollen coat peptide [7 kDa]) and a PC7-like peptide from the self-incompatible *Brassica oleracea* and the self-compatible *Brassica napus*, respectively, interact *in vitro* with two stigmatic proteins, SLG (S[self-incompatibility]-locus glycoprotein) and SLR-1 (S-locus related protein-1), from *Brassica napus* (Doughty *et al.*, 1993; Hiscock *et al.*, 1995). Whether or not these interactions are biologically significant remain to be determined.

Interestingly, the group I allergens from grass pollen share about 20–25% amino acid sequence identity with expansins, cell wall proteins that induce plant cell wall extension *in vitro*, presumably by disrupting the non-covalent bonding between cellulose and hemicellulose in the cell wall matrix. These pollen allergens have potent expansin-like activity that softens the walls of the grass stigma and style (Cosgrove, 1997). Pollen also secretes a variety of cell wall hydrolases. The invasion of pollen tubes into pistil tissues is associated with, and inevitably aided by, a loss in structural integrity of the extracellular matrix of pistil tissues that support the extension process. Pollen grains have a high propensity for pollen tube growth activity. Identification of pollen surface proteins and secretory proteins with the ability to loosen cell walls (Turcich *et al.*, 1983) is another illustration of how pollen is equipped to aid its own destiny in the pollination process.

Potential signal transduction-related proteins, which conceivably can play a role in transducing the pollination signal into pollen activity, have been identified. A calcium-dependent calmodulin-independent kinase cDNA and a receptor kinase gene (PRK-1) in maize and *Petunia inflata* (Mu *et al.*, 1994; Estruch *et al.*, 1994), respectively, have been isolated. Both of these mRNAs accumulate specifically in the late stages of pollen development and during pollen germination, suggesting possible functions in an early pollination event. Specifically, oligonucleotides antisense to the calcium-dependent calmodulin-independent kinase mRNA inhibit *in vitro* maize pollen germination and tube growth, indicating an involvement of the kinase in these processes.

6.3.5 Pollen grain–stigma interactions: the stigma aspect

The *in vivo* pollen tube growth process is in general enhanced in its growth rate and the distances attained by the tubes when compared to the *in vitro* process. The *in vivo* process differs from the *in vitro* process by the assumption of a directionality, from the stigma to the ovary. On germination, the pollen tubes enter the secretory zone and converge from a relatively broad stigmatic region into the stylar transmitting tissue. Therefore a directional bias is established early in the pollen tube growth process. Stigmatic exudates are enriched in sugars, fatty acids and amino acids; many of these are found in more complex forms such as polysaccharides, glycoproteins and glycolipids. The enriched extracellular matrix of the pistils is believed to provide recognition factors, nutrient incentives, adhesion and possibly directional signals (Labarca and Loewus, 1973; Mascarenhas, 1975, 1993; Clarke *et al.*, 1979a,b; Stead *et al.*, 1980; Lord and Sanders, 1992; Sanders and Lord, 1992; Cheung, 1996a,b; Lord *et al.*, 1996). The significance of the stigmatic zone on pollen germination and tube growth is demonstrated by the ablation of this tissue by genetic transformation in tobacco and *Brassica* (Goldman *et al.*, 1994; Kandasamy *et al.*, 1993). In transgenic tobacco plants, which have a wet stigma surface, pollen grains germinate with adequate efficiency, but the extruded tubes are unable to penetrate beyond the secretory zone to invade the transmitting tissue of the style. Application of stigmatic exudates from wild-type pistils to the secretory zone-ablated stigma increases the efficiency of pollen germination and restores pollen tube elongation. In *Brassica*, pollen germination is inhibited on ablated stigma. However, expression of the same cytotoxin gene in its relative, *Arabidopsis*, has little effect on the pollen germination and growth process (Thorsness *et al.*, 1993). The observed differences between *Brassica* and *Arabidopsis* may be the result of inherent properties of pollen germination or the effectiveness of the transgenes in these species.

Glycoproteins are predominant on the stigmatic surface. Two stigmatic glycoproteins, SLG and SLK (S-locus receptor kinase) play fundamental roles in the self-incompatibility response in *Brassica* to mediate allelic-specific rejection of self-incompatible pollen (Stein *et al.*, 1991; Nasrallah *et al.*, 1994; Dickinson, 1996). Another family of proteins, SLR, with sequences related to SLG, is also present in *Brassica* but their location is not limited to the stigma and their precise functions remain to be established, although the stigmatic SLR-1 protein has been shown to interact with two pollen coat proteins *in vitro* (Hiscock *et al.*, 1995). Furthermore, the stigmatic papilla of *Arabidopsis* also expresses a SLR-like glycoprotein (Dwyer *et al.*, 1992). Whether the SLR protein family functions in compatible pollen–pistil interactions remains to be determined.

A predominant class of stigmatic glycoproteins are arabinogalactan proteins (AGPs), which are hydroxyproline-rich proteins with highly branched sugar modifications. AGPs are very sticky proteins and are believed to have a role in the recognition between pollen and stigmas, and to provide adhesion for pollen grains to the stigmatic surface (Clarke *et al.*, 1979a,b; Lord and Sanders, 1992; Du *et al.*, 1996). The exact functional role for these glycoproteins in pollen-stigma interactions remains to be demonstrated.

6.3.6 Pollen tube–transmitting tract interactions

The transmitting tissue of solid styles has a relatively broad extracellular matrix enriched in secretory materials, including an abundance of free sugars, polysaccharides, glycosylated proteins and lipids (Bell and Hicks, 1976), and the transmitting canal in hollow styles is also filled with mucilaginous materials secreted by the epidermal cells lining the transmitting tract (Rosen and Thomas, 1970). On pollination and pollen tube penetration, the carbohydrate contents of this matrix decline and the starch contents in the amyloplasts of the transmitting tissue cells become depleted. Transmitting tissue cells often become devoid of their cytoplasm and in many instances they burst (Herrero and Dickinson, 1979; Wang *et al.*, 1996). The metabolic demand for pollen tube growth is significant, and the transmitting tract is believed to provide nutrients for this process. In addition, directional cues may exist in the transmitting tissue to keep the pollen tubes elongating toward the ovary. Expression of a cytotoxic gene in the transmitting tissue of transgenic tobacco leads to female sterility (Thorsness *et al.*, 1991). The most extensively characterized stylar molecules with a functional role in pollen-tube pistil interactions are the S-RNases found in the transmitting tissue of self-incompatible Solanaceous plants. They function to prohibit pollen tubes from penetrating an incompatible style (Newbingin *et al.*, 1993; Kao and McCubbin, 1996).

It has been shown that the stylar transmitting tissue of three angiosperms can translocate inert latex beads, implicating an extracellular matrix with properties adequate for supporting a process similar to pollen tube growth (Sanders and Lord, 1989). *In vivo* lily pollen tubes adhere to each other and to the outer wall of the epidermal cells in the lily transmitting canal, consistent with the presence of surface adhesive molecules (Jauh and Lord, 1995). Furthermore, an artificial extracellular matrix constituted on a nitrocellulose membrane from lily transmitting tract exudates can support the extension of previously germinated pollen tubes (Jauh *et al.*, 1997). Pollen tubes adhere to this matrix and elongate more rapidly (25% faster) than in the controls. It appears that pollen tubes adhere to this matrix via their tip, with the distal region of the tip detached continuously from the matrix. Since pollen tube tips are enriched in esterified pectins (Li *et al.*, 1994) and the lily transmitting tract is enriched in AGPs, it was suggested that the adhesion of pollen tube tips to this artificial matrix may be mediated by pectin–AGP interactions (Baldwin *et al.*, 1993).

The transmitting tissue extracellular matrix proteins include a collection of different glycoproteins: AGPs (Du *et al.*, 1994; Cheung *et al.*, 1995), extensin-like proteins (Chen *et al.*, 1992; Goldman *et al.*, 1992; Baldwin *et al.*, 1993) and other proline-rich glycoproteins (Lind *et al.*, 1994). While some of these proteins may have structural roles in the transmitting tissue cell walls, others are likely to participate in the pollen tube growth process directly. The high sugar contents and adhesiveness of some of these glycoproteins have led to speculations that they serve as nutrient molecules and surface adhesives for pollen tube growth. The complex protein composition in the extracellular matrix along the pollen tube growth pathway suggests that most probably multiple proteins play over-lapping functions to support pollen tube growth to ensure its success. For instance, a 120 kDa glycoprotein with AGP and extensin properties from *Nicotiana alata* is localized to the intercellular matrix of the transmitting tissue and is taken up by pollen tubes *in vivo*, suggesting a possible participation in the pollen tube growth process (Lind *et al.*, 1994, 1996). Another glycoprotein belonging to the AGP family, the TTS protein from tobacco transmitting tissue, is important for pollen tube growth (Cheung *et al.*, 1995; Wu *et al.*, 1995). Transgenic plants in which TTS protein has been reduced to very low levels by antisense suppres-sion or sense co-suppression have reduced pollen tube growth rates. *In vitro*, TTS protein stimulates pollen tube growth and, when provided at a distance, attracts pollen tubes. Interestingly, TTS proteins added to *in vitro* pollen tube growth cultures are deglycosylated by the pollen tubes and these proteins are incorporated in the walls of pollen tubes. *In vivo*, TTS proteins are also incorpo-rated in the walls of elongating pollen tubes and pollinated transmitting tissue accumulates an underglycosylated TTS protein species that is not detectable in the pre-pollinated styles, suggesting that pollen tubes may also deglycosylate TTS proteins *in vivo*. These observations are consistent with a role for TTS protein as a nutrient resource for pollen tube growth. Surface adhesiveness is believed to provide enhancement for the pollen tube growth process (Sanders and Lord, 1989, 1992; Jauh *et al.*, 1997). TTS protein molecules adhere to the surface and tips of pollen tubes. This pollen tube surface adhesiveness of TTS protein may also contribute to its pollen tube growth-promoting and -attracting activity, perhaps similar to how the *in vitro* adhesive matrix from lily stylar exudates adhere to and promote pollen tube growth rate.

A galactose-rich style glycoprotein protein, GaRSGP, is present in the self-incompatible *Nicotiana alata* (Sommer-Knudsen *et al.*, 1996) and it shares a 97% amino acid homology with the backbone of TTS protein from its relative, the self-compatible tobacco. Curiously, GaRSGP and TTS differ significantly in their biochemical properties and their localization in the transmitting tissue. Both TTS protein and GaRSGP backbones have pIs around 10, whereas the pIs of native TTS protein molecules range from 7.5 to 8.5 and that for the native GaRSGP remains above 10. These indicate that protein modifications, primarily sugar residues, do not influence the pI of GaRSGP backbone but significantly acidify the TTS polypeptide backbone. This difference in biochemical property between TTS protein and GaRSGP already suggests differences in their mode of interactions with other biomolecules in the style and thus different participation in stylar functions. Another significant difference between GaRSGP and TTS protein lies in their relationship with the transmitting tissue extracellular matrix. GaRSGP is tightly bound to the transmitting cell wall matrix and is solublized from pulverized style tissue in buffers containing at least 250 mM NaCl. On the other hand, TTS protein is loosely associated with the transmitting tissue extracellular matrix and is efficiently eluted from hand-bisected style by gentle washing in a variety of salt-free buffers. Furthermore, TTS protein is localized to the intercellular matrix between transmitting tissue cells where pollen tubes elongate; this is consistent with its availability to interact with pollen tubes and having a role in the growth process. Since the GaRSGP antibodies were not reactive in immuno-staining of tissue sections, an indirect analysis of the subcellular localization for GaRSGP by using antibodies against two other proteins led Sommer-Knudsen et al. (1996) to deduce that GaRSGP is localized to the inner cell wall of transmitting tissue cells. This locality would not allow GaRSGP to contact pollen tubes directly, which elongate only in the extracellular space of transmitting tissue cells. The authors concluded that a direct function for GaRSGP in pollen tube growth seems unlikely and they proposed that this protein may be a structural cell wall glycoprotein with a function related to the specialized secretory nature of the transmitting tissue cells or in defense against pathogen attack by immobilizing microbial invasion at the transmitting cell wall. It will be interesting to determine whether differences in TTS protein and GaRSGP are associated with fundamental aspects of pollen tube growth in self-compatible and self-incompatible styles. The characterization of more molecules from closely related self-compatible and -incompatible plant species should produce further insights.

Pollen tube growth is a unique process in plant development and it is likely that the stylar transmitting tissue has evolved special mechanisms to accomplish it. For instance, only the transmitting tissue is apparently capable of producing certain glyco-moieties. In tobacco, the transmitting tissue is uniquely capable of fully glycosylating the TTS protein described above. Ectopically expressed TTS backbone polypeptides are severely underglycosylated, suggesting an arrest in the sugar-polymerization process in non-transmitting tissue (Cheung et al., 1996).

The transmitting tissue extracellular matrix components that contribute to pollen tube growth must be interacting with some pollen tube surface components. The pollen tube wall has been described as having either a two-layered or a three-layered structure. The outer pollen tube wall is pectic in nature and the predominant polysaccharides are arabinans. The inner tube wall is mostly (1,3)-glucan (callose). Proteins, such as the hydroxyproline-rich glycoprotein *PEX1* (Rubenstein et al., 1995b) are also present on the pollen tube surface and in its walls. Pollen tube surface components are potentially available for interactions with pistil components. It has been shown that arabinogalactan epitopes and pectins are deposited along the inner and outer pollen tube wall layers, respectively (Li *et al.*, 1992, 1994). Pectins on the outer pollen tube wall have the potential to interact with molecules in the transmitting tissue extracellular matrix. The pollen tube tip has a single, relatively thin, pectic cell wall predominantly made up of the more pliable methylpectins. Specific surface molecules are conceivably present at this growth apex to interact with pistil extracellular matrix molecules to attain growth.

The pollen tube–pistil extracellular interactions must be transduced to the pollen cytoplasm, possibly via molecules in the pollen extracellular matrix and plasmalemma continuum, to stimulate growth. A potential signal transduction molecule, a rho type GTPase, has been localized to the cortical region of the pollen tube apex and the periphery of the generative cell (Lin *et al.*, 1996). It was suggested that this GTPase may be involved in signalling mechanisms that control the actin-dependent tip growth of pollen tubes.

6.3.7 Pollen tube guidance

Observations suggesting mechanical-, electrical-, chemical- and pollination-induced signals to guide pollen tubes from the stigma to the ovary have all been reported (Rosen, 1961; Mascarenhas, 1975, 1993; Heslop-Harrison, 1987; Cheung, 1995, 1996a; Wilhelmi and Pruess, 1997). A universal mechanism apparently cannot be invoked to explain directional pollen tube growth. It is possible that a combination of mechanisms may exist to ensure the proper pollen tube growth directions, and that directional, architectural, biochemical and possibly also electrical cues exist in multiple foci along the pollen tube growth pathway.

The maize silk is probably the best-described structure in which cellular architecture plus additional factors are involved in pollen tube guidance (Heslop-Harrison *et al.*, 1985). The maize silk hair (equivalent to the stigma) joins the main axis of the style at an angle. It is essential that pollen germinates on silk hair and pollen tubes elongate following the silk hair orientation. Once arrived at the main axis of the style, pollen tubes traverse several rows of cortical cells before entering the transmitting tissue. The pollen tube tips point towards the ovary and they continue their journey in that direction. On the other hand, pollen grains directly germinating in the main axis of the style enter the transmitting tissue in random directions. It was therefore suggested that the architecture of the silk hair guides the pollen tubes towards the ovary. However, additional factors, perhaps chemotropic, must be invoked to keep the pollen tubes targeted to the transmitting tissue instead of turning in different directions prematurely within the cortical tissue. Similar mechanical guidance has been attributed to the trichomes on pearl millet stigma (Heslop-Harrison *et al.*, 1988).

In vitro pollen tubes respond chemotropically to low molecular weight ions and molecules, pistil tissues and their extracts from many plant species (Mascarenhas and Machlis, 1962a,b; Reger *et al.*, 1992a,b). Ca^{2+} has been shown to attract pollen tubes *in vitro* from a number of species, including *Antirrhinun* and pearl millet; the latter also respond chemotropically to glucose. A gradient of increasing Ca^{2+} concentration from the stigma to the ovary has been observed in the *Antirrhinum* style and this has been speculated to play a chemotropic role in pollen tube growth. The tobacco stylar TTS proteins are more highly glycosylated at the ovarian end of the style than at the stigmatic end. Furthermore, two-dimensional SDS-PAGE showed that the more glycosylated TTS proteins have more acidic pIs. Thus, TTS proteins display a gradient of increasing glycosylation and acidity within the tobacco style, coincident with the direction of pollen tube growth. Since TTS protein is a promoter for pollen tube growth, its sugar molecules are hydrolyzed by pollen tubes, and it can attract pollen tubes *in vitro*, it is possible that the differences in sugar and ionic environments associated with these proteins along the style contribute to keeping pollen tubes directed towards where TTS proteins are the most highly glycosylated and most acidic, namely, to the base of the style. Whether this protein-bound sugar and charge gradient is indeed chemotropically active remains to be determined.

Many migrating cells respond to electric currents. Electric currents, partly carried by Ca^{2+}, traverse the growing pollen tubes. Influx of Ca^{2+} across the tube apex, and the presence of a gradient of increasing Ca^{2+} concentration from the base to the tip of elongating tubes suggest that ion currents play an important role in pollen tube growth. Dramatic *in vitro* pollen tube growth oriented towards the cathode has been observed for pollen from *Camellia*, tulip and *Erythrina*. However, pollen tubes from a number of other plants, including tomato and tobacco, grow towards the anode and others show similar preference for both electrodes (Wang *et al.*, 1989; Nakamura *et al.*, 1991). These observed differences may be reflections of the surface properties and cellular physiology of different pollen tubes, enabling them to be most compatible with the extracellular matrix environment they encounter in the styles that they traverse.

The pistil extracellular matrix through which pollen tubes traverse is a highly ionic environment. It has been suggested that the changes in Ca^{2+} concentration around the micropyle due to synergid cell rupture may play an electrotropic role in effecting the growth direction of pollen tubes around the ovules. The gradient of increasing Ca^{2+} from the stigma to the ovary in *Antirrhinum* conceivably could generate an ionic current and contribute electrically to the pollen tube growth process. The increasing acidity gradient associated with TTS protein conceivably could also contribute to the electrical environment in the transmitting tissue. Attempts to measure changes in stylar bioelectric potential due to pollination have also been made in *Lilium longiflorum* (Spanjers, 1981). However, the significance of the differences in this electric parameter induced by self- and cross-pollination has not been further pursued. It is apparent that examination of electropism in pollen tube growth is not a simple task. Development of instrumentation that allows the probing and the perturbation of electrical conditions along the pollen tube growth pathway without affecting the normal pollen tube growth characteristics will be needed to allow better insight into the role of electrical gradients on pollen tube growth *in vivo*.

Extensive mutant screens have been carried out for *Arabidopsis* mutants in which pollen tube growth properties are perturbed but the stigmatic and transmitting tissues, the ovules, and the embryo sacs are developmentally and structurally normal, but few mutants have been recovered. This may be because of gene redundancy and overlapping functions among multiple pollen and pistil components to safeguard against the obliteration of successful gamete fusions. Alternatively, mutations that affect the directional cues for pollen tube growth may also affect essential processes in the vegetative phase, thus precluding the observation of their effects on sexual reproduction. The recent description of an *Arabidopsis* mutant defective in pollen tube guidance illustrates the difficulty in dissecting this important aspect of pollination (Wilhelmi and Pruess, 1996). In this self-sterile mutant, male and female organs and gametophytes are not visibly affected. Pollen tubes do not adhere to the ovular surface as in the wild-type ovary and they do not target the micropyles. Thus pollen tube growth does not result in gamete delivery to the embryo sac. Two genes, *POP2* and *POP3* (pollen–pistil interactions), are simultaneously mutated in these mutants. *POP2* and *POP3* appear to be functionally redundant because self-sterility occurs only in a doubly mutated background although the *pop3* mutation apparently is dominant because homozygous *pop2/pop2* and heterozygous *pop3/POP3* plants are self-sterile. The pollen tube growth phenotype in this mutant suggests that surface adhesion between pollen tubes and ovule surface is an important aspect of pollen tube guidance, as proposed previously. Curiously, both the male and female parents have to carry the mutant alleles in these two loci. This seems to argue against *POP2* and *POP3* directly encoding guidance signal molecules emanating from the ovules because if this were so pollen tubes from the double mutant should be able to target wild-type ovules. A possible explanation may be that *POP2* and *POP3* provide conditions, e.g. surface adhesiveness, that allow pollen tubes to respond to the ovule signals and these may be adequately provided for even if only pollen tubes have the capacity to produce them. Isolation and molecular analysis of *POP2* and *POP3* will be a critical step towards dissecting the components for pollen tube guidance in *Arabidopsis*.

6.3.8   *Pollen tube–ovule interaction in pollen tube guidance*

The most obvious directional guidance for pollen tubes occurs in the ovary where they turn from their basally directed growth direction to enter the ovules (Jensen and Fisher, 1968, 1969; Yan *et al.*, 1991; Chaubal and Reger, 1990, 1992a,b; Russell, 1992, 1994; Plyushch *et al.*, 1995; Willemse *et al.*, 1995;

Hulskamp *et al.*, 1995b; Wilhelmi and Pruess, 1996). Histological studies show that pollen tubes in many plant species need to change elongation direction, sometimes as much as 90°, in order to gain access to ovules. The entrance of pollen tubes into ovules appears to be guided by attractant signals. Ovary fragments and extracts from several species also elicit chemotropic response from *in vitro* grown pollen tubes. The ovule micropyles are covered with exudates that are enriched in glycosylated compounds and glucose has been shown to attract pollen tubes *in vitro*. Ca^{2+} concentration gradients around the micropyle generated by the rupture of a synergid cell before or on pollen tube arrival have been speculated to have tropic effects on pollen tube growth.

Arabidopsis has an elongated ovary with ovules developing along the length of the central septum (Figure 6.1). Pollen tubes enter the ovary and elongate in the transmitting tissue along the central septum. *Arabidopsis* pollen tubes appear to have the tendency to exit the central septum and enter the first available ovules, sequentially fertilizing them (Hulskamp *et al.*, 1995b). Analysis of reproductive mutants in *Arabidopsis* led to the isolation of several ovule and embryo sac developmental mutants in which pollen tubes are no longer directed toward the ovules but grow randomly inside the ovary and no longer show an ordered exit from the central septum. Pollen tubes fail to penetrate these defective ovules. Most dramatically, in an ovary that contains only one functional ovule, a pollen tube has been observed to have elongated past the level of this functional ovule, turned 180° and targetted this ovule. This indicates not only the existence of attractants emanating from the ovule (its sporophytic tissue or the embryo sac) but also that these signals can act at a relatively long distance. Biochemical and molecular comparison of ovarian constituents in the wild-type plants and these mutants may be a formidable task but may help to identify molecules that participate in the guided entrance of pollen tubes into the ovules.

6.3.9 Can pollen–pistil interactions be relaxed?

Normally, productive pollination depends exclusively on the pistil and pollen–pistil interactions and is regulated with stringent specificity. However, these properties are also plastic and subjected to genetic and developmental controls. Characterization of the pollination process during *Arabidopsis* flower development (Kandasamy *et al.*, 1994) showed that immature carpels have highly reduced capacity to support efficient pollen germination or tube growth on their stigmatic surface and transmitting tissue. The pollen tube growth pathway is not well defined and pollen tubes sometimes penetrate parenchymatous tissues. However, these young floral buds have the capacity to support pollen germination on all pistil and other floral organ epidermal surfaces. The species discriminatory mechanism also seems not to be fully acquired in young *Arabidopsis* carpels since pollen tubes from *Brassica* behave in a similar way to *Arabidopsis* pollen tubes in these young buds, although in mature *Arabidopsis* flowers

Brassica pollen tubes do not develop with the same efficacy as *Arabidopsis* tubes and do not target the ovules. These observations indicate that the capacity to support pollen activity exclusively by the carpel is acquired over time. A single recessive mutation in *Arabidopsis*, *fiddlehead*, results in profuse surface adhesions between all contacting epidermal surfaces in the floral bud, and sometimes also between contacting leaves. Pollen germination is no longer restricted to the stigmatic surface of *fiddlehead* mutant plants that have acquired the ability to support pollen germination on all its aerial surfaces. Nevertheless, species recognition seems to be retained in this mutant since only pollen from related Cruciferae plants germinate on the *fiddlehead* tissue surfaces, not Solanaceous pollen (Lolle *et al.*, 1992; Lolle and Cheung, 1993). The biochemical properties necessary to support pollen germination and tube growth also may be acquired by sepal tissues when the carpel developmental pathway controlled by a class C gene, *NAG1* (an *AG* homologue from tobacco) is ectopically expressed in the sepals of transgenic tobacco plants (Cheung *et al.*, 1996). Knowing that flexibility is allowed for pollen–pistil interactive processes may allow designs to broaden or restrict interspecific hybridization, and different aspects of the pollination process may be examined in isolation from all the processes that normally take place simultaneously in the pistil.

6.4   Fertilization and seed development

6.4.1   Double fertilization

Sexual reproduction in angiosperms involves a unique double fertilization process in which one sperm fuses with the egg cell to produce the diploid zygote and the other sperm fuses with the binuclear central cell to produce the triploid endosperm (Russell, 1992, 1993, 1994, 1996). The fertilized ovule develops into the mature seed with an embryo surrounded by the endosperm, which provides nutrition to support embryo maturation and the germination of the seed (West and Harada, 1993; Goldberg *et al.*, 1994). The process of fertilization is not well characterized at the biochemical and molecular level because of the inaccessibility of the female gametophytes. However, histological and cytological analyses in several plant species have provided a good basis upon which future genetic, biochemical and molecular analyses may be built.

Having gained entrance to the embryo sac via one of the two synergid cells, the pollen tube bursts and releases the two sperm cells. Networks of actins are present within the embryo sac (Huang *et al.*, 1993; Huang and Russell, 1994; Huang and Sheridan, 1994). The detection of myosin on sperm cell surfaces has been difficult, but isolated sperm cells track actin if they are introduced into the cytoplasm of streaming *Nitella*, which may provide myosin molecules to coat the sperm cell surfaces (Russell, 1996). This suggests that sperm cells are transported in the embryo sac by an actomysin-based cytoskeletal system. As double fertilization involves two fusion events, whether individual male gametes are programmed or are induced to target either the egg or the central cell has been a topic of considerable investigations. An extreme case of dimorphic sperm cells has been reported for *Plumbago*. After generative cell mitosis, one of the sperms in *Plumbago* inherits almost the entire complement of plastids, while the other sperm inherits most of the mitochondria but no plastids. The plastid-rich sperm cell is not associated with the vegetative nucleus when it migrates in the pollen tube and it fuses with egg cells 94% of the time (Russell, 1984, 1985, 1987), consistent with *Plumbago* being a plant displaying paternal inheritance of the plastid genome. In maize, sperm cells may have differences in their B chromosome contents. The sperm with extra B chromosomes preferentially fuse with the egg cell (Roman, 1948; Carlson, 1969, 1986). Development of fluorescent *in situ* hybridization for B-chromosome DNA fragments in sperm cells (Shi *et al.*, 1996) should be useful towards the further characterization of preferential gametic fusion in maize.

Documentations of preferential gametic fusion have only been made in plants with detectable sperm dimorphism. These are the result of many laborious and careful observations, not easily conducted for other plants which may have more subtle differences in their gametic fusion characteristics. Molecular distinctions between sperm cells may allow the identification of some of the factors that underlie gametic selection by sperm cells. In the last decade, methodologies (Dumas and Mogensen, 1993; Kranz and Duesselhaus, 1996; Rougier *et al.*, 1996) have been developed for the isolation of structurally intact and metabolically active sperm and egg cells from a number of plants, e.g. maize, barley, wheat and *Lolium*. *In vitro* maize gametic fusions have been accomplished by electrofusion and by Ca^{2+}-mediated fusion (Kranz *et al.*, 1991; Kranz and Lorz, 1993, 1994; Faure *et al.*, 1994). Karyogamy (fusion of the male and female haploid nuclei) has also been observed in the gametic fusion products (Faure *et al.*, 1993). Wheat gametes have also been fused by electric pulses (Kovacs *et al.*, 1995). Cell divisions and embryogenesis can be induced in some of these *in vitro*-produced maize and wheat zygotes. Distinct steps in the *in vitro* embryonic developmental pathway mimic those *in vivo* and fertile maize plants have been recovered. Isolated gametes, *in vitro* gamete fusion and early zygotic embryogenesis *in vitro* make available defined systems in which individual components and the early events after gametic fusion can be studied. This should allow the identification of sperm- and egg-specific genes and surface molecules, and fusion-induced genes in addition to the early signalling events in embryogenesis to be dissected.

6.4.2 Sporophytic and gametophytic interactions in seed development

Within the fertilized ovule, the sporophytic integuments, the triploid endosperm and the diploid embryo continue to develop to form the mature seed. Embryogenesis may be induced from somatic cell cultures in response to exogenous hormones (Zimmermann, 1993) and zygotes formed by *in vitro* gamete fusions also undergo embryogenesis and regenerate plants in the absence of the maternal integuments and the endosperm. These phenomena appear to suggest that provided the proper nutritional and hormonal conditions exist embryogenesis can proceed independently of the influence from the tissues that normally surround the zygote. However, many studies have suggested that maternal and endosperm tissues regulate each other's development and that development of the embryo and endosperm is intimately related and each influences the development of the other (Felker *et al.*, 1985; Lopes and Larkins, 1993; West and Harada, 1993; Hong *et al.*, 1996). *sin-1* is a female-sterile mutation in *Arabidopsis* causing defective ovule and embryo sac development and is pleiotropic on flowering times (Robinson-Beers *et al.*, 1992; Ray *et al.* 1996a). Ray *et al.* (1996b) showed that a homozygous mutant (*sin1/sin1*) embryo is phenotypically normal when developed in a heterozygous (*SIN1/sin1*) mother. On the contrary, a wild-type or *sin1/SIN1* heterozygous embryo develops defects when nursed in a homozygous mutant (*sin1/sin1*) mother tissue. These results indicate that *SIN1* has a maternal effect on embryo development and it was suggested that *SIN1* either codes for or controls the production of a diffusable 'morphogen' in the maternal tissue necessary for proper zygotic embryogenesis.

In *Petunia*, Colombo *et al.* (1997) also showed that normal endosperm development requires the expression in the maternal tissues of ovule-specific genes, *FBP7* and *11*, but viable embryos are produced in these degenerative endosperms. The latter observation also suggests that the endosperm is not critical for the later stages of *Petunia* seed development, probably because of the normally minor dependence on the diminutive endosperm in dicot seed development. These results provide support for the inter-dependence of maternal and zygotic tissues in post-fertilization development. On the other hand, mutants in *Arabidopsis* that produce normal embryos and endosperms but defective seed coats have also been described (see Colombo *et al.*, 1997). Whether the difference between these and the observations made in *sin1* mutant plants is the result of different qualitative and quantitative extents of destruction of the maternal tissues or the result of the inherent property of the genes and mutations involved remains to be determined.

6.4.3 Apomixis

Reproduction as a result of gametic fusion and seed formation ensures genetic divergence and produces hybrid rigor. On the other hand, fertilization-independent reproduction via the initiation of embryogenesis from maternal tissues or from unreduced gametophytic cells, a process known as apomixis, also occurs in nature (Hanna and Bashaw, 1987; Aster and Jerling, 1992; Koltunow,

1993; Koltunow *et al.*, 1995; Calzada *et al.*, 1996). Endosperm development is still needed to nourish the developing seeds. Some apomitic plants depend on pollination to fertilize the polar nuclei to initiate an endosperm, while others develop endosperms independent of fertilization. Apomixis permits clonal reproduction through seeds, thus preserving genetic uniformity from the maternal parent to offsprings. If the apomictic trait can be introduced into plants that normally reproduce sexually, hybrid vigor induced by the hybrid crosses will be fixed for perpetuation, bypassing the usual costly and laborious methods of generating hybrid seed sources.

In gametophytic apomixis, reductive meiosis of the megaspore mother cell is either arrested or modified so that $2n$ eggs are produced and allowed to differentiate into embryos with only the maternal genetic constituency. In sporophytic apomixis, sporophytic tissues give rise to $2n$ embryo sacs. Genetic studies suggest that the apomictic trait is controlled by a single or a few genes. The apomictic characteristics suggest that mutations either releasing the repression of the apomictic pathway or activating it can be isolated. Therefore it may be possible to generate mutants that relax the control for zygotic embryogenesis, thus providing both a fundamental understanding of this process and tools to engineer the apomitic trait into normally sexually reproducing plants. Such undertakings are under way in several laboratories, many of them using *Arabidopsis* as the model plant system. Strategies that have been adopted revolve around mutagenizing male-sterile parents and selecting for fruit-producing lines. To date, several fertilization-independent seed-forming mutants have been described (Ohad *et al.*, 1996; Chaudhury *et al.*, 1997). Molecular analyses of the genes identified by these mutations should produce an insight into the controlling mechanism of fertilization-independent embryo development.

6.5 Future perspectives

The molecular and biochemical bases behind sexual differentiation and the reproductive processes are complex and difficult to decipher by any single approach. Genetic studies have already proved to be invaluable in elucidating the regulatory pathways governing reproductive development, especially in organogenesis and embryogensis. Identification of more genetic loci affecting pollination and fertilization will expand the research basis and set the stage for the isolation of genes involved in plant sexual reproduction. On the other hand, substantive descriptions of how development and reproductive processes take place will have to rely on their biochemical, cellular and molecular analyses. The biochemical and molecular constituency required for sexual development and reproduction is likely to be more complex than the genetic mechanisms controlling them since the functions of many gene products rely on post-translational modifications and many gene products probably function in protein complexes rather than in isolation. Sensitive and reliable *in vitro* biochemical and molecular assays for the biological activities of these molecules, coupled with *in vivo* genetic and transgenic analysis of their functional roles, will provide a comprehensive understanding of how these molecules interact to accomplish the reproductive phase in plants.

The rapid advances in understanding basic molecular mechanisms in plant growth and development made by using *Arabidopsis* as a model plant system make it evident that it is crucial to consolidate the studies in sexual reproduction with model plant systems so that a fundamental framework of information may become available. However, plants utilize diversed strategies to accomplish sexual reproduction and many of the mechanisms evolved are both intriguing and uniquely suited for the habits of these plants. Although some of these plants may not currently be suitable for molecular and genetic analysis, it may prove valuable if studies utilizing model systems attempt to both extract from and share with plants displaying their own characteristic sexual reproductive properties information that is relevant.

Studies in plant sexual reproduction have the benefit of a vast repertoire of histological and cytological information concerning the processes involved. With the availability of better tissue preservation methodologies and more refined imaging tools, the pollination and fertilization processes will undoubtedly be captured with a clarity and vividness never before possible. The availability of molecular markers to be used as precise and sensitive detection tools will allow molecular details of the reproductive processes be examined. The concerted utilization of diverse research tools should bring immense advances in our understanding of sexual development and reproduction in the near future.

Acknowledgements

Research conducted in our laboratory was supported by grants from the USDA and NIH.

References

Ainsworth, C., Crossley, S., Buchanan-Wollaston, V., Thangavalu, M. and Parker, J. (1995) Male and female flowers of the dioecious plant sorrel show different patterns of MADS box gene expression. *Plant Cell*, **7** 1593-98.

Angenent, G.C. and Colombo, L. (1996) Molecular control of ovule development. *Trends Plant Sci.*, **1** 228-32.

Angenent, G.C., Franken, J., Busscher, M., Van Dijken, A., Van Went, J.L., Dons, H.J.M. and Van Tunen, A.J. (1995) A novel class of MADS box genes is involved in ovule development in petunia. *Plant Cell*, **7** 1569-82.

Aster, S.E. and Jerling, L. (1992) *Apomixis in Plants*, CRC Press, Boca Raton, FL.

Atal, C.K. (1959) Sex reversal in hemp by application of gibberellin. *Curr. Sci.*, **28** 408-409.

Baker, S.C., Robinson-Beers, K., Villanneva, J.M., Gaiser, J.C. and Gasser, C.S. (1997) Interactions among genes regulating ovule development in *Arabidopsis thaliana*. *Genetics,* **145** 1109-24.

Baldwin, T.C., McCann, M.C. and Roberts, K. (1993) A novel hydroxyproline-deficient arabino-galactan protein secreted by suspension cultured cells of *Daucus carota*. *Plant Physiol.,* **103** 115-23.

Bedinger, P.A. (1992) The remarkable biology of pollen. *Plant Cell,* **4** 879-87.

Bedinger, P.A., Hardeman, K.J. and Loukides, C.A. (1994) Travelling in style: the cell biology of pollen. *Trends Cell Biol.,* **4** 132-38.

Bell, J. and Hicks, G. (1976) Transmitting tissue in the pistil of tobacco. Light and electron micro-scopic observations. *Planta,* **131** 187-200.

Bensen, R.J., Johai, G.S., Crane, V.C., Tossberg, J.T., Schnable, P.S., Meeley, R.B. and Briggs, S.P. (1995) Cloning and characterization of the maize An1 gene. *Plant Cell,* **7** 75-84.

Bernier, G., Havelange, A., Houssa, C., Petitjean, A. and Lejeune, P. (1993) Physiological signals that induce flowering. *Plant Cell,* **5** 1147-55.

Bowman, J.L., Sakai, H., Jack, T., Weigel, D., Mayer, U. and Meyerowitz, E.M. (1992) SUPERMAN, a regulator of floral homeotic genes in *Arabidopsis* development. *Development,* **114** 599-615.

Bracale, I., Caporali, E., Galli, M.G., Longo, C., Marziani-Longo, G., Rossi, G., Spada, A., Soave, E., Falavigina, A., Raffaldi, F., Maestri, E., Restivo, F.M. and Tassi, F. (1991) Sex determination and differentiation in *Asparagus officinals* L. *Plant Sci.,* **80** 67-77.

Burbulis, I.E., Iacobucci, M. and Shirley, B.W. (1996) A null mutation in the first enzyme of flavonoid biosynthesis does not affect male fertility in *Arabidopsis*. *Plant Cell,* **8** 1013-25.

Cai, G., Moscatelli, A. and Cresti, M. (1997) Cytoskeletal organization and pollen tube growth. *Trends Plant Sci.,* **2** 86-91.

Calzada, J.-P.V., Crane, C. and Stelly, D.M. (1996) Apomixis: the asexual revolution. *Science,* **274** 1322-23.

Carlson, W.R. (1969) Factors affecting preferential fertilization in maize. *Genetics,* **62** 543-54.

Carlson, W.R. (1986) The B-chromosome of maize. *CRC Crit. Rev. Plant Sci.,* **3** 201-26.

Chailakhyan, M.K. (1979) Genetic and hormonal regulation of growth, flowering and sex expression in plants. *Am. J. Bot.,* **66** 717-36.

Champault, A. (1973) Effet de quelques regulateurs de la croissance sur des noeuds isoles de *Mercurialis annua* L. (2n=16) cultives *in vitro*. *Bull. Soc. Bot. Fr.,* **120** 87-100.

Champault, A., Guerin, B. and Teller, G. (1985) Cytokinin contents and specific characteristics of tissue strains from three sexual genotypes of *Mercurialis annua*: evidence for sex gene involvement at callus-tissue level. *Planta,* **166** 429-37.

Chaubal, R. and Reger, B.J. (1990) Relatively high calcium is localized in synergid cells of wheat ovaries. *Sex Plant Reprod.,* **3** 98-102.

Chaubal, R. and Reger, B.J. (1992a) Calcium in the synergids and other regions of pearl millet ovaries. *Sex Plant Reprod.,* **5** 34-46.

Chaubal, R. and Reger, B.J. (1992b) The dynamics of calcium distribution in the synergid cells of wheat after pollination. *Sex Plant Reprod.,* **5** 206-13.

Chaudhury, A.M., Ming, L., Miller, C., Craig, S., Dennis, E.S. and Peacock, W.J. (1997) Fertiliza-tion independent seed development in *Arabidopsis thaliana*. *Proc. Natl Acad. Sci. USA,* **94** 4223-28.

Chen, C.-G., Cornish, E.D. and Clarke, A.E. (1992) Specific expression of an extensin-like gene in the sytle of *Nicotiana alata*. *Plant Cell,* **4** 1053-62.

Cheung, A.Y. (1995) Pollen–pistil interactions in compatible pollen tube growth. *Proc. Natl Acad. Sci. USA,* **92** 3077-80.

Cheung, A.Y. (1996a) Pollen-pistil interactions during pollen tube growth. *Trends Plant Sci.,* **1** 45-51.

Cheung, A.Y. (1996b) The pollen tube growth pathway: its molecular and biochemical contributions and responses to pollination. *Sex. Plant Reprod.,* **9** 330-36.

Cheung, A.Y., Wang, H. and Wu, H.-M. (1995) A floral transmitting tissue-specific glycoprotein attracts pollen tubes and stimulates their growth. *Cell,* **82** 383-93.

Cheung, A.Y., Zhan, Z-Y., Wang, H. and Wu, H.-M. (1996) Organ-specific and agamous-regulated expression and glycosylation of a pollen tube growth-promoting protein. *Proc. Natl Acad. Sci. USA,* **93** 3853-58.

Clarke, A.E., Anderson, R.L. and Stone, B.A. (1979a) Form and function of arabinogalactans and arabinogalactan-proteins. *Phytochemistry,* **18** 521-40.

Clarke, A.E., Gleeson, P., Harrison, S. and Knox, R.B. (1979b) Pollen-stigma interactions: identification and characterization of surface components with recognition potential. *Proc. Natl Acad. Sci. USA,* **76** 3358-62.

Clouse, S.D. (1996) Molecular genetic studies confirm the role of brassinosteroids in plant growth and development. *Plant J.,* **10** 1-8.

Coe, E.H., McCormick, S.M. and Modena, S.A. (1981) White pollen in maize. *J. Hered.,* **72** 318-20.

Coen, E.C. and Meyerowitz, E.M. (1991) The war of the whorls: genetic interactions controlling flower development. *Nature,* **353** 31-37.

Colombo, L., Franken, J., Koetze, E., Vanwent, J., Dons, H.J.M., Angenent, G.C. and Van Tunen, J. (1995) The petunia MADS box gene FBP11 determines ovule identity. *Plant Cell,* **7** 1859-68.

Colombo, L., Franken, J., Vanderkrol, A., Wittich, P.E., Dons, H.J.M. and Angenent, G.C. (1997) Down-regulation of ovule-specific MADS box genes from *Petunia* results in maternally controlled defects in seed development. *Plant Cell,* **9** 703-15.

Cosgrove, D.J. (1997) Creeping walls, softening fruit, and penetrating pollen tubes: the growing roles of expansins. *Proc. Natl Acad. Sci. USA,* **94** 5504-5505.

Dauphin, B., Teller, G. and Durand, B. (1979) Identification and quantitative analysis of cytokinins from shoot apices of *Mercurialis ambigua* by gas chromatography–mass spectrometry computer system. *Planta,* **144** 113-19.

Dauphin-Guerin, B., Teller, G. and Durand, B. (1980) Different endogenous cytokinins between male and female *Mercurialis annua* L. *Planta,* **148** 124-29.

Dellaporta, S.L. and Calderon-Urrea, A. (1993) Sex determination in flowering plants. *Plant Cell,* **5** 1241-51.

Dellaporta, S.L. and Calderan-Urrea, A. (1994) The sex determination process in maize. *Science,* **266** 1501-1505.

DeLong, A., Calderan-Urrea, A. and Dellaporta, S.L. (1993) Sex determination gene tasselseed2 of maize encodes a short-chain alcohol dehydrogenase required for stage-specific floral organ abortion. *Cell,* **74** 757-68.

Dickinson, H.G. (1995) Dry stigmas, water and self-incompatibility in *Brassica. Sex Plant Reprod.,* **8** 1-10.

Dickinson, H.G. (1996) Plant signalling comes of age: identification of self-pollination in *Brassica* involves a transmembrane receptor kinase. *Trends Plant Sci.,* **1** 136-38.

Donnison, I.S. Siroky, J., Vyskot, B., Saedler, H. and Grant, S.R. (1996) Isolation of Y chromosome-specific sequences from *Silene latifolia* and mapping of male sex determining genes using representational difference analysis. *Genetics,* **144** 1891-99.

Doughty, J., Hedderson, F., McCubbin, A. and Dickinson, H.G. (1993) Interaction between a coating-borne peptide of the *Brassica* pollen grain and stigmatic S (incompatibility)-locus-specific glycoproteins. *Proc. Natl Acad. Sci. USA,* **90** 467-71.

Du, H., Simpson, R.J., Moritz, R.L., Clarke, A.E. and Bacic, A. (1994) Isolation of the protein backbone of an arabinogalactan-protein from the styles of *Nicotiana alata* and characterization of a corresponding cDNA. *Plant Cell,* **6** 1643-53.

Du, H., Simpson, R.J., Clarke, A.E. and Bacic, A. (1996) Molecular characterization of a stigma-specific gene encoding an arabinogalactan protein (AGP) from *Nicotiana alata. Plant J.,* **9** 313-23.

Dumas, C. and Mogensen, H.L. (1993) gametes and fertilization: maize as a model system for experimental embryogenesis in flowering plants. *Plant Cell,* **5** 1337-48.

Durand, B. (1967) L'expression du sexe chez les *Mercuriales annuelles. Bull. Soc. Fr. Physiol. Veg.,* **13** 195-202.

Durand, R. and Durand, B. (1984) Sexual differentiation in higher plants. *Physiol. Plant.,* **60** 267-74.

Dwyer, K.D., Lalonde, B.A., Nasrallah, J.B. and Nasrallah, M.E. (1992) Structure and expression of AtS1, an *Arabidopsis thaliana* gene homologous to the S-locus related genes of *Brassica. Mol. Gen. Genet.,* **231** 442-48.

Echlin, P. (1971) The role of the tapetum during microsporogenesis of angiosperms, in *Pollen and Development and Physiology* (ed. J. Heslop-Harrison), Butterworths, London, pp. 41-61.

Elleman, C.J., Franklin-Tong, V. and Dickinson, H.G. (1992) Pollination in species with dry stigmas: the nature of the early stigmatic response and the pathway taken by pollen tubes. *New Phytol.,* **121** 413-24.

Estruch, J.J., Kadwell, S., Merlin, E. and Crossland, L. (1994) Cloning and characterization of a maize pollen-specific calcium-dependent clamodulin-independent protein kinase. *Proc. Natl Acad. Sci. USA,* **91** 8837-41.

Faure, J.E., Mogensen, H.L., Dumas, C., Lorz, H. and Kranz, E. (1993) Karyogamy after electrofusion of single egg and sperm cell protoplasts from maize: cytological evidence and time course. *Plant Cell,* **5** 747-55.

Faure, J.E., Digonnet, C. and Dumas, C. (1994) An *in vitro* system for adhesion and fusion of maize gametes. *Science,* **263** 1598-1600.

Felker, F.C., Peterson, D.M. and Nelson, O.C. (1985) Anatomy of immature grains of eight maternal effect shrunken endosperm mutants. *Am. J. Bot.,* **72** 248-56.

Franken, A.A. (1970) Sex characteristics and inheritence of sex in asparagus (*Asparagus officinalis* L.). *Euphytica,* **19** 277-87.

Franklin-Tong, V.E., Ride, J.P., Read, N.D., Trewavas, A.J. and Franklin, F.C.H. (1993) The self-incompatibility response in *Papaver rhoeas* is mediated by cytosolic free calcium. *Plant J.,* **4** 163-77.

Franklin-Tong, V.E., Ride, J.P. and Franklin, F.C. (1995) Recombinant stigmatic self-incompatibility (S) protein elicits a Ca^{++} transient in pollen of *Papaver rhoeas. Plant J.,* **8** 299-307.

Gaiser, J.C., Robinson-Beers, K. and Gasser, C.S. (1995) The *Arabidopsis* SUPERMAN gene mediates asymmetric growth of the outer integument of ovules. *Plant Cell,* **7** 333-45.

Galun, E. (1961) Study of the inheritance of sex expression in the cucumber: the interaction of major genes with modifying genetic and non-genetic factors. *Genetica,* **32** 134-63.

Gasser, C.S. and Robinson-Beers, K. (1993) Pistil development. *Plant Cell,* **5** 1231-39.

Goldberg, R., Beals, T.P. and Sanders, P.M. (1993) Anther development: basic principles and practical applications. *Plant Cell,* **5** 1217-29.

Goldberg, R.B., dePaiva, G. and Yadejari, R. (1994) Plant embryogenesis: zygote to seed. *Science,* **266** 605-14.

Goldman, de M.H.S, Pezzotti, M., Seurinck, J. and Mariani, C. (1992) Developmental expression of tobacco pistil-specific genes encoding novel extensin-like proteins. *Plant Cell,* **4** 1041-51.

Goldman, de M.H.S., Goldberg, R.B. and Mariani, C. (1994) Female sterile tobacco plants are produced by stigma-specific cell ablation. *EMBO J.,* **13** 2976-84.

Goto, K. and Meyerowitz, E.M. (1994) Function and regulation of the *Arabidopsis* floral homeotic gene pistilla. *Genes Dev.,* **8** 1548-60.

Grant, S., Houben, A., Vyskot, B., Siroky, J., Pan, W., Macas, J. and Saedler, H. (1994) Genetics of sex determination in flowering plants. *Rev. Genet.,* **15** 214-30.

Gray, J.E., McClure, B.A., Bonïg, I., Anderson, M.A. and Clarke, A.E. (1991) Action of the style product of the self-incompatibility gene of *Nicotiana alata* (S-RNase) on *in vitro* grown pollen tubes. *Plant Cell,* **3** 271-83.

Hamdi, S., Teller, G. and Louis, J.P. (1987) Master regulatory genes, auxin levels, and sexual organogeneses in the dioecious plant *Mercurialis annua. Plant Physiol.,* **85** 393-99.

Hanna, W.W. and Bashaw, E.C. (1987) Apomixis: its identification and use in plant breeding. *Crop Sci.,* **27** 1136-39.

Hansen, D.J., Bellman, S.K. and Sacher, R.M (1976) Gibberellic acid-controlled sex expression in corn tassels. *Crop Sci.*, **16** 371-74.

Hardenack, S., Ye, D., Saedler, H. and Grant, S. (1994) Comparison of MADS box gene expression in developing male and female flowers of the dioecious plant white campion. *Plant Cell*, **6** 1775-87.

Hepler, P. (1997) Tip growth in pollen tubes: calcium leads the way. *Trends Plant Sci.*, **2** 79-80.

Herrero, M. (1992) From pollination to fertilization in fruit trees. *Plant Growth Reg.*, **11** 27-32.

Herrero, M. and Dickinson, H.G. (1979) Pollen–pistil incompatibility in petunia hybrida: changes in the pistil following compatible and incompatible intraspecific crosses. *J. Cell Sci.*, **36** 1-18.

Heslop-Harrison, J. (1960) The experimental control of sexuality and inflorescence structure in *Zea mays* L. *Proc. Linn. Soc. Lond.*, **172** 108-24.

Heslop-Harrison, J. (1987) Pollen germination and pollen tube growth. *Intl Rev. Cytol.*, **107** 1-78.

Heslop-Harrison, Y. and Reger, B.J. (1988) Tissue organization, pollen receptivity and pollen tube guidance in normal and mutant stigmas of the grass *Pennisetum pyphorides* (Burm.) Stapf et Hubb. *Sex Plant Reprod.*, **1** 182-93.

Heslop-Harrison, Y., Heslop-Harrison, J. and Reger, B.J. (1985) Pollen tube guidance and the regulation of tube number in *Zea mays*. *Acta Bot. Neèrl.*, **34** 193-211.

Hill, J.P. and Lord, E.M. (1989) Floral development in *Arabidopsis thaliana*—a comparison of the wild type and the homeotic pistillata mutant. *Can. J. Bot.*, **67** 2922-36.

Hiscock, S.J., Doughty, J., Willis, A.C. and Dickinson, H.G. (1995) A 7kDa pollen coating-borne peptide from *Brassica napus* interacts with S-locus glycoprotein and S-locus-related glycoprotein. *Planta*, **196** 367-74.

Hong, S.K., Kitano, H., Satoh, H. and Nagato, Y. (1996) How is embryo size genetically regulated in rice? *Development*, **122** 2051-58.

Huang, B.Q. and Russell, S.D. (1994) Fertilization in *Nicotiana tabacum*: cytoskeletal modifications in the embryo sac during synergid degeneration. A hypothesis for short distance transport of sperm cells prior to gamete fusion. *Planta*, **194** 200-214.

Huang, B.Q. and Sheridan, W.F. (1994) Female gametophyte development in maize: microtubular organization and embryos sac polarity. *Plant Cell*, **6** 845-61.

Huang, B.Q., Strout, G.W. and Russell, S.D. (1993) Fertilization in *Nicotiana tabacum*: ultrastructural organization of propane jet-frozen embryo sacs *in vivo*. *Planta*, **191** 256-64.

Hulskamp, M., Kopczak, S.D., Horejsi, T.F., Kihl, B.K. and Pruitt, R.E. (1995a) Identification of genes required for pollen–stigma recognition in *Arabidopsis thaliana*. *Plant J.*, **8** 703-14.

Hulskamp, M., Schneitz, K. and Pruitt, R.E. (1995b) Genetic evidence for a long range activity that directs pollen tube guidance in *Arabidopsis thaliana*. *Plant Cell*, **7** 57-64.

Irish, E. and Nelson, T. (1989) Sex determination in monoecious and dioecious plants. *Plant Cell*, **1** 737-44.

Jack, T., Brockman, L.L. and Meyerowitz, E.M. (1992) The homeotic gene APETALA3 of *Arabidopsis thaliana* encodes a MADS box and is expressed in petals and stamens. *Cell*, **68** 683-92.

Jack, T., Fox, G. and Meyerowitx, E.M. (1994) *Arabidopsis* homeotic gene Apetela 3 ectopic expression: transcriptional and posttranscriptional regulation determine floral organ identity. *Cell* **76**, 703-706.

Jauh, G.Y. and Lord, E.M. (1995) Movement of the tube cell in the lily style in the absence of the pollen grain and the spent pollen tube. *Sex. Plant Reprod.*, **8** 168-72.

Jauh, G.Y., Eckard, K.J., Nothnagel, E.A. and Lord, E.M. (1997) Adhesion of lily pollen tubes on an artificial matrix. *Sex. Plant Reprod*, **10** 173-80.

Jensen, W.A. and Fisher, D.B. (1968) Cotton embryogenesis: the entrance and discharge of the pollen tube in the embryo sac. *Planta*, **78** 158-83.

Jensen, W.A. and Fisher, D.B. (1969) Cotton embryogenesis: the tissues of the stigma and style and their relation to the pollen tube. *Planta*, **84** 97-121.

Kandasamy, M.K., Thorsness, M.K., Rundle, S.J., Goldberg, M.L., Nasrallah, J.B. and Nasrallah,

M.E. (1993) Ablation of papillar cell function in *Brassica* flowers results in the loss of stigma receptivity to pollination. *Plant Cell,* **5** 263-75.

Kandasamy, M.K., Nasrallah, J.B. and Nasrallah, M.E. (1994) Pollen–pistil interactions and developmental regulation of pollen tube growth in *Arabidopsis. Development,* **20** 3405-18.

Kao, T.-H. and McCubbin, A.G. (1996) How flowering plants discriminate between self and non-self pollen to prevent inbreeding. *Proc. Natl Acad. Sci. USA,* **93** 12059-65.

Kempin, S.A., Mendel, M.A. and Yanofsky, M.F. (1993) Conversion of perianth into reproductive organs by ectopic expression of the tobacco floral homeotic gene NAG1. *Plant Physiol.,* **103** 1041-46.

Khono, T. and Shimmen, T. (1988) Accelerated sliding of pollen tube organelles along Characeae actin bundles regulated by Ca^{++}. *J. Cell Biol.,* **106** 1539-43.

Knox, R.B. (1984) Pollen–pistil interactions. *Encyclopedia Plant Physiol.,* **17** 508-608.

Koltunow, A.M. (1993) Apomixis embryo sacs and embryos formed without meiosis or fertilization in ovules. *Plant Cell,* **5** 1425-37.

Koltunow, A.M., Bicknell, R.A. and Chaudbury, A.M. (1995) Apomixis: molecular strategies for the generation of genetically identical seeds without fertilization. *Plant Physiol.,* **108** 1345-52.

Kovacs, M., Barnabas, B. and Kranz, E. (1995) Electro-fused isolated wheat (*Triticum aestivum* L.) gametes develop into multicellular structures. *Plant Cell Rep.,* **15** 178-80.

Kranz, E. and Lorz, H. (1993) *In vitro* fertilization with isolated, single gametes results in zygote embrogenesis and fertile maize plants. *Plant Cell,* **5** 739-43.

Kranz, E. and Lorz, H. (1994) *In vitro* fertilization of maize by single egg and sperm cell protoplast fusion mediated by high calcium and high pH. *Zygote,* **2** 125-28.

Kranz, E. and Duesselhaus, T. (1996) *In vitro* fertilization with isolated higher plant gametes. *Trends Plant Sci.,* **1** 82-89.

Kranz, E., Bautor, J. and Lorz, H. (1991) *In vitro* fertilization of single isolated gametes of maize by electrofusion. *Sex Plant Reprod.,* **4** 12-16.

Krishnamoorthy, H.N. and Talukdar, A.R. (1976) Chemical control of sex expression in *Zea mays* L. *Z. Pflanzenphysiol.,* **79** 91-94.

Kunz, C., Chang, A., Faure, J.-D., Clarke, A.E., Polya, G.M. and Anderson, M.A. (1996) Phosphorylation of style S-RNases by Ca^{++}-dependent protein kinases from pollen tubes. *Sex Plant Reprod.,* **9** 25-34.

Labarca, C. and Loewus, F. (1973) The nutritional role of pistil exudate in pollen tube wall formation in *Lilium longiflorum*. II. Production and utilization of exudate from the stigma and stylar canal. *Plant Physiol.,* **52** 87-92.

Lebel-Hardenack, S. and Grant, S.R. (1997) Genetics of sex determination in flowering plants. *Trends Plant Sci.,* **2** 130-36.

Li, Y. Q., Bruun, L., Pierson, E.S. and Cresti, M. (1992) Periodic deposition of arabinogalactan epitopes in the cell wall of pollen tubes of *Nicotiana tabacum* L. *Planta,* **188** 532-38.

Li, Y.Q., Chen, F., Linsken, H.F. and Cresti, M. (1994) Distribution of unesterified and esterified pectins in cell walls of pollen tubes of flowering plants. *Sex Plant Reprod.,* **7** 145-52.

Lin, Y., Wang, Y., Zhu, J.-K. and Yang, Z. (1996) Localization of a Rho GTPase implies a role in tip growth and movement of the generative cell in pollen tubes. *Plant Cell,* **8** 293-303.

Lincoln, C., Long, J., Yamaguchi, J., Serikawa, K. and Hake, S. (1994) A knotted1-like homeobox gene in *Arabidopsis* is expressed in the vegetative meristem and dramatically alters leaf morphology when expressed in transgenic plants. *Plant Cell,* **6** 1859-76.

Lind, J.L., Bacic, A., Clarke, A. and Anderson, M.A. (1994) A style-specific hydroxyproline-rich glycoprotein with properties of both extensins and arabinogalactan proteins. *Plant J.,* **6** 491-502.

Lind, J.L., Bonig, I., Clarke, A.E. and Anderson, M.A. (1996) A style-specific 120 kDa glycoprotein enters pollen tubes of *Nicotiana alata* in vivo. *Sex Plant Reprod.,* **9** 75-86.

Lolle, S.J. and Cheung, A.Y. (1993) Promiscuous germination and growth of wildtype pollen from *Arabidopsis* and related species on the shoot of the *Arabidopsis* mutant, *Fiddlehead. Dev. Biol.,* **155** 250-58.

Lolle, S.J., Cheung, A.Y. and Sussex, I.M. (1992) *Fiddlehead*—an *Arabidopsis* mutant constitutively expressing an organ fusion program that involves interactions between epidermal cells. *Dev. Biol.*, **152** 383-92.

Lopes, M.A. and Larkins, B.A. (1993) Endosperm origin, development and function. *Plant Cell*, **5** 1383-99.

Lord, E.M. and Sanders, L.C. (1992) Roles for the extracellular matrix in plant development and pollination: a special case of cell movement in plants. *Dev. Biol.*, **153** 16-28.

Lord, E.M., Walling, L.L. and Jauh, G.Y. (1996) Cell adhesion in plants and its role in pollination, in *Membranes: Specialized Functions in Plants* (eds M. Smallwood, J.P. Knox and D.J. Bowles), Bios, Oxford, pp. 21-37.

Louis, J.P. (1989) Genes for the regulation of sex differentiation and male fertility in *Mercurialis annua* L. *J. Hered.*, **80** 104-11.

Louis, J.P. and Durand, B. (1978) Studies with the dioecious angiosperm *Mercurialis annua* L. ($2n=16$): correlation between genic and cytoplasmic male sterility, sex segregation and feminizing hormones (cytokinins). *Mol. Gen. Genet.*, **165** 309-22.

Ma, H. (1994) The unfolding drama of flower development: recent results from genetic and molecular analyses. *Genes Dev.*, **8** 745-56.

Malho, R. and Trewavas, A.J. (1996) Localized apical increases of cytosolic free calcium control pollen tube orientation. *Plant Cell*, **8** 1935-49.

Malho, R., Fejio, J.A. and Pais, M.A. (1992) Effect of electrical fields and external ionic currents on pollen tube orientation. *Sex Plant Reprod.*, **5** 57-63.

Malho, R., Read, N.D., Pais, M.S. and Trewavas, A.J. (1994) Role of cytosolic free calcium in the reorientation of pollen tube growth. *Plant J.*, **5** 331-41.

Malho, R., Read, N.D., Trewavas, A.J. and Pais, M.S. (1995) Calcium channel activity during pollen tube growth and reorientation. *Plant Cell*, **7** 1173-84.

Mariani, C., De Beuckeleer, M., Truettner, J., Leemans, J. and Goldberg, R.B. (1990) Induction of male sterility in plants by a chimaeric ribonuclease gene. *Nature*, **347** 737-41.

Mascarenhas, J.P. (1975) The biochemistry of angiosperm pollen development. *Bot. Rev.*, **41** 259-314.

Mascarenhas, J.P. (1993) Molecular mechanisms of pollen tube growth and differentiation. *Plant Cell*, **5** 1303-14.

Mascarenhas, J.P. and Machlis, L. (1962a) Chemotropic response of *Antirrhinum majus* pollen to calcium. *Nature*, **196** 292-93.

Mascarenhas, J.P. and Machlis, L. (1962b) The pollen tube chemotropic factor from *Antirrhinum majus*: bioassay, extraction and partial purification. *Am. J. Bot.*, **49** 482-89.

McConn, M. and Browse, J. (1996) The critical requirement for linolenic acid is pollen development, not photosynthesis, in an *Arabidopsis* mutant. *Plant Cell*, **8** 403-516.

Mena, M., Ambrose, B.A., Meeley, R.B., Briggs, S.B., Yanofsky, M.F. and Schmidt, R.J. (1996) Diversification of C-function activity in maize flower development. *Science*, **274** 1537-40.

Miller, D.D., Scordilis, S.P. and Hepler, P.K. (1995) Identification and localization of three classes of myosins in pollen tubes of *Lilium longiflorum* and *Nicotiana alata*. *J. Cell Sci.*, **108** 2549-63.

Mizukami, Y. and Ma, H. (1992) Ectopic expression of the floral homeotic gene AGAMOUS in transgenic *Arabidopsis* plants alters floral organ identity. *Cell*, **71** 119-31.

Mo, Y., Nagel, C. and Taylor, L. (1992) Biochemical complementation of chalcone synthase mutants defines a role for flavonols in functional pollen. *Proc. Natl Acad. Sci. USA*, **89** 7213-17.

Mu, J.-H., Lee H.-S., and Kao, T.-H. (1994) Characterization of a pollen-expressed receptor-like kinase gene of *Petunia inflata* and the activity of its encoded kinase. *Plant Cell*, **6** 709-21.

Nakamura, N., Fukushima, A., Iwayama, H. and Suzuki, H. (1991) Electrotropism of pollen tubes of camellia and other plants. *Sex Plant Reprod.*, **4** 138-43.

Nasrallah, J.B., Stein, J.C., Kandasamy, M.K. and Nasrallah, M.E. (1994) Signalling the arrest of pollen tube development in self-incompatible plants. *Science*, **266** 1505-1508.

Negri, S.S. and Olmo, H.P. (1966) Sex-conversion in a male *Vitis vinifera* L. by a kinin. *Science,* **152** 1624-25.

Newbingin, E., Anderson, M.A. and Clarke, A.E. (1993) Gametophytic self-incompatibility systems. *Plant Cell,* **5** 1315-24.

Nickerson, N.H. (1957) Sustained treatment with gibberellic acid of five different kinds of maize. *Ann. Mo. Bot. Gard.,* **46** 19-37.

Ohad, N.O., Margossian, L., Hsu, Y.-C., Williams, C., Repetti, P. and Fischer, R.L. (1996) A mutation that allows endosperm development mutant fertilization. *Proc. Natl Acad. Sci. USA,* **93** 5319-24.

Okamuro, J.K., den Boer, B.G.W., Lotys-Prass, C., Szeto, W. and Jofuku, K.D. (1996) Flowers into shoots: photo and hormonal control of a meristem identity switch in *Arabidopsis. Proc. Natl Acad. Sci. USA,* **93** 13831-36.

Okamuro, J.K., Szeto, W., Lotys-Prass, C. and Jofuku, K.D, (1997) Photo and hormonal control of meristem identity in the *Arabidopsis* flower mutants apetala1 and apetala2. *Plant Cell,* **9** 37-47.

Parker, J.S. and Clark, M.S. (1991) Dosage sex-chromosome systems in plants. *Plant Sci.,* **80** 79-92.

Phinney, B.O. (1956) Growth response of single-gene dwarf mutants in maize to gibberellic acid. *Proc. Natl Acad. Sci. USA,* **42** 185-89.

Phinney, B.O. (1961) Dwarfing genes in *Zea mays* and their relation to the gibberellins, in *Plant Growth Regulation* (ed. R.J. Klein), Ames, I.A., Iowa State University Press, pp. 489-501.

Phinney, B.O. (1984) Gibberellin A1, dwarfism and the control of short elongation in higher plants, in *SEB Seminar Series,* vol. 3 (eds A. Crozier and J.R. Hillman), Cambridge, Cambridge University Press, pp. 17-41.

Pierce, L.C. and Currence, T.M. (1962) The inheritance of hermaphroditism in *Asparagus officinalis* L. *Proc. Am. Soc. Hort. Sci.,* **80** 368-76.

Pierce, L.K. and Wehner, T.C. (1990) Review of genes and linkage groups in cucumber. *Hort. Sci.,* **25** 605-15.

Pierson, E.S. and Cresti, M. (1992) Cytoskeleton and cytoplasmic orgnaization of pollen and pollen tubes. *Intl Rev. Cytol.,* **140** 73-125.

Pierson, E.S., Miller, D.D., Callaham, D.A., Shipley, A.M., Rivers, B.A., Cresti, M. and Hepler, P.K. (1994) Pollen tube growth is coupled to the extracellular calcium ion fluxes and the intracellular calcium gradient: effect of BAPTA-type buffers and hypertonic media. *Plant Cell,* **6** 1815-28.

Pierson, E.S., Miller, D.D., Callaham, D.A., van Aken, J., Hackett, G. and Hepler, P.K. (1996) Tip-localized calcium entry fluctuates during pollen tube growth. *Dev. Biol.,* **174** 160-73.

Plyushch, T.A., Willemse, M.T.M., Franssen-Verheijen, M.A.W. and Reinders, M.C. (1995) Structural aspects of *in vitro* pollen tube growth and micropylar penetration in *Gasteria verrucosa* (mill) H. Duval and *Lilium longiflorum* thumb. *Protoplasma,* **187** 13-21.

Preuss, D., Lemiewx, B., Yen, G. and Davis, R. (1993) A conditional sterile mutation eliminates surface components from *Arabidopsis* pollen and disrupts cell signalling during fertilization. *Genes Dev.,* **7** 974-85.

Ray, S., Lang, J.D., Golden, T. and Ray, S. (1996a) SHORT INTEGUMENT (SIN1), a gene required for ouvle development in *Arabidopsis*, also controls flowering time. *Development,* **122** 2631-38.

Ray, S., Golden, T. and Ray, A. (1996b) Maternal effects of the short integument mutation on embryo development in *Arabidopsis. Dev. Biol.,* **180** 365-69.

Reger, B.J., Chaubal, R. and Pressey, R. (1992a) Chemotropic responses by pearl millet pollen tubes. *Sex Plant Reprod.,* **5** 47-56.

Reger, B.J., Pressey, R. and Chaubal, R. (1992b) *In vitro* chemotropism of pearl millet pollen tubes to stigma tissue: a response to glucose produced in the medium by tissue-bound invertase. *Sex. Plant Reprod.,* **5** 201-205.

Robinson-Beers, K., Pruitt, R.E. and Gasser, C.S. (1992) Ovule development in wildtype *Arabidopsis* and two female sterile mutants. *Plant Cell,* **4** 1237-50.

Roman, H. (1948) Directed fertilization in maize. *Proc. Natl Acad. Sci USA,* **34** 46-52.

Rood, S.B., Pharis, R.P. and Major, D.J. (1980) Changes of endogenous gibberellin-like substances with sex reversal of the apical inflorescence of corn. *Plant Physiol.*, **66** 793-96.

Rosen, W.G. (1961) Studies on pollen tube chemotropisim. *Am. J. Bot.* **48** 889-95.

Rosen, W.G. and Thomas, H.R. (1970) Secretory cells of lily pistils. I. Fine structure and function. *Am. J. Bot.*, **57** 1108-14.

Rougier, M., Antoine, A.F., Alden, D. and Dumas, C. (1996) New lights in early steps of *in vitro* fertilization in plants. *Sex Plant Reprod.*, **9** 324-29.

Rubenstein, A.L., Broadwater, A.H., Lowrey, K.B. and Bedinger, P.A. (1995a) *Proc. Natl Acad. Sci. USA*, **92** 3086-90.

Rubenstein, A.L., Marquez, J., Suarez-Cervera, M. and Bedinger, P.A. (1995b) Extensin-like glycoproteins in the maize pollen tube wall. *Plant Cell,* **7** 2211-25.

Russell, S.D. (1984) Ultrastructure of the sperm of *Plumbago zeylancia*: 2. Quantitative cytology and three-dimensional reconstruction. *Planta,* **162** 385-91.

Russell, S.D. (1985) Preferential fertilization in *Plumbago*: ultrastructural evidence for gamete-level recognition in an angiosperm. *Proc. Natl Acad. Sci. USA*, **82** 6129-32.

Russell, S.D. (1987) Quantitative cytology of the egg and central cell of *Plumbago zeylanica* and its impact on cytoplasmic inheritance patterns. *Theor. Appl. Genet.*, **74** 693-99.

Russell, S.D. (1992) Double fertilization. *Intl Rev. Cytol.*, **140** 357-88.

Russell, S.D. (1993) The egg cell development and role in fertilization and early embryogenesis. *Plant Cell,* **5** 1349-59.

Russell, S.D. (1994) Fertilization in higher plants, in *Pollen–Pistil Interactions and Pollen Tube Growth* (eds A.G. Stephenson and T.H. Kao), American Society of Plant Physiologists, Rockville, MD, pp. 140-52.

Russell, S.D. (1996) Attraction and transport of male gametes for fertilization. *Sex Plant Reprod.*, **9** 337-42.

Sakai, H., Medrano, L.J. and Meyerowitz, E.M. (1995) Role of SUPERMAN in maintaining *Arabidopsis* floral whorl boundaries. *Nature,* **378** 199-203.

Sanders, L.C. and Lord, E.M. (1989) Directed movement of latex particles in the gynoecia of three species of flowering plants. *Science,* **243** 1606-1608.

Sanders, L.C. and Lord, E.M. (1992) A dynamic role for the stylar matrix in pollen tube extension. *Intl Rev. Cytol.*, **140** 297-318.

Schmidt, R.J., Veit, B., Mandel, M.A., Mena, M., Hake, S. and Yanofsky, M.F. (1993) Identification and molecular characterization of ZAG1, the maize homologue of the *Arabidopsis* floral homeotic gene AGAMOUS. *Plant Cell,* **5** 729-37.

Schultz, E.A., Pickett, F.B. and Haugh, G.W. (1991) The FLO10 gene product regulates the expression domain of homeotic genes AP3 and PI in *Arabidopsis* flowers. *Plant Cell,* **3** 1221-37.

Shi, L., Zhu, T., Mogensen, H.L. and Keim, P. (1996) Sperm identification in maize by fluorescence *in situ* hybridization. *Plant Cell,* **8** 815-21.

Shifriss, O. (1961) Sex control in cucumber. *J. Hered.,* **52** 5-12.

Sinha, N.R., Williams, R.E. and Hake, S. (1993) Overexpression of the maize homeobox gene, Knotted-1, casues a switch from determinate to indeterminate cell fates. *Genes Dev.,* **7** 787-95.

Sommer-Knudsen, J., Clarke, A.E. and Bacic, A. (1996) A galactose-rich, cell wall glycoprotein from styles of *Nicotiana alata. Plant J.*, **9** 71-83.

Spanjers, A.W. (1981) Bioelectric potential changes in the style of *Lilium longiflorum* thumb. after self- and cross-pollination of the stigma. *Planta,* **153** 1-5.

Stead, A.D., Roberts, N. and Dickinson, H.G. (1980) Pollen-stigma interaction in *Brassica oleracea*: the role of stigmatic proteins in pollen grain adhesion. *J. Cell Sci.,* **42** 417-23.

Steer, M.W. and Steer, J.M. (1989) Pollen tube tip growth. *New Phytol.,* **111** 323-58.

Stein, J.C., Howlett, B., Boyer, D.C., Nasrallah, M.E. and Nasrallah, J.B. (1991) Molecular cloning of a putative receptor protein kinase gene encoded at the self-incompatible locus of *Brassica oleracea. Proc. Natl Acad. Sci. USA*, **88** 8816-20.

Sun, T-P., Goodman, H.M. and Ausubel, F.M. (1992) Cloning the *Arabidopsis* GA1 locus by genomic substraction. *Plant Cell,* **4** 119-28.

Tang, X., Hepler, P.K. and Scordilis, S.P. (1989) Immunochemical and immunocytochemical identification of a myosin heavy chain polypeptide in *Nicotiana* pollen tube. *J. Cell Sci.,* **92** 569-74.

Taylor, L.P. and Jorgensen, R. (1992) Conditional male fertility in chalcone synthase deficient petunia. *J. Hered.,* **83** 11-17.

Thorsness, M.K., Kandasamy, M.K., Nasrallah, M.E. and Nasrallah, J.B. (1991) A *Brassica* S-locus gene promoter targets gene expression and cell death to the pistil and pollen of transgenic *Nicotiana. Dev. Biol.,* **143** 173-84.

Thorsness, M.K., Kandasamy, M.K., Nasrallah, M.E. and Nasrallah, J.B. (1993) Genetic ablation of floral cells in *Arabidopsis. Plant Cell,* **5** 253-61.

Trebitsh, T., Staub, J.E. and O'Neill, S.D. (1997) Identification of a 1-aminocyclopropane/carboxylic acid synthase gene linked to the female (F) locus that enhances female sex expression in cucumber. *Plant Physiol.,* **113** 987-95.

Turcich, M.P., Hamilton, D.A. and Mascarenhas, J.P. (1983) Isolation and characterization of pollen-specific maize genes with sequence homology to ragweed allergens and pectate lyases. *Plant Mol. Biol.,* **23** 1061-65.

Venglat, S.P. and Sawhney, V.K. (1996) Benzylaminopurine induces phenocopies of floral meristem and organ identity mutants in wild-type *Arabidopsis* plants. *Planta,* **198** 480-87.

Verbeke, J.A. (1992) Fusion events during floral morphogenesis. *Ann. Rev. Plant Physiol. Plant Mol. Biol.,* **43** 583-98.

Walker, D.B. (1978) Morphogenetic factors controlling differentiation and dedifferentiation of epidermal cells in the gynoecium of Catharanthus roseus. *Planta,* **142** 181-86.

Wang, C., Rathore, K.S. and Robinson, K.R. (1989) The responses of pollen to applied electrical fields. *Dev. Biol.,* **136** 405-10.

Wang, H., Wu, H.-M. and Cheung, A.Y. (1996) Pollination induces mRNA poly(A) tail-shortening and cell deterioration in flower transmitting tissue. *Plant J.,* **9** 715-27.

Weigel, D. and Meyerowitz, E.M. (1994) The ABCs of floral homeotic genes. *Cell,* **78** 203-209.

Welk, Sr. M., Millington, W.F. and Rosen, W.G. (1965) Chemotropic activity and the path of the pollen tube in lily. *Am J. Bot.,* **52** 774-81.

West, M.A.L. and Harade, J.J. (1993) Embryogenesis in higher plants: an overview. *Plant Cell,* **5** 1361-69.

Westergaard, M. (1958) The mechanism of sex determination in dioecious flowering plants. *Adv. Genet.,* **9** 217-81.

Wilhelmi, L.K. and Preuss, D. (1996) Self-sterility in *Arabidopsis* due to defective pollen tube guidance. *Science,* **274** 1535-37.

Wilhelmi, L.K. and Preuss, D. (1997) Blazing new trails. *Plant Physiol.,* **113** 307-12.

Willemse, M.T.M., Plyushch, T.A. and Reinders, M.C. (1995) *In vitro* micropylar penetration of the pollen tube in the ovule of *Gasteria verrucosa* (mill) H Duval, *Lilium longiflorum* thumb: conditions, attraction and application. *Plant Sci.,* **108** 201-208.

Wu, H.-M., Wang, H. and Cheung, A.Y. (1995) A pollen tube growth stimulatory glycoprotein is deglycosylated by pollen tubes and displays a glycosylation gradient in the flower. *Cell,* **82** 393-403.

Yan, H., Yang, H.-Y. and Jensen, W.A. (1991) Ultrastructure of the micropyle and its relationship to pollen tube growth and synergid degeneration in sunflower. *Planta,* **184** 166-75.

Yanofsky, M.F., Ma, H., Bowman, J.L., Drews, G.N., Feldmann, K.A. and Meyerowitz, E.M. (1990) The protein encoded by the *Arabidopsis* homeotic gene agamous resembles transcription factors. *Nature,* **346** 35-39.

Ye, D., Oliveira, M., Veuskens, J., Wu, Y., Installe, P., Hinnisdaels, S., Truong, A.T., Brown, S., Mouras, A. and Negrutiu, I. (1991) Sex determination in the dioecious Melandrium. The X/Y chromosome system allows complementary cloning strategies. *Plant Sci.,* **80** 93-106.

Yin, T. and Quinn, J.A. (1995) Tests of a mechanistic model of one hormone regulating both sexes in *Cucumis sativas* (Cucurbitaceae). *Am. J. Bot.* **82** 1537-46.

Ylstra, B., Muskens, M. and van Tunen, A.J. (1996) Flavonols are not essential for fertilization in *Arabidopsis thaliana. Plant Mol. Biol.*, **32** 1155-58.

Yokota, T. (1997) The structure, biosynthesis and function of brassinosteroids. *Trends Plant Sci.*, **2** 137-43.

Yokota, E. and Shimmen, T. (1994) Isolation and characterization of plant myosin from pollen tubes of lily. *Protoplasma,* **177** 153-62.

Zimmerman, V.L. (1993) Somatic embryogenesis: a model for early development in higher plants. *Plant Cell,* **5** 1411-23.

7 Embryogenesis

Ramón A. Torres Ruiz

7.1 Introduction

In recent years molecular biologists have paid increasing attention to plant embryogenesis because of the undoubted progress in animal developmental biology and the question of whether plants make use of similar and/or different mechanisms to establish the blueprint of their body organization (Wardlaw, 1955; Johri, 1984; Raghavan, 1986; Meinke, 1991; Johri *et al.*, 1992; Lindsey and Topping, 1993; West and Harada, 1993; Goldberg *et al.*, 1994; Jürgens *et al.*, 1994; Laux and Jürgens, 1997). This chapter is organized in two parts. The first provides an overview of the main processes associated with embryo development while the second focuses on recent progress that has been made in advancing our knowledge of the genetic and molecular events regulating this phenomenon in Angiosperms. Much of the work that is described is based on studies of embryogenesis in *Arabidopsis thaliana* but other examples examine maize and rice, which are increasingly contributing to this field.

Angiosperms exhibit some intriguing features that are tightly linked to their evolutionary success (Esau, 1977; Ehrendorfer, 1978; Crane *et al.*, 1995). One of these is the structure of the flower itself, which can exhibit an enormous degree of complexity, often with astonishing adaptations. Flowers produce seeds that protect the embryo and are responsible for the spread of a species over considerably disparate areas. After shedding, the seed coat may degenerate and allow the seed to germinate. Germination is essentially brought about or maintained by the activity of (terminally located) meristems, often supported by highly elaborate endosperm structures or cotyledons that deliver nourishing substances (Steeves and Sussex, 1989). The young seedling is formed during embryogenesis and passes through a dormant stage interrupted by the onset of germination (i.e. embryogenesis can be considered as the phase between fertilization and germination). Exceptions to this scheme exist in orchids and some other plants where the embryo remains arrested at an early developmental stage and may undergo a period of after-ripening (Natesh and Rau, 1984). At the opposite end of the spectrum are monocotyledoneous embryos such as maize that exhibit advanced development and produce several leaf primordia prior to germination (Esau, 1977). With the exception of these examples, it seems that the basic body organization of the seedling is elaborated during embryogenesis. From the comparative genetic and molecular studies that have already been undertaken it seems that all Angiosperms, and perhaps other plants, share the same basic principles of development.

7.2   Descriptive studies of zygotic embryogenesis and apomixis

Seedlings of dicotyledoneous plants generally exhibit a stereotypic body organization (Figure 7.1). Exceptions to this 'core' body plan exist. For instance, anisocotyledoneous or (pseudo) monocotyledoneous species such as some Gesneriaceae, develop cotyledons of different size, or only one cotyledon (Rosenblum and Basile, 1984; Sanchez-Burgos and Dengler, 1988). This seems, at least in *Monocotylea*, to be the result of developmental competition between two cotyledon primordia (Tsukaja, 1997).

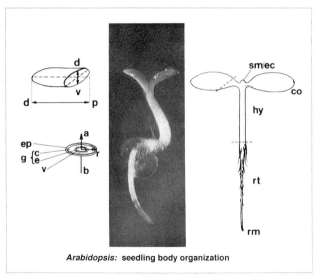

Arabidopsis: seedling body organization

Figure 7.1 Schematic representation of the main body axes in *Arabidopsis thaliana*. Middle: *Arabidopsis* seedling. Right: apical–basal axis with shoot meristem/epicotyl (sm, ec), cotyledons (co), hypocotyl (hy), root (rt) and root meristem (rm). Upper left: dorso–ventral (d, v) and proximo–distal (p, d) axes of the cotyledon (or subsequent leaves). Lower left: radial axis (r) with epidermis (ep), cortex (c) and endodermis (e), which together constitute ground tissue (g) and vascular tissue (v) with respect to apical (a)–basal (b) axis. For details see text.

7.2.1   Principal body organization of the seedling

In essence, in dicotyledoneous plants one can describe two principal axes along which organs and tissues are organized (Jürgens *et al.*, 1991, 1994; Jürgens, 1995; Laux and Jürgens, 1997). An apical–basal pattern is characterized by the succession of a shoot apical meristem, small epicotyl, in most cases two cotyledons, hypocotyl, root and root meristem. Superimposed on this is a radial pattern that is most clearly visible in the hypocotyl: epidermis, ground tissue (including endodermis) and vascular tissue. In addition, a proximo–distal axis and a dorso–ventral axis underlie the organization of the cotyledon and subsequent leaves. The petiole is proximal while the leaf blade is more distal to the main axis. At the same time the cotyledon has a dorsal (adaxial) and ventral (abaxial) surface. Internally the position of xylem and phloem delineates the dorsal/adaxial and ventral/abaxial sides of the cotyledon. In leaves this asymmetry can be more pronounced because of the differentiation of palisade parenchyma tissue versus ground tissue cells, mesophyll cells, the unequal distribution of stomata, etc. For instance, the pattern of this axis can be disturbed by single gene mutations as shown by *lop1* which causes partial rotation of xylem and phloem out of the normal axis (Carland and McHale, 1996).

The analysis of monocots has revealed that the apical shoot meristem does not originate laterally with respect to the cotyledon but adopts this position because during initial embryogenesis the primordium elaborating the cotyledon overgrows the shoot meristem and enlarges laterally (Natesh and Rau, 1984, and references therein). However, the principal body organization of a monocotyledoneous plant is very similar to that of a dicot (Jürgens *et al.*, 1994).

7.2.2 *Basic processes in plant embryogenesis*

The determination of the correct spatial organization of tissues and organs, which is best described by the term 'pattern formation', is only one of the main processes that takes place during the embryogenesis of the plant (Jürgens *et al.*, 1994). Seedlings exhibit specific dimensions and their different organs occupy precise proportions such that from generation to generation a characteristic body shape is achieved. Work with mutants suggests that, at least in part, this is a consequence of the shape of the cells that comprise the body of the plant (see below). Growth processes including cell division and/or cell elongation (which also influences shape) presumably occur during the last stages of embryogenesis. The most interesting point in this respect is how different tissues coordinate their growth in relation to each other and how their absolute size is determined. Progress in this field would help to understand phenomena connected with allometry, i.e. the diverse growth of tissues between related species that leads to contrasting physical appearances although the body pattern is not altered. Allometry seems to be of major importance for some aspects of plant and animal breeding (Leibenguth, 1982). The early establishment of distinct tissues and organs points to processes of cell commitment and differentiation where preceding positional information has been sensed and correctly interpreted.

7.2.3 *Important events preceding zygotic embryogenesis*

When considering embryogenesis some aspects of the life cycle of a plant have to be recognized. Of greatest importance is the alternation of gametophytic and sporophytic generation in angiosperms. It is clear that the gametophytic development by itself involves important steps and fundamental events preceding zygotic embryogenesis (see chapter 6).

In the nineteenth century two major botanical discoveries were made by Hofmeister, Strasburger and Nawaschin. Firstly, gametophytes represent an extremely reduced but partly autonomous (haploid) generation in angiosperms and, secondly, double fertilization generates the zygote and endosperm (see Ehrendorfer, 1978; Johri *et al.*, 1992). The male gametophyte consists of one vegetative and two generative cells generated through mitotic divisions of one of the meiotic products of the microspore mother cell (Rutishauser, 1969; Johri *et al.*, 1992; McCormick, 1993; Reiser and Fischer, 1993). The mitotic division of the generative cell can be postponed in some plants to a time when the pollen tube reaches the egg (Rutishauser, 1969; Johri *et al.*, 1992). In contrast to microspore development, where all meiotic products lead to gametophytes, the development of macrospores often includes the degeneration (or programmed cell death, see chapter 11) of all but one meiotic product (Rutishauser, 1969; Johri *et al.*, 1992; Reiser and Fischer, 1993; Russell, 1993). The resulting female gametophyte, the embryo sac, often develops through three mitotic divisions of the remaining nucleus leading to an eight-nucleated, seven-celled embryo sac (Figure 7.2). In monosporic embryo sac development all nuclei are genetically identical. This may be different in disporic (developed from two meiotic products) or tetrasporic (participation of all meiotic products) embryo sacs. One commonly known 'prototype' of embryo sac, the *Polygonum* type, occurs in mono- and dicotyledoneous plants (Rutishauser, 1969; Reiser and Fischer, 1993). For example, this type has been described in detail in *Capsella bursa-pastoris* development (Schulz and Jensen, 1973, 1986) and is essentially also followed in *A. thaliana* (Mansfield and Briarty, 1991; Schneitz *et al.*, 1995) although the latter seems to show a rare deviation from this type (Webb and Gunning, 1990).

The embryo sac of *Arabidopsis* displays a well-defined and polar organization (Figure 7.2). At the micropylar pole the egg cell is accompanied by two further cells, the synergids that nurture the egg cell and support fertilization, for which the so-called filiform apparatus has been developed (Johri *et al.*, 1992; Reiser and Fischer, 1993). On the opposite chalazal pole the three antipodal cells are localized and these may also provide nourishment (Johri *et al.*, 1992; Reiser and Fischer, 1993). The centre is occupied by the diploid central cell which, after double fertilization, leads to the triploid endosperm. The embryo sac is not only polarized by its organization but also by its position within the ovule, i.e. egg and synergids are in contact with the most distally located nucellus and integument cells whereas antipodes are in contact with the chalazal tissue. The ovule arises from placental tissue, which also bears the macrospore mother cell, and often develops one or two integument cell layers, which surround the embryo sac, leaving the micropyle 'open' (Figure 7.2; see Rutishauser, 1969; Mansfield *et al.*, 1991; Johri *et al.*, 1992; Schneitz *et al.*, 1995; Angenent and Colombo, 1996).

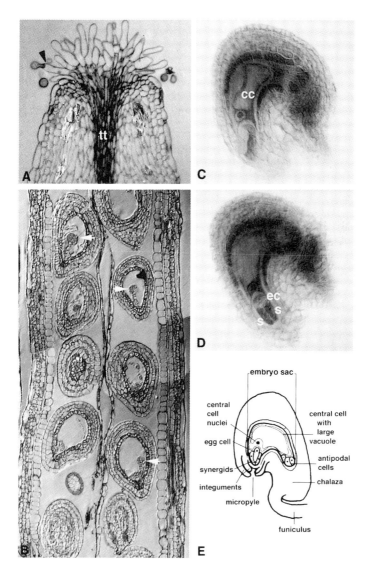

Figure 7.2 *A. thaliana*: silique, ovule and embryo sac organization. A: Upper (stigma and stylus) region of the silique with enlarged papillar cells and transmitting tissue (tt). The arrowhead points to a pollen tube invading a papillar cell. B: Lower part (ovar) of the silique separated internally into two chambers. Ovules are organized in rows. The arrowheads point to developing embryos (16 cell to early globular stage). C and D: View of an embryo sac in a young ovule (cc = central cell; ec = egg cell; s = synergide). E: Schematic presentation of ovule and embryo sac. (Parts C and D kindly provided by K. Schneitz, University of Zürich).

Ovules are organized in rows within the two chambers of the silique (Figure 7.2).

In other plants significant deviations from this type of embryo sac occur. Thus the number of embryo sac cells may be lowered or elevated considerably; some embryo sacs lack antipodes, others possess central or antipodal cells with higher ploidy numbers (Rutishauser, 1969; Johri *et al.*, 1992). Deviations occur even within species, e.g. for *Arabidopsis* variation in the time point when antipodes degenerate has been reported (Murgia *et al.*, 1993).

Embryologists have paid much attention to polarization and have noticed that in numerous cases the egg cell exhibits internal polarization, which is characterized by unequal distribution of cytoplasm, (m)RNA, proteins and organelles (Schulz and Jensen, 1968; Natesch and Rau, 1984; Mansfield *et al.*, 1991). Thus polarity in zygotes is often preceded by polarity in egg cells.

On pollination the microspore develops (if compatible) an outgrowth that enters the stigma and grows within a special tissue (tela conductrix, Rutishauser, 1969; transmitting tissue, Mascarenhas, 1993) towards the egg cell, possibly conducted by chemical signals (discussed in Mansfield *et al.*, 1991; Mascarenhas, 1993; Wilhelmi and Preuss, 1996, and references therein). It leaves the gynoecial tissue at the base of the funiculus (the connection between ovule and placenta), grows further and enters the micropyle. The pollen tube intrudes into the filiform aparatus of the degenerating synergids and releases the two generative cells, which fuse with their targets, the egg and central cell (Russell, 1993). After fusion of the plasma membranes the nuclei of the male gametes migrate towards the egg and central cell nuclei to give rise to the zygote and first triploid endosperm cell. The latter undergoes further mitoses and a complex developmental program before finally acting as a source of nutrients for the embryo (Mansfield and Briarty, 1990a,b) or the seedling (Lopes and Larkins, 1993).

Double fertilization leading to embryo and endosperm has generally been considered as a special feature of angiosperms. However, in *Ephedra*, which is the most basal member of the Gnetales, double fertilization of an egg nucleus and a second (ventral canal) nucleus is known (Friedman, 1990, 1992). Both diploid nuclei undergo further divisions that could potentially lead to eight embryos. In *Ephedra trifurca* only one embryo, which typically originates from the true zygote nucleus, survives (Friedman, 1992). Since Gnetales represent the most closely related extant group of seed plants, this finding supports the view that endosperm originates from a supernumerary specialized or reprogrammed embryo that has nutritive functions. The endosperm can be considered as a modified evolutionary homologue of the embryo and double fertilization a synapomorphy of angiosperms and their sister group Gnetales (Friedman, 1990, 1992).

7.2.4   Critical stages of embryo development

7.2.4.1   Classification of cell division routes leading to plant embryos
After fertilization the zygote does not immediately enter the next division. It has been noted that zygote dimensions may be altered (Natesh and Rau, 1984). In

Arabidopsis this results in elongation along the micropylar chalazal axis (which corresponds with the apical–basal axis of the incipient embryo) and correlates with the rearrangement of microtubules (Webb and Gunning, 1991). Most remarkably, in some cases polarization may be achieved or enhanced by (dramatic) post-fertilization shifts that rearrange the distribution of cytoplasm, organelles and vacuoles while in other plants the unequal distribution of cytoplasm and organelles in the egg is also found to extend into the zygote (Schulz and Jensen, 1968; Natesh and Rau, 1984).

Generally the zygote undergoes a transverse and unequal division thought to be connected with the observed polarization. However, it is important to note that some plants seem to deviate from this route by performing vertical or oblique zygotic divisions (Rutishauser, 1969; Natesh and Rau, 1984; Johri *et al.*, 1992). This has led to a separate classification of this group of plants with respect to their embryogenesis: the Piperaceen-type of embryogenesis (Rutishauser, 1969; Johri *et al.*, 1992). Most notably, plants of this group, such as *Korthasella dacrydii*, *Rafflesia* and *Balanophora*, seem to start with unpolarized zygotes (Rutishauser, 1935 in Rutishauser, 1969). This fact, as well as the development of somatic embryos (see below), must be considered when the significance of polarization is discussed. Apart from these exceptions the zygote divides into a large and highly vacuolated basal cell (oriented towards the micropylar pole; often designated as *cb, cellula basale*) and a cytoplasmically dense apical cell (towards the chalazal pole; *ca, cellula apicale*). This difference in cell fate is also highlighted by the unequal distribution of the transcript of the *ATML* gene (Lu *et al.*, 1996). The *cb* generally gives rise to the suspensor, an extraembryonic filamentous organ that probably promotes the continued growth of the embryo proper (Yeung and Meinke, 1993). Plants differ with respect to the fates of *ca* and *cb. ca* may totally contribute to the embryo proper and even part of the suspensor (Caryophyllaceen-type) while *cb* solely leads to (parts of) the suspensor. At the other extreme *cb* may contribute to the suspensor and a considerable part of the embryo proper while *ca* only provides part of the embryo (see Rutishauser, 1969; Johri *et al.*, 1992). A further diagnostic feature for the classification of plants is the pattern of division taken by *ca* (and *cb*) which is limited to certain variants. This and the contribution of *ca* and *cb* to embryo and suspensor has led to several variants of embryo type classification. The most common classification subdivides embryos that start with a transversal zygotic division into five types: Onagrad, Asterad, Solanad, Chenopodiat and Caryophyllad. The sixth type is the Piperad. An alternative (simplified) classification emphasizes the contribution of *ca* and *cb* to the embryo proper (see Esau, 1977). Although a classification is quite useful to compare different plants the tight adherence to one type of classification may be problematical. This is best demonstrated by those species that realize more than one type of developmental route as is the case, for example, in *Gossypium* (Pollock and Jensen, 1964). For more detailed information see Maheshwari (1950), Rutishauser (1969), Esau (1977), Raghavan (1986) and references therein).

Despite differences in division patterns it should be kept in mind that most, if not all, embryos develop very similarly through the more advanced globular, heart and torpedo stages, and reach the basic organization described at the beginning of this chapter. It should be noted that mono- and dicotyledoneous embryos develop similarly (see below). The detailed course of *Arabidopsis* embryogenesis is discussed below. For comparison an overview of embryo development in a monocotyledenous plant is included.

7.2.4.2 Dicot versus monocot development

Descriptive analysis has been especially intensive in cruciferous plants such as *Capsella bursa-pastoris*, *Brassica* sp. and, more recently, *Arabidopsis* (Hanstein, 1870; Schulz and Jensen, 1968; Tykarska, 1976, 1979; Mansfield and Briarty, 1991; Jürgens and Mayer, 1994). All three develop similarly and follow the Onagrad type of development (see Figure 7.3). The vacuolated basal cell (*cb*) that results from division of the zygote (Figure 7.3A) divides until a filamentous row of seven to nine cells is reached (at this stage the embryo proper is at the globular stage). The suspensor cells retain characteristic vacuolization and are assumed to serve as a bridge across which nutritive substances can pass from maternal tissue to the embryo (Mansfield and Briarty, 1991). This assumption is also supported by observations of other species where suspensors develop into conspicuous, haustorial-like organs (Johri *et al.*, 1992; Yeung and Meinke, 1993). In *Arabidopsis* the suspensor, i.e. the basal cell *cb*, partly contributes to the embryo proper (Onagrad type of embryo development). The uppermost cell, the hypophysis (Figure 7.3E) intrudes into the upper cell group where it undergoes a characteristic sequence of divisions that results in the development of a conspicuous lens-shaped cell (hypophyseal cell group). At the globular stage the precursor of that root meristem is clearly distinguishable from the rest of the embryo (Figure 7.3F,G). The apical cell resulting from the division of the zygote (*ca*) leads to the larger part of the embryo. Certain stages during the course of the following divisions seem to be of special significance. Through two vertical divisions of *ca* the quadrant stage is achieved. Embryologists have traced back the origin of cotyledon primordia in mono- as well as dicotyledonous plants back to this early stage (Natesh and Rau, 1984). A subsequent transverse partition of the quadrant produces the octant stage which again seems to mark a critical step (Figure 7.3C,D). The boundary between the upper and lower group of (four) cells (upper tier and lower tier, respectively) is often termed the O'-boundary (Tykarska, 1976, 1979; West and Harada, 1993). According to this observation, the cotyledons, epicotyl and shoot meristem primordia originate from the upper four cells (upper tier) while the lower group (lower tier) and the hypophysis constitute the rest of the embryo. This is largely, but not entirely, correct. Clonal analysis has shown that the shoulder region of the cotyledons originates from

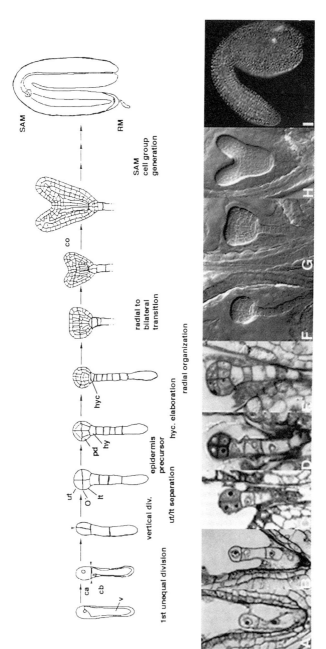

Figure 7.3 Arabidopsis embryo development from zygote to mature embryo. Top: schematic representation of *A. thaliana* development with important events indicated. Cellule apicale (*ca*) and cellule basale (*cb*) give rise to the main part of the embryo (*ca*) and the suspensor, including the hypophyseal cell group (*cb*). v, vacuole; ut, upper tier; O, O' boundary; lt, lower tier; pd, protoderm; hy, hypophysis cell (uppermost suspensor cell); hyc, hypophyseal cell group; co, cotyledon; SAM, shoot apical meristem; RM, root meristem (SAM and RM indicated by small spotted regions). Bottom: sections and whole mount preparations of different stages. A: generation of *ca* and *cb* after first zygotic division. B: after additional division of *cb*. C: young embryo (densely stained cells) between quadrant and octant stage. D: octant stage embryo. E: 16-cell stage. F: early globular stage. G: triangular stage. H: heart stage embryo. I: advanced torpedo stage embryo. A–E: semi-thin sections (3.5 μm); F–I: whole mounts, not to scale.

heart stage cells located below the O'-boundary (Scheres *et al.*, 1994). A set of tangential divisions of the octant stage cells separates the precursor of the epidermis (protoderm) from a set of internally located cells (which lose any contact to the external environment, dermatogen stage; Figure 7.3E). In the following stages the protoderm increases by anticlinal divisions. The centrally located cells undergo a series of longitudinal and transverse divisions such that precursors of cortex and vascular cells separate and vascular precursor cells become recognizable in the centre by their (vertically) elongated shape. At this mid-globular stage (Figure 7.3F) the basic radial pattern is achieved (see above). The globular shape of the embryo persists for additional divisions until its organization is significantly changed. At the late globular stage the radial symmetry of the embryo is transformed into a bilateral symmetry by the enhanced activity of cells at opposite sites of the apical region. A locally increased frequency of divisions produces incipient cotyledon primordia, which reach the level of the most apically located cells; this gives the embryo a triangular-like appearance (triangular or transition stage; Figure 7.3G). The incipient primordia overgrow the centrally located cells, leading to characteristic heart-shaped embryos (early heart stage). The heart stage embryo (Figure 7.3H) undergoes further divisions and exhibits cell growth, especially in the lower tier. This enables the hypocotyl to be recognized. Thus, by this stage the embryo exhibits (nearly) all the elements of the apical–basal pattern. One exception, pointed out by Barton and Poethig (1993) is the group constituting the precursor of the shoot apical meristem. This is morphologically distinguishable only at later torpedo stages (Barton and Poethig, 1993). However, *in situ* analyses by Long *et al.* (1996) using the *SHOOT MERISTEM* gene indicate that a group of subepidermal, central cells might be already committed to develop into shoot apical meristem (SAM) precursors, which means that SAM is already present at the globular stage as anlage. In contrast to the SAM the root (apical) meristem (RM) is developed and distinguishable at the heart stage. With the exception of the SAM, subsequent development does not add qualitatively to this basic organization. Rather the embryo divides further and grows by cell elongation (torpedo stage, Figure 7.3I). Physiologically it prepares for desiccation and germination, e.g. the storage of protein and starch in the hypocotyl and cotyledons (Goldberg *et al.*, 1989; Mansfield and Briarty, 1992). In late torpedo stages cotyledons bend, giving the seed its characteristic shape (cotyledon stage). The suspensor has degenerated and the endosperm has been hydrolysed. At the end of development the embryo enters desiccation.

Whether or not mono- and dicotyledoneous embryogenesis are comparable has been a much debated issue (Rutishauser, 1969; Natesh and Rau, 1984, Johri *et al.*, 1992). In particular, the origin of the SAM and the cotyledons remains unclear. One hypothesis that has received some support is that separate modes of development take place in both groups (see Natesh and Rau, 1984; Johri *et al.*, 1992). According to this view in monocots the apical meristem and cotyledons originate from different 'tiers', i.e. the cotyledon from a tier of apical cells and the shoot meristem precursor from cells beneath this tier. More recent investigations of different species have revealed similarities in early embryo stages between both groups (see Natesh and Rau, 1984, and references therein). In both mono- and dicotyledoneous plants the apex 'behaves' similarly and is organized late in development. However, in contrast to monocotyledoneous plants it is not displaced in dicots as, in general, both cotyledons grow symmetrically.

Morphological and embryological comparisons with dicots are less problematical if monocotyledoneous plants like onion (*Allium cepa*) are used in the comparison rather than grasses, which exhibit a more complex organization. Nevertheless in this section the development of a grass embryo is considered, as maize and rice are the foci of combined genetical and embryological studies. As mentioned previously, the first division stages seem to be similar in their development between monocots and dicots. Early patterns of cell division can be more (barley) or less (maize) regular (Randolph, 1936; Norstog, 1972). The resulting embryo is club-shaped; the upper dome-shaped part represents the embryo proper and the lower part the suspensor (Clark, 1996). Cells of a region located beneath the most apical part of the club-shaped embryo begin to divide and form the incipient SAM. This is visible as a small lateral bulge that is separated by an indentation from the rest of the embryo. At some distance beneath the SAM, on the same side, the RM begins to develop internally. The region laterally and above the SAM expands and forms the scutellum, which is thought to originate from the cotyledon or to be itself an elaborated cotyledon (see Esau, 1977, and references therein). This organ connects the embryo with the well-developed endosperm, which is greatly enlarged in grasses and supports the germination of the seed. The SAM is further surrounded by a special organ called the coleoptile. In contrast to dicots and some monocots (e.g. *Allium cepa*) the SAM often develops within the coleoptile additional leaf primordia prior to embryo maturity (i.e. the epicotyl consists of SAM and leaf primordia). Maturation of the embryo ends with the entrance into dormancy phase (Clark, 1996).

Within the monocots the grass embryo appears more evolutionarily advanced and exhibits considerable morphological complexity and this situation has led to controversial discussions on the origin of the different organs (see Esau, 1977 and references therein).

7.2.5 Apomixis and polyembryony

In the past apomixis was a term that covered all forms of asexual reproduction. This definition dates back to 1908 (Winkler, 1908, cited in Rutishauser, 1969) and states that, as in vegetative reproduction, the fertile seeds resulting from apomixis are genetically identical to the female parent (except in rare cases). This definition's disadvantage is that it includes all events of vegetative reproduction in plants that embrace a considerable number of variants (Rutishauser,

1969). A more restricted definition stresses the asexual formation of seed, i.e. asexual reproduction in the ovule in angiosperms (Rutishauser, 1969; Nogler, 1984; Koltunov, 1993). It emphasizes the development in ovules and the occurrence of different generations and does not exclude fertilization, which occurs in pseudogamous species (see below). It also separates apomixis from somatic embryogenesis since it includes embryo formation close to gametophytic structures without passing through a callus phase (see Koltunov, 1993). In addition, one should bear in mind that occasionally unreduced egg cells may be fertilized such that a sometimes viable zygote with altered ploidy arises.

Apomixis mimics the processes of sexual reproduction, i.e. female gametogenesis, fertilization and embryo development. It may actually have developed from them. For instance it is conspicuous that the mechanism of apomictic embryo sac development in one species is generally similar to that of its sexual relatives (Rutishauser, 1969). Three forms are often differentiated: apospory, diplospory and adventitious embryogeny arising from nucellus or inner integument tissue (Koltunov, 1993). The three variants can be described as starting at different stages of gametophytic development.

Aposporic initials develop from nucellar cells that are different from the megaspore mother cell (sometimes they arise from integumental cells). They are often conspicuous by having dense cytoplasmic staining and prominent nucleoli (Rutishauser, 1969). They lead to unreduced embryo sacs by mitotic divisions. Interestingly, they can coexist in one ovule with 'normal' sexual/reduced embryo sacs (Koltunov, 1993). Two forms are known: Hieracium type and Panicum type. In the former a bipolar two-nucleated embryo sac is formed. Two further mitoses produce an embryo sac that is comparable to the Polygonum type of embryo sac (see above) except that the nuclei are diploid. In the latter only two mitoses are performed and lead to a monopolar four-celled embryo sac with one egg cell, two synergids and one central cell with one polar nuclei (see Rutishauser, 1969). All sexual relatives of the apomictic Hieracium species exhibit the Polygonum type of monosporic embryo sac development. The Panicum type only occurs in the Panicoidae and is possibly derived from the Hieracium type. For historical reasons it should be remembered that Mendel, after having done most of his famous work in peas, worked, strongly encouraged and influenced by the botanist Nägeli, with Hieracium. It was apomixis that caused him serious difficulties and frustation since his results were not interpretable at that time (Czihak, 1984).

In diplosporic apomixis development starts with the megaspore mother cell (Rutishauser, 1969; Koltunov, 1993). There are different variants known that demonstrate alterations of meiosis, all of which aim to produce embryo sacs with diploid nuclei. It is most common for deviations to occur in meiosis I, which is often interrupted in anaphase or metaphase (leading to restitution nucleus and meioitic diplospory) or may be completely reduced (mitotic diplospory). There is also mention of at least one further type that starts with premeiotic endomitosis;

this is *Allium nutans* type (Rutishauser, 1969).

Adventitious embryogenesis starts from cells of the nucellus or inner integument of the ovule (Lakshmanan and Ambegaokar, 1984). One of the best analysed plants in this respect is *Citrus*, which is increasingly becoming a model system for analysis of nucellar embryony (Koltunov, 1993). As in apospory those cells that are committed to become embryos are recognizable by virtue of their large nuclei and dense cytoplasm. In *Citrus* nucellar initial and zygotic cells develop at the same time. As more than one embryo develops, in some cases more than 50 initials are reported (Wakana and Uemoto, 1988; Koltunov, 1993) and the ovule is enlarged. However, only the initials at the micropylar region form embryos and grow into the embryo sac. This observation is interesting for two reasons: firstly, it suggests an inhibitory effect by the chalazal region; secondly, the proembryo is directed towards the embryo sac, possibly under the influence of the endosperm. Nucellar and zygotic embryos develop similarly or are indentical. Not all initiated embryos can fully maturate (especially in the chalazal region) but at the end seeds result that harbour several (nucellar and sexual) embryos (Koltunov, 1993). This type of apomixis has therefore also been considered to be like polyembryony.

Little is known about the events that initiate apomixis. However, it is known for several plants that entering this route of embryogenesis is dependent on fertilization or pollination of polar cells and, moreover, apomictic embryo development can be dependent on the endosperm (Rutishauser, 1969). The term used in this case is pseudogamy, for seed development with unreduced gametophytes, or induced nucellar embryony, in the case of adventitious embryony. If pollination is not necessary seed development is autonomous. Thus, unreduced gametophytes show diploid parthenogenesis and vegetative nucellar cells show autonomous nucellar embryony.

As discussed above, apomixis may lead to polyembryony. However, polyembryony can also arise by events that are not connected to apomixis (Johansen, 1950; Lakshmanan and Ambegaokar, 1984). Johansen (1950) has proposed a strict definition for polyembryony. This states that either eggs give rise to a multicellular body from which several embryos arise or additional embryos arise on a normal embryo by epidermal budding (observed in monocots, *Erythronium americanum*), or the zygote divides and two daughter cells develop more or less separately (*Cymbidium bicolor*).

Experimentally induced (chemical, *in vitro*) polyembryony is always excluded (e.g. *Eranthis*, Haccius, 1960). Further possibilities are given by embryos that develop from synergids and asexual embryos (gametophytic with haploid, diploid and sporophytic forms; see Johri, 1984). Polyembryos can occasionally arise in certain *Arabidopsis* lines (Figure 7.4B,C). However, in most cases it is unclear how and from which cells they originate. In some cases polyembryony can lead to the production of twins. These may possess different ploidy and/or different genetic identity (Figure 7.4B,C). Despite the cases counted as apomictic (i.e. adventitious embryogeny) in general very little is known about the inductive and mechanistic reasons for polyembryony. The advent of *A. thaliana* mutants that are related to apomixis and polyembryony show that this system might be suitable for analysing such processes at the molecular level (see below).

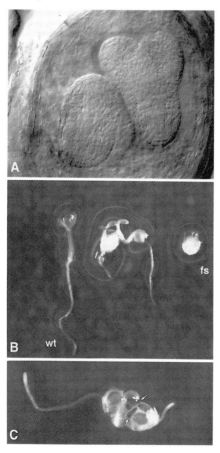

Figure 7.4 *A. thaliana* polyembryos. A: Ovule harbouring two embryos at late heart stage. B and C: Triple seedling. In this case a mutant *fass* seedling is fused with two wild-type seedlings (B: side view; C: top view; small arrows point to *fass* cotyledons).

7.3   Genetic and experimental analysis of embryo development in *Arabidopsis* and other plants

Direct experimental analysis of plant embryos is difficult since they develop in ovules that are in turn enclosed by fruits. Numerous approaches have been undertaken to analyse plant embryos. Attempts have been made to isolate and cultivate embryos in sterile culture, aiming to alleviate experimental manipulation (Raghavan and Torrey, 1963; Wu *et al.*, 1992; Liu *et al.*, 1993a). Young embryos have turned out to be sensitive to this approach but more advanced stages could be shown to respond to auxin transport blockers in a characteristic way (Liu *et al.*, 1993b). In certain cases the difficulties faced with the 'explant' approach can be bypassed by generating embryos artificially. The best-characterised systems are carrot somatic embryogenesis, embryos generated from pollen and the generation of embryos by *in vitro* fertilization (Reinert, 1959; Kranz and Lörz, 1993; Kranz and Dresselhaus, 1996; see Mordhorst *et al.*, 1997; Reynolds, 1997). For space limitations this chapter focuses on embryogenesis *in vivo*. Special aspects originating from artificial embryogenesis, which are of immediate significance for zygotic embryo development, are integrated where appropiate.

In recent years the approach of undertaking a genetic and molecular analysis has been applied to plant embryogenesis. The power of this strategy has already resulted in a considerable enhancement of our understanding of pattern formation and morphogenesis in the fruitfly *Drosophila melanogaster* (Nüsslein-Volhard, 1991) and has been successfully applied to the study of flower development (Yanofsky, 1995).

7.3.1 *Uncovering important gene functions for embryogenesis*

Numerous mutagenesis screens have provided embyro defect/lethal and seedling mutants, especially in *Arabidopsis* (Müller, 1963; Meinke and Sussex, 1979; Meinke 1985, 1996 and references therein; Jürgens *et al.*, 1991, 1994; Feldman, 1991; Benfey *et al.*, 1993; Scheres *et al.*, 1995; Laux *et al.*, 1996), maize (Neuffer and Sheridan, 1980; Sheridan and Neuffer, 1982; Clark and Sheridan, 1991; Sheridan and Clark, 1993) and rice (Nagato *et al.*, 1989; Hong *et al.*, 1995). To a lesser extent the search for embryo mutants has also taken place in other species, such as *Petunia* and pea, and these studies have added important imformation to the field (Dubois *et al.*, 1996; Liu *et al.*, 1996). Once embryo mutants have been characterized, the next stage is commonly to clone the mutant gene responsible. Positional cloning methods, transposon tagging with homologous/heterologous systems (*Petunia*-transposon/En-I/Spm, Ac/Ds) and insertional mutagenesis with T-DNA (Shevell *et al.*, 1994; Souer *et al.*, 1996; Lukowitz *et al.*, 1996; Busch *et al.*, 1996; Di Laurenzio *et al.*, 1996) have all been employed to achieve this. Additionally, in screens designed to find mutants affecting the adult plant some have been found whose defects can be traced back to the embryonic stages, for instance the *pinoid* mutants with supernumerary cotyledons (Bennett *et al.*, 1995). Further interesting developmental mutants are identified by gene and promoter trap tagging experiments, which also provide cell and tissue specific markers which are particularly valuable for studies of embryo mutants (Topping *et al.*, 1991; Springer *et al.*, 1995; Topping and Lindsey, 1997).

The mutants found cover all aspects of embryogenesis, as mentioned above, as well as basic cell functions. It is therefore convenient to discuss them under the different aspects of embryo development.

Conspicuous morphological mutants have been found that affect body organization and shape in *Arabidopsis*. One class that has attracted considerable attention is characterized by missing, altered or defective pattern elements. Mutants in the gene *GNOM*, which turned out to be allelic to the formerly found mutant *emb30* (Meinke, 1985), eliminate apical and basal structures (Figure 7.5; Mayer *et al.*, 1991, 1993a). A group represented by *shoot meristemless* (*stm*; Barton and Poethig, 1993; Endrizzi *et al.*, 1996), *pinhead* (*pnh*; McConnell and Barton, 1995) and *zwille* (Figure 7.6; *zll*; Jürgens *et al.*, 1994; Endrizzi *et al.*, 1996) exhibit partial or perfect obliteration of the shoot apical meristem (SAM). The mutant *wuschel* possesses a reduced SAM and grows by periodic proliferation of stem and rosette leaves (Laux *et al.*, 1996). Similar screens in maize have revealed several cases where the shoot apical meristem primordium is underdeveloped or missing (Sheridan and Clark, 1993). In petunia the *no apical meristem* mutant is likely to fall into this class (Souer *et al.*, 1996). A further group of mutants, *gurke* (*gk*), *pepino* (*pep*) and *R63-11*, exhibit defects in the shoot apical region (shortly apical region), including cotyledons, and in the hypocotyl (Figure 7.7; Mayer *et al.*, 1991; Torres *et al.*, 1996a,b). In weaker *gk* alleles cortical cells are inappropriately expanded along the radial axis. Strong alleles segregate extreme phenotypes which have obliterated a considerable part of the hypocotyl and mainly consist of a root that in all alleles appears essentially normal by morphological criteria (Torres Ruiz *et al.*, 1996b). *pepino* and *R63-11* resemble weak *gk* alleles (Torres Ruiz *et al.*, 1996a). The mutant laterne differs from the former in that cotyledons are precisely deleted (or sometimes supernumerary) while the hypocotyl is elaborated as in wild-type (Mayer *et al.*, 1993b; Torres Ruiz *et al.*, 1996a; Torres Ruiz, unpublished). Comparable to this is *lanceolata* reported from tomato (Caruso, 1968). The most severe defect of *fackel* mutants resides in the central part of the seedling (Mayer *et al.*, 1991). Thus obviously one group of genes has a spatially restricted task in elaborating the SAM. A second group is probably involved in organizing the apical region and hypocotyl. It is striking that this apparent relation between apical region and hypocotyl has an inverse counterpart at the opposite pole represented by *monopteros* mutants, which lack root and hypocotyl (Figure 7.8). Moreover, as opposed to those genes specifying the SAM, the mutants *hobbit*, *bombadil*, *gremlin* and *orc* display localized defects in the hypophyseal cell region (Figure 7.9; Scheres *et al.*, 1996). Root expansion mutants such as *pompom*, *cobra*, *lion's tail*, etc. have to be separated from the aforementioned as they do not obliterate (pattern) elements of the root (Hauser *et al.*, 1995).

A genetic screen performed to detect radial pattern mutants was based on the rationale that such mutants could cause retardation of root growth (Scheres *et al.*,

Figure 7.5 Comparison of wild-type *Arabidopsis* (right) and *gnom* mutant (left).

Figure 7.6 *Arabidopsis* shoot meristem mutants. A: Wild-type seedling. B and D: *zwille* mutant. In D the *zwille* seedling has developed one terminal leaf between the cotyledons. C: *wuschel* mutant (not to scale). (Kindly provided by Th. Laux and B. Moussian, University of Tübingen.)

1995). This idea arose from the phenotype of the mutant *short root* (*shr*), which had been detected before (Benfey *et al.*, 1993). This and the subsequently found mutants *scarecrow* (*scr*) and *pinocchio* interfere with cortex versus endodermis

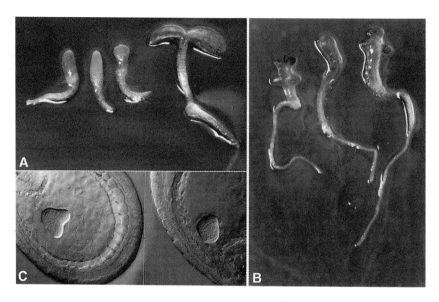

Figure 7.7 *A. gurke* mutants. A: *gurke* (strong allele) in comparison to wild-type. B: Weak *gurke* allele. C: Comparison of wild-type (left) and *gurke* embryo (right), not to scale.

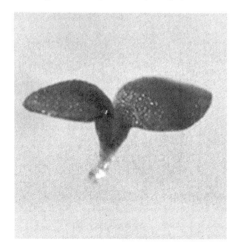

Figure 7.8 *A. monopteros* mutant. Note cotyledons insert in a basal peg. Hypocotyl and root are missing. (Kindly provided by Th. Berleth and Chr. Hardke, Ludwig-Maximilians University, München).

specification while *wooden leg* and *gollum* affect vascular tissue (Benfey *et al.*, 1993; Scheres *et al.*, 1995, 1996; Di Laurenzio *et al.*, 1996). In contrast *knolle*

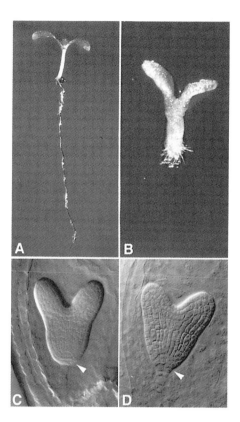

Figure 7.9 *Arabidopsis hobbit* mutant. A: Wild-type, B: *Hobbit* mutant. C: Wild-type heart stage embryo. D: *Hobbit* heart stage embryo. The arrowheads point to wild-type (C) and the abnormal (D) hypophyseal cell group (Parts A, B and D from Willemsen *et al.* (1998), *Development*, **125** 521-33, Company of Biologists Ltd., with permission.)

(*kn*) and *keule* (*keu*) seem to disturb cellular differentiation but not radial pattern (Mayer *et al.*, 1991; Lukowitz *et al.*, 1996; Assaad *et al.*, 1996). This collection of mutants immediately leads to the question on their possible mechanistic interconnection in apical–basal versus radial specification.

Additional aspects of body organization and organ identity were uncovered with quite different phenotypes. For instance, the mutants *häuptling* and *amp* develop supernumerary cotyledons but have also other defects (Figure 7.10; Jürgens *et al.*, 1991; Chaudhury *et al.*, 1993). Others have striking phenotypes that are connected with embryo organ identity or polarity as well as cell identity (Jürgens *et al.*, 1991; Meinke, 1992; Vernon and Meinke, 1994). Duplications of the root, as seen in the doppelwurzel-phenotype (Jürgens *et al.*, 1991), have also been observed in rice (Hong *et al.*, 1995).

Figure 7.10 *Arabidopsis häuptling* mutant. A: Wild-type (the arrowheads point to trichomes on the primary leaves; cotyledons do not develop trichomes). B to D: *häuptling* mutants with different numbers of cotyledons (four in B, five in C and three in D). (Parts B to D kindly provided by S. Ploense.)

Quite different effects were displayed by those mutants that from early on were classified as affecting the body shape evoked by striking cell shape alterations as seen in *fass* (Figure 7.11), *knopf, mickey* and *enano* (Mayer *et al.,* 1991; Torres Ruiz and Jürgens, 1994; Traas *et al.,* 1995; Torres Ruiz *et al.,* 1996a). In addition, these mutants led to arrested seedlings that are extremely small in comparison with wild-type (Torres Ruiz and Jürgens, 1994; Torres Ruiz *et al.,* 1996a). Again similar, if not identical, mutants have been found in other systems. The *tra1* mutant from *Petunia* exhibits an identical body and cellular phenotype in comparison to *fass* (Figure 7.11; Dubois *et al.,* 1996). The mutants *knolle* and *keule* (Figure 7.12) have been classified as mutations in genes involved in cytokinesis (Mayer *et al.,* 1991; Assaad *et al.,* 1996; Lukowitz *et al.,* 1996). Several rice mutants are significantly reduced in their body size (Hong *et al.,* 1996). Many more were assigned from early on to represent more general cell functions. Such defects were shown to affect variable cotyledon and hypocotyl distortions, to arrest embryos variably at different developmental stages

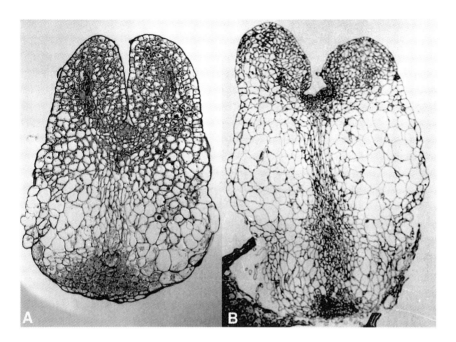

Figure 7.11 Comparison of morphogenesis mutants of *Arabidopsis* and *Petunia*. A: *Arabidopsis fass* mutant (from Torres Ruiz and Jürgens (1994), *Development*, **120** 2967-78, Company of Biologists Ltd., with permission). B: *Petunia tra1* mutant (from Dubois *et al.*, 1996, The Plant Journal, Vol. 10, pp. 47-59, Blackwell Science, with permission). Semi-thin sections, not to scale.

leading to embryo abortion or to alter pigmentation (Müller, 1963; Meinke and Sussex, 1979; Meinke, 1985; Jürgens *et al.*, 1991). For instance, the mutant *bio1* belongs to this class and has been particularly well analysed (Patton *et al.*, 1996; Meinke, 1996 and references therein). The initial and rough classification on pigment phenotypes has been refined. Among those exhibiting anthocyanin over-expression (*fusca*) some reflect physiological disturbments in (late) torpedo stages with deleterious consequences for the germinating seedling (Müller, 1963; Miséra *et al.*, 1994). Many of these turned out to represent genes encoding important functions involved in hypocotyl elongation in the dark (skoto- and photomorphogenesis) and several found in independent screens were found to be allelic (e.g. *fusca*, *cop*, *det*; see Miséra *et al.*, 1994, and references therein).

7.3.2 Molecular analysis of genes influencing important steps in plant embryogenesis

7.3.2.1 Polarity and asymmetric division

Of the mutants that have been characterized in this class perhaps *gnom/emb30* is one of particular interest as it has such a striking phenotype being either cone-

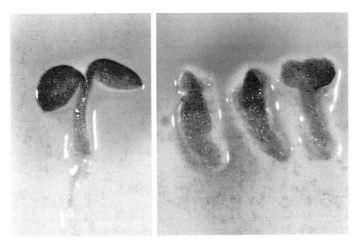

Figure 7.12 Comparison of *Arabidopsis* wild-type (left) and *keule* mutant (right).

shaped without a root and with reduced cotyledon structures or ball-like (Figure 7.5; Mayer *et al.*, 1991, 1993a). A further curiosity was detected after extensive analysis of transheterozygous seedlings that revealed partial or full intragenic complementation between certain allelic groups from the same and different backgrounds (Mayer *et al.*, 1993a; Busch *et al.*, 1996). A possible significance for promoting asymmetric division was deduced from the fact that *gnom* embryos fail to execute unequal division of the zygote, instead the newly established cell wall partitions the zygote more or less equally (Mayer *et al.*, 1993a). In addition, the apical–basal expression pattern of the *AtLTP1* gene deviates from the usual wild-type pattern, which is confined to the apical end (Vroemen *et al.*, 1996). Instead, *gnom* embryos can have no or reversed *AtLTP1* expression. Culturing of allelic *emb30* mutants revealed that they were not able to develop roots (Franzman *et al.*, 1989). Cloning revealed significant similarity in a stretch of 157 amino acids and additionally limited similarity throughout the protein to *SEC7p*, a yeast gene with secretory function in the Golgi-apparatus (Shevell *et al.*, 1994). The transcript was found to be present in all tissues and a role for the encoded peptide in normal cell division as well as cell expansion and cell adhesion was proposed after histological analysis (Shevell *et al.*, 1994). These findings suggest that *GNOM/EMB30* has a general function (Meinke, 1996). On the other hand, there remain some important features to be considered. For instance, *sec7* mutants are (cell) lethal in yeast (Achstetter *et al.*, 1988) whereas *Arabidopsis* seedlings demonstrate absence of cell lethality. As yet we do not know whether the overall distribution of its transcript necessarily reflects that of the protein. Additionally, a further yeast protein, Yec2p, of unknown function, exhibits more extended homology than Sec7p. Gene disruption of *YEC2* does not lead to a phenotype in yeast (Busch *et al.*, 1996). Sequencing revealed that even alleles with premature stop codon mutations can partially complement and this observation has lead to the proposal that there is inefficient translational readthrough. A possible linkage to the Golgi apparatus seems possible in the light of (false/ectopic) cell plate formation but certainly *GNOM* is not a prerequisite for every unequal division performed in a plant since *gnom/emb30* mutants develop normal stomata (Mayer *et al.*, 1993a). Topping and Lindsey (1997) suggest an indirect role of *GNOM/EMB30* in regulating polarity. Thus the precise function of the *GNOM/EMB30* gene product remains a mystery and further work at the protein level will be necessary in order to elucidate its function.

Polarity (and unequal division) is a fundamental process that enables a seemingly uniform cell to develop into a complex organism (Bünning, 1952). The understanding of how polarity might be established or maintained comes from a number of studies on systems such as *Fucus* and *Equisetum* (Haupt, 1962), with the former proving to be particularly valuable. *Fucus* zygotes occur as 'free' stages in water where they can be easily manipulated and observed with respect to their further fate (see Quatrano and Shaw, 1997). Unpolarized zygotes can be polarized by a number of external stimuli (Haupt, 1962; Quatrano, 1990; Brownlee and Berger, 1995; Quatrano and Shaw, 1997). Experiments have shown that a phase in which polarization can be reversed is followed by one in which it is irreversibly fixed. Fixation of the axis is paralleled by the localization of F-actin at the pole of future rhizodermis. This first stable asymmetry in the zygote is accompanied by intensive vesicle transport directed towards the same site to support rhizoid outgrowth (Kropf *et al.*, 1989; Bouget *et al.*, 1996). At the same time actin mRNA is localized at the opposite end (Bouget *et al.*, 1996). This distribution is later changed when the zygote has undergone division. On zygotic division actin mRNA and F-actin are co-localized at the cell plates of the first two divisions of the rhizoid.

Polarity depends on the asymmetric distribution and localized activity of proteins (Horvitz and Herskowitz, 1992). Several cases in animals have been identified where localized mRNAs have been shown to be responsible for gradients of specific proteins (St. Johnston and Nüsslein-Volhard, 1992; St Johnston, 1995) or even for polar accumulation of cytoskeletal proteins (Wilhelm and Vale, 1993). Thus *Fucus* mRNA localization may also be a mechanism that operates to accumulate or maintain proteins at specific sites (Bouget *et al.*, 1996). However, it cannot be the only mechanism as localized F-actin at the rhizoid tip is not accompanied by localized actin mRNA. Moreover, ablation experiments in *Fucus* have elegantly shown that essential positional information is somehow stored in its cell walls (Berger *et al.*, 1994). This adds credence to alternative models that envisage the fixation of position-dependent fate-determining factors in the cell wall (Brownlee and Berger, 1995).

The question that might be posed is whether higher plants make use of this mechanism too. This is not clear yet although it has been shown that the pattern of early cell divisions and their fate is very similar between *Fucus* and *Arabidopsis* (Brownlee and Berger, 1995; Quatrano and Shaw, 1997). It should be noted at this point that findings are coming from experiments with (carrot) somatic embryos that identify the influence of proteins and non-protein substances, such as carbohydrates and others, on plant tissue development. For instance, De Jong *et al.* (1992) demonstrated that chitinase could rescue embryo development of a temperature-sensitive mutant. Kreuger and Van Holst (1993) showed that a certain class of proteins, arabinogalactan proteins (AGPs; see Kreuger and Van Holst, 1996), promote embryogenic cell formation in carrot suspension cultures (see Mordhorst *et al.*, 1997 for a discussion on *in vitro* embryogenesis).

From elegant ablation studies in *Arabidopsis* a different mechanism (to that outlined in *Fucus*) has been proposed that invokes signalling between living cells rather than 'dead' cell walls and cells. This is the case in specifying the identity of cortex/endodermis initials, which use their daughters as a template (van den Berg *et al.*, 1995). Extension of these ablation experiments has analysed the role of the quiescent centre, a distinct central region of mitotically inactive cells in root meristems that are surrounded, for example, by initial cells of cortex/endodermis and columella. These experiments show that the quiescent centre inhibits the differentiation of surrounding cells (van den Berg *et al.*, 1997). Thus a balance of short-range signals inhibiting differentiation and signals that reinforce cell fate decisions control pattern formation in root meristems. The initiation and maintenance of the quiescent centre itself might be directly linked to the accumulation of high levels of auxin, as shown in maize (Kerk and Feldman, 1995). The observations made in ablation and other experiments point to inductive signals between symplastically connected cells. Following dye-coupling experiments in *Arabidopsis* epidermis the situation may be similar to that in animals, i.e. undifferentiated cells are symplastically connected and become gradually isolated with progressing differentiation (Duckett *et al.*, 1994). The importance of plasmodesmata with respect to morphogenesis has also been emphasized for other plants, such as ferns (Tilney *et al.*, 1990). Nucleic acid and protein cell-to-cell transport through plasmodesmata has been shown for transcription factors involved in adult plant development (see Mezitt and Lucas, 1996, and references therein). Thus the formative processes in the embryo may make use of 'symplastical signalling'.

This discussion started with asymmetric division in the zygote. However, unequal divisions have to be made several times in early embryogenesis. One example is given by the unequal division of the daughter cell of the cortex/endodermis initial, which is dependent on the action of the putative transcription factor *SCARECROW* (Di Laurenzio *et al.*, 1996) while the establishment of the identity of the resulting daughter cells seems to depend on *SHORTROOT*. Does this exclude the cell wall as a source for positional information in higher plants? There seems to be at least one important precedent for this. Mutants in *FASS* lead to oblique cell walls from first zygotic division onwards (Torres Ruiz and Jürgens, 1994). *fass* mutants have been shown to lack preprophasebands (PPBs), which predict the position of the future cell wall (Traas *et al.*, 1995). Therefore it is tempting to speculate that some 'information' is left behind after the PPB vanishes. However, this does not touch the process of polarity because even if first cell walls are extremely oblique in *fass*, cytoplasmic content and vacuolization are asymmetric, as in wild-type *ca* and *cb* (Torres Ruiz and Jürgens, 1994). Obviously internal polarity has to be achieved at some (zygotic) stage to form a plant. In the ovule or embryo sac environment it is possibly influenced by physical restriction and nourishing, maternal factors driving the embryo development in the known basal to apical direction. However, this external induction is probably not essential, as evidenced by numerous variants of artificial embryogenesis. With respect to this theme one should pay attention to 'upside-down' mutants which exhibit a 180° reverted axis, as observed in rice (Hong *et al.*, 1995), as well as to cases of polarity reversion occurring in organs of the adult plant, for instance reversion of extra flowers in the *Hooded* mutant (a homeobox gene mutation; Müller *et al.*, 1995).

7.3.2.2 Apical–basal versus radial organization

Following zygotic division a number of discrete decisions (including further asymmetries) have to be performed, the most important are the separation of upper and lower tier, establishment of the protoderm, the establishment of the embryonic root and root meristem, the transition from radial to bilateral symmetry with elaboration of cotyledon primordia and the elaboration of the shoot meristem primordium (and epicotyl). The corresponding mutants exhibit deviations from wild-type embryo development at stages where the respective organ/region seedling defect is organized or earlier, e.g. the mutants *stm* and *zll* deviate in late torpedo stages by obliterating the shoot meristem cell group (Barton and Poethig, 1993; Jürgens *et al.*, 1994; Endrizzi *et al.*, 1996), *gurke* and *pepino* at transition stage, *monopteros* at an octant stage (Berleth and Jürgens, 1993) and so on. Actually, the correspondence of seedling defect with embryo defect has been a criterion for discriminating between this and other mutants (Jürgens *et al.*, 1991). Starting from the apex the characterization and cloning of *SHOOT MERI-STEMLESS* has provided important information in this field (Barton and Poethig, 1993; Long *et al.*, 1996). *stm* mutants fail to develop a morphologically visible SAM as embryos and stems if grown in sterile culture. *STM* has been shown to encode a potential transcription factor harbouring a homeo box and having significant homology to the earlier characterized gene *KNOTTED1* that seems to play an important role in maize meristem identity with comparable temporal and spatial expression patterns (Smith *et al.*, 1995). *KNAT1* and other *KN1*-like homeobox genes in *Arabidopsis* are probably involved in SAM determination together with *STM*. For instance, lobed leaves are formed with ectopic meristems in *Arabidopsis* when *KNAT1* is overexpressed (Chuck *et al.*, 1996). Moreover early *STM* expression indicates the presence of anlage SAM in globular stage embryos as a SAM primordium is only morphologically recognizable in late torpedo stages (Barton and Poethig, 1993; Long *et al.*, 1996). Two further genes operate in the SAM *ZWILLE* and *WUSCHEL*, leading to defects in its organization and premature termination (Figure 7.6; Jürgens *et al.*, 1994; Laux *et al.*, 1996; Endrizzi *et al.*, 1996). Double mutant analysis suggests that *STM* acts upstream of *ZLL* and *WUS*. From this and more detailed phenotypic analysis it is concluded that *STM* keeps central meristem cells undifferentiated while *ZLL* and *WUS* are required for their proper function (Endrizzi *et al.*, 1996). *ZLL* has an embryonic but poor adult phenotype (Jürgens *et al.*, 1994; Endrizzi *et al.*, 1996) that resembles the situation in *pinhead* (McConnell and Barton, 1995) and the maize mutant *dek ed*41* (see Mordhorst *et al.*, 1997). *pinhead* mutants develop adventitious shoots *in vitro* and similarly *dek ed*41* develops normal-looking plants. In contrast to these mutants another class exhibits defects in the cotyledons, hypocotyl and, at least partially, the SAM. Thus the strongest alleles of *gurke* embrace variable defects in the apical region which becomes visible at an early heart stage (Figure 7.7). Extreme phenotypes from these lines develop limited hypocotyl tissue and do not show SAM by morphological criteria while the root tissues essentially display wild-type elements (Torres Ruiz *et al.*, 1996b). Weak alleles of *gurke* as well as *pepino* alleles possess SAMs but those tested in *gurke* cannot fully develop normal flowers. It is therefore probable that genes like *GURKE* and *PEPINO* are required in adult stages together with those mentioned before. From work in *Petunia* it is known that mutations in the *NO APICAL MERISTEM* (*NAM*) result in SAM-less embryos (Souer *et al.*, 1996). From occasional 'escape shoot' formation in a fraction of seedlings and its subsequent development (every organ is normal except the flower), a function in the adult plant is assumed in determining positions of meristems and primordia. Cloning and molecular analysis have revealed a protein of unknown function that shares homology with proteins in *Arabidopsis* and rice, thus establishing a novel class of proteins with a possible capacity for transcriptional control (Souer *et al.*, 1996). Interestingly, two genes (*CUP-SHAPED COTYLEDON1* and *2, CUC1* and *2*) act in concert to control organ separation, for example, of cotyledons during transition stage (Aida *et al.*, 1997). *CUC2* has been cloned and shows homology to *NAM* (Aida *et al.*, 1997). Thus, we have so far a small glimpse of the factors that control development of the apical region in general and in particular that of the SAM. A third class of mutants affecting the apical region lead to variable obliteration of cotyledons or elevation of cotyledon number without altering the principle organization and shape of the hypocotyl and the root (*pinoid*; *pinformed*, *laterne*). The defect can be traced back to the heart stage and the adult phenotype points to their late requirement (Mayer *et al.*, 1993b; Bennet *et al.*, 1995; Torres Ruiz, unpublished). At least *laterne* seems not to 'code' for the ability to make cotyledons *per se* since this line frequently segregates tricots or fusecots (Torres Ruiz, unpublished). Thus the (possibly indirect) defect seen in this mutant might be the perturbation of a factor whose threshold might be critical for initiating cotyledons or leaves. Interestingly, it has been shown for *pinformed* and *pinoid* that auxin transport is disturbed (Okada *et al.*,

1991; Bennett *et al.*, 1995). This link between cotyledon formation and hormones is reminiscent of experiments with auxin and its transport blockers (and other substances). Haccius (1960) showed that real monocotyly can be induced with certain concentrations of phenylboric acid in *Eranthis hiemalis* while the application of 2,4-dichlorophenoxyacetic acid (2,4-D) induces syncotyly (Haccius and Trompeter, 1960). This experiment has also been carried out in *Brassica* with a similar result, suggesting that differentially transported auxin might be involved in establishing bilateral symmetry and thus positioning of cotyledons (Liu *et al.*, 1993b). Older experiments with species such as *Phaseolus* demonstrated the potential of basipetal (but not acropetal) transport of auxin in late torpedo stages (Fry and Wangerman, 1976). The same perturbing effect of auxins on the elaboration of bilateral symmetry has been demonstrated in wheat (Fischer and Neuhaus, 1996). What is needed now is an unambiguous proof of whether hormones play a direct or indirect role in embryo/seedling pattern formation, especially the transition of embryo symmetry. Besides position, the identity of cotyledons has also to be considered and this seems to be temporally controlled. For instance, it has been shown that cotyledons in the *leafy* mutant can adopt features of primary leaves (Meinke, 1992; West *et al.*, 1994) whereas subsequent leaves in precociously germinating *Brassica* embryos retain features of cotyledon (Finkelstein and Crouch, 1984; Fernandez, 1997).

Genetic complexity in the formation of the more basal body does not lag behind that of the apical part. *monopteros* (*mp*) mutants delete hypocotyl and the root (Figure 7.8; Berleth and Jürgens, 1993), giving rise to a complementary *gurke* phenotype. Surprisingly, *MONOPTEROS*, which encodes a putative transcription factor (C. Hardtke and T. Berleth, personal communication), seems not to be responsible for root formation *per se*, as culturing of *mp* seedlings allows correct root formation. Rather, a role for the *mp* gene product in the formation of embryonic hypocotyl/root by generating correctly oriented axialization/elongation of cells is proposed (Przemeck *et al.*, 1996). *pinformed*-like adult *mp* phenotypes (including distortions in auxin transport and vascularization) suggest a postembryonic function in the same process (Przemeck *et al.*, 1996), which is thought to depend on orienting signals and cellular response mechanisms (Sachs, 1991). At least four to six genes have been identified which interfere with hypophyseal cell group formation: *BASAL DELETION*, *MÖWE*, *HOBBIT* (Figure 7.9), *BOMBADIL*, *ORC* and *GREMLIN* (Berleth *et al.*, 1996; Scheres *et al.*, 1996). This implies that the embryonic development of the root pole is highly complex. Moreover the action of these genes has to be coordinated with an additional set including *SCARECROW (SCR)*, *SHORT ROOT (SHR)*, *PINOCCHIO*, *GOLLUM* and *WOODEN LEG*, all of which were uncovered by their root phenotype but are intimately involved in radial pattern organization of the embryonic hypocotyl and root (Benfey *et al.*, 1993; Scheres *et al.*, 1995). Like many of the other genes mentioned previously in this section the activity of this group is probably required for both early and later stages. Such a hypothesis is supported by examining defects in secondary and regenerated roots in *hobbit* and *scarecrow* (Scheres *et al.*, 1996; Di Laurenzio *et al.*, 1996). As outlined above genes like *SCARECROW* take over specifications in the cortex or vascular tissue, partly by influencing corresponding asymmetric divisions. Both *scr* and *shr* mutants fail to execute the periclinal and asymmetric division of the cortex/ endodermis initials but do not share the same tasks. This is deduced from the fact that in *shr* solely cortex-specific cell surface markers are expressed in the cortex cell row while in *scr* characteristics of both cortex and endodermis are demonstrated. Seemingly *SCR* does not participate, at least initially, in specifying cortex and/or endodermis cell identity. *SCARECROW* has been cloned and encodes a putative novel transcription factor, as judged by the presence of motifs that are associated with potential activation, nuclear localization and protein–protein interactions (Di Laurenzio *et al.*, 1996). It is expressed in the heart stage (first ground tissue then endodermis) and in adult stages, where it is restricted to the endodermis, perhaps to maintain its identity.

7.3.2.3 Cytokinesis

Two mutants *knolle* and *keule* (Figure 7.12) were initially classified as exhibiting radial defects because of their bloated surface layer (Mayer *et al.*, 1991). Embryo analysis revealed that in both mutants protoderm-forming periclinal divisions are not appropriately performed (Mayer *et al.*, 1991; Lukowitz *et al.*, 1996; Assaad *et al.*, 1996). At first glance, in *knolle* mis-expression of the epidermis specific marker *AtLTP1* also points to a failure to produce a correct protoderm as does double-mutant analysis with a *fusca* mutant, which normally does not express anthocyanin in epidermal cells (Vroemen *et al.*, 1996; Lukowitz *et al.*, 1996). However, detailed analysis shows that *knolle* and *keule* are primarily defective in cytokinesis, which leads to continued connection between protodermal and internal cells (Lukowitz *et al.*, 1996; Assaad *et al.*, 1996). Thus, cell files can be discontinuous in their fates, appearing as if radial pattern is incorrect (Lukowitz *et al.*, 1996; Assaad *et al.*, 1996; Vroemen *et al.*, 1996). Molecular cloning of *KNOLLE* revealed similarity to syntaxins, proteins that are involved in vesicular transport. In accordance with this, *in situ* hybridization of embryos exhibits expression patterns in single cells or groups in a 'pacht'-like manner (Lukowitz *et al.*, 1996). Thus *KNOLLE* does not seem to be involved in the specification of radial pattern. A similar function is probable for *KEULE* since *keule* mutants display a similar cellular phenotype, i.e. cells with interrupted cell walls and cell wall stubs, slow cell division and mis-oriented cell division planes (Assaad *et al.*, 1996). Moreover in *keule* embryos *AtLTP1* mRNA is localized in the abnormal outer cell layer; this has also been seen in mutant *raspberry* embryos (Vroemen *et al.*, 1996; Yadegari *et al.*, 1994). Taken together this strongly indicates that *KNOLLE* and *KEULE* exert their function in cytokinesis. The *cytokinesis-defective* (*cyd*) mutant of pea displays a similar (cellular) phenotype to *keule* (Liu *et al.*, 1996). *knolle*, *keule* and *cytokinesis-defective* can be phenocopied by treatments with caffeine (Assaad *et al.*, 1996; Liu *et al.*, 1996). However, it is not known whether *cyd* encodes an orthologous or additional function.

7.3.2.4 Cell shape and size

Of particular interest are those mutants that have dramatic effects on the shape and/or size of the seedling without principal alteration of the body pattern. Several mutants have been found that fit into this class in *Arabidopsis*, *Petunia*, rice and maize. Special focus has been placed on mutants of the gene *FASS* (or *TON* in the screen of Traas *et al.*, 1995) that have been found in different screens in different allelic strength (Mayer *et al.*, 1991; Torres Ruiz and Jürgens, 1994; Traas *et al.*,1995; Fisher *et al.*, 1996). *fass* mutants (Figure 7.11) are unable to correctly direct the cell division plane from first zygotic division onwards and do not generate PPBs (see above; Torres Ruiz and Jürgens, 1994; Traas *et al.*, 1995). As a consequence, all cells of embryos and seedlings adopt an irregular shape and seedlings are extremely compressed along the longitudinal axes of their organs (Figure 7.11); this holds true for adult plants (and their organs) which can be grown on soil and exhibit extreme miniaturization in comparison to wild-type (Torres Ruiz and Jürgens, 1994). At the same time seedlings have a correctly elaborated pattern (apical–basal and radial) which means that cell division pattern and cell shape are not instrumental for pattern formation in plants. It should be noted that *Petunia* mutants in *TRA1* have a similar (histological) phenotype (Figure 7.11; Dubois *et al.*, 1996). Whether *FASS* is a structural element of the cell wall or indirectly involved in its orientation is not clear. Fisher *et al.* (1996) discuss a possible involvement of *FASS* in the regulation of hormone levels (auxin). Similarly Dubois *et al.* (1996) discuss involvement in hormone response or microtubule formation for *TRA1*. A more detailed insight will depend on further molecular work, including the cloning of this and related gene(s) such as an additional (*TON*) gene leading to fass phenotype (Traas *et al.*, 1995) and genes like *KNOPF*, *ENANO* and *MICKEY* that display compression along their longitudinal axes (Mayer *et al.*, 1991; Torres Ruiz *et al.*, 1996a). Both *knopf* and *enano* mutations lead to elongation of cortical and epidermal cells in a radial rather than an apical–basal direction. This is especially pronounced in *knopf*. This radial versus apical–basal elongation is reminiscent of opposed elongation defects when the mutants *angustifolia* and *rotundifolia* are considered (Tsuge *et al.*, 1996). The former leads to proximo–distal elongation and the latter to lateral elongation. It will be interesting to compare the responsible molecules once they have been cloned. The mutant *enano* is not only characterized by radial expansion but also by occasional irregular growth of cells in any of the tissues in young seedlings (Spiess and Torres Ruiz, unpublished). Whether this is a primary defect or a consequence of the former phenomenon remains to be determined. Mutants in the gene *HYDRA1* are also considered to be primarily defective in the regulation of cell shape although *hydra1* mutants exhibit some phenotypic variability (Topping *et al.*, 1997). Taken together these observations concerning shape mutants suggest that cell shape elaboration is influenced by quite different factors or mechanisms. For the cytoskeleton this fits into the fact that its establishment is based and controlled by numerous and disparate factors (Seagull 1989; Staiger and Lloyd, 1991). This notion is also supported by double-mutant analyses. For instance double mutants of *fass* and *enano* are additive (Spiess and Torres Ruiz, unpublished).

While these mutations lead to size reduction by altering cell shape (at least partly), size can generally be changed by elevating or lowering the number of all cells and/or their overall dimensions (without altering their shape). It can be regularly observed that if growth conditions are suboptimal plants produce seeds that germinate to produce considerably smaller seedlings. The influence of physiological and physical factors is evident. Hong *et al.* (1996) have provided several examples, e.g. *REDUCED EMBRYO 1, 2* and *3* (*RE 1, 2, 3*) and *GIANT EMBRYO 1* (*GE 1*), for embryo size determination by physical restriction. Thus enlarged endosperm limits the size of embryos in *re 1, 2, 3* mutants while its reduction allows the embryo to enlarge. One of the tasks lying ahead will be to unravel how these processes are integrated in order to build up the plant embryo.

7.3.2.5 Apomixis and polyembryonic mutants

The success of the combined genetic and molecular approach has led to considerable progress in analysing zygotic embryogenesis. Increasingly, attempts are being made to answer other and related questions connected to apomixis (and polyembryony; Koltunov, 1993; Vielle Calzada *et al.*, 1996).

Physiological conditions (temperature, inorganic salts and nutrients) can influence apomixis, particularly in facultative apospory (Brown and Empry, 1958). However, interesting observations (as mentioned above) and experimental data have already pointed to the genetic basis of these asexual reproduction forms. For instance, crosses with the (sexual) *Ranunculus cassubicifolius* and (apomictic) *R. megacarpus* demonstrated that apomixis can be genetically transferred through pollen and may represent a quantitative effect (see Rutishauser, 1969). The alteration of the genetic basis could account for the assumed origin of the Panicum from Hieracium type of aposporic apomixis (see above). In this view simple mutations might have played a role in establishing apomictic mechanisms (Koltunov *et al.*, 1995; Ohad *et al.*, 1996) as well as in refining some apomictic variants. This is supported by genetic (mutation) analysis in the *Arabidopsis* system. For instance, mutations in the *FIE* (*FERTILIZATION INDE-PENDENT ENDOSPERM*) gene of *Arabidopsis* lead to initiation of endosperm development in unfertilized ovules (Ohad *et al.*, 1996), thus showing that embryo and endosperm development can be uncoupled processes. Vernon and Meinke (1994) have found and analysed an *Arabidopsis* mutant leading to twin embryos, which arise from suspensor cells. It is proposed that the embryo proper normally suppresses the embryogenic potential of the suspensor (Schwartz *et al.*, 1994), a function that might be disturbed in twin mutants (Schwartz *et al.*,

1994). Considering the progress of the genetic approach to zygotic embryo-genesis an increasing number of genetic and molecular reports on apomixis and polyembryony can be expected.

7.4 Concluding remarks

The studies and progress discussed in this chapter show that patterns of cell division and position-dependent cell fate determination play decisive roles in plant embryo development. Of considerable importance are the cloning and analysis of molecules and mechanisms that control (unequal) cell division and it will be interesting to see at which point these processes are linked to cell cycle mechanisms. As we have seen, disturbed cell division *per se* does not necessarily lead to altered body organization (but altered morphogenesis). Thus it seems obvious that unequal cell divisions are connected with additional 'information' (or 'quality'). The spectrum of mutants and genes known, suggests that different cell division modes take place in embryos (and adult plants). It has also been shown that there is already a precedent in that the development of initial and daughter cells is co-ordinated by cell-to-cell signalling. Embryogenesis will profit from the progress of analysis of flower development, which provides cases of macromolecular transport through plasmodesmata. The precise informational capacity and role of hormones in the embryo and the significance of signalling molecules bound to (or coming from) the cell wall has to be further clarified.

In the context of cell division patterns it should be noted that biologists might learn from mathematicians, who have demonstrated the capability of Linden-mayer (L) systems to simulate the generation of plants of different species (see Prusinkiewicz and Hanan, 1989). Interestingly, some algorithms (context-sensi-tive L systems) allow for cellular interactions (Prusinkiewicz and Hanan, 1989).

Additional work will follow to answer questions linked to cytokinesis and cell wall growth. Of particular interest will be how positions of newly formed cell walls are determined and which mechanisms control cell wall growth and elongation along the apical–basal and radial axes of the embryo. It is evident from the preceding discussion that we do not yet understand the control of the absolute cell number (numbers of cell cycles) determining the size of organs and the complete body of the plant. An analysis of the basic aspects of embryogene-sis will be of significant importance to the agricultural and horticultural indus-tries, and *Arabidopsis* offers, together with other plants, a valuable tool for expanding our understanding of this fascinating area of plant development.

Acknowledgements

I thank Fahrah Assaad, Monika Frey, Georg Haberer and Wolfgang Haupt for reading this manuscript. I thank Wolfgang Haupt for guidance through the labyrinths of our software, all colleagues for providing information prior to publication and Alfons Gierl for generous assistance. I am especially indebted to my colleagues K. Schneitz (Zürich), T. Laux and B. Mousssian (Tübingen), T. Berleth and C. Hardke (München), B. Scheres and co-workers (Utrecht), S. Ploense (Minneapolis), J. Durand and R.S. Sangwan (Orsay, Amiens) for providing photographs. Support for the work cited in the text from my laboratory was provided by the Deutsche Forschungsgemeinschaft, the Leonhard-Lorenz-Stiftung and the Technische Universität München, and this is gratefully acknowledged.

References

Achstetter, T., Franzusoff, A., Field, C. and Schekman, R. (1988) SEC7 encodes an unusual, high molecular weight protein required for membrane traffic from the yeast Golgi apparatus. *J Biol. Chem.*, **263** 11711-17.

Aida, M., Ishida, T., Fukaki, H., Fujisawa, H. and Tasaka, M. (1997) Genes involved in organ separation in *Arabidopsis*: an analysis of the *cup-shaped cotyledon* mutant. *Plant Cell*, **9** 841 57.

Angenent, G.C. and Colombo, L. (1996) Molecular control of ovule development. *Trends Plant Sci.*, **1** 228-32.

Assaad, F.F., Mayer, U., Wanner, G. and Jürgens, G. (1996) The *KEULE* gene is involved in cytokinesis in *Arabidopsis*. *Mol. Gen. Genet.*, **253** 267-77.

Barton, M.K. and Poethig, R.S. (1993) Formation of the shoot apical meristem in *Arabidopsis thaliana*: an analysis of development in the wild-type and in the *shoot meristemless* mutant. *Development*, **119** 823-31.

Benfey, P.N., Linstead, P.J., Roberts, K., Schiefelbein, J.W., Hauser, M.-Th. and Aeschbacher, R.A. (1993) Root development in *Arabidopsis*: four mutants with dramatically altered root morphogenesis. *Development*, **119** 57-70.

Bennett, S.R.M., Alvarez, J., Bossinger, G. and Smyth, D.R. (1995) Morphogenesis in pinoid mutants of *Arabidopsis thaliana*. *Plant J.*, **8** 505-20.

Berger, F., Taylor, A. and Brownlee, C. (1994) Cell fate determination by the cell wall in early *Fucus* development. *Science*, **263** 1421-23.

Berleth, T. and Jürgens, G. (1993) The role of the *monopteros* gene in organising the basal body region of the *Arabidopsis* embryo. *Development*, **118** 575-87.

Berleth, T., Hardtke, C.S., Przemeck, G.K.H. and Müller, J. (1996) Mutational analysis of root initiation in the *Arabidopsis* embryo. *Plant and Soil*, **187** 1-9.

Bouget, F.-Y., Gerttula, S., Shaw, S.L. and Quatrano, R.S. (1996) Localization of actin mRNA during the establishment of cell polarity and early cell divisions in *Fucus* embryos. *Plant Cell*, **8** 189-201.

Brown, W.V. and Empry, W.H.P. (1958) Apomixis in the Gramineae: Panicoidae. *Am. J. Bot.*, **45** 253-63.

Brownlee, C. and Berger, F. (1995) Extracellular matrix and pattern in plant embryos: on the lookout for developmental information. *Trends Gen.*, **11** 344-48.

Bünning, E. (1952) Morphogenesis in plants. *Surv. Biol. Progr.*, **2** 105-40.

Busch, M., Mayer, U. and Jürgens, G. (1996) Molecular analysis of the *Arabidopsis* pattern formation gene GNOM: gene structure and intragenic complementation. *Mol. Gen. Genet.*, **250** 681-91.

Carland, F.M. and McHale, N.A. (1996) LOP1: a gene involved in auxin transport and vascular patterning in *Arabidopsis*. *Development*, **122** 1811-19.

Caruso, J.L. (1968) Morphogenetic aspects of a leafless mutant in tomato. I. General patterns in development. *Am. J. Bot.*, **55** 1169-76

Chaudhury, A.M., Letham, S., Craig, S. and Dennis, E.S. (1993) *amp1*—a mutant with high cytokinin levels and altered embryonic pattern, faster vegetative growth, constitutive photomorphogenesis and precocious flowering. *Plant J.*, **4** 907-16.

Chuck, G., Lincoln, C. and Hake, S. (1996) *KNAT1* induces lobed leaves with ectopic meristems when overexpressed in *Arabidopsis*. *Plant Cell*, **8** 1277-89.

Clark, J.K. (1996) Maize embryogenesis mutants, in *Embryogenesis: The Generation of a Plant* (eds E. Cummings and T. Wang), Bios Scientific Publishers, Oxford, pp. 89-112.

Clark, J.K. and Sheridan, W.F. (1991) Isolation and characterization of 51 embryo-specific mutations of maize. *Plant Cell*, **3** 935-51.

Crane, P.R., Friis, E.M. and Pedersen, K.R. (1995) The origin and early diversification of angiosperms. *Nature*, **374** 27-33.

Czihak, G. (1984) *Johann Gregor Mendel*. Dokumentierte Biographie und Katalog zur Gedächtnisausstellung anläßlich des hundertsten Todestages mit Facsimile seines Hauptwerkes, G. Czihak, Salzburg

De Jong, A.J., Cordewener, L., Loschiavo, F., Terzi, M., Vandekerckhove, J., Van Kammen, A. and De Vries, S.C. (1992) A carrot somatic embryo mutant is rescued by chitinase. *Plant Cell*, **4** 425-33.

Di Laurenzio, L., Wysocka-Diller, J., Malamy, J.E., Pysh, L., Helariutta, Y., Freshour, G., Hahn, M.G., Feldmann, K.A. and Benfey, P.N. (1996) The *SCARECROW* gene regulates an asymmetric cell division that is essential for generating the radial organization of the *Arabidopsis* root. *Cell*, **86** 423-33.

Dubois, F., Bui Dang Ha, D., Sangwan, R.S. and Durand, J. (1996) The *Petunia tra1* gene controls cell elongation and plant development, and mediates responses to cytokinins. *Plant J.*, **10** 47-59.

Duckett, C.M., Oparka, K.J., Prior, D.A.M., Dolan, L. and Roberts, K. (1994) Dye-coupling in the root epidermis of *Arabidopsis* is progressively reduced during development. *Development*, **120** 3247-55.

Ehrendorfer, F. (1978) Spermatophyta, in *Strasburger Lehrbuch der Botanik* (eds. D. von Denffer, F. Ehrendorfer, K. Mägdefrau and H. Ziegler), Gustav Fischer, Stuttgart, pp. 699-855.

Endrizzi, K., Moussian, B., Haecker, A., Levin, J.Z. and Laux, T. (1996) The *SHOOT MERISTEMLESS* gene is required for maintenance of undifferentiated cells in *Arabidopsis* shoot and floral meristems and acts at a different regulatory level than the meristem genes *WUSCHEL* and *ZWILLE*. *Plant J.*, **10** 967-79.

Esau, K. (1977) *Anatomy of Seed Plants*. 2nd edn, John Wiley, New York.

Feldmann, K.A. (1991) T-DNA insertion mutagenesis in *Arabidopsis*: mutational spectrum. *Plant J.*, **1** 71-82

Fernandez, D.E. (1997) Developmental basis of homeosis in precociously germinating *Brassica napus* embryos: phase change at the shoot apex. *Development*, **124** 1149-57.

Finkelstein, R.R. and Crouch, M.L. (1984) Precociously germinating rapeseed embryos retain characteristics of embryogeny. *Planta*, **162** 125-31.

Fischer, C. and Neuhaus, G. (1996) Influence of auxin on the establishment of bilateral symmetry in monocots. *Plant J.*, **9** 659-69.

Fisher, R.H., Barton, M.K., Cohen, J.D. and Cooke, T.J. (1996) Hormonal studies of *fass*, an *Arabidopsis* mutant that is altered in organ elongation. *Plant Physiol.*, **110** 1109-21.

Franzmann, L., Patton, D.A. and Meinke, D.W. (1989) *In vitro* morphogenesis of arrested embryos from lethal mutants of *Arabidopsis thaliana*. *Theor. Appl. Genet.*, **77** 609-16.

Friedman, W.E. (1990) Double fertilization in *Ephedra*, a nonflowering seed plant: its bearing on the origin of Angiosperms. *Science*, **247** 951-54.

Friedman, W.E. (1992) Evidence of a pre-angiosperm origin of endosperm: implications for the evolution of flowering plants. *Science*, **255** 336-39.

Fry, S.C. and Wangermann, E. (1976) Polar transport of auxin through embryos. *New Phytol.*, **77** 313-17.

Goldberg, R.B., Barker, S.J. and Perez-Grau, L. (1989) Regulation of gene expression during plant embryogenesis. *Cell,* **56** 149-60.

Goldberg, R.B., de Paiva, G. and Yadegari, R. (1994) Plant embryogenesis: zygote to seed. *Science,* **266** 605-14.

Haccius, B. (1960) Experimentell induzierte Einkeimblättrigkeit bei *Eranthis hiemalis.* II Monokotylie durch Phenylborsäure. *Planta,* **54** 482-97.

Haccius, B. and Trompeter, G. (1960) Experimentell induzierte Einkeimblättrigkeit bei *Eranthis hiemalis.* I Synkotylie durch 2,4-Dichlorphenoxyessigsäure. *Planta,* **54** 466-81.

Hanstein, J. (1870) Die Entwicklungsgeschichte der Monocotylen und Dicotylen. *Bot. Abhandl Bonn,* **1** 1-112.

Haupt, W. (1962) Die Entstehung der Polarität in pflanzlichen Keimzellen, insbesondere die Induktion durch Licht. *Ergebnisse der Biologie,* **25** 1-32.

Hauser, M.-T., Morikami, A. and Benfey, P.N. (1995) Conditional root expansion mutants of *Arabidopsis. Development,* **121** 1237-52.

Hong, S.K., Aoki, T., Kitano, H., Satoh, H. and Nagato, Y. (1995) Phenotypic diversity of 188 rice embryo mutants. *Dev. Genet.,* **16** 298-310.

Hong, S.K., Kitano, H., Satoh, H. and Nagato, Y. (1996) How is embryo size genetically regulated in rice? *Development,* **122** 2051-58.

Horvitz, H.R. and Herskowitz, I. (1992) Mechanisms of asymmetric cell division: two Bs or not two Bs, that is the question. *Cell,* **68** 237-55.

Johansen, D.A. (1950) *Plant Embryology*, Chronica Botanica Company, Waltham, Mass.

Johri, B.M. (1984) *Embryology of Angiosperms*, Springer, Berlin.

Johri, B.M., Ambegaokar, K.B. and Srivastava, P.S. (1992) *Comparative Embryology of Angiosperms*, Vol 1, Springer, Berlin.

Jürgens, G. (1995) Axis formation in plant embryogenesis: cues and clues. *Cell,* **81** 467-70

Jürgens, G. and Mayer, U. (1994) *Arabidopsis*, in *EMBRYOS: Colour Atlas of Development* (ed. JBL Bard), Wolfe Publishing, London, pp. 7-21

Jürgens, G., Mayer, U., Torres Ruiz, R.A., Berleth, T. and Miséra, S. (1991) Genetic analysis of pattern formation in the *Arabidopsis* embryo. *Dev. Suppl.,* **1** 27-38.

Jürgens, G., Torres Ruiz, R.A. and Berleth, T. (1994) Embryonic pattern formation in flowering plants. *Ann. Rev. Genet.,* **28** 351-71.

Kerk, N.M. and Feldman, L.J. (1995) A biochemical model for the initiation and maintenance of the quiescent center: implications for organization of root meristems. *Development,* **121** 2825-33.

Koltunov, A.M. (1993) Apomixis: embryo sacs and embryos formed without meiosis or fertilization in ovules. *Plant Cell,* **5** 1425-37.

Koltunov, A.M., Bicknell, R.A. and Chaudhurry, A.M. (1995) Apomixis: molecular strategies for the generation of genetically identical seeds without fertilization. *Plant Physiol.,* **108** 1345-52.

Kranz, E. and Dresselhaus, T. (1996) *In vitro* fertilization with isolated higher plant gametes. *Trends Plant Sci.,* **1** 82-89.

Kranz, E. and Lörz, H. (1993) *In vitro* fertilization with isolated, single gametes results in zygotic embryogenesis and fertile maize plants. *Plant Cell,* **5** 739-46.

Kreuger, M. and Van Holst, G.-J. (1993) Arabinogalactan proteins are essential in somatic embryogenesis of *Daucus carota* L. *Planta,* **189** 243-48.

Kreuger, M. and Van Holst, G.-J. (1996) Arabinogalactan proteins and plant differentiation. *Plant Mol. Biol.,* **30** 1077-86.

Kropf, D.L., Berge, S.K. and Quatrano, R.S. (1989) Actin localization during *Fucus* embryogenesis. *Plant Cell,* **1** 191-200.

Lakshmanan, K.K. and Ambegaokar, K.B. (1984) Polyembryony, in *Embryology of Angiosperms* (ed. B.M. Johri) Springer, Berlin, pp. 445-74.

Laux, T. and Jürgens, G. (1997) Embryogenesis: a new start in life. *Plant Cell,* **9** 989-1000.

Laux, T., Mayer, K.F.X., Berger, J. and Jürgens, G. (1996) The *WUSCHEL* gene is required for shoot and floral meristem integrity in *Arabidopsis. Development*, **122** 87-96.

Leibenguth, F. (1982) *Züchtungsgenetik*, Georg Thieme, Stuttgart.

Lindsey, K. and Topping, J.F. (1993) Embryogenesis: a question of pattern. *J. Exp. Bot.*, **44** 359-74.

Liu, C.M., Xu, Z.H. and Chua, N.H. (1993a) Proembryo culture: *in vitro* development of early globular-stage zygotic embryos from *Brassica juncea. Plant J.*, **3** 291-300.

Liu, C.M., Xu, Z.H. and Chua, N.H. (1993b) Auxin polar transport is essential for the establishment of bilateral symmetry during early plant embryogenesis. *Plant Cell*, **5** 621-30.

Liu, C.-M., Johnson, S., Hedley, C.L. and Wang, T.L. (1996) The generation of a legume embryo: morphological and cellular defects in pea mutants, in *Embryogenesis: The Generation of a Plant* (eds T.L. Wang and A. Cuming), Bios Scientific Publishers, Oxford.

Long, J.A., Moan, E.I., Medford, J.I. and Barton, M.K. (1996) A member of the KNOTTED class of homeodomain proteins encoded by the STM gene of *Arabidopsis. Nature*, **379** 66-69.

Lopes, M.A. and Larkins, B.A. (1993) Endosperm origin, development and function. *Plant Cell*, **5** 1383-99.

Lu, P., Porat, R., Nadeau, J.A. and O'Neill, S.D. (1996) Identification of a meristem L1 layer-specific in *Arabidopsis* that is expressed during embryonic pattern formation and defines a new class of homeobox genes. *Plant Cell*, **8** 2155-68.

Lukowitz, W., Mayer, U. and Jürgens, G. (1996) Cytokinesis in the *Arabidopsis* embryo involves the syntaxin-related *KNOLLE* gene product. *Cell*, **84** 61-71.

Maheshwari, P. (1950) *An Introduction to the Embryology of Angiosperms*, McGraw-Hill, London.

Mansfield, S.G. and Briarty, L.G. (1990a) Development of free-nuclear endosperm in *Arabidopsis thaliana* L. *Arabidopsis Inf. Serv.*, **27** 53-65.

Mansfield, S.G. and Briarty, L.G. (1990b). Endosperm cellularization in *Arabidopsis thaliana* L. *Arabidopsis Inf. Serv.*, **27** 65-72.

Mansfield, S.G. and Briarty, L.G. (1991) Early embryogenesis in *Arabidopsis thaliana*. II. The developing embryo. *Can. J. Bot.*, **69** 461-76.

Mansfield, S.G. and Briarty, L.G. (1992) Cotyledon cell development in *Arabidopsis thaliana* during reserve deposition. *Can. J. Bot.*, **70** 151-64.

Mansfield, S.G., Briarty, L.G. and Erni, S. (1991) Early embryogenesis in *Arabidopsis thaliana*. I. The mature embryo sac. *Can. J. Bot.*, **69** 447-60.

Mascarenhas, J.P. (1993) Molecular mechanisms of pollen tube growth and differentiation. *Plant Cell*, **5** 1303-14.

Mayer, U., Torres Ruiz, R.A., Berleth, T., Miséra, S. and Jürgens, G. (1991) Mutations affecting body organization in the *Arabidopsis* embryo. *Nature*, **353** 402-407.

Mayer, U., Büttner, G. and Jürgens, G. (1993a) Apical–basal pattern formation in the *Arabidopsis* embryo: studies on the role of the *GNOM* gene. *Development*, **117** 149-62.

Mayer, U., Berleth, T., Torres Ruiz, R.A., Misera, S. and Jürgens, G. (1993b). Pattern formation during *Arabidopsis* embryo development, in *Cellular Communication in Plants* (ed. R.M. Amasino), Plenum, New York, pp. 93-98.

McConnell, J.R. and Barton, M.K. (1995) Effect of mutations in the *PINHEAD* gene of *Arabidopsis* on the formation of shoot apical meristems. *Dev. Genet.*, **16** 358-66.

McCormick, S. (1993) Male gametophyte development. *Plant Cell*, **5** 1265-75.

Meinke, D.W. (1985) Embryo-lethal mutants of *Arabidopsis thaliana*: analysis of mutants with a wide range of lethal phases. *Theor. Appl. Genet.*, **69** 543-52.

Meinke, D.W. (1991) Perspectives on genetic analysis of plant embryogenesis. *Plant Cell*, **3** 857-66.

Meinke, D.W. (1992) A homeotic mutant of *Arabidopsis thaliana* with leafy cotyledons. *Science*, **258** 1647-50.

Meinke, D.W. (1996) Embryo-defective mutants of *Arabidopsis*: cellular functions of disrupted genes and developmental significance of mutant phenotypes, in *Embryogenesis: The Generation of a Plant* (eds E. Cummings and T. Wang), Bios Scientific Publishers, Oxford, pp. 35-50.

Meinke, D.W. and Sussex, I.M. (1979) Embryo-lethal mutants of *Arabidopsis thaliana*. A model system for genetic analysis of plant embryo development. *Dev. Biol.*, **72** 50-61.

Mezitt, L.A. and Lucas, W.J. (1996) Plasmodesmatal cell-to-cell transport of proteins and nucleic acids. *Plant Mol. Biol.*, **32** 251-73.

Miséra, S., Müller, A., Weiland-Heidecker, U. and Jürgens, G. (1994) The *FUSCA* genes of *Arabidopsis*: negative regulators of light responses. *Mol. Gen. Genet.*, **244** 242-52.

Mordhorst, A.P., Toonen, M.A.J. and de Vries, S.C. Plant embryogenesis. *Crit. Rev. Plant Sci.*, **16** 535-76.

Müller, A.J. (1963) Embryonentest zum Nachweis rezessiver Letalfaktoren bei *Arabidopsis thaliana*. *Biol. Zentralbl.*, **82** 133-63.

Müller, K.J., Romano, N., Gerstner, O., Garcia-Maroto, F., Pozzi, C., Salamini, F. and Rohde, W. (1995) The barley *Hooded* mutation caused by a duplication in a homeobox gene intron. *Nature*, **374** 727-30.

Murgia, M., Huang, B.Q., Tucker, S. and Musgrave, M.E. (1993) Embryo sac lacking antipodal cells in *Arabidopsis thaliana* (Brassicaceae). *Am. J. Bot.*, **80** 824-38.

Nagato, Y., Kitano, H., Kamijima, O., Kikuchi, S. and Satoh, H. (1989) Developmental mutants showing abnormal organ differentiation in rice embryos. *Theor. Appl. Genet.*, **78** 11-15.

Natesh, S. and Rau, M.A. (1984). The embryo, in *Embryology of Angiosperms* (ed. B.M. Johri) Springer, Berlin, pp. 377-443.

Neuffer, M.G. and Sheridan, W.F. (1980) Defective kernel mutants of maize. I. Genetic and lethality studies. *Genetics*, **95** 929-44.

Nogler, G.A. (1984) Gametophytic apomixis, in *Embryology of Angiosperms* (ed. B.M. Johri), Springer, Berlin, pp. 475-518.

Norstog, K. (1972) Early development of the barley embryo: fine structure. *Am. J. Bot.*, **59** 123-32.

Nüsslein-Volhard, C. (1991) Determination of the embryonic axes of *Drosophila*. *Dev. Suppl.*, **1** 1-10.

Ohad, N., Margossian, L., Hsu, Y.-C., Williams, C., Repetti, P. and Fischer, R.L. (1996) A mutation that allows endosperm development without fertilization. *Proc. Natl Acad. Sci. USA*, **93** 5319-24.

Okada, K., Ueda, J., Komaki, M.K., Bell, C.J. and Shimura, Y. (1991) Requirement of the auxin polar transport system in early stages of *Arabidopsis* floral bud formation. *Plant Cell*, **3** 677-84.

Patton, D.A., Volrath, S. and Ward, E.R. (1996) Complementation of an *Arabidopsis* thaliana biotin auxotroph with an *Escherichia coli* biotin biosynthetic gene. *Mol. Gen. Genet.*, **251** 261-66.

Pollock, E.G. and Jensen, W.A. (1964) Cell development during early embryogenesis in *Capsella* and *Gossypium*. *Am. J. Bot.*, **51** 915-21.

Prusinkiewicz, P. and Hanan, J. (1989) Lindenmayer systems, fractals, and plants. *Lecture Notes in Biomathematics* (ed. S. Levin), Springer, Berlin.

Przemeck, G.K.H., Mattsson, J., Hardtke, C.S., Sung, Z.R. and Berleth, T. (1996) Studies on the role of the *Arabidopsis* gene *MONOPTEROS* in vascular development and plant cell axialization. *Planta*, **200** 229-37.

Quatrano, R.S. (1990) Polar axis fixation and cytoplasmic localization in *Fucus*, in *Genetics of Pattern Formation and Growth Control* (ed. A. Mahowald), Alan R. Liss, New York, pp. 31-46

Quatrano, R.S. and Shaw, S.L. (1997) Role of the wall in the determination of cell polarity and the plane of cell division in *Fucus* embryos. *Trends Plant Sci.*, **2** 15-21.

Raghavan, V. (1986) *Embryogenesis in Angiosperms. A developmental and experimental study*. Cambridge University Press, Cambridge.

Raghavan, V. and Torrey, J.G. (1963) Growth and morphogenesis of globular and older embryos of *Capsella* in culture. *Am. J. Bot.*, **50** 540-51.

Randolph, L.F. (1936) Developmental morphology of the caryopsis in maize. *J. Agric. Res.*, **53** 821-916.

Reinert, J. (1959) Über die Kontrolle der Morphogenese und die Induktion von Adventivembryonen in Gewebekulturen aus Karotten. *Planta*, **53** 318-33.

Reiser, L. and Fischer, R.L. (1993) The ovule and the embryo sac. *Plant Cell*, **5** 1291-1301.

Reynolds, T.L. (1997) Pollen embryogenesis. *Plant Mol. Biol.*, **33** 1-10.

Rosenblum, I.M. and Basile, D.V. (1984) Hormonal regulation of morphogenesis in *Streptocarpus* and its relevance to evolutionary history of the Gesneriaceae. *Am. J. Bot.*, **71** 52-64.

Russell, S.D. (1993) The egg cell: development and role in fertilization and early embryogenesis. *Plant Cell*, **5** 1349-59.

Rutishauser, A (1969). *Embryologie und Fortpflanzungsbiologie der Angiospermen*, Springer, New York.

Sachs, T. (1991). Cell polarity and tissue patterning in plants. *Dev. Suppl.*, **1** 83-93.

Sanchez-Burgos, A.A. and Dengler, N.G. (1988) Leaf development in isophyllous and facultatively anisophyllous species of *Pentadenia* (Gesneriaceae). *Am. J. Bot.*, **75** 1472-84.

Scheres, B., Wolkenfelt, H., Willemsen, V., Terlouw, M., Lawson, E., Dean, C. and Weisbeek, P. (1994) Embyonic origin of the *Arabidopsis* primary root meristem initials. *Development*, **120** 2475-87.

Scheres, B., Di Laurenzio, L., Willemsen, V., Hauser, M.-T., Janmaat, K., Weisbeek, P. and Benfey, P. (1995) Mutations affecting the radial organisation of the *Arabidopsis* root display specific defects throughout the embryonic axis. *Development*, **121** 53-62..

Scheres, B., McKhann, H., van den Berg, C., Willemsen, V., Wolkenfelt, H., de Vrieze, G. and Weisbeek, P. (1996) Experimental and genetic analysis of root development in *Arabidopsis thaliana*. *Plant and Soil*, **187** 97-105.

Schneitz, K., Hülskamp, M. and Pruitt, R.E. (1995) Wild-type ovule development in *Arabidopsis thaliana*: a light microscope study of cleared whole-mount tissue. *Plant J.*, **7** 731-49.

Schulz, P. and Jensen, W.A. (1968) *Capsella* embryogenesis: the egg, zygote and young embryo. *Am. J. Bot.*, **55** 807-19.

Schulz, P. and Jensen, W.A. (1973) *Capsella* embryogenesis: the central cell. *J. Cell. Sci.*, **12** 741-63.

Schulz, P. and Jensen, W.A. (1986) Prefertilization ovule development in *Capsella*: the dyad, tetrad, developing megaspore, and two-nucleate gametophyte. *Can. J. Bot.*, **64** 875-84.

Schwartz, B.W., Yeung, E.C. and Meinke, D.W. (1994) Disruption of morphogenesis and transformation of the suspensor in abnormal *suspensor* mutants of *Arabidopsis*. *Development*, **120** 3235-45.

Seagull, R.W. (1989). The plant cytoskeleton. *CRC Crit. Rev. Plant Sci.*, **8** 131-67.

Sheridan, W.F. and Clark, J.K. (1993) Mutational analysis of morphogenesis of the maize embryo. *Plant J.*, **3** 347-58.

Sheridan, W.F. and Neuffer, M.G. (1982) Maize developmental mutants. Embryos unable to form leaf primordia. *J. Hered.*, **73** 318-29.

Shevell, D.E., Leu, W.-M., Gillmor, C.S., Xia, G., Feldmann, K.A. and Chua, N.-H. (1994) *EMB30* is essential for normal cell division, cell expansion, and cell adhesion in *Arabidopsis* and encodes a protein that has similarity to Sec7. *Cell*, **77** 1051-62.

Smith, L.G., Jackson, D. and Hake, S. (1995) Expression of *knotted1* marks shoot meristem formation during maize embryogenesis. *Dev. Genet.*, **16** 344-48.

Souer, E., van Houwelingen, A., Kloos, D., Mol, J. and Koes, R. (1996) The *NO APICAL MERISTEM* gene of *Petunia* is required for pattern formation in embryos and flowers and is expressed at meristem and primordia boundaries. *Cell*, **85** 159-70.

Springer, P.S., McCombie,W.R., Sundaresan, V. and Martienssen, R.A. (1995) Gene trap tagging of *Prolifera*, an essential *MCM2-3-5*-like gene in *Arabidopsis*. *Science*, **268** 877-80.

St Johnston, D. (1995) The intracellular localization of messenger RNAs. *Cell*, **81** 161-70.

St Johnston, D. and Nüsslein-Volhard, C. (1992) The origin of pattern and polarity in the *Drosophila* embryo. *Cell*, **68** 201-19.

Staiger, C.J. and Lloyd, C.W. (1991) The plant cytoskeleton. *Curr. Opin. Cell Biol.*, **3** 33-42.

Steeves, T.A. and Sussex, I.M. (1989). *Patterns in Plant Development.*. Cambridge University Press, Cambridge.

Tillich, H.-J. (1992) Bauprinzipien und Evolutionslinien bei monokotylen Keimpflanzen. *Bot. Jahrb. Syst.*, **114** 91-132.

Tilney, L.G., Cooke, T.J., Connelly, P.S. and Tilney, M.S. (1990) The distribution of plasmodesmata and its relationship to morphogenesis in fern gametophytes. *Development*, **110** 1209-21.

Topping, J.F. and Lindsey, K. (1997) Promoter trap markers differentiate structural and positional components of polar development in *Arabidopsis*. *Plant Cell*, **9** 1713-25.

Topping, J.F., Wi, W. and Lindsey, K. (1991) Functional tagging of regulatory elements in the plant genome. *Development*, **112** 1009-19.

Topping, J.F., May, V.J., Muskett, P.R. and Lindsey, K. (1997) Mutations in the *HYDRA1* gene of *Arabidopsis* perturb cell shape and disrupt embryonic and seedling morphogenesis. *Development*, **124** 4415-24.

Torres Ruiz, R.A. and Jürgens, G. (1994) Mutations in the *FASS* gene uncouple pattern formation and morphogenesis in *Arabidopsis* development. *Development*, **120** 2967-78.

Torres Ruiz, R.A., Fischer, T. and Haberer, G. (1996a). Genes involved in the elaboration of apical pattern and form in *Arabidopsis thaliana*: genetic and molecular analysis, in *Embryogenesis: The Generation of a Plant* (eds E. Cummings and T. Wang), Bios Scientific Publishers, Oxford, pp. 15-34.

Torres Ruiz, R.A., Lohner, A. and Jürgens, G. (1996b) The *GURKE* gene is required for normal organization of the apical region in the *Arabidopsis* embryo. *Plant J.*, **10** 1005-16.

Traas, J., Bellini, C., Nacry, P., Kronenberger, J., Bouchez, D. and Caboche, M. (1995) Normal differentiation patterns in plants lacking microtubular preprophase bands. *Nature*, **375** 676-77.

Tsuge, T., Tsukaya, H. and Uchimiya, H. (1996) Two independent and polarized processes of cell elongation regulate leaf blade expansion in *Arabidopsis thaliana* (L.) Heynh. *Development*, **122** 1589-1600.

Tsukaja, H. (1997) Determination of the unequal fate of cotyledons of a one-leaf plant, *Monophyllaea*. *Development*, **124** 1275-80.

Tykarska, T. (1976) Rape embryogenesis. I. The proembryo development. *Acta Soc. Bot. Pol.*, **45** 3-15.

Tykarska, T. (1979) Rape embryogenesis. II. Development of the embryo proper. *Acta Soc. Bot. Pol.*, **48** 391-421.

van den Berg, C., Willemsen, V., Hage, W., Weisbeek, P. and Scheres, B. (1995) Cell fate in the *Arabidopsis* root meristem determined by directional signalling. *Nature*, **378** 62-65.

van den Berg, C., Willemsen, V., Hage, W., Weisbeek, P. and Scheres, B. (1997) Short-range control of cell differentiation in the *Arabidopsis* root meristem. *Nature*, **390** 287-89.

Vernon, D.M. and Meinke, D.W. (1994) Embryogenic transformation of the suspensor in twin, a polyembryonic mutant of *Arabidopsis*. *Dev. Biol.*, **165** 566-73.

Vielle Calzada, J.-P., Crane, C.F. and Stelly, D.M. (1996) Apomixis: the asexual revolution. *Science*, **274** 1322-23.

Vroemen, C.W., Langeveld, S., Mayer, U., Ripper, G., Jürgens, G., Van Kammen, A. and de Vries, S. (1996) Pattern formation in the *Arabidopsis* embryo revealed by position-specific lipid transfer protein gene expression. *Plant Cell*, **8** 783-91.

Wakana, A. and Uemoto, S. (1988) Adventive embryogenesis in (Rutaceae). *Citrus* II. Postfertilization development. *Am. J. Bot.*, **75** 1031-47.

Wardlaw, C.W. (1955) *Embryogenesis in Plants*, Methuen, London.

Webb, M.C. and Gunning, B.E.S. (1990) Embryo sac development in *Arabidopsis thaliana*. I. Megasporogenesis, including the microtubular cytoskeleton. *Sex. Plant Reprod.*, **3** 244-56.

Webb, M.C. and Gunning, B.E.S. (1991) The microtubular cytoskeleton during development of the zygote, proembryo, and free-nuclear endosperm in *Arabidopsis thaliana* (L.) Heynh. *Planta*, **184** 187-95.

West, M.A.L. and Harada, J.J. (1993) Embryogenesis in higher plants: an overview. *Plant Cell*, **5** 1361-69.

West, M.A.L., Yee, K.M., Danao, J., Zimmerman, J.L., Fischer, R.L., Goldberg, R.B. and Harada, J.J. (1994) *LEAFY COTYLEDON1* is an essential regulator of late embryogenesis and cotyledon identity in *Arabidopsis*. *Plant Cell*, **6** 1731-45.

Wilhelm, J. and Vale, R. (1993) RNA on the move: the mRNA localization pathway. *J. Cell Biol.*, **123** 269-74.

Wilhelmi, L.K. and Preuss, D. (1996) Self-sterility in *Arabidopsis* due to defective pollen tube guidance. *Science*, **274** 1535-37.

Wu, Y., Haberland, G., Zhou, C. and Koop, H.U. (1992) Somatic embryogenesis, formation of morphogenetic callus and normal development in zygotic embryos of *Arabidopsis thaliana* cultured *in vitro*. *Protoplasma,* **169** 35-39.

Yadegari, R., de Pavia, G.R., Laux, T., Koltunow, A.M., Apuya, N., Zimmerman, J.L., Fischer, R.L., Harada, J.J. and Goldberg, R.B. (1994) Cell differentiation and morphogenesis are uncoupled in *Arabidopsis* raspberry embryos. *Plant Cell*, **6** 1713-29.

Yanofysky, M.F. (1995) Floral meristems to floral organs: genes controlling early events in *Arabidopsis* flower development. *Ann. Rev. Plant Physiol. Plant Mol. Biol.*, **46** 167-88.

Yeung, E.C. and Meinke, D.W. (1993) Embryogenesis in angiosperms: development of the suspensor. *Plant Cell*, **5** 1371-81.

8 Patterns in vegetative development

Rob Martienssen and Liam Dolan

8.1 Introduction

Patterns of vegetative development are among the most ancient in the plant kingdom. Primitive plants underwent branching, lateral organogenesis and photosynthetic and non-photosynthetic cellular differentiation well before the emergence of flowers and seeds. The basic principles by which shoots, leaves and roots are patterned might thus be considered a blueprint for floral and embryonic organs that have evolved more specialized roles. This view of plant development has been held to some degree since botanists first began to systematically catalogue plant form, but recently genetic studies have lent considerable support to this hypothesis.

Classically, the origin of the root has been less clear, but in principle roots might be considered to be derived shoots or leaves. The earliest vascular plants, including the Cooksoniopsida, Rhyniopsida and Zosterophyllopsida, lacked leaves and roots (Kenrick and Crane, 1997). Roots evolved once in the Lycopsid lineage and a number of times in the derived Eutracheophyte lineage (Gensel, 1992), which includes Spenopsida, Pteropsida and the Spermatophytes. Lycopsid roots arise adventitiously from shoots, have an exarch xylem organization (protoxylem differentiating on the outside of the metaxylem) and are considered to be derived from microphylls, the primitive leaf-like structures borne on the shoots of this group of plants. Eutracheophyte roots, with their endarch xylem (protoxylem differentiating on the inside of the metaxylem), arise either adventitiously (from shoot structures) or from bipolar embryos. In extant seed plants (gymnosperms and angiosperms) the basal end of the embryo contributes to the primary root meristem. The fate of this meristem is variable but the mature root 'system' comprises derivatives of primary roots and lateral roots, and may be supplemented in certain instances with roots of adventitious origin (Steeves and Sussex, 1989).

The pattern of vegetative development in the shoot is largely determined by the positioning of branches and lateral organs along the primary body axis. Lateral organs include leaves, which are distinguished from axillary shoots by their flattened appearance and determinate growth pattern. The primary root, on the other hand, is radially symmetric and indeterminate. Lateral roots reiterate this pattern and are positioned along the main axis via reorganization of the pericycle layer. Although the process of organogenesis appears to be very different in each system, general principles of vegetative and embryonic patterning emerge from an examination of them both. Of primary importance are the signalling interactions by which lateral organs and tissues arise from stem cell populations, and the positional cues by which they are placed and oriented with respect to the main axis. Cell–cell interactions, cell cycle control and asymmetric cell division are examples of mechanisms common to each process. These mechanisms are being studied from a genetic perspective in *Arabidopsis*, and are the subject of this chapter.

8.2 Shoot development

The shoot apical meristem (SAM) arises very early during embryogenesis. Indeed, the entire apical half of the globular embryo can be considered a SAM. According to this view, the cotyledon primordia at the heart stage are the first pair of lateral organs (Kaplan, 1969; Laux and Jurgens, 1997). Mutations that disrupt the differentiation of the embryonic SAM and the cotyledons include *gurke, topless, zwille* and *pinhead*, and are discussed in more detail in chapter 7. *TOPLESS* regulates the expression of *SHOOTMERISTEMLESS (STM)* which in turn may regulate *ZWILLE* (Evans and Barton, 1997; Endrizzi *et al.*, 1996). However, of these three genes, only *STM* is thought to have a role in the vegetative SAM, and this role may differ from that in the embryo (Evans and Barton, 1997). Nonetheless, important parallels can be drawn between these two phases of plant development (Kaplan and Cooke, 1997), some of which will be explored below.

8.2.1 Organization of the shoot apical meristem

The SAM of *Arabidopsis* has a typical organization for a higher plant (Satina *et al.*, 1940), comprising a central and a peripheral zone, each divided into two tunica layers (L1 and L2), and a corpus (L3). The central zone of vacuolated cells comprises only 6–10 cells in each layer at the time of germination, while the peripheral zone comprises a ring 2–4 cells wide surrounding the central zone and interleaving between incipient organ primordia (Figure 8.1). Labelling studies have revealed that the cell cycle time in the central zone is 2–3 times longer than in the peripheral zone (Brown *et al.*, 1963). The *Arabidopsis* meristem adopts its characteristic layered organization by the torpedo stage of embryogenesis (Barton and Poethig, 1993). The outer (L1) layer is maintained as a separate lineage by restricting cell division to within the layer. The subepidermal, or L2 lineage, is also maintained as a single cell layer within the meristem, but contributes to the underlying tissue in the leaves via periclinal (out-of-plane) divisions during organogenesis. The inner layer, L3, divides in all directions and contributes to the rib meristem below the shoot apex.

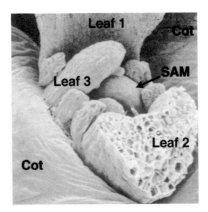

Figure 8.1 The *Arabidopsis* shoot apical meristem (SAM). A scanning electron micrograph of an 8-day-old seedling of the Landsberg ecotype. Cot (cotyledons), leaves 1, 2 and 3, and the SAM are indicated. The second leaf has been removed to expose the SAM just prior to the floral transition. The third leaf is in plastochron 5. Scanning electron micrograph kindly provided by V. Irish (Yale University).

After germination, the cells at the summit of the SAM behave in many respects as stem cells do (Vaughan, 1955; Brown *et al.*, 1963). That is to say, they divide very slowly, and their progeny fall into two classes. The first class replenishes the meristem, equivalent to 'mother' cells in a stem cell lineage. The second class, equivalent to 'daughter cells', emerges into the peripheral zone of the apex, where division rates are much higher and organogenesis occurs (Figure 8.1). However, clonal analysis in *Arabidopsis* (Furner and Pumfrey, 1992; Irish and Sussex, 1992; Schnittger *et al.*, 1996) and in other plants (Poethig, 1989) has revealed that both products of stem cell division frequently adopt stem cell fates (resulting in marked sectors of increasing size) or peripheral fates (resulting in sector termination). These results show that stem cell fate does not depend on asymmetric division, so that positional cues are likely to be more important.

8.2.2   Growth of the SAM and its genetic regulation

The vegetative apex increases steadily in size during growth, more than doubling in volume when the transition to flowering is delayed in short days (Vaughan, 1955; Lincoln *et al.*, 1994; Medford *et al.*, 1994). Importantly, organogenesis is unaffected by this large increase. A number of mutations are thought to have a role in controlling stem cell versus peripheral cell division patterns (Meyerowitz, 1996; Laux and Jurgens, 1997) with consequent effects on meristem size (Weigel and Clark, 1996) *shootmeristemless (stm)* and *wuschel (wus)* were both selected as seedling mutants that failed to initiate leaves, although the cotyledons still emerged (Barton and Poethig, 1993; Laux *et al.*, 1996). The SAM in *stm* and *wus* is considerably reduced in size and may be absent altogether in *stm*. In contrast, the meristem in *clavata1 (clv1)* mutants is greatly increased in size. *clavata* mutants are fasciated, which means that they have enlarged meristems that initiate additional organs with aberrant phyllotaxis. Internodes are also affected resulting in fused, ribbed stems (Leyser and Furner, 1992; Clark et al., 1993).

Null alleles of stm can be suppressed in a heterozygous clv1/+ background, resulting in viable vegetative shoots lacking STM (Clark et al., 1996). This shows that stm is not required for meristem establishment (Clark, 1997) nor, strictly speaking, for maintenance of the stem cell population itself (Endrizzi et al., 1996). Similarly, clv1 mutants are largely suppressed in stm/+ heterozygous plants. It has been proposed that these additive interactions represent the bypass of one system for size control via another, so that stm and clv1 have opposite effects on cell proliferation and thus on meristem size (Weigel and Clarke, 1996; Meyerowitz, 1996; Clarke, 1997).

The *WUSCHEL (WUS)* gene also affects meristem function, resulting in a shoot meristem-less phenotype. Mutant wus seedlings have an enlarged flattened apex that fails to form leaves but continually initiates new defective axillary meristems instead (Laux et al., 1996). clv1 is completely epistatic to wus, so that wus,clv1 double mutants have wus-like apices (Laux et al., 1996). It is tempting to propose, therefore, that there are two pathways that regulate meristem size: one involving STM and another involving WUS and CLV. However, wus,stm double mutants resemble stm rather than wus, so they may also be involved in a common pathway. It has been proposed that STM and CLV independently regulate WUS, presumably in opposite directions, and that WUS is a key regulator of meristem size and function (Laux and Schoof, 1997). Alternatively, STM may simply be required to maintain cell types affected by wus, resulting in indirect rather than direct interaction (Evans and Barton, 1997).

8.2.3 Genes involved in signalling within the apex

STM is a homeodomain protein related to the KNOTTED family of transcriptional regulators (Long et al., 1996). This family of proteins was first discovered in maize as dominant mutations that altered leaf cell fate (Vollbrecht et al., 1991). When these genes are ectopically expressed in *Arabidopsis* they generate new meristems from the dorsal (adaxial) surface of greatly reduced leaves (Sinha et al., 1993; Lincoln et al., 1994; Chuck et al., 1996). Knotted homeodomain proteins, therefore, appear to be required for meristem fate. There are at least two homologues of STM in *Arabidopsis*, KNAT1 and KNAT2, which could, in principle, have redundant overlapping functions with STM. However, sequence comparisons suggest that these three classes of genes also exist in maize, indicating that the family is very ancient and that novel functions may have arisen over evolutionary time (Schneeburger et al., 1995; M. Freeling, personal communication). KNAT1, for example, is down-regulated in cells at the summit of the apex on the transition to flowering in *Arabidopsis* (Lincoln et al., 1994). Expression is restored in floral meristems that arise later. This expression pattern suggests a role in floral induction, or perhaps in organogenesis.

CLAVATA1 encodes a receptor-like membrane kinase that has a predicted extracellular domain consisting of multiple leucine-rich repeats (Clark *et al.*, 1997). These features are also found in a large and expanding family of proteins that are required for gene-for-gene pathogen recognition and disease resistance (reviewed by Boyes *et al.*, 1996). Similar repeats in the *Drosophila* transmembrane receptor kinase Toll are known to interact with extracellular peptide signals that control axis specification in the early embryo (Belvin and Anderson, 1996). The same pathway has also been co-opted for cellular immunity to bacterial and fungal pathogens in *Drosophila*, suggesting that it functions in much the same way as in plants. Thus it is likely that CLV1 is a receptor for a peptide ligand that somehow controls meristem structure.

Genetically, a good candidate for a ligand of the CLV1 kinase is the CLV3 gene product (Clark *et al.*, 1995). *clv1* and *clv3* are epistatic, and *trans* heterozygous combinations give partially mutant phenotypes. These are classical properties of receptor–ligand genetic interactions, as reduction in the concentration of signal at the source reduces the range of its action, and it is considerably exacerbated by a reduction in the level of the cognate receptor (e.g. Baker *et al.*, 1990).

The CLV1 receptor kinase is expressed in a broad domain in the internal layers of the apex and yet exerts its influence on all layers of the SAM (Clark *et al.*, 1997). Presumably, some downstream effector of the CLV1 signal must be able to pass between these layers to co-ordinate their growth. The KNOTTED homeodomain regulatory protein has been shown to pass directly between cells in this way (Lucas *et al.*, 1995; reviewed by Hake and Char, 1997; Jackson and Hake, 1997). Although *STM* itself is not downstream of *CLV1*, it is possible that the effects of *CLV1* may be mediated by transcription factors that pass between cells in a similar way (Clark, 1997).

An alternative view of the role of *CLAVATA1* in meristem function is that it is required for lateral inhibition of central zone identity. Evidence for lateral inhibition comes from surgical isolation of portions of the SAM in *Trachymene* (Ball, 1980), lupin (Pilkington, 1929; Ball, 1950) and potato (Sussex, 1952). Isolated pieces first enlarge by unregulated cell division and then reorganize into new apices (Steeves and Sussex, 1989). In many plants, irradiation often results in twinned embryos and meristems, apparently because of the induction of apical cell death and consequent reorganization of the meristem into multiple apices (Poethig, 1989). It has been argued from these experiments that the central cells at the summit of the apex emit a signal that inhibits surrounding cells from forming another apex. If the *CLAVATA* gene products regulate this inhibition, mutant cells would be free to adopt a central cell fate. This could result in the elaboration of new apices and ultimately in a greatly enlarged meristem. This would be consistent with the negative regulation of *WUS*, which is required for central cell fate (Laux *et al.*, 1996).

8.2.4 Regulation of phyllotaxis

Lateral organs arise from small groups of founder cells in the peripheral zone of the shoot apex (Lyndon, 1983). The first two leaves are initiated in the embryo after the cotyledons and SAM have differentiated (Barton and Poethig, 1993). The third and successive leaves arise sequentially following germination (Figure 8.2). Vegetative development can be conveniently divided into stages called 'plastochrons' that correspond to the period of time between the initiation of one organ and the initiation of the next. The incipient leaf primordia is considered to be in plastochron 0, while the first stage of leaf development following initiation is called plastochron 1 (P1), and leaf development is complete by plastochron 6 or 7 (Pyke *et al.*, 1991). In *Arabidopsis*, each plastochron lasts 24–48 hours, meaning that a new leaf is initiated almost every day. The number of cells in each primordium increases approximately tenfold during this time (Pyke *et al.*, 1991).

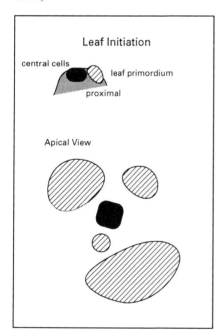
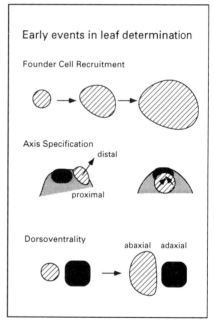

Figure 8.2 A schematic drawing of early events in leaf determination. The central zone (stem cells) are black, while leaf founder cells are hatched. Cells in the peripheral zone are in grey. Axis specification occurs in plastochrons 0, 1 and 2, while founder cell recruitment is probably not complete until plastochron 3.

The sites of leaf initiation in *Arabidopsis* and many other species arise in a precise spiral pattern around the apex (called the phyllotactic pattern), except for the first two leaves, which initiate opposite one another, in between the cotyledons. This pattern can be explained if each primordium is initiated at the maximum possible distance from neighbouring, older primordia (Richards, 1948). These observations imply the existence of an inhibitory signal that regulates phyllotaxis. When individual primordia are surgically isolated from the meristem, ectopic leaves initiate, supporting this idea (Snow and Snow, 1931, 1933).

Profound alterations in primordia number and spacing are found in weak alleles of *stm* and *clv* (Endrizzi *et al.*, 1996; Clark *et al.*, 1993). Signals that regulate phyllotaxis must pass through the meristem in order to be received by neighbouring primordia, and it is possible that *clv* and *stm* mutant apices have a reduced ability to transduce these signals. In some alleles this results in additional, more crowded floral and vegetative organs (Endrizzi *et al.*, 1996). It is unlikely that altered meristem size is directly responsible for these effects as environmental changes in meristem size have no effect (see earlier).

Other mutants in phyllotaxis in *Arabidopsis* include *altered meristem programming1 (amp1)*, which produces small leaves more rapidly than wild-type (Chaudhury *et al.*, 1993). Several alleles have been independently isolated, including *constitutive photomorphogenesis2 (cop2)* (Telfer *et al.*, 1997). Interestingly, flowering time and the time of onset of adult leaf characteristics are relatively unaffected in these mutants, resulting in plants with additional juvenile leaves (Telfer *et al.*, 1997). Cytokinin levels are substantially increased, which may account for some of the phenotypic effects (Chaudhury *et al.*, 1993). The *FOR EVER YOUNG (FEY)* gene encodes an oxidoreductase that also affects phyllotaxis (Callos *et al.*, 1994). In *fey* shoots the phyllotactic spiral, which is constant in wild-type plants, switches between clockwise and anticlockwise and the meristem is often consumed in necrotic tissue. One interpretation is that individual primordia abort in these mutants, leading to a loss of phyllotactic information and misplacement of new primordia, reminiscent of the surgical manipulations described above.

8.2.5 Development of axillary meristems

Axillary meristems arise in the axils of rosette and cauline leaves, and clonal analysis suggests they are related to the upper (adaxial) portion of the incipient leaf primordia (Snow and Snow, 1942; Irish and Sussex, 1992). In *Arabidopsis*, axillary buds give rise to an inflorescence meristem composed of a few secondary basal branches and flowers, but do not grow out until the transition to flowering (Grbic and Bleecker, 1996). Mutations and environmental effects that alter flowering time can induce a reversion to a vegetative state and aerial rosettes of vegetative leaves on these inflorescence branches. Aerial rosettes are also found in weak alleles of *stm* (Endrizzi *et al.*, 1996) but their relationship to floral induction, if any, is not clear. Axillary outgrowth is controlled by auxin, so that auxin-resistant mutants have a bushy appearance in which axillary branches are released from apical dominance prematurely (Hobbie *et al.*, 1994). *pinhead* and *revoluta* mutants have the opposite phenotype, with fewer axillary branches than normal (McConnell and Barton, 1995; Talbert *et al.*, 1995). In *revoluta*, axillary branches and flowers are sometimes reduced to filaments and this is thought to reflect an altered distribution of mitotic activity between axillary, lateral and primary meristems, but it is not known if this is the primary defect. The differentiation of axillary branches as flowers, inflorescences or vegetative shoots is discussed in chapter 9.

8.3   Organogenesis of the leaf

8.3.1   Recruitment of founder cells

Once initiated, a group of founder cells for each organ is recruited from the meristem. There are between four and ten leaf founder cells in each layer in *Arabidopsis*, as determined by clonal analysis (Furner and Pumfrey, 1992; Irish and Sussex, 1992). Telfer and Poethig (1994) have argued that founder cells continue to be recruited from the meristem after the first two leaves have been initiated (plastochron 3). This means that the primary axes of growth and symmetry are imposed on the primordium before the boundary between leaf and meristem cells has been established (Figure 8.2). Mutations in genes that regulate this boundary might thus be expected to result in fused leaves with partially determined axes of symmetry when recruitment encroaches on neighbouring primordia.

Mutants with an *stm* genotype have partially fused or missing organs at all stages of development, consistent with a role in regulating founder cell recruitment. The *stm-2* allele, for example, allows the outgrowth of two or three leaves that arise in aberrant phyllotaxis, and are fused at the base, as are floral primordia and floral organs found on axillary shoots (Barton and Poethig, 1993; Endrizzi *et al.*, 1996). In strong alleles, cotyledons are also fused, so the residual SAM is completely enclosed (Endrizzi *et al*, 1996). *STM* is first expressed as a stripe separating the presumptive cotyledon primordia in wild-type globular embryos consistent with a role in cotyledon separation (Long *et al.*, 1996). *STM* is subsequently expressed throughout the vegetative apex, being down-regulated in leaves only when they are initiated (Jackson *et al.*, 1994; Long *et al.*, 1996; Evans and Barton, 1997). If *STM* is required to restrict founder cell recruitment, mutant apices and neighbouring organs might be consumed by the first few primordia, accounting for meristem defects (Endrizzi *et al.*, 1996). These could be rescued by *clv* mutants if central cell identity were restored to some cells preventing their incorporation into leaves (see above). Thus restricting founder cell recruitment may be the primary role of *STM*, with downstream consequences for meristem maintenance and indeterminancy.

Apart from *stm*, very few other mutants have been described that are involved in founder cell recruitment in the vegetative phase. Two genes whose expression pattern indicates a role in leaf founder cell recruitment are *NO APICAL MERISTEM* (*NAM*) in petunia (Souer *et al.*, 1996) and *FIMBRIATA* (*FIM*) in *Antirrhinum* (Simon *et al.*, 1994; Ingram *et al.*, 1997). Both genes have homologues in *Arabidopsis*: *CUP-SHAPED COTYLEDONS2* (*CUC2*) and *UNUSUAL FLORAL ORGANS* (*UFO*), respectively (Levin and Meyerowitz, 1995; Ingram *et al.*, 1995; Aida *et al.*, 1997) and are expressed at the boundaries of cotyledons, leaves and floral organs (Souer *et al.*, 1996; Lee *et al.*, 1997). Mutations in these genes result in fusions between cotyledons in the embryo (*cuc, nam*) and between stamens, petals and other organs in the flower (*cuc, nam* and *ufo*). In addition, *cuc* and *nam* mutants cause meristem arrest in the seedling, while *ufo* has secondary effects on organ identity in the flower. However, neither mutant has any effect on leaves.

The *UFO/FIM* and *NAM/CUC* genes were initially thought to be required for correct positioning and separation of organs within their respective whorls (Levin and Meyerowitz, 1995; Souer *et al.*, 1996; Aida *et al.*, 1997). However, an alternative interpretation is that these genes restrict founder cell recruitment. In loss-of-function mutants, additional cells might be recruited into primordia at the expense of the meristem and neighbouring organs, resulting in fused organs and meristem arrest. The absence of a phenotype in leaves may be attributable to their spiral rather than whorled phyllotaxis (Aida *et al.*, 1997; Ingram *et al.*, 1997; Souer, 1997). Alternatively, redundant factors may be expressed during vegetative growth: the *CUC1* and *CUC2* genes are known to be redundant in the embryo and in the flower (Aida *et al.*, 1997) and additional *nam*-like genes may be expressed in the vegetative phase (Souer, 1997). Interestingly, *UFO* transcripts are found in wild-type embryos but are undetectable in *stm* mutants (Evans and Barton, 1997). It is possible that *STM* may restrict founder cell recruitment by regulating *UFO* and related genes in the apex. Only when all of these hypothetical genes are down-regulated, as in *stm* mutants, would a phenotype become apparent. A number of genes that are expressed at the boundaries of leaf primordia have been detected in *Arabidopsis* using enhancer and gene trap reporter genes, and an example is shown in Figure 8.3 (Springer and Martienssen, unpublished). Many of these expression patterns are altered in *stm* and other mutants and transgenic backgrounds, consistent with this idea.

8.3.2 *Boundaries and the regulation of cell proliferation*

Organ primordia undergo rapid cell proliferation, demonstrated by radiolabelling with tritiated thymidine (Brown *et al.*, 1963; Sussex and Rosenthal, 1973; Van't Hof *et al.*, 1985; Lyndon, 1990) or by expression of cell cycle-regulated genes, such as the *CDC47* homolog *prolifera* (Springer *et al.*, 1995) and others (Fobert *et al.*, 1993; Hemerly *et al.*, 1993). However, cells immediately surrounding the primordium divide much less frequently, resulting in a 'buffer' zone of cells that

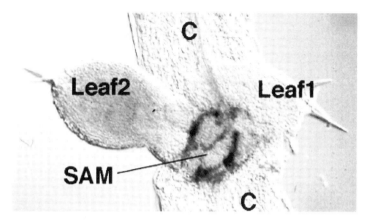

Figure 8.3 Reporter gene expression at leaf primordia boundaries. A 4-day-old Landsberg seedling was stained for β-glucuronidase expression and cleared in ethanol before differential interference contrast microscopy. A gene trap transposon has inserted in a gene expressed at leaf primordia boundaries. C (cotyledon petioles); leaf 1 (folded over), leaf 2 and the SAM are indicated (Springer and Martienssen, unpublished).

stop dividing around the base of the leaf. It is thus possible for clonally related sectors to extend to the tip of one leaf without penetrating the next (Furner and Pumfrey, 1992; Irish and Sussex, 1992). This zone may correspond to the region of expression of boundary genes such as *FIM (UFO)*. and *NAM (CUC)*. Evidence supporting this has been obtained in *Antirrhinum*, where *FIMB* has been shown to interact with a homolog of the yeast protein SKP1, which controls the proteolysis of cell cycle regulatory molecules in yeast (Ingram *et al.*, 1997). It is plausible then that the regulation of cell proliferation is an important property of primordial boundaries in plants, just as it is an important property of compartment boundaries in animals (Pardee, 1989; Diaz-Benjumea and Cohen, 1993; Couso *et al.*, 1994).

In order to establish a boundary between meristems and lateral organs, primordial fate must be determined within the primordia itself. Mutations in genes that effect determination might thus be expected to result in leafless plants or plants in which the identity of leaves has been changed (e.g. Caruso, 1968; Caruso and Cutter, 1970). The *Arabidopsis* mutants *embryonic flower1(emf1)* and *embryonic flower2 (emf2)* do not produce vegetative leaves, although rudimentary primordia are sometimes observed (Yang *et al.*, 1995). These mutants are viable and progress directly from seedling to floral organs in the space of a few weeks. *EMF* genes might thus determine rosette leaf identity, although they are usually interpreted as inflorescence meristem identity genes (Chen *et al.*, 1997).

Three regulatory genes expressed in leaves as well as flowers may have a role in leaf determination. *AINTEGUMENTA (ANT)* is a homologue of *APETALA2 (AP2)* and is required for integument formation in the ovule (Elliott *et al.*, 1996; Klucher *et al.*, 1996; Okamuro *et al.*, 1997). *AP2* itself is required for floral organ identity (see chapter 9) but *ap2,ant* double mutants have severe reductions in sepals, petals and stamens, suggesting a redundant role in floral determination (Elliott *et al.*, 1996). *AP2* and *ANT* are both expressed in leaves but double mutants have no vegetative phenotype. *GLOSSY15* is a close homologue of *AP2* in maize and confers juvenile identity to epidermal leaf cells, fuelling speculation that leaf and floral organ identity might be specified in the same way (Moose and Sisco, 1996; Poethig, 1997).

LEAFY (LFY) is expressed in leaves (Kelly *et al.*, 1995; Blazquez *et al.*, 1997) as well as in floral primordia, where it transforms indeterminate shoots into determinate flowers with additional effects on petal and stamen identity (Weigel *et al.*, 1992). *FLORICAULA (FLO)* is a functional homolog of *LFY* in snapdragon (Coen *et al.*, 1990) and *flo* leaves sometimes fail to separate, forming a continuous double spiral (Carpenter *et al.*, 1995). This is exactly the phenotype expected for a defect in founder cell recruitment. In pea, which has compound leaves, *UNIFOLIATA (UNI)* is the apparent orthologue of *FLO* (Hofer *et al.*, 1997). In addition to the predicted floral phenotypes, leaves are entire in *unifoliata* mutants, suggesting *UNI* acts at an early stage of leaflet and tendril determination in compound leaves (Sachs, 1969; Young 1983; Gould *et al.*, 1994). In flowers, *LFY* acts together with *UFO* to determine organ identity (Simon *et al.*, 1994; Ingram *et al.*, 1995, 1997; Lee *et al.*, 1997). Plants that ectopically express *UFO* under the control of the 35S promoter have lobed upper leaves, but only in the presence of wild-type *LEAFY* (Lee *et al.*, 1997). This is at least consistent with the role of these two genes in boundary specification and determination proposed above, and may be relevant to the evolution of determinate flowers and leaves from indeterminate axillary shoots (Friis *et al.*, 1987; Hofer *et al.*, 1997).

8.3.3 Determination of leaf identity

The identity of leaves can be switched between adult and juvenile phases through the action of *PAUSED*, *HASTY* and other genes (Telfer *et al.*, 1997). Mutations in these genes are usually considered heterochronic on the basis that the affected primordia arise sequentially over time rather than being determined simultaneously. Changes in phase result in changes in shape as well as in the appearance of trichomes on the adaxial (upper) surface of juvenile leaves, on both surfaces of adult leaves and on the abaxial (lower) surface of cauline leaves. These changes are sensitive to the phytohormone gibberellin as well as to environmental effects on flowering time (Chien and Sussex, 1996; Telfer *et al.*, 1997).

Cotyledons and floral organs can be converted into leaves by combinations of homeotic mutations (Bowman *et al.*, 1991; Meinke, 1992; Meinke *et al.*, 1994) and are classically considered modified leaves (Goethe, 1790; Steeves and Sussex, 1989). Typically, loss-of-function mutations result in transformations into more leaf-like organs, although mutations that convert leaves into cotyledons have been described (Conway and Poethig, 1997). Thus the leaf has been modified successively during evolution to provide specialized functions in storage and reproduction in embryos and flowers, respectively (Coen and Meyerowitz, 1991). Organ identity in the flower is conferred by a set of MADS box proteins, expressed in overlapping whorls of floral organs (see Chapter 9). Each of these genes has additional roles in organogenesis and meristem determinancy, however, suggesting that this simple picture may not be complete (Bowman *et al.*, 1991; Weigel and Clarke, 1996). *CURLY LEAF (CLF)* is required to regulate negatively the expression of floral organ identity genes in the leaf and in the flower. *CLF* encodes a homologue of the polycomb group genes in *Drosophila*, which regulate homeotic genes via chromatin structure (Goodrich *et al.*, 1997; Jurgens, 1997). Despite the global nature of the mutation, the only effect of *clf* on leaf morphogenesis is an upward curling also observed in leaves that overexpress the homeotic MADS box genes *AGAMOUS* or *AP3* (Mizukami and Ma, 1992; Krizek and Meyerowitz, 1996). Thus additional genes must be required for floral organ identity that are not regulated by *CLF*.

8.3.4 Development of leaf axes

Lateral organs in plants resemble limbs in animals in that they are patterned as secondary fields, which arise from the primary body plan and must be properly oriented with respect to it (Wardlaw, 1965; Bryant, 1975). Lateral organ primordia have three primary axes (Figure 8.3): the proximodistal (PD) axis runs along the length of the leaf and comprises the axis of outgrowth. It lies perpendicular to the primary apical–basal axis of the shoot. The dorsoventral (DV) axis between the upper (adaxial) surface and the lower (abaxial) surface of the leaf lies parallel to the primary apical–basal axis of the shoot. Patterning of the organ in this axis gives the lateral organ its characteristic flattened shape. Finally, the left–right, or medio–lateral, axis lies along the surface of the meristem but is perpendicular to the primary axis of shoot growth. As in animals, mutations in axis specification are likely to be important in understanding the earliest events in organogenesis in plants (Steeves and Sussex, 1989).

The proximodistal axis of the leaf is marked via the differentiation of ground, vascular and epidermal cell types. The midvein arises as a leaf trace that differentiates acropetally from the vascular bundles in the stem into the early leaf primordia. Only a single midvein is formed in each leaf, with successively minor veins branching from it and each other to form a reticulate network (Vaughan, 1955; Nelson and Dengler, 1997). At around the same time, subepidermal cells at the tip of the leaf differentiate to form a hydathode and associated epithem (the subtending airspace). The hydathode consists of a series of specialized stomata known as water pores that allow fluids to pass to the surface of the leaf (Zeiger *et al.*, 1987). In true leaves, the vasculature is characterized by numerous freely ending veinlets as well as multiple secondary hydathodes along the margin of the leaf. In contrast, the cotyledon has much reduced vasculature, with secondary veins branching from the midvein in large loops and few or no freely ending veinlets. A single hydathode is found at the tip. Trichomes, stomata and other cell types differentiate with respect to the proximodistal axis, with distal structures ceasing cellular proliferation and differentiating first (Pyke *et al.*, 1991). The basal (proximal) regions of the leaf are marked by the development of stipules, which are reduced to small multicellular structures in *Arabidopsis* and arise on either side of the dorsal surface at the base of the leaf (Bowman, 1994).

The dorsoventral axis is marked by epidermal cell types such as trichomes on the adaxial side of juvenile leaves, and elongated pavement cells over the midvein on the abaxial side (Telfer and Poethig, 1994). Internally, the adaxial layer of mesophyll cells comprises densely packed, elongated palisade cells, while the abaxial layer comprises spongy cells (Pyke *et al.*, 1991). The vasculature is also polar with xylem at the adaxial side and phloem abaxial, continuous with the concentric arrangement in the stem (Vaughan, 1955; Dharmawardhana *et al.*, 1992; Nelson and Dengler, 1997). Molecular markers for most tissues and cell types exist (e.g. Baima *et al.*, 1995; Sundaresan *et al.*, 1995; Lu *et al.*, 1996; Springer *et al.*, unpublished).

Plants in which either of the homeobox genes *KNAT1* (from *Arabidopsis*) or *KNOTTED1* (from maize) is expressed under control of the 35S promoter from CaMV have deeply lobed, asymmetric leaves that are considerably reduced in size (Lincoln *et al.*, 1994). Each lobe is subtended by stipules on either side (Chuck *et al.*, 1996). The reiteration of stipules supports the interpretation that each lobe represents a reiterated proximodistal axis. Similarly, the pattern of the vasculature in these plants is drastically altered, adopting a more distributed pattern as if there were multiple primary (mid) veins. Parallels can be drawn with compound leaves in tomato, which express *KNOTTED* homologues that are restricted to the meristem in other species (Hareven *et al.*, 1996).

Ectopic expression of KNAT1 and other homeodomain proteins also results in the appearance of shoots on the upper (adaxial) surface of the leaf, primarily at the base and in the sinuses between the lobes. Each new meristem is associated with a vascular bundle, which is where *KNAT1* transcripts accumulate in transgenic plants (Chuck *et al.*, 1996). Adventitious shoots on leaves (epiphylly) are known in many species (reviewed by Kerstetter and Hake, 1997) and can be induced on the surface of tobacco leaves by ectopic expression of the cytokinin biosynthesis gene *isopentenyl transferase* (Li *et al.*, 1992). This may be related to aberrations in shoot development observed in cytokinin-requiring mutants (Deikman and Ulrich, 1995). The role of hormones in development is reviewed in chapter 4. The extent to which misregulation of hormones and key regulatory genes like *KNAT1* can explain the variety of angiosperm forms awaits the molecular analysis of some of these more exotic species (Cronk and Moeller, 1997). If *KNOTTED*-like genes (such as *STM*) function to inhibit leaf founder cell recruitment (see earlier), ectopic expression in the vasculature might induce the production of ectopic axes as well as organize new meristems.

The PD axis is also disrupted in *Arabidopsis* plants homozygous for the recessive mutation, *asymmetric leaf1 (as1)* (Redei, 1965; Curtis and Martienssen, unpublished). Homozygous plants have lobed leaves and the number and depth of these lobes increases in successive leaves until cauline leaves, which are heavily incised and straplike. Cotyledons are unaffected in the mutant but the stamens are reduced (Barabas and Redei, 1971). The arrangement of marginal and surface structures is altered in *as1* mutant leaves, but the most striking aberration is in the vasculature, which is more distributed than wild-type, although not as extreme as in KNAT1 transgenic plants. Interestingly, *as1* leaves and *35S::KNAT1* leaves still emerge in a regular phyllotaxis, suggesting that the signals responsible for the phyllotactic arrangement of leaves do not correspond to those that specify the PD axis.

One of the most interesting mutants in organogenesis to be described in recent years is the snapdragon mutant *phantastica (phan)* (Waites and Hudson, 1995). Strong and weak alleles at this locus are acutely cold sensitive and have short radially symmetric leaves at the restrictive temperature. Cell types in the epidermis and ground tissue of mutant leaves indicate that adaxial structures have been lost. At intermediate temperatures, transformation of the adaxial to the abaxial domain is incomplete, resulting in patches of abaxial tissue bordered by ectopic leaf margins. Waites and Hudson (1995) have proposed that laminal outgrowth in wild-type leaves is the consequence of the juxtaposition of adaxial and abaxial identity, and that varying phenotypic effects of *phan* reflect varying degrees of adaxial identity. Petals are affected in *phan* mutants, but the cotyledons and sepals are normal.

The *Arabidopsis* mutant *leaf morphology7 (lem7)* bears many resemblances to *phan* (Meisel *et al.*, 1996). Firstly, it is cold sensitive, being completely rescued at 30°C and showing intermediate phenotypes at 23°C. Secondly, dorsal cell types are selectively eliminated, including trichomes in the juvenile epidermis and palisade cells in the ground tissue. Thirdly, the leaves and floral organs are much narrower than normal, but the cotyledons are relatively unaffected. In its most extreme phenotype, *lem7* mutants fail to produce leaves, producing a sort of callus in place of the SAM.

Another class of mutants that may affect axis specification are the temperature-sensitive *arrested development1 (add1)* mutants described by Pickett *et al.* (1996). At the permissive temperature (16°C) they produce leaves slowly, but eventually flower and set seed. At the restrictive temperature (29°C) these mutants undergo developmental arrest. At intermediate temperatures, or following post-germination temperature shifts, fingerlike leaves are produced by *add1* plants that appear to have lost dorsoventrality (Pickett *et al.*, 1996). Leaves shifted late in development had intermediate 'spoon-shaped' phenotypes reminiscent of the *lem7* and *phan* leaves described above. Narrow spoon-shaped leaves that may be compromised in proximal dorsoventral characters are also found in weak alleles of *stm*, though the origin of these leaves with respect to residual or adventitious meristems is unclear (Barton and Poethig, 1993; Endrizzi *et al.*, 1996).

Although little can be concluded in the absence of further molecular information, all these results are consistent with the classical notion that dorsoventrality is imposed on the incipient primordia from the meristem above (Sussex, 1955). A number of genes are required to interpret this information and it is tempting to draw parallels with animal genes that specify dorsal identity in limb primordia, such as the selector gene *apterous*, and the boundary gene *vestigial* (Diaz-Benjumea and Cohen, 1993; Williams *et al.*, 1994; Waites and Hudson, 1995). Expression of *apterous* in the dorsal compartment defines the DV boundary. Interaction with positional determinants like *wingless* and *decapentaplegic* establishes and then interprets this boundary, which then becomes the wing margin (Cohen *et al.*, 1991; Campbell *et al.*, 1993). The margin of the leaf lamina might be similarly determined (Waites and Hudson, 1995). Waites and Hudson (1995) point out, however, that although growth is severely reduced, completely ventralized *phan* leaves still grow out, unlike wings in *apterous* null mutants. This may mean that the PD axis arises independently of the DV axis in plants or that additional factors are involved.

8.3.5 Sensitivity of leaf mutants to temperature

Temperature-sensitive mutants are commonly encountered in screens for mutations that affect leaf morphogenesis in *Arabidopsis* (see above), *Antirrhinum* and maize (Miles, 1989). The paucity of temperature-sensitive mutants in plants probably reflects an inherently temperature-sensitive process in leaf development rather than the nature of the alleles themselves. Pickett *et al.* (1996) have proposed that *polycomb*-like chromatin effects may be involved. Mutations in chromatin proteins in *Drosophila* are acutely temperature sensitive, and often lead to variegation because changes in chromatin can be mitotically inherited. The discovery that polycomb group proteins regulate organ identity genes in leaves (Goodrich *et al.*, 1997) lends some support to this idea, although *curly-leaf* mutants have not been reported to be temperature sensitive.

Another possibility is that axis specification in leaves depends on diffusion of signalling molecules, as it does in animals (e.g. Irvine and Weischaus, 1994). Diffusion is acutely temperature sensitive, even when facilitated by carrier proteins. The discovery that protein and RNA signals can pass between cells through plasmodesmata raises this interesting possibility. KNOTTED protein and RNA has been shown to pass between cells in tobacco by microinjection of labelled derivatives (Lucas *et al.*, 1995; Jackson and Hake, 1997). As these proteins may have a role in establishing and orienting the primary axes of growth, this might explain the temperature sensitivity of these mutations. Plasmodesmatal trafficking of homeodomain and other transcription factors has been extensively reviewed elsewhere (Lucas *et al.*, 1993; Hake and Char, 1997; Jackson and Hake, 1997) and will not be explored further here.

8.3.6 Determination of leaf shape

Once the primary axes of growth have been laid down, the major determinant of leaf shape is the polarity and extent of cellular expansion (Poethig, 1997). This has been demonstrated by treatment of apices with high doses of radiation to block cell division (Foard, 1971) or with molecules that promote cell expansion (Fleming *et al.*, 1997). In each case, primordial outgrowth is achieved independently of cell division. Expansion is thought to depend on the orientation of cellulose microfibrils and is facilitated by expansins, extracellular proteins that intercalate with the cell wall to loosen and disrupt non-covalent interactions that restrict cell growth (Cosgrove, 1997). A huge variety of leaf shapes can be generated by this relatively simple mechanism. In *Arabidopsis*, mutants in cell expansion include *diminuto (dim), rotundifolia (rot)* and *angustifolia(an). rot* and *an* have essentially independent and opposite effects on the orientation of leaf cell expansion and result in more rounded and narrower leaves, respectively (Tsuge *et al.*, 1996). *an* also affects trichome branching, a phenotype known to depend on cell expansion (Hulskamp *et al.*, 1994). *diminuto*, in contrast, has a general defect in cell elongation and results in leaves and cotyledons that open in the dark. The *DIM* gene encodes a novel protein that appears to regulate tubulin (Takahashi *et al.*, 1995).

Another gene with a major effect on cell elongation is *ERECTA*. Mutant plants have a compact inflorescence, rosette and fruit. The *ERECTA* gene encodes an LRR receptor kinase closely related to *CLAVATA1,* indicating a role for cell–cell communication in the control of organ shape (Torii *et al.*, 1996). The interaction of cell expansion and organogenesis has been examined in detail in several other plant species, for example the lobed leaf primordia of *Tropaeoleum* are effectively rescued by increased proliferation of the sinuses, resulting in round rather than lobed leaves (reviewed by Poethig, 1997). It is possible that genes that effect cell shape and cellular proliferation in *Arabidopsis*, such as *ERECTA*, will have major modifying effects on mutants that alter or reiterate the primary axes of growth. This is likely to confound many studies in the Landsberg ecotype because this is typically marked with this seemingly harmless mutation.

8.3.7 Spacing of cell types and their determination

Trichomes, vascular bundles and perhaps stomatal complexes are evenly spaced in mature leaves. This suggests the existence of mechanisms that inhibit the specification of similar cellular identities in nearby cells (Nelson and Dengler, 1997; Larkin *et al.*, 1997). In animals, secreted factors and their receptors have been found to play key roles in this type of spacing differentiation, being involved in both short- and long-range interactions that specify cell type (e.g. Baker *et al.*, 1990). In *Arabidopsis* a number of mutants have been isolated that may have a role in this process. These include *tryptychon (try)* and *reduced trichome number (rtn)*, which are required for trichome spacing and for temporal regulation of trichome formation in the leaf (reviewed by Larkin *et al.*, 1997). A number of more general defects that affect spacing differentiation of both epidermal and vascular cell types have also been identified. These include *patch-work (pwk)* (Gu and Martienssen, unpublished) *tousled* (Roe *et al.*, 1993) and *lopped1 (lop)* (Carland and McHale, 1996). It is important to realize, however, that these global defects may affect organogenesis at an earlier stage, resulting in major alterations in cellular differentiation. For example, they may influence hormone biosynthesis and response (Sachs, 1981; Carland and McHale, 1996). Many terminally differentiated cell types in *Arabidopsis* are highly endopoly-ploid (Galbraith *et al.*, 1991; Melarango *et al.*, 1993) but the significance of this is unclear. The role of pattern formation in cellular differentiation has been extensively reviewed recently and so these issues are not reiterated here (Nelson and Dengler, 1997; Larkin *et al.*, 1997).

Stomatal complexes in cotyledons are separated by only one or two subsidi-ary epidermal cells. This spacing is maintained by the stereotypical series of symmetric and asymmetric cell divisions that give rise to stomatal complexes (Zeiger *et al.*, 1987; Croxdale *et al.*, 1992; Sachs, 1994). Mutants in this process include *fourlips (flp)*, which alters guard cell specification, and *too many mouths (tmm)*, which alters guard cell mother cell specification, resulting in multiple reiterated guard cell pairs (Yang and Sack, 1995). These mutants may be deficient in symmetric guard mother cell division or in asymmetric meristemoid division, respectively. In contrast, trichomes arise as single cells that may not share the same lineage as the subsidiary cells surrounding them (Larkin *et al.*, 1996). Mutants that fail to form trichomes include *glabra1 (gl1)*, *glabra2 (gl2)* and *transparent testa glabra (ttg)*. In weak alleles of *gl1* and *ttg*, dose-dependent trichome clustering is observed (Larkin *et al.*, 1994), which could mean that lateral inhibition is a component of trichome identity. These genes are likely to act in a hierarchy with spacing differentiation genes like *rtn* and *try*, as well as with other genes involved in specifying trichome shape (reviewed by Larkin *et al.*, 1997; Folkers *et al.*, 1997). All of these mutants appear to encode transcrip-tion factors and some affect the patterning of root hairs as well as trichomes. However, they appear to have different functions related to division and differen-tiation in these different cell types. The function of these genes in the root is discussed in the following section and so is not reviewed here.

8.4   Organogenesis of the root

8.4.1   Derivation of the root

The *Arabidopsis* primary root meristem (radicle) is patterned in the embryo and emerges with the initiation of cell division and elongation in the germinating seedling (Dolan *et al.*, 1993) (Figure 8.4). The seedling root is derived from three tiers of cells that are already in place in the heart stage embryo (Scheres *et al.*, 1994a). Clonal and histological analyses indicate that the quiescent centre and columella (central root cap) are derived from the hypophyseal cell and the promeristem is derived from the tier of cells abutting these hypophyseal derivatives (Scheres *et al.*, 1994a). The next apical-most tier gives rise to the part of the root that lies between the collet and the future meristem but undergoes no further cell division on germination.

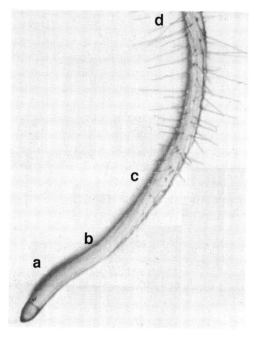

Figure 8.4 Arabidopsis root zones: (a) meristem, (b) elongation, (c) differentiation and (d) mature zone where lateral primordia are first apparent and where secondary thickening is initiated in the stele.

The establishment of pattern in the embryonic root is reviewed elsewhere in this book (see chapter 7) and is not dealt with in detail here. Briefly, *MONOP-TEROS* is required for the development of the embryonic root since loss-of-function alleles are devoid of root tissue in the embryo (Berleth and Jurgens, 1993). Mutations in *SCARECROW* (*SCR*) and *SHORTROOT* (*SHR*) incorrectly specify the cortex endodermal lineage in both the embryo and the seedling root meristem, indicating that pattern formation and the maintenance of pattern utilize some of the same gene products (Scheres *et al.*, 1994b).

The characteristic symmetry of the mature root is also in place by the heart stage of embryogenesis (Scheres *et al.*, 1994a) (Figure 8.5). The root is composed of concentric layers of tissues: epidermis, ground tissue and stele. The epidermis is composed of 16–22 cell files and in the seedling root is derived from a ring of approximately 16 initials that also give rise to the lateral root cap on the outside of the root (Dolan *et al.*, 1994; Scheres *et al.*, 1994a). The ground tissue is composed of an eight-cell file endodermis inside the eight-cell file cortex. The cortex and endodermis are derived from a common set of eight initials. The pericycle forms a layer of variable number of cell files and is the site of origin of the future lateral roots. The stele comprises a diarch xylem interspersed with phloem and parenchyma (thin-walled cells). During secondary growth (secondary thickening), these parenchyma cells divide and form a cambium, producing xylem on the inside and phloem to the outside while the pericycle proliferates to form a periderm (Dolan *et al.*, 1993; Dolan and Roberts, 1995) (Figure 8.5C).

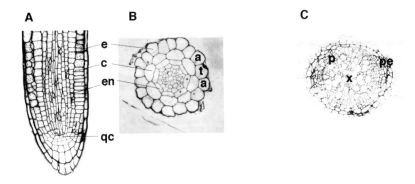

Figure 8.5 Cellular organization of the *Arabidopsis* root. A, Median longitudinal section of a primary root in which the cell walls were labelled with an anti-pectin antibody, imaged by immuno-fluorescence and reverse contrast printed. B, Cross-section through the elongation zone of the root showing the radial organization of cell types. C, Secondary thickened root after loss of epidermis and ground tissue. a, trichoblasts; c, cortex; e, epidermis; en, endodermis; p, phloem; pr, periderm; qc, quiescent centre; t, trichoblast; x, xylem.

8.4.2 Stability of initials in the root meristem

Higher plant meristems are characterized by a population of slowly dividing cells at their core, which give rise to a population of rapidly dividing cells (Steeves and Sussex, 1989). The central zone of the shoot meristem also divides relatively slowly while division in the peripheral zone is more rapid. BrdU and tritiated thymidine incorporation into *Arabidopsis* root indicate that the four central cells are relatively quiescent (7.7% of cells incorporated BrdU in 24 hours) (Dolan *et al.*, 1993; Fujie, *et al.*, 1993). Above these is the initial zone in which cells divide more rapidly (38% of cells incorporating BrdU in 24 hours) while cells in the meristem proper divide most rapidly (84% cells incorporated precursor over a 24-hour period).

It is well established that initial cells in the central zone of shoot meristems are unstable, i.e. initials are continuously being replaced (see earlier). By inducing genetically marked clones in the root meristem of three-day-old seedlings, it has been shown that root initials are also unstable (Kidner, 1997). Observations of clones induced in epidermal initials predict that all 16 initials would be replaced over a three-day period, while it was predicted that the cortical endodermal initials have a turnover time of approximately 10 days.

Clonal analyses of shoot meristems have indicated that cell division patterns exhibit some variability (for a review see Irish, 1991). While the three meristematic lineages of the shoot (L1, L2 and L3) give rise to stereotypical regions of the leaf, invasions of cells from one layer into another occur. For example, layer invasions in which an epidermal cell is observed to divide periclinally into the underlying mesophyll occur once in approximately every 2000 divisions in the tobacco shoot epidermis (Stewart and Burk, 1970). The clonal analysis of the three-day-old seedling root also reveals that cells divide from one tissue layer (lineage) into another with a frequency similar to that observed in the tobacco shoot epidermis (Kidner, 1997). This variability in the pattern of cell divisions (cell lineage) highlights the flexibility in cell fate in the root and suggests that positional cues must play a major role in the specification of cell fate in the root.

8.4.3 Specification of cell identity

Clonal analysis in a number of systems has highlighted the importance of positional information in the development of plants. Little is known about the nature of these positional cues although recent microinjection experiments indicate that RNA and protein can move from cell to cell in the maize shoot meristem (Lucas *et al.*, 1995). Furthermore recent laser microsurgical experiments suggest that the extracellular matrix in the *Fucus* embryo plays a central role in the specification of the fate of the rhizoid and thallus cells (Berger *et al.*, 1994). In a similar fashion, direct evidence that positional information is important in the development of the root comes from similar microsurgical investigations (Berger and Dolan, unpublished).

On ablation of a ground tissue cell in the *Arabidopsis* root a neighbouring cell from an adjacent layer can move into the space and assume the fate of the dead cell (Van den Berg *et al.*, 1995). In a similar fashion, ablation of a ground tissue initial results in a neighbouring cell moving into the space and assuming the role of the dead initial. If the new cell is devoid of contacts with more mature cells by killing the three cells above the new initial, it fails to undergo its stereotypical set of divisions, indicating that a signal from mature cells is required for the specification of initial fate. The molecular nature of this signal remains unknown.

Cell fate in the epidermis is also at least partially specified by positional cues (Berger and Dolan, unpublished). Hair-forming cells (derived from trichoblasts) are located over the anticlinal walls of underlying cortical cells, while non-hair cells (derived from atrichoblasts) are located over the periclinal walls of these cortical cells (Figure 8.5B). The majority of cell divisions in the epidermis are transverse and consequently increase the number of cells in a file (Berger, *et al.*, 1997). Occasionally longitudinal anticlinal divisions occur in the trichoblast file resulting in one daughter cell located in the same relative location as the mother cell and another in the atrichoblast position. The daughter cell in the atrichoblast location differentiates as a non-hair epidermal cell in two cell clones, indicating that cell fate is flexible until at least the last cell division. Further surgical experiments confirm the importance of position in this system and indicate that the spatial cues may be located in discrete domains in the extracellular matrix (cell wall) (Berger and Dolan, unpublished). It is hypothesized that the pattern is established during embryogenesis and that cues are imprinted in the extracellular matrix at that time. These cues are maintained during subsequent development and functioning of the meristem.

Local cell signalling is also involved in the maintenance of the balance between cell division and differentiation (Van den Berg *et al.*, 1997). Ablation of the central cells in the quiescent centre during root growth results in the progressive differentiation of columella initial cells, which normally show no signs of differentiation but continue to divide. Ablation of central cells in mutants in which no (post-germination) cell division takes place illustrates that the primary defect is in differentiation and not cell division. This suggests that the central cells negatively regulate the differentiation of neighbouring cells during development and may thereby serve to maintain the meristematic state of the cells in this location rather than regulating cell division *per se*. CLAVATA1 may have a similar role in lateral inhibition of central cell fate in the shoot (see earlier).

Screening for mutations with abnormal cellular patterns of differentiation has identified genes that encode products involved in the specification of cell fate in the root. *SHORT ROOT (SHR)* is defined by recessive alleles that were identified in a screen for reduced root growth. In addition to their decreased stature, *shr* roots lack an endodermis and further investigation indicates that *SHR* activity is required for the asymmetric periclinal division of the cortical endodermal initial daughter cell (Benfey *et al.*, 1993; Scheres *et al.*, 1994b). That the endodermal layer is missing (and not the cortex) was shown using a monoclonal antibody that recognizes an AGP epitope in the wild-type endodermis. Double mutants of *shr* with *fass*, a mutant in which extranumerary cell divisions take place, failed to restore the missing endodermis (Scheres *et al.*, 1994b). The lack of an endodermis in this double mutant background indicates that the SHR product is not only required for the execution of the division in the cortical endodermal initial daughter cell but may also be needed for the specification of endodermal identity.

SCARECROW (SCR) is another gene required for the establishment of the radial pattern in the root (Di Laurenzio *et al.*, 1996). Cortical endodermal initial daughter cells fail to undergo the formative asymmetric periclinal division in *scr* roots and result in the development of a single cell layer (mutant cell layer) in the position normally occupied by both endodermis and cortex in wild-type. This observation indicates that *SCR* is required for the execution of this asymmetric cell division. The so-called mutant layer is present in *scr, fass* double mutants, indicating that the *scr* phenotypes do not simply result from an inability of the initial daughter to divide (Scheres *et al.*, 1994b). The mutant cell layer also exhibits characteristics of both cortex and endodermis as revealed by the presence of a casparian strip (an endodermal marker) and cell surface epitopes normally expressed in the cortex. Since the mutant cell layer exhibits identities of both cortex and endodermis, it has been proposed that cell specification is not altered in *scr* roots but that the primary defect is in the execution of the asymmetric cell division (Di Laurenzio *et al.*, 1996). Consistent with this conclusion is the observation that *SCR* is expressed in the endodermal/cortical initials. Since *SCR* is also expressed in the older endodermal cells, the authors suggest that *SCR* expression in these cells may act as a source of polarizing signal for the basal initial daughter. It has already been shown that signals may move from older to younger cells in the cell specification in the initial zone. Nevertheless it is still formally possible that *SCR* may play a role in the specification of endodermal and/or cortical cell fate. Since *SCR* is expressed throughout the endodermis, *SCR* may be a negative regulator of cortical identity in cells in the endodermal location.

In addition to the difference in morphology, the cell division patterns of trichoblasts differ from atrichoblasts (Berger *et al.*, 1997). Young trichoblasts divide faster than atrichoblasts and this difference is responsible for the differences in cell length between hair and non-hair cells apparent at maturity. Genetic analysis indicates that the differential rates of cell division are regulated by *TTG* since all epidermal cells in *ttg* roots cycle as trichoblasts (with a relatively rapid cell cycle) (Koornneef, 1981; Galway *et al.*, 1994). Differential cell cycle time is independent of *GL2* activity, suggesting either that cell cycle is regulated in an independent pathway that shares *TTG* or that the cell cycle differences are established before *GL2* activity is required (Berger *et al.*, 1997).

Loss-of-function *ttg* alleles develop hairs on every root epidermal cell. In addition, the difference in the rates of cell division between the two cell types is absent in *ttg* root, suggesting that *TTG* acts early (Galway *et al.*, 1994; Di Cristina *et al.*, 1996; Berger *et al.*, 1997). Loss-of-function mutations in *GL2* also result in the formation of hairs on all epidermal files (Masucci *et al.*, 1996; Di Cristina *et al.*, 1996). *GL2* encodes a homeodomain protein expressed predominantly in the non-hair cell files from close to the initials through meristematic, elongation and differentiation zones, suggesting that *GL2* is a positive regulator of non-hair cell development (Rerie *et al.*, 1994; Masucci *et al.*, 1996). *GL2* transcription is greatly reduced in *ttg* backgrounds, indicating that *TTG* activity is required for the correct spatial expression of *GL2* (Di Cristina *et al.*, 1996).

The pattern of cell divisions is normal in *gl2* roots consistent with the conclusion that *TTG* acts upstream of *GL2* in the pathway (Berger *et al.*, 1997).

A further gene involved in the process of epidermal development is *CAPRICE (CPC)*, which is defined by a loss-of-function mutation with an almost hairless phenotype (Wada *et al.*, 1997). *CPC* encodes a *myb*-like protein that lacks a transcriptional activation domain. Consequently it is postulated that this protein is a transcriptional repressor, competing for *myb* binding sites. Since *GL2* is activated by *GL1*, a *myb* protein, in the development of the shoot epidermis, it is possible that *CPC* negatively regulates *GL2* function *in vivo* (Oppenheimer *et al.*, 1991; di Cristina *et al.*, 1996). *ttg,cpc* double mutants exhibit an intermediate phenotype, an observation that has been interpreted to indicate that *CPC* and *TTG* act either independently or in different cell types (Wada *et al.*, 1997). *gl2* is epistatic to *cpc* in the double mutant combination. It is therefore possible that *TTG* positively regulates *GL2* in atrichoblasts while *CPC* negatively regulates *GL2* in trichoblast files. The absence of *CPC* in atrichoblasts would allow a GL1-like factor to activate *GL2* transcription, although a GL1-like activator remains to be found in the root. A prediction of this model is that ectopic expression of *CPC* would result in the ectopic repression of *GL2* transcription, which would in turn result in the development of ectopic root hairs (*gl2* phenotype). *CPC* expression driven by a 35S promoter does indeed resemble the *gl2* phenotype (Wada *et al.*, 1997). Future cloning of candidate genes and the examination of gene expression patterns will be instructive in expanding this model.

Three further genes that mutate to a hairy phenotype have been defined by recessive mutations. *extra root hair1 (erh1)*, *erh3* and *pompom1 (pom1)* produce extranumerary root hairs (Schneider *et al.*, 1997). *erh3* also produces additional floral organs in whorls one and two. All three exhibit defects in cell expansion since mutant roots are shorter than wild-type. While their interaction with other genes involved in root epidermal development has not been ascertained, the absence of early differentiation between trichobasts in *erh1* and *erh3* backgrounds indicates that these genes may act early. *erh1* and *erh3* would be candidates for genes encoding a possible GL1-like factor that activates *GL2* transcription in atrichoblasts.

8.4.4 *Role of plant hormones in epidermal development*

The genetic dissection of the ethylene and auxin signalling pathways has provided tools for the investigation of the roles played by these phytohormones in root epidermal development since anecdotal evidence indicated that ethylene was a regulator of hair growth (Tanimoto *et al.*, 1995). Genetic and molecular analyses indicate that *ETHYLENE RESISTANT1 (ETR1)* encodes the ethylene receptor and acts upstream of *CONSTITUTIVE TRIPLE RESPONSE1 (CTR1)*, a negative regulator of ethylene signalling (for a review see Kieber, 1997). Loss-of-function *ctr1* alleles develop ectopic root hairs, indicating that ethylene is a positive regulator of root hair development. *etr1,auxin resistant1 (aux1)* double mutants develop fewer hairs than wild-type (Masucci and Schiefelbein, 1996). Since neither mutant alone exhibits a root hair phenotype, the novel phenotype of the *etr1,aux1* double mutant is consistent with these genes acting at the same point in the pathway. *AUX1* encodes a protein similar to amino acid permeases, indicating that it may be involved in auxin transport but is also known to be ethylene resistant in the root. *auxin resistant2 (axr2)* (Wilson *et al.*, 1990) plants develop fewer root hairs while the auxin resistant *dwarf (dwf)* (Mirza *et al.*, 1984) and *axr3* (Leyser *et al.*, 1996) fail to develop hairs, supporting the conclusion that auxin is involved in the process.

Evidence is mounting that auxin and ethylene act late in the process of hair cell development. *axr2* is epistatic to *ttg* and expression of the *GL2* promoter GUS fusion is indistinguishable from wild-type in the *axr2* background, leading Masucci and Schiefelbein (1996) to conclude that auxin acts downstream of *GL2* in the process. *GL2* promoter activity was unchanged in the *ctr1* background and by modulators of ethylene synthesis, supporting the view that ethylene acts after *GL2* (Masucci and Schiefelbein, 1996).

It has been possible to precisely determine the point at which ethylene is sensed in the epidermis using two approaches. By transferring marked roots from conventional growth media to media containing the ethylene precursor 1-aminocyclopropane carboxylic acid (ACC), Masucci and Schiefelbein (1996) showed that cells in the elongation zone could respond to ethylene. Examination of the phenotype of *ethylene overproducer3 (eto3)* indicate that ethylene is perceived in the early elongation zone (Cao *et al.*, unpublished).

In conclusion, it appears that during pattern formation a prepattern is set up in the embryo that is subsequently maintained by positional cues through the cell division zone. Once cell division has ceased ethylene and auxin are required for the establishment of tip growth during the final stages of hair cell differentiation.

8.4.5 Development of lateral roots

Lateral roots are derived from approximately 11 G2-arrested pericycle cells located next to the xylem pole (Laskowski *et al.*, 1995). The first stages of lateral root development involve the initiation of transverse anticlinal divisions forming 8–11 short cells, which undergo radial expansion (stage I) (Malamy and Benfey, 1997). These cells undergo a periclinal division forming an inner layer (IL) and an outer layer (OL) (stage II). The outer layer undergoes a periclinal division to form OL1 and OL2 (stage III). Subsequently IL divides periclinally to form IL1 and IL2 (stage IV). In subsequent stages central regions of OL1 give rise to the columella initials, root cap and epidermis, and derivatives of OL2 give rise to the future central cells of the QC, endodermal cortical cells initials and stele initials. The remainder of the root is derived from the inner layers. Clonal analyses are required to test this model and to determine the variability in cell lineage in the emerging lateral.

Since *SCR* is required for the establishment of radial pattern in the embryo, it might be predicted to have a role in the development of lateral roots. Consistent with this suggestion is the observation that *SCR* is expressed in OL at stage II and becomes progressively restricted to those cells that will constitute the endodermal lineage (OL2 at stage V) (Malamy and Benfey, 1997). By stage VII, when mature meristem cellular organization is established, its expression is restricted to the endodermal lineage although central cells of the future quiescent centre express *SCR*. *SCR* expression disappears from central cells with subsequent growth of the lateral root.

In contrast *GL2* is expressed relatively late, at stage VI, after the epidermis has been established as a distinct lineage (Malamy and Benfey, 1997). This suggests that *GL2* does not play a role in the generation of the pattern in the developing lateral root but may play a role in its maintenance. This is consistent with the proposed role of *GL2* acting relatively late in the specification of cell fate in the epidermis.

These observations indicate that patterning mechanisms are initiated early in the development of the lateral root meristem (*SCR* is expressed in the two-layered primordium) and that the mature cellular organization is achieved by stage VI. Surgical experiments indicate that two-layered lateral root meristems are unable to maintain growth when removed from the site of origin and placed onto nutrient medium (Laskowski *et al.*, 1995). 100% of six-layered meristems (stage V or VI above) on the other hand survive and grow *in vitro*. Occasionally a three-layered meristem (stage III) will survive. This indicates that meristem function as defined by the 'ability of a group of cells to form an organ' (sic) meristem is established between stage III and stage V/VI, i.e. just before the final cellular organisation is established.

Auxin is a potent inducer of lateral root development and screens for mutations in which laterals fail to develop have identified genes involved in auxin regulated lateral root development. These studies indicate that auxin is required early for the initiation of lateral root primordium and subsequently for the maturation of the primordium (around the time of emergence) (Celenza *et al.*, 1995).

ALTERED LATERAL ROOT1 (ALF1) is defined by recessive mutations that result in the development of supernumerary lateral roots (Celenza *et al.*, 1995). Such roots develop from the hypocotyl as well as the root proper and on occasion can develop from petioles and leaves. Since *alf1* plants produce excessive amounts of free IAA, *ALF1* is considered to be a negative regulator of auxin production. The development of additional roots in *alf1* background indicates that auxin is required for the initiation of laterals.

alf3 plants initiate laterals that cease growth soon after emergence from the parent root, indicating that *ALF3* activity is required late in lateral root development. *alf3* seedlings generally develop twice as many lateral root primordia than wild-type. Treatment of *alf3* roots with IAA rescues the mutant phenotype, indicating that *ALF3* may be involved in an auxin-related process. *alf3,alf1* roots develop beyond the size of *alf1* roots, suggesting that the excess auxin produced by the *alf1* mutant can stimulate lateral root maturation in the *alf3* background.

alf4 roots form virtually no lateral roots, indicating that the gene is required at the earliest stages of root initiation (Celenza *et al.*, 1995). Treatment of roots with IAA does not rescue the phenotype and *alf4* is epistatic to *alf1*, indicating that the *alf4* mutant is insensitive to the excess auxin produced in the *alf1* background (Celenza *et al.*, 1995). Furthermore, *alf4* is epistatic to *alf3*.

Taken together these interactions suggest that *ALF1* is a negative regulator of IAA production and may serve to regulate the concentration of free IAA in the root. *ALF4* may act in the sensing of such free IAA. *ALF4* acts before *ALF3* in lateral root formation, so *ALF3* activity is probably required to maintain a high IAA concentration in the lateral root primordium during the later stages of lateral root emergence.

alf mutants do not display embryonic root phenotypes, i.e. the primary root develops normally (although some may abort early), suggesting that embryonic root development is independent of *ALF1*, *ALF3* and *ALF4* function. In a similar, though opposite, manner *mp* embryos fail to initiate a normal embryonic root but develop normal laterals in tissue culture, indicating that *MP* is required for root development in the embryo but not in laterals (Berleth and Jurgens, 1993). *rml1* and *rml2* roots fail to initiate cell division on germination and consequently produce a short root that fails to grow more than a few millimetres. *rml1* and *rml2* lateral roots also fail to develop after emergence from parent roots (Cheng *et al.*, 1995). Therefore there appear to be genes specific to either lateral or embryonic root development and a suite that is common to both.

8.5   Conclusions

The pattern of vegetative growth is influenced at a number of fundamental levels, from initiation and specification of the primary axes to determination of organ identity and cellular differentiation. The molecular genetic determinants that confer pattern to the shoot and root meristems and their products are beginning to emerge from mutant studies in *Arabidopsis*, and common as well as distinct mechanisms are found. Distinct elements include mechanisms of axis specification and organ determination in the shoot. Meristem function and cellular differentiation may have common components, including lateral inhibition, asymmetric division, spacing differentiation and cellular expansion. The root and shoot epidermis, for example, arise as a commmon lineage in the embryo, and many of the genes involved in cellular differentiation in the root epidermis also play a major role in the shoot epidermis (*TTG, GL2, CPC*). It has also been shown that the balance between proliferation and differentiation in the shoot meristem is at least partly mediated by cell communication involving the CLAVATA1 protein. In a parallel set of experiments it has been demonstrated that cell communication events are central to the maintenance of the meristematic state (or suppression of differentiation) in the root (Van den Berg, 1997). In principle these similarities might represent evolutionary convergence, i.e. shoot and root meristems have independent origins and both co-opt similar suites of genetic activities during their evolution. The distinct cell types affected by these genes in shoots and roots lend some support to this view.

Alternatively, it is possible that the root is derived from a shoot structure— leaf or meristem. Since the first vascular plants were essentially leafless shoots, devoid of roots, it is at least chronologically possible that root meristems could have evolved from shoot structures. Furthermore, it is now clear that lycopsid roots are derived from primitive leaves (microphylls) and emerge in radial arrays from modified rhizome-like structures known as rhizomorphs (see Stewart, 1983). That the angiosperm root might be derived from the shoot (Wilson, 1953) is not so outlandish since it has been proposed (though not universally accepted) that the megaphyll, the leaf of higher vascular plants, is derived from a modified shoot system. In this case it has been proposed that the progressive modification of shoot trusses by 'overtopping', 'planation' and 'webbing' resulted in the formation of a leaves (Zimmermann, 1930). Consequently the (megaphyllous) leaf is considered to be a modified system of shoot units (telomes). It is anticipated that an understanding of the genetic mechanisms underpinning the development of vegetative structures in the shoot and root will be instructive in identifying which of these mechanisms most nearly reflects reality.

In the shoot, it is becoming clear that mechanisms that determine the structure and organization of floral and embryonic organs are probably shared by lateral organs in the vegetative phase, and may have arisen first in this context. One way this may have happened is by wholesale duplication of gene families involved in vegetative pattern formation and their recruitment, via alterations in expression pattern, in floral and seed development as these specialized structures evolved in the angiosperm radiation (Friis *et al.*, 1987; Hofer *et al.*, 1997). This might account for the apparently higher levels of genetic redundancy in basic patterning mechanisms in vegetative organs. Alternatively, many of these genes are likely to have lethal consequences when they are disrupted, so their role cannot readily be determined by classical genetics in *Arabidopsis*. The advent of the *Arabidopsis* genome project (Goodman *et al.*, 1995), and the prospect of a complete catalogue of all the genes in *Arabidopsis*, will go a long way to answering issues of genetic redundancy during development.

Acknowledgements

We would like to thank Vivian Irish for providing and Marja Timmermans for preparing Figure 8.1, and Patricia Springer for providing Figure 8.3. We would also like to thank Marja Timmermans, Detlef Weigel and Kathy Barton for stimulating discussions, as well as past and present members of our respective laboratories.

References

Aida, M., Ishida, T., Fukaki, H., Fujisawa, H. and Tasaka, M. (1997) Genes involved in organ separation in *Arabidopsis*: an analysis of the cup-shaped cotyledon mutant. *Plant Cell,* **9** 841-57.

Baima, S., Nobili, F., Sessa, G., Lucchetti, S., Ruberti, I. and Morelli, G. (1995) The expression of the Athb-8 homeobox gene is restricted to provascular cells in *Arabidopsis thaliana. Development,* **121** 4171-82.

Baker, N., Mlodzik, M. and Rubin, G.M. (1990) Spacing differentiation in the developing *Drosophila* eye: a fibrinogen-related lateral inhibitor encoded by *Scabrous. Science,* **250** 1370-77.

Ball, E. (1950) Isolation, removal and attempted transplants of the central portion of the shoot apex of *Lupinus albus* L. *Am. J. Bot.,* **37** 117-36.

Ball, E. (1980) Regeneration from isolated portions of the shoot apex of *Trachymene coerulea* R.C. Grah. *Ann. Bot.,* **45** 103-12.

Barabas, Z. and Redei, G.P. (1971) Facilitation of crossing by the use of appropriate parental stocks. *Arabidopsis Inf. Serv.* **8** 7-8.

Barton, M.K. and Poethig, R.S. (1993) Formation of the shoot apical meristem in *Arabidopsis thaliana*: an analysis of development in the wild-type and in the *shootmeristemless* mutant. *Development,* **119** 823-31.

Belvin, M.P. and Anderson, K.V. (1996) A conserved signaling pathway: the *Drosophila* toll-dorsal pathway. *Ann. Rev. Cell Dev. Biol.,* **12** 393-416.

Benfey, P.N., Linstead, P.J., Roberts, K., Schiefelbein, J.W., Hauser, M.T. and Aeschbacher, R.A. (1993) Root development in *Arabidopsis*: four mutants with dramatically altered root morphogenesis. *Development,* **119** (1) 57-70.

Berger, F., Taylor, A. and Brownlee, C. (1994) Cell fate determination by the cell wall in early *Fucus* development. *Science,* **263** 1421-23.

Berger, F., Hung, C.Y., Dolan, L. and Schiefelbein, J.W. (1997) Control of cell division in the root epidermis of *Arabidopsis thaliana. Dev. Biol.,* in press.

Berleth, T. and Jurgens, G. (1993) The role of MONOPTEROUS gene in organising the basal body region of *Arabidopsis. Development,* **118** 575-87.

Blazquez, M.A., Soowal, L.N., Lee, I. and Weigel, D. (1997) *LEAFY* expression and flower initiation in *Arabidopsis. Development,* **124** 3835-44.

Bowman, J.L. (ed.) (1994) *Arabidopsis: an atlas of morphology and Development,* Springer-Verlag, Berlin.

Bowman, J.L., Smyth, D.R. and Meyerowitz, E.M. (1991) Genetic interactions among floral homeotic genes of *Arabidopsis. Development,* **112** 1-20.

Boyes, D.C., McDowell, J.M. and Dangl, J.L. (1996) Plant pathology: many roads lead to resistance. *Curr. Biol.,* **6** 634-37.

Brown, J.A.M., Miksche, J.P. and Smith, H.H. (1963) An analysis of ^{3}H-thymidine distribution throughout the vegetative meristem of *Arabidopsis thaliana* (L.) Heynh. *Radiation Bot.,* **4** 107-13.

Bryant, P.J. (1975) Pattern formation in the imaginal wing disc of *Drosophila* melanogaster: fate map, regeneration and duplication. *J. Exper. Zool.,* **193** 49-78.

Callos, J.D., DiRado, M., Xu, B., Behringer, F.J., Link, B.M. and Medford, J.I. (1994) The forever young gene encodes an oxidoreductase required for proper development of the *Arabidopsis* vegetative shoot apex. *Plant J.,* **6** 835-47.

Campbell, G., Weaver, T. and Tomlinson, A. (1993) Axis specification in the developing *Drosophila* appendage: the role of wingless, decapentaplegic and the homeobox gene aristaless. *Cell,* **74** 1113-23.

Carland, F.M. and McHale, N.A. (1996) *LOP1*: a gene involved in auxin transport and vascular patterning in *Arabidopsis*. *Development,* **122** 1811-19.

Carpenter, R., Copsey, L., Vincent, C., Doyle, S., Magrath, R. and Coen, E. (1995) Control of flower development and phyllotaxy by meristem identity genes in *Antirrhinum*. *Plant Cell,* **7** 2001-11.

Caruso, J.L. (1968) Morphogenetic aspects of a leafless mutant in tomato. I. General patterns of *Development. Am. J. Bot.,* **55** 1169-76.

Caruso, J.L. and Cutter, E.G. (1970) Morphogenetic aspects of a leafless mutant in tomato. II. Induction of vascular cambium. *Am. J. Bot.,* **57** 420-29.

Celenza, Jr, J.L. Grisafi, P.L. and Fink, G.R. (1995) A pathway for lateral root formation in *Arabidopsis thaliana. Genes Dev.,* **9** 2131-42.

Chaudhury, A.M., Letham, S., Craig, S. and Dennis, E.S. (1993) *amp1*—a mutant with high cytokinin levels and altered embryonic pattern, faster vegetative growth, constitutive photo-morphogenesis and precocious flowering. *Plant J.,* **4** 907-16.

Chen, L., Cheng, J.C., Castle, L. and Sung, Z.R. (1997) EMF genes regulate *Arabidopsis* inflorescence development. *Plant Cell,* **9** 2011-24.

Cheng, J.-C., Seeley, K.A. and Sung, R.S. (1995) RML1 and RML2, *Arabidopsis* genes required for cell proliferation at the root tip. *Plant Physiol.,* **107** 365-76.

Chien, J.C. and Sussex, I.M. (1996) Differential regulation of trichome formation on the adaxial and abaxial leaf surfaces by gibberellins and photoperiod in *Arabidopsis thaliana* (L.) Heynh. *Plant Physiol.,* **111** 1321-28.

Chuck, G., Lincoln, C. and Hake, S. (1996) *KNAT1* induces lobed leaves with ectopic meristems when overexpressed in *Arabidopsis*. *Plant Cell,* **8** 1277-89.

Clark, S.E. (1997) Organ formation at the vegetative shoot meristem. *Plant Cell,* **7** 1067-76.

Clark, S.E., Running, M.P. and Meyerowitz, E.M. (1993) *CLAVATA1*, a regulator of meristem and flower development in *Arabidopsis*. *Development,***119** 397-418.

Clark, S.E., Running, M.P. and Meyerowitz, E.M. (1995) CLAVATA3 is a specific regulator of shoot and floral meristem development affecting the same processes as CLAVATA1. *Development,* **121** 2057-67.

Clark, S.E., Jacobsen, S.E., Levin, J.Z. and Meyerowitz, E.M. (1996) The *CLAVATA* and *SHOOT MERISTEMLESS* loci competitively regulate meristem activity in *Arabidopsis*. *Development,* **122** 1567-75.

Clark, S.E., Williams, R.W. and Meyerowitz, E.M. (1997) The *CLAVATA1* gene encodes a putative receptor kinase that controls shoot and floral meristem size in *Arabidopsis*. *Cell,* **89** 575-85.

Coen, E.S. and Meyerowitz, E.M. (1991) The war of the whorls: genetic interactions controlling flower development *Nature,* **353** 31-37.

Coen, E.S., Romero, J.M., Doyle, S., Elliott, R., Murphy, G. and Carpenter, R. (1990) *floricaula*: a homeotic gene required for flower development in *Antirrhinum majus*. *Cell,* **63** 1311-22.

Cohen, B., Wimmer, E.A. and Cohen, S.M. (1991) Early development of leg and wing primordia in the *Drosophila* embryo. *Mech. Dev.,* **33** 229-40.

Conway, L.J. and Poethig, R.S. (1997) Mutations of *Arabidopsis thaliana* that transform leaves into cotyledons. *Proc. Natl Acad. Sci. USA,* **94** 10209-14.

Cosgrove, D. (1997) Relaxation in a high-stress environment: the molecular bases of extensible cell walls and cell enlargement. *Plant Cell,* **9** 1031-41.

Couso, J.P., Bishop, S.A. and Martinez-Arias, A. (1994) The wingless signalling pathway and the patterning of the wing margin in *Drosophila*. *Development,* **120** 621-36.

Cronk, Q. and Moeller, M. (1997) Strange morphogenesis—organ determination in Monophyllaea. *Trends Plant Sci.,* **2** 327-28.

Croxdale, J., Smith, J., Yandell, B. and Johnson, J.B. (1992) Stomatal patterning in Tradescantia, an evaluation of the cell lineage theory. *Dev. Biol.,* **149** 158-67.

Deikman, J. and Ulrich, M. (1995) A novel cytokinin-resistant mutant of *Arabidopsis* with abbreviated shoot development. *Planta,* **195** 440-49.

Dharmawardhana, D.P., Ellis, B.E. and Carlson, J.E. (1992) Characterization of vascular lignification in *Arabidopsis thaliana. Can. J. Bot.,* **70** 2238-44.

Diaz-Benjumea, F.J. and Cohen, S. (1993) Interaction between dorsal and ventral cells in the imaginal disc directs wing development in *Drosophila. Cell,* **75** 741-52.

Di Cristina, M., Sessa, G., Dolan, L., Linstead, P., Baima, S., Ruberti, I. and Morelli, G. (1996) The *Arabidopsis* ATHB-10 (GLABRA2) is a HD-ZIP protein required for the repression of ectopic root hair formation. *Plant J.,* **10** 393-402.

Di Laurenzio, L.D., Wysocka-Diller, J., Malamy, J.E., Pysh, L., Helariutta, Y., Freshour, G., Hahn, M.G., Feldman, K.A. and Benfey, P.N. (1996) The SCARECROW gene regulates an asymmetric cell division that is essential for generating radial organization of the *Arabidopsis* root. *Cell,* **86**, 423-33.

Dolan, L. and Roberts, K. (1995) Secondary thickening in the roots of *Arabidopsis thaliana*: anatomy and cell surface changes. *New Phytol.,* **131** 121-28.

Dolan, L., Janmaat, K., Willemsen, V., Linstead, P., Poethig, S., Roberts, K. and Scheres, B. (1993) Cellular organisation of the *Arabidopsis thaliana* root. *Development,* **119** 71-84.

Dolan, L., Duckett, C.M., Grierson, C., Linstead, P., Schneider, K., Lawson, E., Dean, C. and Roberts, K. (1994) Clonal relationships and cell patterning in the root epidermis of *Arabidopsis. Development,* **120** 2465-74.

Dolan, L., Linstead, P., Kidner, C., Boudonck, K., Cao, X.F. and Berger, F. (1997) Cell fate in plants: lessons from the *Arabidopsis* root, Society for Experimental Biology Symposium, Society for Experimental Biology, London.

Elliott, R.C., Betzner, A.S., Huttner, E., Oakes, M.P., Tucker, W.Q., Gerentes, D., Perez, P. and Smyth, D.R. (1996) AINTEGUMENTA, an APETALA2-like gene of *Arabidopsis* with pleio-tropic roles in ovule development and floral organ growth. *Plant Cell,* **8** 155-68.

Endrizzi, K., Moussian, B., Haecker, A., Levin, J.Z. and Laux, T. (1996) The *SHOOT MERISTEM-LESS* gene is required for maintenance of undifferentiated cells in *Arabidopsis* shoot and floral meristems and acts at a different regulatory level than the meristem genes *WUSCHEL* and *ZWILLE. Plant J.,* **10** 967-79.

Evans, M.M.S. and Barton, M.K. (1997) Genetics of angiosperm shoot apical meristem development. *Ann. Rev. Plant Physiol. Mol. Biol.,* **48** 673-702.

Fleming, A.J., McQueen-Mason, S., Mandel, T. and Kuhlemeier, C. (1997) Induction of leaf primordia by the cell wall protein expansin. *Science,* **276** 1415-18.

Foard, D.E. (1971) The initial protrusion of a leaf primordium can form without concurrent periclinal cell divisions. *Can. J. Bot.,* **49** 1601-1603.

Fobert, P.R., Coen, E.S., Murphy, G.J.P. and Doonan, J.H. (1993) Patterns of cell division revealed by transcriptional regulation of genes during the cell cycle in plants. *EMBO J.,* **13** 616-24.

Folkers, U., Berger, J. and Hulskamp, M. (1997) Cell morphogenesis of trichomes in *Arabidopsis*: differential control of primary and secondary branching by branch initiation regulators and cell growth. *Development,* **124** 3779-86.

Friis, E.M., Chaloner, W.G. and Crane P.R. (eds) (1987) *The Origins of Angiosperms,* Cambridge University Press, Cambridge.

Fujie, M., Kuriowa, H., Suzuki, T., Kwano, S. and Kuriowa, T. (1993) Organelle DNA synthesis in the quiescent centre of *Arabidopsis thaliana* (Col). *J. Exper. Bot.,* **44** 689-93.

Furner, I.J. and Pumfrey, J.E. (1992) Cell fate in the shoot apical meristem of *Arabidopsis thaliana. Development,* **115** 755-64.

Galbraith, D.W., Harkins, K.R. and Knapp, S. (1991) Systemic endopolyplody in *Arabidopsis thaliana. Plant Physiol.,* **96** 985-89.

Galway, M.E., Masucci, J.D., Lloyd, A.M., Walbot, V., Davis, R.W. and Schiefelbein, J.W. (1994) The TTG gene is required to specify epidermal cell fate and cell patterning in the *Arabidopsis* root. *Dev. Biol.,* **166** 740-54.

Gensel, P.G. (1992) Phylogenetic relationships of the Zosterophylls and Lycopsids: evidence from morphology paleoecology, and clasistic methods of inference. *Ann. Missouri Bot. Gard.*, **79** 450-73.

Goethe, J.W. (1790) *Versuch die Metamorphose der Pflanzen zu erklaren*, C.W. Ettinger, saxe-Gotha.

Goodman, H., Ecker, J. and Dean, C. (1995) The genome of *Arabidopsis thaliana. Proc. Natl Acad. Sci. USA*, **92** 10831-35.

Goodrich, J., Puangsomlee, P., Martin, M., Long, D., Meyerowitz, E.M. and Coupland, G. (1997) A Polycomb-group gene regulates homeotic gene expression in *Arabidopsis*. *Nature*, **386** 44-51.

Gould, K.S., Cutter, E.G. and Young, J.P.W. (1994) The determination of pea leaves, leaflets and tendrils. *Am. J. Bot.*, **81** 352-60.

Grbic, B. and Bleecker, A.B. (1996) An altered body plan is conferred on *Arabidopsis* plants carrying dominant alleles of two genes. *Development*, **122** 2395-2403

Hake, S. and Char, B.R. (1997) Cell–cell interactions during plant development. *Genes Dev.*, **11** 1087-97.

Hareven, D., Gutfinger, T., Parnis, A., Eshed, Y. and Lifschitz, E. (1996) The making of a compound leaf: genetic manipulation of leaf architecture in tomato. *Cell*, **84** 735-44.

Hemerly, A.S., Ferreira, P., Engler, J., van Montagu, M., Engler, G. and Inze, D. (1993) cdc2a expression in *Arabidopsis* is linked with competence for cell division. *Plant Cell*, **5** 1711-23.

Hobbie, L., Timpte, C. and Estelle, M. (1994) Molecular genetics of auxin and cytokinin. *Plant Mol. Biol.*, **26** 1499-1519.

Hofer, J., Turner, L., Hellens, R., Ambrose, M., Matthews, P., Michael, A. and Ellis, N. (1997) UNIFOLIATA regulates leaf and flower morphogenesis in pea. *Curr. Biol.*, **7** 581-87.

Hulskamp, M., Misera, S. and Jurgens, G. (1994) Genetic dissection of trichome cell development in *Arabidopsis*. *Cell*, **76** 555-66.

Ingram, G.C., Goodrich, J., Wilkinson, M.D., Simon, R., Haughn, G.W. and Coen, E.S. (1995) Parallels between *UNUSUAL FLORAL ORGANS* and *FIMBRIATA*, genes controlling flower development in *Arabidopsis* and *Antirrhinum*. *Plant Cell*, **7** 1501-10.

Ingram, G.C., Doyle, S., Carpenter, R., Schultz, E.A., Simon, R. and Coen, E.S. (1997) Dual role for *fimbriata* in regulating floral homeotic genes and cell division in *Antirrhinum*. *EMBO J.*, **16** 6521-34.

Irish, V.F. (1991) Cell lineage in plant development. *Curr. Opinions Genet. Devel.*, **1** 169-73.

Irish, V.F. and Sussex, I.M. (1992) A fate map of the *Arabidopsis* embryonic shoot apical meristem. *Development*, **115** 745-53.

Irvine, K.D. and Wieschaus, E. (1994) Fringe, a boundary-specific signalling molecule, mediates interactions between dorsal and ventral cells during *Drosophila* development. *Cell*, **79** 595-606.

Jackson, D. and Hake, S. (1997) Morphogenesis on the move: cell-to-cell trafficking of plant regulatory proteins. *Curr. Opin. Genet. Dev.*, **7** 495-500.

Jackson, D., Veit, B. and Hake, S. (1994) Expression of maize KNOTTED1 related homeobox genes in the shoot apical meristem predicts patterns of morpogenesis in the vegetative shoot. *Development*, **120** 405-13.

Jofuku, K.D., den Boer, BG., Van Montagu, M. and Okamuro, J.K. (1994) Control of *Arabidopsis* flower and seed development by the homeotic gene *APETALA2*. *Plant Cell*, **6** 1211-25.

Jurgens, G. (1997) Memorizing the floral ABC. *Nature*, **386** 17.

Kaplan, D.R. (1969) Seed development in *Downingia*. *Phytomorphology*, **19** 253-78.

Kaplan, D.R. and Cooke, T.J. (1997) Fundamental concepts in the embryogenesis of dicotyledons: a morphological interpretation of embryo mutants. *Plant Cell*, **9** 1903-19.

Kelly, A.J., Bonnlander, M.B. and Meeks-Wagner, D.R. (1995) NFL, the tobacco homolog of FLORICAULA and LEAFY, is transcriptionally expressed in both vegetative and floral meristems. *Plant Cell*, **7** 225-34.

Kenrick, P. and Crane, P.R. (1997) The origin and early evolution of plants on land. *Nature*, **389** 33-39.

Kerstetter, R. and Hake, S. (1997) Shoot meristem formation in vegetative development. *Plant Cell,* **9** 1001-10.

Kidner, C.A. (1997) Clonal analysis of the *Arabidopsis* primary root. Ph.D. thesis, University of East Anglia.

Kieber, J.J. (1997) The ethylene signal transduction pathway in *Arabidopsis. J. Exper. Bot.,* **307** 211-18.

Klucher, K.M., Chow, H., Reiser, L. and Fischer, R.L. (1996) The AINTEGUMENTA gene of *Arabidopsis* required for ovule and female gametophyte development is related to the floral homeotic gene APETALA2. *Plant Cell,* **8** 137-53.

Koornneef, M. (1981) The complex syndrome of ttg mutants. *Arabid. Inform. Serv.,* **18** 45-51.

Krizek, B.A. and Meyerowitz, E.M. (1996) The *Arabidopsis* homeotic genes *APETALA3* and *PIS-TILLATA* are sufficient to provide the B class organ identity function. *Development,* **122** 11-22.

Larkin, J.C., Oppenheimer, D.G., Lloyd, A., Paparozzi, E. and Marks, D. (1994) Roles of the GLABROUS1 and TRANSPARENT TESTA GLABRA genes in *Arabidopsis* trichome development. *Plant Cell,* **6** 1065-76.

Larkin, J.C., Young, N., Prigge, M. and Marks, M.D. (1996) The control of trichome spacing and number in *Arabidopsis. Development,* **122** 997-1005.

Larkin, J.C., Marks, D., Nadeau, J. and Sack, F. (1997) Epidermal cell fate and patterning in leaves. *Plant Cell,* **9** 1109-20.

Laskowski, M.J., Williams, M.J., Nusbaum, C. and Sussex, I. (1995) Formation of lateral root meristems is a two-stage process. *Development,* **121** 3303-10.

Laux, T. and Jurgens, G. (1997) Embryogenesis: a new start in life. *Plant Cell,* **9** 989-1000.

Laux, T. and Schoof, H. (1997) Maintaining the shoot meristem—the role of *CLAVATA1. Trends Plant Sci.,* **2** 325-27.

Laux, T., Mayer, K.F., Berger, J. and Jurgens, G. (1996) The *WUSCHEL* gene is required for shoot and floral meristem integrity in *Arabidopsis. Development,* **122** 87-96.

Lee, I., Wolfe, D.S., Nilsson, O. and Weigel, D. (1997) A *LEAFY* co-regulator encoded by *UNUSUAL FLORAL ORGANS. Curr. Biol.,* **7** 95-104.

Levin, J.Z. and Meyerowitz, E.M. (1995) UFO: an *Arabidopsis* gene involved in both floral meristem and floral organ development. *Plant Cell,* **7** 529-48.

Leyser, O. and Furner, I. (1992) Characterization of three shoot apical meristem mutants of *Arabidopsis thaliana. Development,* **116** 397-403.

Leyser, H.M.O., Pickett, F.B., Dharmasiri, S. and Estelle, M. (1996) Mutations in the AXR3 gene of *Arabidopsis* result in altered auxin response including ectopic expression from the SAUR-AC1 promoter. *Plant J.,* **10** 403-13.

Li, Y., Hagen, G. and Guilfoyle, T.J. (1992) Altered morphology in transgenic tobacco plants that overproduce cytokinins in specific tissues and organs. *Dev. Biol.,* **153** 386-95.

Lincoln, C., Long, J., Yamaguchi, J., Serikawa, K. and Hake, S. (1994) A *knotted1*-like homeobox gene in *Arabidopsis* is expressed in the vegetative meristem and dramatically alters leaf morphology when overexpressed in transgenic plants. *Plant Cell,* **6** 1859-76.

Long, J.A., Moan, E.I., Medford, J.I. and Barton, M.K. (1996) A member of the *KNOTTED* class of homeodomain proteins encoded by the *STM* gene of *Arabidopsis. Nature,* **379** 66-69.

Lu, P., Porat, R., Nadeau, J.A. and O'Neill, S.D. (1996) Identification of a meristem L1 layer-specific gene in *Arabidopsis* that is expressed during embryonic pattern formation and defines a new class of homeobox genes. *Plant Cell,* **8** 2155-68.

Lucas, W.J., Ding, B. and Van der Schoot, C. (1993) Plasmodesmata and the supracellular nature of plants. *New Phytol.,* **125** 435-76.

Lucas, W.J, Bouche-Pillon, S, Jackson, D.P., Nguyen, L., Baker, L., Ding, B. and Hake, S. (1995) Selective trafficking of KNOTTED1 homeodomain protein and its mRNA through plasmodesmata. *Science,* **270** 1980-83.

Lyndon, R.F. (1983) The mechanism of leaf initiation, in *The Growth and Functioning of Leaves* (eds J.E. Dale and F.L. Milthorpe), Cambridge University Press, Cambridge, pp. 3-24.

Lyndon R.F. (1990) Plant development: the cellular basis, in *Topics in Plant Physiology*: 3 (eds M. Black and J. Chapman) Unwin Books, London, pp. 253-85.

Malamy, J.E. and Benfey, P. (1997) Organization and cell differentiation in lateral roots of *Arabidopsis thaliana. Development,* **124** 33-44.

Masucci, J.D. and Schiefelbein, J.W. (1996) Hormones act downstream of TTG and GL2 to promote root hair outgrowth during epidermis development in the *Arabidopsis* root. *Plant Cell,* **5** 1505-17.

Masucci, J.D., Rerie, W.G., Foreman, D.R., Zhang, M., Galway, M.E., Marks, M.D. and Schiefelbein, J.W. (1996) The homeobox gene *GLABRA2* is required for position dependent cell differentiation in the root epidermis of *Arabidopisis thaliana. Development,* **122** 1253-60.

McConnell, J.R. and Barton, M.K. (1995) Effect of mutations in the *PINHEAD* gene of *Arabidopsis* on the formation of shoot apical meristems. *Dev. Genet.,* **16** 358-66.

Medford, J.I., Callos, J.D., Behringer, F.J. and Link, B.M. (1994) Development of the vegetative shoot apical meristem, in *Arabidopsis* (eds E. Meyerowitz and C. Somerville), Cold Spring Harbor Press, Cold Spring Harbour, New York, pp. 355-78.

Meinhardt, H. (1984) Models of pattern formation and their application to plant development, in *Positional Controls in Plant Development* (eds P.W. Barlow and D.J. Carr), Cambridge University Press, Cambridge, pp. 1-32.

Meinke, D.W. (1992) A homoeotic mutant of *Arabidopsis thaliana* with leafy cotyledons. *Science,* **258** 1647-50.

Meinke, D.W., Franzmann, L.H., Nickle, T.C. and Yeung, E.C. (1994) Leafy cotyledon mutants of *Arabidopsis. Plant Cell,* **6** 1049-64.

Meisel, L., Xie, S. and Lam, E. (1996) *lem7,* a novel temperature-sensitive *Arabidopsis* mutation that reversibly inhibits vegetative development. *Dev. Biol.,* **179** 116-34.

Melarango, J.E., Mehrotra, B. and Coleman, A. (1993) Relationship between endopolyploidy and cell size in epidermal tissue of *Arabidopsis. Plant Cell,* **5**, 1661-68.

Meyerowitz, E.M. (1996) Plant development: local control, global patterning. *Curr. Opin. Gen. Dev.,* **6** 475-79.

Miksche, J.P. and Brown, J.A.M. (1965) Development of vegetative and floral meristems of *Arabidopsis thaliana. Am. J. Bot.,* **52** 533-37.

Miles, D. (1989) An interesting leaf developmental mutant from a Mu active line that causes the loss of leaf blade. *Maize Genet. Coop. Newsl.,* **63** 66-67.

Mirza, J.I., Olsen, G.M., Iversen, T.-H. and Maher, E.P. (1984) The growth and gravitropic responses of wildtype and auxin resistant mutants of *Arabidopsis thaliana. Physiol. Plant.,* **60** 516-22.

Mizukami, Y. and Ma, H. (1992) Ectopic expression of the floral homeotic gene AGAMOUS in transgenic *Arabidopsis* plants alters floral organ identity. *Cell,* **71** 119-31.

Moose, S.P. and Sisco, P.H. (1996) Glossy15, an APETALA2-like gene from maize that regulates leaf epidermal cell identity. *Genes Dev.,* **10** 3018-27.

Nelson, T. and Dengler, N. (1997) Leaf vascular pattern formation. *Plant Cell,* **9** 1121-35.

Okamuro, J.K., Caster, B., Villarroel, R., Van Montagu, M. and Jofuku. K.D. (1997) The *AP2* domain of *APETALA2* defines a large new family of DNA binding proteins in *Arabidopsis. Proc. Natl Acad. Sci. U S A,* **94** 7076-81.

Oppenheimer, D.G., Herman, P.L., Sivakumaran, S., Esch, J. and Marks, M.D. (1991) A myb gene required for leaf trichome differentiation in *Arabidopsis* is expressed in stipules. *Cell,* **67** 483-93.

Pardee, A.B. (1989) G1 events and regulation of cell proliferation. *Science,* **246** 603-608.

Pickett, F.B., Champagne, M.M. and Meeks-Wagner, D.R. (1996) Temperature-sensitive mutations that arrest *Arabidopsis* shoot development. *Development,* **122** 3799-3807.

Pilkington, M. (1929) The regeneration of the stem apex. *New Phytol.,* **28** 37-53.

Poethig, R.S. (1989) Genetic mosaics and cell lineage analysis in plants. *Trends Genet.,* **5** 273-77.

Poethig, R.S. (1997) Leaf morphogenesis in flowering plants. *Plant Cell,* **9** 1077-87.

Przemeck, G.K., Mattsson, J., Hardtke, C.S., Sung, Z.R. and Berleth, T. (1996) Studies on the role of the *Arabidopsis* gene MONOPTEROS in vascular development and plant cell axialization. *Planta,* **200** 229-37.

Pyke, K.A., Marrison, J.L. and Leech, R.M. (1991) Temporal and spatial development of the cells of the expanding first leaf of *Arabidopsis thaliana. J. Exper. Bot.* **42** 1407-16.

Redei, G.P. (1965) Non-mendelian megagametogenesis in *Arabidopsis. Genetics,* **51** 857-72.

Rerie, W.G., Feldmann, K. and Marks, M.D. (1994) The *GLABRA2* gene encodes a homeodomain protein required for normal trichome development in *Arabidopsis. Genes Dev.,* **8** 1388-99.

Richards, F.J. (1948) The geometry of phyllotaxis and its origin. *Symp. Soc. Exper. Biol.,* **2** 217-45.

Roe, J.L., Rivin, C.J., Sessions, R.A., Feldmann, K.A. and Zambryski, P.C. (1993) The *Tousled* gene in *A. thaliana* encodes a protein kinase homolog that is required for leaf and flower development. *Cell,* **75** 939-50.

Sachs, T. (1969) Regeneration experiments on the determination of the form of leaves. *Isr. J. Bot.,* **18** 21-30.

Sachs, T. (1981) The control of patterned differentiation of vascular tissues. *Adv. Bot. Res.,* **9** 151-262.

Sachs, T. (1994) Both cell lineages aand cell interactions contribute to stomatal patterning. *Int. J. Plant Sci.,* **155** 245-47.

Satina, S., Blakeslee, A.F. and Avery, A.G. (1940) Demonstration of the three germ layers in the shoot apex of *Datura* by means of induced polyploidy in periclinal chimeras. *Am. J. Bot.,* **27** 895-905.

Scheres, B., Wolkenfelt, H., Willemsen, V., Terlouw, M., Lawson, E., Dean, C. and Weisbeek, P. (1994a) Embryonic origin of the *Arabidopsis* primary root and root meristem initials. *Development,* **120** 2475-87.

Scheres, B., Dilaurenzio, L. Willemsen, V., Hauser, M.T., Janmaat, K., Weisbeek, P. and Benfey, P.N. (1994b) Mutations affecting the radial organisation of the *Arabidopsis* root display specific defects in the embryonic axis. *Development,* **121** 53-62.

Schneeberger, R.G., Becraft, P.W., Hake, S. and Freeling, M. (1995) Ectopic expression of the knox homeo box gene rough sheath1 alters cell fate in the maize leaf. *Genes Dev.,* **9** 2292-2304.

Schneider, K., Wells, B., Dolan, L. and Roberts, K. (1997) Structural and genetic analysis of epidermal differentiation in *Arabidopsis* primary roots. *Development,* **124** 1789-98.

Schnittger, A., Grini, P.E., Folkers, U. and Hulskamp, M. (1996) Epidermal fate map of the *Arabidopsis* shoot meristem. *Dev. Biol.,* **175** 248-55.

Simon, R., Carpenter, R., Doyle, S. and Coen, E. (1994) *Fimbriata* controls flower development by mediating between meristem and organ identity genes. *Cell,* **78** 99-107.

Sinha, N.R., Williams, R.E. and Hake, S. (1993) Overexpression of the maize homeo box gene, Knotted-1, causes a switch from determinate to indeterminate cell fates. *Genes Dev.,* **7** 787-95.

Snow, M. and Snow, R. (1931) Experiments on phyllotaxis I. The effect of isolating a primordium. *Phil. Trans. Roy. Soc. Lond. B,* **221** 1-43.

Snow, M. and Snow, R. (1933) Experiments on phyllotaxis II. The effect of displacing a primordium. (1933) *Phil. Trans. Roy. Soc. Lond. B,* **222** 353-400.

Snow, M. and Snow, R. (1942) The determination of axillary buds. *New Phytol.,* **41** 13-22.

Souer, E. (1997) Genetic analysis of meristem and organ-primordium formation in petunia hybrids, Ph.D. Dissertation, Free University of Amsterdam, The Netherlands.

Souer, E., van Houwelingen, A., Kloos, D., Mol, J. and Koes, R. (1996) The no apical meristem gene of petunia is required for pattern formation in embryos and flowers and is expressed at meristem and primordia boundaries. *Cell,* **85** 159-70.

Springer, P., McCombie, W., Sundaresan, V. and Martienssen, R.A. (1995) Gene trap transposon tagging of prolifera, an MCM2-3-5 like gene in *Arabidopsis. Science,* **268** 877-80.

Steeves, T.A. and Sussex, I.M. (1989) *Patterns in Plant Development,* Cambridge University Press, Cambridge.

Stewart, R.N. (1983) *Paleobotany and the Evolution of Plants,* Cambridge University Press, Cambridge.

Stewart, R.N. and Burke, L.G. (1970) Independence of tissues derived from apical layers in ontogeny of tobacco leaf and ovary. *Am. J. Bot.,* **57** 1010-16.

Sundaresan, V., Springer, P., Haward, S., Dean, C., Jones, J., Ma, H. and Martienssen, R.A. (1995) Gene trap and enhancer trap transposon mutagenesis in *Arabidopsis thaliana*. *Genes Dev.*, **9** 1797-1810.

Sussex, I.M. (1952) Regeneration of the potato shoot apex. *Nature*, **170** 755-57.

Sussex, I.M. (1955) Experimental investigation of leaf dorsiventrality and orientation in the juvenile shoot. *Phytomorphology*, **5** 286-300.

Sussex, I.M and Rosenthal, D. (1973) Differential H-thymidine labelling of nuclei in the shoot apical meristem of *Nicotiana*. *Bot. Gaz.*, **134** 295-301.

Takahashi, T., Gasch, A., Nishizawa, N. and Chua, N.H. (1995) The *DIMINUTO* gene of *Arabidopsis* is involved in regulating cell elongation. *Genes Dev.*, **9** 97-107.

Talbert, P.B., Adler, H.T., Parks, D.W. and Comai, L. (1995) The *REVOLUTA* gene is necessary for apical meristem development and for limiting cell divisions in the leaves and stems of *Arabidopsis thaliana*. *Development*, **121** 2723-35.

Tanimoto, M., Roberts, K. and Dolan, L. (1995) Ethylene is a positive regulator of root hair development in *Arabidopsis*. *Plant J.*, **8** 943-48.

Telfer, A. and Poethig, R.S. (1994) Leaf development in *Arabidopsis*, in *Arabidopsis* (eds E. Meyerowitz and C. Somerville), Cold Spring Harbor Press, Cold Spring Harbour, New York, pp. 379-403.

Telfer, A., Bollman, K.M. and Poethig, R.S. (1997) Phase change and the regulation of trichome distribution in *Arabidopsis thaliana*. *Development*, **124** 645-54.

Torii, K.U., Mitsukawa, N., Oosumi, T., Matsuura, Y., Yokoyama, R., Whittier, R.F. and Komeda, Y. (1996) The *Arabidopsis ERECTA* gene encodes a putative receptor protein kinase with extracellular leucine-rich repeats. *Plant Cell*, **8** 735-46.

Tsuge, T., Tsukaya, H. and Uchimiya, H. (1996) Two independent and polarized processes of cell elongation regulate leaf blade expansion in *Arabidopsis thaliana* (L.) Heynh. *Development*, **122** 1589-1600.

Van den Berg, C., Willemsen, V., Hage, W., Weisbeck, P. and Scheres, B. (1995). Determination of cell fate in the root meristem by directional signalling. *Nature*, **378** 62-65.

Van den Berg, C., Willemsen, V., Hedricks, G., Weisbeek, P. and Scheres, B. (1997) Short-range control of cell differentiation in the *Arabidopsis* root meristem. *Nature*, **390** 287-89.

Van't Hof, J., Kuniyuki, A. and Bjernes, C.A. (1978) The size and number of replicon families of chromosomal DNA of *A. thaliana*. *Chromosoma*, **68** 269.

Vaughan, J.G. (1955) The morphology and growth of the vegetative and reproductive apices of *Arabidopsis*, Capsella and Anagallis. *Lind. Soc. Lond. Bot.* **55** 279-301.

Vollbrecht, E., Veit, B., Sinha, N. and Hake, S. (1991) The developmental gene Knotted-1 is a member of a maize homeobox gene family. *Nature*, **350** 241-43.

Wada, T., Tachibana, T., Shimura, T. and Okada, K. (1997) Epidermal cell differentiation in *Arabidopsis* determined by a Myb homolog. *CPC. Science*, **277** 1113-16.

Waites, R. and Hudson, A. (1995) *phantastica*: a gene required for dorsoventrality of leaves of *Antirrhinum majus*. *Development*, **121** 2143-54.

Wardlaw, C.W. (1965) *Organization and Evolution in Plants*, Longmans, Green and Co., pp. 17-54.

Weigel, D. and Clark, S.E. (1996) Sizing up the floral meristem. *Plant Physiol.*, **112** 5-10.

Weigel, D., Alvaez, J., Smyth, D., Yanofsky, M.F. and Meyerowitz, E.M. (1992) LEAFY controls floral meristem identity in *Arabidopsis*. *Cell*, **69** 843-59.

Williams, J.A., Paddock, S.W., Vorwerk, K. and Carroll, S.B. (1994) Organization of wing formation and induction of a wing-patterning gene at the dorsal/ventral compartment boundary. *Nature*, **368** 299-305.

Wilson, C.L. (1953) The telome theory. *Bot. Rev.*, **19** 417-37.

Wilson, A.K., Pickett, F.B., Turner, J.C. and Estelle, M. (1990) A dominant mutation in *Arabidopsis* confers resistance to auxin, ethylene and abscisic acid. *Mol. Gen. Genet.*, **222** 377-83.

Yang, M. and Sack, F. (1995) The too many mouths and fourlips mutations affect stomatal production in *Arabidopsis*. *Plant Cell*, **7** 2227-39.

Yang, C.H., Chen, L.J. and Sung, Z.R. (1995) Genetic regulation of shoot development in *Arabidopsis*: role of the EMF genes. *Dev. Biol.*, **169** 421-35.

Young, J.P.W. (1983) Pea leaf morphogenesis: a simple model. *Ann. Bot.*, **52** 311-16.

Zeiger, E., Farquhar, G.D. and Cowan, I.R. (eds) *Stomatal Function*, Stanford University Press, Stanford.

Zimmermann, W. (1930) *Die Phylogenie der Pflanzen. Erste Teil*, G. Fischer, Jena.

9 Genetic control of floral induction and floral patterning

Ilha Lee, Detlef Weigel and François Parcy

9.1 Introduction

As with several other fields that are reviewed in this book, *Arabidopsis* has been adopted by many researchers as a model species for the study of floral induction in the hope that the key factors of floral induction are conserved in other species and that only the regulation of these key factors has changed during evolution. Initial results are promising. For example, the floral regulator *LEAFY (LFY)* has been found to promote flowering not only in *Arabidopsis* but also in divergent species such as tobacco and even aspen trees. However, the analysis of *LFY* and its orthologues in other species has also demonstrated that regulatory details are bound to differ: expression of the *LFY* orthologue in snapdragon is restricted to flowers, while it is expressed constitutively in tobacco. *Arabidopsis* represents an intermediate case, with low expression during the vegetative phase and up-regulation of *LFY* on floral induction.

In addition to floral induction, the development of individual flowers has been subject of intensive analysis in *Arabidopsis*. In contrast to floral induction, these studies have been motivated less by the desire to solve a classical problem of plant physiology than by the wish to understand basic concepts of pattern formation in plants. For example, morphologists have traditionally been inclined to emphasize the role of cell lineage in plant development, as opposed to positional information. Although our understanding of pattern formation in individual flowers is still superficial, the analysis of homeotic mutants has already demonstrated that position is far more important than lineage in the patterning of flowers.

This chapter sets out to review the current situation in research in the areas of floral induction and floral patterning. Both these areas are intellectually and conceptually related, and recent findings on the role of meristem-identity genes have allowed the two areas to be connected in a concrete fashion. The reader is encouraged to consult a number of recent reviews for additional information and different view points (Okamuro *et al.*, 1993; Ma, 1994; Martínez-Zapater *et al.*, 1994; Weigel and Meyerowitz, 1994; Okada and Shimura, 1995; Weigel, 1995; Yanofsky, 1995; Aukerman and Amasino, 1996; Peeters and Koornneef, 1996; Simon and Coupland, 1996; Wilson and Dean, 1996).

9.2   Floral induction and flowering time

Flowering is under the control of both external and endogenous signals, and this has been recognized for a long time. For example, most plants growing in temperate climates produce flowers only during a specific season, indicating their ability to use external cues to determine the right time for flowering. Conversely, most trees start to produce flowers only when they have reached a certain age, although they have been exposed to the same external cues for several years, indicating that endogenous signals also play important roles.

Before the subject of the genetics of floral induction in *Arabidopsis* is discussed, the morphological events associated with the transition to flowering are described. In addition, a background to what is known about physiological and hormonal factors controlling floral induction in general is provided.

9.2.1   Morphology of the floral transition in Arabidopsis

Both vegetative and reproductive structures, including leaves, lateral shoots and flowers, are derived from a group of undifferentiated cells contained in the shoot apical meristem. On floral induction, the shoot meristem ceases to produce leaves with associated shoots, called paraclades, and generates flowers instead. As is typical for many members of the Brassicaceae, the flowers of *Arabidopsis* are not subtended by leaf-like bracts. The flowers themselves are rather stereotypical in organization, with four major types of floral organs arranged in four concentric rings or whorls (Figure 9.1). The outermost or first whorl is occupied by four green sepals, the most leaf-like type of floral organ. The second whorl is formed by four white petals, which are often showy in other species, where they serve to attract pollinators. The third whorl is occupied by six yellow stamens, which are the male reproductive organs. Among the four whorls, organ

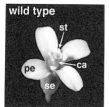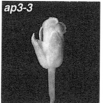

Figure 9.1 Floral morphology of wild type and of three homeotic mutants, representing A function (*ap2*), B function (*ap3*), and C function (*ag*). In wild type, three major floral organ types, including sepals (se), petals (pe), stamens (st) and carpels (ca) can be distinguished, while *ap2* mutants have only stamens and carpels, *ap3* mutants only sepals and carpels, and *ag* mutants only sepals and petals. In addition, *ag* mutant flowers are indeterminate and their floral formula is (sepals, petals, petals)n.

number in the third whorl is most variable, and wild-type flowers with five or even four stamens are not uncommon. The fourth and central-most whorl is taken up by the female reproductive structure, the gynoecium, which is formed by two congenitally fused carpels (Smyth *et al.*, 1990).

In addition to the morphological differences between floral organs and leaves, there are two other major features that distinguish flowers from shoots, namely determinacy and phyllotaxis. While the *Arabidopsis* wild-type shoot meristem can potentially produce an unlimited number of primordia, and is thus indeterminate, the wild-type floral meristem is normally consumed by the production of a limited number of floral organs (typically sixteen), and is thus determinate. Furthermore, leaves arise one at a time in a phyllotactic spiral on the shoot, while the floral organs in each whorl arise at about the same time.

When one looks at a mature *Arabidopsis* plant, it appears that there are at least three obvious phases through which a plant passes: a vegetative phase, during which the rosette leaves are produced, an early inflorescence phase, during which paraclades subtended by reduced leaves called bracts are produced, and a late inflorescence phase, during which bractless flowers are produced. The rosette part of the plant is characterized by the absence of internode elongation between leaves, while internodes elongate in the inflorescence part. From the appearance of the mature plant it would thus appear that the shoot meristem undergoes at least two transitions, first from a vegetative to an early inflorescence, and then from an early to a late inflorescence state.

In contrast to these macroscopic observations, a careful microscopic examination of the morphological events associated with the transition from the vegetative to the reproductive phase strongly suggests that there is no distinct early inflorescence phase. Hempel and Feldman (1994) observed that all leaf primordia that had been initiated before plants are induced to flower are initially indistinguishable, and that no paraclades are initiated before the transition to flowering is made. Once plants are induced to flower, the shoot apical meristem switches from the production of leaf primordia to flower primordia, and at the same time paraclades are released in the axils of previously initiated leaf primordia. This occurs in a basipetal fashion, that is the first paraclade to be released is the one in the axil of the last leaf produced. The differentiation of paraclades is accompanied by a reduction in growth of the subtending leaf primordia, which develop into bracts instead of rosette leaves.

That the morphology of leaves is plastic, and that bracts are not determined at the time of their inception, is corroborated by studies of trichome formation (Chien and Sussex, 1996; Telfer *et al.*, 1997). In addition to the transition to flowering, there are several other, more gradual, changes in organ morphology that occur during the plant life cycle and that are collectively known as phase change. One example is the distribution of trichomes on the abaxial and adaxial surfaces of leaves. The first set of leaves has only adaxial trichomes, and their density increases in subsequent leaves. Later arising leaves start to have an increasing number of abaxial trichomes, while the number of adaxial trichomes declines, such that the most apical bracts are devoid of adaxial trichomes. Importantly, before adaxial trichomes disappear they become restricted to the apical part of later-formed leaves. Conversely, when abaxial trichomes are formed, they are first found on the basal part of the leaf. In other words, the apical part of a particular leaf is more similar to the basal part of the previously initiated leaf than to the basal part of the same leaf.

These observations reconcile with the known gradient of leaf maturation, in which the apical part of the leaf matures first. Thus, a leaf does not behave as a unit, as one would expect if the character of a leaf was determined by the shoot apical meristem at the inception of the leaf primordium, but rather a leaf behaves as a mosaic in which each part behaves accordingly to the overall maturation state of the plant.

Trichome formation, along with changes in leaf size and shape, is affected by similar physiological and genetic pathways as flowering. For example, both flowering and abaxial trichome formation are delayed when *Arabidopsis* plants are grown in short days (Chien and Sussex, 1996; Telfer *et al.*, 1997). Thus, while we will focus on describing the effects of various environmental, hormonal and genetic factors on flowering time, one should remember that most of these factors also affect other aspects of phase change. However, despite the parallels between phase change in general and the transition to flowering in particular, an important difference is that the changes in leaf morphology are gradual, while the switch from leaf with paraclade to bractless flower is a precipitous one.

9.2.2 *Environmental factors affecting floral induction*

The two environmental signals that have been studied in most detail are vernalization and photoperiod. Photoperiodism, or the effect of changes in day length, was first discovered by Garner and Allard (1920) in studies of a tobacco variety, Maryland Mammoth, which did not flower in summer but quickly flowered once days got shorter in the fall.

According to their requirement for photoperiods of different lengths, plants can be divided into three major groups: long-day and short-day plants (Doorenbos and Wellensiek, 1959), and a third group, comprising plant species in which flowering is not affected by photoperiod, known as day-neutral. In contrast to what their name implies, most short-day plants actually measure the length of the uninterrupted dark period instead of the daily light period, and they require a dark period exceeding a critical length of night before they will flower.

Photoperiodism is mediated mainly by the phytochrome class of photorecep-tors (see chapter 10), which constitute a reversible pigment system that detects red and far-red light, as indicated by the observation that the effect of red light can be completely reversed if exposure to red light is immediately followed by exposure to far-red light. However, other pigment systems are likely to be involved in photoperiodism since blue and far-red light can affect the flowering of long-day plants in a manner inconsistent with phytochrome absorption (Lang, 1965; Zeevaart, 1976).

The site where photoperiod is perceived is the leaf, as demonstrated by applying inductive photoperiods to various plant organs while maintaining the rest of the plant under non-inductive photoperiods (Bocchi *et al.*, 1956). The physical separation of the site of photoperiod perception and response suggests the existence of transmissible floral signals, for which the term 'florigen' was first proposed by Chailakhyan (1936). Zeevaart (1958) subsequently demonstrated the existence of a transmissible floral promoter by a series of elegant grafting experiments. Similar experiments have provided evidence for transmissible floral inhibitors, since a late-flowering donor can cause delay of flowering in an otherwise early-flowering receptor (Paton and Barber, 1955; King and Zeevaart, 1973; Lang *et al.*, 1977).

Vernalization is the promotion of flowering by transient exposure to cold temperature, and was first observed by Gaßner (1918), who studied the flowering of winter annual rye. Later work has shown that many species have a vernalization requirement for flowering, including winter annuals, biennials and perennial herbaceous plants (Chouard, 1960). A few species have an absolute requirement for vernalization (qualitative response; true biennials), but the flowering of other species is just accelerated by vernalization and will occur even in unvernalized plants (quantitative response).

9.2.3 *Hormones and floral induction*

Although floral induction can be influenced by almost all plant hormones, including gibberellins, cytokinins, auxin, abscisic acid and ethylene, the effect of plant hormones on floral induction varies from species to species. The literature on the role of hormones tends to be very confusing as almost any combination of promoting and inhibitory effects can be found in different species.

Gibberellins (GAs) can induce flowering in many long-day- and cold-requiring rosette plants under non-inductive conditions, and this is often accompanied by internode elongation (Pharis and King, 1985) (see chapter 4). However, internode elongation and flowering are independently controlled processes in other species and in these GAs cannot substitute for vernalization to induce flowering. In addition, not all rosette plants can be induced to flower by GAs, and in some perennial plants GAs even show inhibitory effects on flowering (Bernier, 1988). Abscisic acid (ABA), which antagonizes the role of GAs during, for example, seed germination, inhibits flowering in several short- and long-day plants grown under inductive conditions, although the endogenous level does not show any consistent correlation with floral induction (Zeevaart, 1976) (see Chapter 4). Exogenous cytokinins affect flower initiation in some species, with promotive effects being more frequent than inhibitory ones. Exogenous auxin generally inhibits floral initiation, although promotion has been observed in some species, such as apple, mango or pineapple (Bernier, 1988).

9.2.4   Arabidopsis *flowering-time mutants*

Unfortunately, attempts to isolate the hypothetical florigen have been largely unsuccessful, although there are a few reports on the preparation of floral inductive extracts, mostly from two species, morning glory and duckweed (Takimoto and Kaihara, 1990; Ishioka *et al.*, 1991; Kozaki *et al.*, 1991). Similarly, complex mixtures of oligosaccharides have been implicated in floral induction (Tran Thanh Van *et al.*, 1985).

Several groups have undertaken systematic searches for mutations that interfere with the normal floral induction process in *Arabidopsis*, as indicated by changes in time to flowering. Wild-type *Arabidopsis* is a facultative long-day plant, meaning that it will flower much earlier under long days than under short days. Typical long-day conditions used in the laboratory are sixteen hours of light, although many researchers use continuous light, while typical short-day conditions range from eight to ten hours of light. As with most other long-day plants, *Arabidopsis* reacts to night breaks, and relatively brief interruptions of the dark period will significantly accelerate flowering. Vernalization accelerates the flowering of many geographical races (ecotypes) of *Arabidopsis*, although the commonly used laboratory strains such as Landsberg *erecta* or Columbia show only a very small response to vernalization (Napp-Zinn, 1985). Light quality is another environmental factor that controls flowering time in *Arabidopsis*. Blue light and far-red light accelerate flowering, whereas red light is inhibitory (Brown and Klein, 1971; Eskins, 1992; Bagnall, 1993).

The pioneer in the field of flowering-time genes is Koornneef, who first isolated a large set of late-flowering mutants (Koornneef *et al.*, 1991). He also selected a large set of photomorphogenetic mutants because of their abnormal light responses during the seedling stage, and later showed that many of these are also defective in the control of flowering time (Goto *et al.*, 1991a). Additional flowering-time mutants were identified from ecotypes other than the Landsberg *erecta* strain used by Koornneef (Ray *et al.*, 1996; Sanda *et al.*, 1997), and by specific screens for early-flowering mutants (Sung *et al.*, 1992; Zagotta *et al.*, 1992). A number of mutants isolated for reasons such as defects in phytohormone biosynthesis can also be added to the growing list of flowering-time loci.

Although a coherent picture of floral induction in *Arabidopsis* is not available yet, and will take a while to construct, progress in the analysis of the different flowering-time genes has been made on several fronts. Double mutant analyses have allowed many genes to be grouped according to common functions, while physiological analyses have put the genes into context with known floral-induction processes. Finally, cloning of flowering-time genes has allowed the first gain-of-function tests.

9.2.4.1 *Light perception*

Light is used not only as the energy source for photosynthesis but also as signal for developmental processes, including seed germination, stem elongation, phototropism and flowering (see chapter 10). Perception and transduction of light signals are mediated by red/far-red, blue and UV photoreceptors. Phytochromes, which absorb red and far-red light, are composed of a 120 kD soluble apoprotein that is covalently linked to a tetrapyrrole chromophore. *Arabidopsis* contains at least five distinct genes encoding the apoprotein components of phytochrome. Classically, two types of phytochrome have been distinguished: the light-labile type I or etiolated-tissue phytochrome, which is most abundant in dark-grown tissue, and the light-stable type II or green-tissue phytochrome, which is constitutively present at equal levels in both dark- and light-grown tissues. In *Arabidopsis*, the apoprotein of the type I phytochrome is encoded by the *PHYTOCHROME A (PHYA)* gene, while that of the type II phytochrome is heterogeneous and encoded by at least four genes, *PHYB–E* (Sharrock and Quail, 1989). Point mutations in *PHYA* and *PHYB*, and in genes that are required for chromophore biosynthesis, were initially isolated because of their failure to suppress hypocotyl elongation in newly germinated seedlings, and they were therefore known as *LONG HYPOCOTYL (HY)* mutations. *HY3* has since been renamed *PHYB*, and *HY8* has been renamed *PHYA*.

phyA mutants flower normally under both long and short days, but fail to respond to the extension of an eight-hour short day with eight hours of low-fluence light, or to a one-hour night break (Johnson *et al.*, 1994; Reed *et al.*, 1994). In wild-type, either treatment accelerates flowering almost as effectively as regular long days. A phenotype that is the opposite of *phyA* mutants is seen in *PHYA* overexpressing plants, which are day-neutral and flower as early under short days as wild-type grown under long days (Bagnall *et al.*, 1995). These results show that *PHYA* mediates the photoperiod response in *Arabidopsis*, although its function is partially redundant, as indicated by the mild loss-of-function phenotype. In other species, however, the requirement for *PHYA* activity is higher. For example, the phenotype of pea *phyA* mutants is considerably more severe than that of *Arabidopsis phyA* mutants, and the pea mutants behave under regular long days as if they were grown under short days (Weller *et al.*, 1997).

In contrast to *phyA* mutants, *phyB* mutants flower early, under both long and short days. However, they are still photoperiod sensitive and it has been suggested that *PHYB* mediates the so-called shade-avoidance syndrome and thus acts independently of the photoperiodic response (Goto *et al.*, 1991b; Reed *et al.*, 1994). The early flowering syndrome of *phyB* mutants is further enhanced by a mutation in *PHYD*, whose structure is very similar to that of *PHYB* but which is expressed at much lower levels (Aukerman *et al.*, 1997). *PHYB* deficiency in other plants such as *Sorghum*, *Brassica* and cucumber also causes an early-flowering phenotype, indicating that the inhibitory effect of type II phytochrome on the floral transition is common to many plants, including monocots (Adamse *et al.*, 1988; Beall *et al.*, 1991; Devlin *et al.*, 1992; Childs *et al*, 1995). In addition, the *Sorghum* and *Brassica* mutants have been reported to overproduce GAs, suggesting a link between phytochrome and GA. However, *phyB* mutants of *Arabidopsis* show primarily increased responsiveness to GAs instead of a change in GA levels (Reed *et al.*, 1996).

In addition to red and far-red light, floral induction in *Arabidopsis* is influenced by blue light (Brown and Klein, 1971; Goto *et al.*, 1991a; Eskins, 1992). A night break with blue light accelerates flowering of *Arabidopsis* and when the ratio of red to far-red light is held constant, increasing ratios of blue light lead to a proportionate acceleration of flowering. The blue-light photoreceptor *CRYPTOCHROME 1 (CRY1)* is encoded by the *HY4* locus, which was originally identified through mutations in which blue light does not inhibit hypocotyl elongation (Koornneef *et al.*, 1980; Ahmad and Cashmore, 1993). Unfortunately, the flowering-time phenotype of *hy4* mutants is not consistent: one allele, *hy4-105*, has been reported to cause early flowering in short days, while other alleles have been reported to delay flowering in long and short days (Goto *et al.*, 1991a; King and Bagnall, 1996; Zagotta *et al.*, 1996). On the other hand, *hy4* mutants are insensitive to a night break with blue light, while night breaks using red and far-red light are of similar effectiveness in *hy4* mutants as in wild-type (King and Bagnall, 1996). The specific insensitivity of *hy4* mutant to a blue-light night break suggests that *CRY1* affects floral induction independently of the phytochrome system.

Mutants of the *de-etiolated (det)* and *constitutive photomorphogenic (cop)* classes show phenotypes opposite to those of *hy* mutants and identify genes that negatively regulate photomorphogenesis. In *det* and *cop* mutants, hypocotyl elongation in the dark is inhibited, and dark-grown mutant seedlings resemble light-grown wild-type seedlings. Interestingly, at least two mutants of this class, *det1* and *cop1*, are day-neutral and flower at the same time under long and short days (McNellis *et al.*, 1994; Pepper and Chory, 1997). The connection between *det,cop* mutants and floral induction has been further strengthened through the recent isolation of genetic suppressors of *det1* (Pepper and Chory, 1997). Although these mutations were initially isolated because of their ability to suppress the seedling phenotype of *det1*, two mutations, *ted1* and *ted2* (*ted* standing for reversal of the *det* phenotype), restore photoperiod sensitivity and thus delay flowering under short days in a *det1* mutant background, and the *ted1,det1* double mutant is late flowering under both long and short days. Importantly, *TED1* does not correspond to any of the previously identified late-flowering loci. These results demonstrate that some of the downstream components involved in the photomorphogenetic response of seedlings are also involved in the initiation of flowering.

In order for a plant to measure the length of day or night, it has to integrate the environmental input with an endogenous circadian rhythm. This connection has been substantiated by the analysis of *det1* and *cop1* mutants, in which the activity period of a circadian-regulated promoter is shortened in the light and in which the shorter periods are maintained in constant darkness (Millar *et al.*,

1995). In addition, another day-neutral mutant, *elf3*, lacks rhythmicity in two distinct circadian responses in constant light, although rhythmicity in constant dark and during the light-to-dark transition is almost normal (Hicks *et al.*, 1996). A further connection between circadian rhythm and the phytochrome response is revealed by the long-hypocotyl phenotype of *elf3* mutants. As one might expect, mutations that affect phytochrome function, such as *hy2*, are also defective in circadian rhythm, and the period of *CAB2* expression in continuous red light is lengthened in *hy2* mutants (Millar *et al.*, 1995).

   Other screens for early-flowering mutants have led to the identification of the *phytochrome-signaling early-flowering (pef)* mutants. Analysis of the hypocotyl phenotype has shown that *pef1* is defective in both *PHYA-* and *PHYB*-mediated responses, while *pef2* and *pef3* are specifically defective in *PHYB*-mediated responses (Ahmad and Cashmore, 1996).

9.2.4.2 Hormone signalling

Arabidopsis behaves in a similar way to many other rosette plants, and flowering is accelerated by application of exogenous GAs, although the effect is much more pronounced in short than in long days (Langridge, 1957; Wilson *et al.*, 1992; Jacobsen and Olszewski, 1993). Conversely, treatment of imbibed seeds with a GA inhibitor delays flowering (Napp-Zinn, 1969).

   The role of endogenous GA in flowering has been confirmed with a mutant, *ga1-3*, which lacks the GA biosynthetic enzyme *ent*-kaurene synthetase A (Sun and Kamiya, 1994). While flowering is only slightly delayed in long days in *ga1-3* mutants, they never flower in short days, indicating an absolute requirement for GA under short days (Wilson *et al.*, 1992). Although the flowering time of *ga1-3* mutants in long days is largely normal, it is presently unclear whether GA has indeed no role for flowering in long days, or whether the requirement for GA is merely much lower in long than in short days, because *ga1-3* mutants still contain low levels of GA (Zeevaart and Talón, 1992). Consistent with the role of GA in floral induction, flowering of *gibberellic acid insensitive (gai)* mutants is also delayed strongly in short days. In addition, the *spindly* mutation, which constitutively activates GA signal transduction, causes early flowering (Jacobsen and Olszewski, 1993).

   The requirement of GA for flowering under non-inductive photoperiods cannot be overcome by vernalization as cold treatment does not induce flowering of *ga1-3* in short days. However, this does not necessarily mean that the vernalization effect on flowering is mediated by GA, as *ga1-3*, when combined with another late-flowering mutant, *fca*, is still vernalization-sensitive (Wilson and Dean, 1996).

   The effect of GA in other processes such as seed germination is often antagonized by abscisic acid (ABA); *ABA deficient 1 (aba1)* and *ABA insensitive 1 (abi1)* mutants flower slightly early under short days, a phenotype that can be considered the opposite of *ga1* and *gai* mutants (Wilson and Dean, 1996).

9.2.4.3 Genes specifically affecting flowering time

Apart from mutations that were initially identified because of their effects on light or hormone signalling, a large number of mutations that seem to specifically affect phase change, including flowering time, have been identified in *Arabidopsis*. Most of these have been isolated in early ecotypes such as Landsberg *erecta* and Columbia (McKelvie, 1962; Redeí, 1962; Hussein, 1968; Vetrilova, 1973; Koornneef *et al.*, 1991). Mutants that flower late because they grow slowly were excluded, hence *bona fide* late-flowering mutants are both temporally and developmentally late, and at least twelve loci have been identified in this class: *CONSTANS (CO), GIGANTEA (GI), LUMINIDEPENDENS (LD),* and *FCA, FD, FE, FHA, FPA, FT, FVE, FWA* and *FY*.

On the basis of the response to environmental stimuli such as photoperiod, vernalization and far-red enriched light, late-flowering mutants have been classified into three different groups (Martínez-Zapater and Somerville, 1990; Koornneef *et al.*, 1991; Bagnall, 1993). The first group, *fca, fve, fpa, fy* and *ld*, shows a stronger response to environmental stimuli than wild-type, and vernalization abolishes the lateness of these mutants almost completely. The second group, *ft, fe, fwa* and *fd*, shows a strong response to photoperiod, but responds only weakly to vernalization and far-red enriched light. This response is similar to that of wild-type, although these mutants flower later than wild-type even under optimal conditions. The third group, *co, gi* and *fha*, does not respond to long photoperiods, vernalization or increased far-red light.

Although the Landsberg *erecta* wild-type strain responds only marginally to vernalization, it has been possible to isolate mutations that identify the vernalization pathway in the background of the vernalization-sensitive *fca* mutation (Chandler *et al.*, 1996). *vernalisation requiring (vrn)* mutations confer insensitivity to vernalization not only in the *fca* background but also in other late-flowering backgrounds and to short-day grown wild-type. Interestingly, the acceleration of flowering by GA is not affected by *vrn1*, indicating that GA acts on a different pathway than vernalization.

A third environmental input, which has not been studied in detail, is that of growth temperature. Growing *Arabidopsis* at 16°C instead of the customary 22 to 25°C considerably delays flowering, and the effect is similar to that of growing plants in short days. The effects of the various flowering-time mutations under different temperature regimens have not been studied, and the relation to the other pathways is thus unknown.

Unfortunately, an exhaustive analysis of all possible double mutant combinations among late-flowering mutants has not yet been published. However, transgression analyses, in which the flowering time of double mutants is compared with that of single mutants, have identified at least two pathways, one defined by *fe, ft, fd, fwa, co* and *gi*, and one by *fca, fpa, fve* and *fy* (Koornneef *et al.*, 1991; Martínez-Zapater *et al.*, 1994). Importantly, none of these double mutant combinations eliminates flowering. However, flowering in both long and short days is blocked in a double mutant carrying the *gal-3* mutation, which eliminates flowering in short days, along with a late-flowering mutation such as *co*, which eliminates the acceleration of flowering caused by long days (Putterill *et al.*, 1995).

The picture that emerges from these analyses is that of a slow pathway that relies on GA and operates in both long and short days, but whose effects in long days are normally masked by the more efficient *CO/GI* pathway. One prediction that follows from this scenario is that constitutive activation of the *CO/GI* pathway in short days should override the failure of *gal-3* mutants to flower.

In addition to late-flowering mutants, early-flowering mutants have been isolated (Shannon and Meeks-Wagner, 1991; Zagotta *et al.*, 1992; Coupland, 1995). Many of the early-flowering mutants have pleiotropic phenotypes, including the *elf3* mutant discussed above and the *terminal flower 1 (tfl1)* mutant, which is discussed in more detail below.

A unique phenotype is seen in *embryonic flower (emf)* mutants, exemplified by at least two loci (Sung *et al.*, 1992; Bai and Sung, 1995; Yang *et al.*, 1995). In the strongest mutant, *emf1-2*, all leaves, including cotyledons, exhibit carpelloid character, indicating an extreme acceleration of the floral program.

9.2.4.4 Molecular analysis of flowering-time genes

So far, three genes identified by late-flowering mutations have been cloned. The gene product encoded by *LD* is likely to be a transcriptional regulator since the predicted amino acid sequence shows similarity to homeo domains and contains a glutamine-rich region, reminiscent of transcriptional activation domains (Lee *et al.*, 1994a; Aukerman and Amasino, 1996). Consistent with the proposed role of *LD* in constitutive floral induction, the abundance of *LD* mRNA does not differ between long and short days. *LD* is expressed primarily in rapidly growing regions including shoot apices, and could thus have a direct role in regulating genes responsible for flower initiation.

CO also encodes an apparent transcriptional regulator, a member of the zinc-finger protein superfamily (Putterill *et al.*, 1995). Long days increase the abundance of *CO* RNA compared to short days, suggesting that *CO* levels are important to determine flowering time. This is confirmed by the dose-dependent effect of *CO* wild-type copies on flowering, and by the day-length-independent early flowering observed in transgenic plants that overexpress *CO* constitutively (Putterill *et al.*, 1995; Simon *et al.*, 1996).

In contrast to *CO* and *LD*, *FCA* seems to function in post-transcriptional regulation of other genes, since the FCA protein contains two RNA-binding domains and FCA can bind to RNA *in vitro*, indicating that FCA is involved in RNA processing (Macknight *et al.*, 1997).

9.2.5 Natural variation of flowering time in Arabidopsis

Apart from mutations induced in the laboratory, variation in flowering time between natural populations has been a valuable source for the identification of flowering-time genes. The ecotypes most commonly used in the laboratory, such as Columbia and Landsberg *erecta*, are early ecotypes and show only a small response to vernalization. However, winter-annual ecotypes, such as Stockholm, flower very late and are exquisitely responsive to vernalization. Most of the natural variation is due to a single locus, *FRI*, and naturally occurring dominant alleles of this locus confer a strong vernalization requirement in most of the winter annual ecotypes (Karlovska, 1974; Napp-Zinn, 1985; Lee *et al.*, 1993; Clarke and Dean, 1994; Kowalski *et al.*, 1994; Clarke *et al.*, 1995; Lee and Amasino, 1995; Sanda *et al.*, 1997). The activity of the dominant *FRI* alleles requires a functional allele at another locus, *FLC*, and it turned out that some of the earliest laboratory strains, such as Landsberg *erecta*, carry inactive alleles at both loci (Koornneef *et al.*, 1994; Lee *et al.*, 1994b).

9.3 Floral induction and flower initiation

9.3.1 Genes controlling the initiation of individual flowers

One of the consequences of floral induction is the up-regulation of several genes that control the initiation of individual flowers. Most genes in this class have a positive role in flower development and their inactivation causes a complete or partial transformation of flower into shoot meristems. A notable feature of these flower-meristem-identity genes is their redundancy, either at the molecular level, that is between genes that encode closely related proteins, or at the genetic level, that is between genes that do not encode closely related proteins but have apparently similar activities at the cellular or organismal level.

9.3.1.1 Flower-meristem-identity genes LEAFY and APETALA1

LEAFY (LFY) and *APETALA1 (AP1)* can be considered the cardinal flower-meristem-identity genes. Both genes encode transcription factors that are expressed at high levels in floral anlagen and floral primordia, before any organs differentiate (Mandel *et al.*, 1992; Weigel *et al.*, 1992). *LFY* is apparently a unique gene that has no close relatives except for its orthologs in other plant species. The nuclear localization of LFY protein, along with the presence of acidic and glutamine/proline-rich regions reminiscent of transcriptional activation domains in other proteins, suggests that it functions as a transcription factor (Weigel *et al.*, 1992; Levin and Meyerowitz, 1995). *AP1* encodes a member of the MADS domain family of transcription factors, which are found throughout all eukaryotic kingdoms. The plant MADS domain proteins themselves fall into several subfamilies, and *AP1* belongs to a clade that is represented by at least two other closely related genes, *CAULIFLOWER (CAL)* and *AGAMOUS-LIKE 8 (AGL8)* (Purugganan *et al.*, 1995; Theißen and Saedler, 1995). *CAL*, whose normal expression pattern is very similar to that of *AP1*, has been shown to act redundantly with *AP1* (Bowman *et al.*, 1993; Kempin *et al.*, 1995). It is possible that *AGL8* also acts redundantly with these two genes. Although the *AGL8* expression domain normally does not overlap with that of *AP1* and *CAL*, it invades the *AP1* domain in *ap1* mutants, raising the possibility that AGL8 protein can functionally substitute for AP1 activity in *ap1* mutants (Mandel and Yanofsky, 1995a).

All *lfy* alleles cause the transformation of at least a few basal flowers into bracts with associated paraclades. In contrast, the phenotype of more apical flowers differs greatly between different *lfy* alleles. In a typical null allele, the apical flowers are subtended by a leaf-like bract, while the flowers themselves exhibit various characteristics of shoots, such as partially spiral phyllotaxis and formation of secondary flowers (Schultz and Haughn, 1991; Huala and Sussex, 1992; Weigel *et al.*, 1992). In addition, petals and stamens are almost eliminated, and most organs are either bract-like, sepals or carpels, or of mixed identity in any combination of these three organ types. In contrast, weaker alleles can form apical flowers that have all the normal organ types, and in the weakest allele, *lfy-22*, apical flowers can even be completely normal (Levin and Meyerowitz, 1995).

The difference between basal and apical flowers in *lfy* mutants can be attributed to a delay in the activation of the other flower-meristem-identity gene, *AP1*. In both weak and strong *lfy* mutants, *AP1* RNA is expressed normally in more apical nodes but it is not expressed in basal nodes, which correspond to flowers that are replaced by bracts with paraclades (Mandel and Yanofsky, 1995b; Gustafson-Brown, 1996). That *AP1* expression in apical nodes is responsible for their partial flower character in *lfy* mutants is confirmed by the phenotype of *lfy,ap1* double mutants, in which the distinction between basal and apical positions largely disappears, and in which all flowers are replaced by shoot-like structures (Huala and Sussex, 1992; Weigel *et al.*, 1992). However, not only is *LFY* required for normal *AP1* expression, but *AP1* is also required for high levels of *LFY* expression in flowers, as *LFY* expression is reduced in *ap1,cal* double mutants (Bowman *et al.*, 1993).

Interestingly, the phenotypic differences between *lfy* and *ap1* alleles of different strengths are largely eliminated in double mutant combinations, and combination of a weak *lfy* with a weak *ap1* allele produces a phenotype very similar to that seen in the combination of two strong alleles, confirming the redundancy of these two genes (Bowman *et al.*, 1993). What could be the reason for this situation, in which *LFY* and *AP1* positively regulate each other but also act redundantly? A clue might come from the observation that the sharp transition between the production of bracts with paraclades and the production of bractless flowers is attenuated in *lfy* and *ap1* mutants, indicating that the mutual reinforcement of *LFY* and *AP1*, together with their redundant action, is critical for the precipitous transition from leaf/paraclade to flower.

Despite their redundancy, the individual functions of *LFY* and *AP1* are rather different. Similarly to *lfy*, the *ap1* single mutant phenotype can be interpreted as a partial conversion of flowers into shoots, as several flowers are borne on a single pedicel, with additional flowers typically arising in the axils of first-whorl organs (Irish and Sussex, 1990; Bowman *et al.*, 1993; Schultz and Haughn, 1993; Shannon and Meeks-Wagner, 1993). In the most extreme case, a flower meristem is initiated but then behaves as a shoot meristem and produces additional flower meristems in a phyllotactic spiral, a process that can be repeated a few times, before flowers finally differentiate. This effect of *ap1* mutations is aggravated in combination with *cal* mutations, which on their own have no phenotypic defects, indicating the redundant action of these two genes, which are very similar both in their sequence and in their early expression pattern (Bowman *et al.*, 1993; Kempin *et al.*, 1995). As expected from the observation that the differences between *ap1* mutant alleles of different strength disappear in a *lfy* mutant background, the *lfy,ap1* phenotype is not further enhanced by a *cal* mutation (Bowman *et al.*, 1993).

9.3.1.2 *Flower-meristem-identity genes* UFO *and* APETALA2

Other genes with positive roles in flower initiation are *UNUSUAL FLORAL ORGANS (UFO)* and *APETALA2 (AP2)*. *ufo* mutants resemble weak *lfy* alleles in several ways, although the organ-identity defects are more pronounced than the meristem-identity defects when compared with weak *lfy* alleles (Levin and Meyerowitz, 1995; Wilkinson and Haughn, 1995). Similarly, a weak allele of *AP2*, *ap2-1*, causes defects that resemble those seen in *ap1* mutants, including the production of secondary flowers in the axils of bract-like first-whorl organs, particularly in short days (Okamuro *et al.*, 1997a). That these two genes are likely to function in either the *LFY* or *AP1* pathway is further demonstrated by the observation that either *ufo,ap1* or *lfy,ap2-1* double mutants are similar to *lfy,ap1* double mutants (Huala and Sussex, 1992; Levin and Meyerowitz, 1995; Wilkinson and Haughn, 1995).

AP2 encodes a member of the large class of plant-specific transcription factors that are characterized by the AP2 domain. Among these, AP2 belongs to a subclass that contains two copies of the AP2 domain (Jofuku *et al.*, 1994; Ohme-Takagi and Shinshi, 1995; Okamuro *et al.*, 1997b). Another gene encoding a protein with two AP2 domains is the *AINTEGUMENTA (ANT)* gene; mutations in this gene were originally identified because of their effects on ovule development. As with *AP2*, *ANT* turned out to be more widely expressed, and double mutant analysis shows that *ANT* acts at least partially redundantly with *AP2* (Elliott *et al.*, 1996; Klucher *et al.*, 1996).

The exact biochemical function of UFO protein is unknown, but it shares with a number of other proteins from animals, fungi and plants a region of sequence similarity called the F-box (Bai *et al.*, 1996; Samach *et al.*, 1997). The F-box was functionally identified as a domain through which the yeast cell-cycle regulators cyclin F, Cdc4p and Skp2p interact with the same protein, Skp1p (Bai *et al.*, 1996). It has been proposed that F-box proteins target other proteins for ubiquitin-mediated proteolysis through their interaction with Skp1p. This hypothesis is further supported by the finding that many F-box proteins have additional protein-interaction domains, such as WD40 repeats and leucine-rich repeats. While UFO does not contain any additional obvious protein-interaction domains, regions that are downstream of the F-box are functionally important, since stop codons that leave more than half of the protein intact cause a null mutant phenotype, and it is possible that these regions interact with other, as yet unidentified, proteins (Ingram *et al.*, 1995; Lee *et al.*, 1997).

Despite three cell-cycle regulators being the founding members of the F-box family, the functions of F-box proteins and of Skp1p are not necessarily cell-cycle-specific. For example, the yeast F-box protein Grr1p was originally identified as a positive regulator of glucose-induced gene expression, and Grr1p acts by inactivating a transcription factor, Rgt1p, which in turn represses glucose-induced expression of *HXT* genes (Flick and Johnston, 1991; Özcan and Johnston, 1995). Subsequently, Grr1p has been linked genetically to the cell cycle, suggesting that Grr1p integrates nutrient uptake with cell division (Barral *et al.*, 1995). Conversely, Skp1p is required not only for cell-cycle progression, but also for glucose-induced *HXT* expression (Li and Johnston, 1997).

9.3.1.3 Other roles of flower-meristem-identity genes
All the genes discussed have roles in addition to their early function in the initiation of flowers. *LFY*, *AP1* and *AP2*, as well as *AGAMOUS (AG)*, are required for the maintenance of a determinate flower meristem, although this function is not always obvious in the single mutants. *LFY*'s role in maintaining flower-meristem identity is revealed when heterozygous *lfy* mutants are grown in short days, where new inflorescence shoots will initiate from the center from otherwise normal flowers (Okamuro *et al.*, 1996). A similar effect is seen in short-day grown *ag* mutants. The roles of *AP1* and *AP2* in preventing reversion of the flower to a shoot are uncovered in *ap1,ag* double mutants, which show a higher incidence of reversion than *ag* single mutants, and in *ap1,ag,ap2* triple mutants, in which 'flowers' often produce an indeterminate number of primordia, which either remain undifferentiated or develop into bract-like structures in whose axils lateral 'flowers' arise (Bowman *et al.*, 1993). Whether the roles of *LFY* and of *AP1/AP2/AG* in maintaining flower-meristem-identity are equivalent is difficult to judge, for *AG* is expressed normally in the center of reverting flower meristems of *lfy* heterozygotes, suggesting that the effects of *LFY* are not mediated by down-regulation of *AG* expression (Okamuro *et al.*, 1996). In addition to their function in establishing and maintaining meristem identity, *AP1*, *AP2* and *UFO* seem to have direct roles in controlling organ identity.

9.3.2 Regulation of flower-meristem-identity gene activity

9.3.2.1 Flower-meristem-identity genes and floral induction
Since flower-meristem-identity genes affect the initiation of flowers, it is obvious that their activity should be regulated by floral inductive cues. This question has

been addressed by two types of experiments. First, the RNA expression patterns of flower-meristem-identity genes have been compared with floral induction and floral determination (Figure 9.2). Second, the normal expression patterns have been manipulated using transgenic plants to determine whether altered expression levels or patterns affect flower initiation.

The expression of *UFO* and *AP2* is not restricted to the flower, indicating that their flower-specific activity results from interaction with other factors whose expression is more restricted. *AP2* RNA is expressed widely, including in leaves, and there is no evidence for its expression levels changing with plant age or floral inductive cues (Jofuku *et al.*, 1994; Putterill *et al.*, 1995). Although the expression pattern of *UFO* is more specific than that of *AP2*, *UFO* is expressed in a similar pattern in shoot meristems and in stage 2 flower meristems, which resemble shoot meristems in morphology. Consistent with such an expression pattern, ectopic expression of *UFO* does not affect flower initiation (Lee *et al.*, 1997).

In contrast to *UFO* and *AP2*, *AP1* expression is tightly linked to flower formation and is a marker for floral determination—its expression indicates that an *Arabidopsis* plant has been irrevocably determined to produce flowers (Hempel *et al.*, 1997). Conversely, forced overexpression of *AP1* in transgenic plants causes early flowering along with the conversion of lateral and main shoots into flowers (Mandel and Yanofsky, 1995b; Liljegren and Yanofsky, 1996).

A case that is intermediate between the quasi-constitutive expression of *UFO* and *AP2* and the flower-specific expression of *AP1* is observed for *LFY*, whose RNA levels increase from the vegetative to the reproductive phase (Blázquez *et al.*, 1997). Before the transition to flowering, *LFY* is expressed in the anlagen and primordia of leaves, which arise on the shoot apical meristem in positions equivalent to those where flowers arise after floral induction. *LFY* RNA levels in the leaf primordia increase continuously during the vegetative phase, which is particularly obvious when plants are grown in short days, where it takes about eight weeks before flowers are initiated. Since morphological studies suggest that floral inductive cues may act directly to control the fate of lateral primordia produced by the shoot apical meristem (Hempel and Feldman, 1994), it has been proposed that the *LFY* activity levels are critically involved in controlling the fate of these lateral primordia (Blázquez *et al.*, 1997). This conjecture has been tested by changing *LFY* expression levels, by increasing the number of wild-type copies or by overexpressing the gene from a constitutive viral promoter (*35S::LFY*). In either case, the number of total leaves produced before the first bractless flower appears is reduced when compared to wild-type, indicating that increased *LFY* expression levels lead to an earlier conversion of newly arising primordia into flowers. In addition, high-level overexpression of *LFY* in *35S::LFY* plants causes all lateral shoots to be replaced by solitary flowers (Weigel and Nilsson, 1995; Blázquez *et al.*, 1997; Nilsson and Weigel, 1997).

While the transgenic experiments demonstrate that the regulation of *LFY* and *AP1* transcript accumulation is critical for normal flower formation, post-tran-

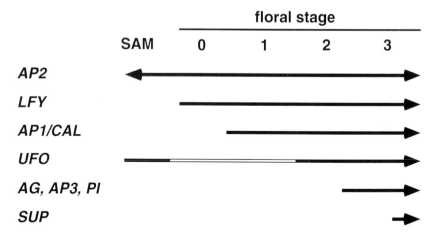

Figure 9.2 Temporal expression patterns of floral regulatory genes. Stage 0 represents the floral anlage, before a floral primordium becomes morphological distinct. Stage 3 is when the first organ primordia, those of sepals, are initiated, and when floral patterning is considered to have occurred. This stage has been called the floritypic stage in *Antirrhinum* (Carpenter *et al.*, 1995). SAM, shoot apical meristem.

scriptional regulation is important as well since the phenotypes of *35S::LFY* and *35S::AP1* plants are still affected by floral inductive conditions. The phenotypes of both *35S::LFY* and *35S::AP1* are attenuated in short days, although transcription rates are presumed to be the same as in long days (Mandel and Yanofsky, 1995b; Weigel and Nilsson, 1995; Liljegren and Yanofsky, 1996). Similarly, the phenotypes of *lfy* and *ap1* loss-of-function mutants are enhanced under conditions that delay floral induction, such as short days or lower growing temperatures (Huala and Sussex, 1992; Weigel *et al.*, 1992; Bowman *et al.*, 1993; Schultz and Haughn, 1993). These effects have been attributed to the competence of meristems to respond to *LFY* and *AP1* activities, although the level at which competence acts is unknown.

9.3.2.2 Genetic factors controlling flower-meristem-identity gene activity
Unfortunately, we are only at the brink of understanding the mechanisms that underlie the connection between flower-meristem-identity gene activity and the control of flowering time. In this respect, it is important to note that at least one of the flower-meristem-identity genes, *LFY*, can be said to have properties of a flowering-time gene as well. Increasing the number of *LFY* wild-type copies, thereby presumably increasing expression levels without changing the spatial expression pattern, leads to an earlier onset of flowering. Conversely, lowering

the number of *LFY* wild-type copies (in heterozygous mutant plants) leads to an increase in the number of total leaves produced before the first bractless flower appears, and thus a delay in flowering (Blázquez *et al.*, 1997).

Obvious candidates for direct upstream regulators of flower-meristem-identity genes are the flowering-time genes; several groups have started to explore the relationship between expression of flower-meristem-identity genes and flowering-time gene activity. The best understood case is that of *CO*, for which it has been shown that inducing its activity in short days, where *CO* normally is inactive, leads to rapid up-regulation of flower-meristem-identity genes *LFY* and *AP1*. Importantly, the *LFY* response is more rapid than the *AP1* response when compared to the expression profiles seen in plants that are induced to flower by shifting them from non-inductive short days to inductive long days, suggesting that *CO* acts preferentially upstream of *LFY* (Simon *et al.*, 1996). Since the *CO* expression domain overlaps that of *LFY*—during vegetative development, *CO* is expressed in the shoot apical meristem and leaf primordia—and since *CO* encodes a transcription factor, it is a very good candidate for a direct trans-activator of *LFY* expression. This argument is further strengthened by the observation that the initial up-regulation of *LFY* on transfer from short to long days is not restricted to newly arising primordia but extends into previously formed leaf primordia, where *CO* is also found (Simon *et al.*, 1996; Blázquez *et al.*, 1997).

One of the questions that arise from this possible interaction between *CO* and *LFY* is why *LFY* is excluded from the shoot apical meristem proper. This is important because ectopic expression of *LFY* in the shoot apical meristem leads to its conversion into a flower meristem (Weigel and Nilsson, 1995). Part of the solution to this problem comes from the finding that *CO* not only upregulates expression of *LFY* but also expression of a negative regulator of *LFY*, *TERMINAL FLOWER 1 (TFL1)*, whose loss of function causes a phenotype somewhat similar to that of *35S::LFY*. In *tfl1* mutants, *LFY* and *AP1* are ectopically expressed in shoot meristems, with the consequence that the main shoot terminates prematurely in a compound flower and the most apical lateral shoots are transformed into solitary flowers (Shannon and Meeks-Wagner, 1991; Alvarez *et al.*, 1992; Weigel *et al.*, 1992; Bowman *et al.*, 1993). As with *LFY*, *TFL1* expression is specific to the shoot apex, where it is expressed in a patch of cells below the shoot apical meristem. The *TFL1* expression domain does not directly overlap with that of *LFY*, indicating that *TFL1* acts non-autonomously (Bradley *et al.*, 1997). This is consistent with the predicted structure of the TFL1 protein, which shares similarity with a group of proteins termed phosphatidylethanolamine binding proteins, which have been implicated in signal transduction events in other organisms.

A further parallel between *TFL1* and *LFY* is that *TFL1* expression and function are not restricted to the reproductive phase. Similarly to *LFY*, *TFL1* is expressed at low levels during the vegetative phase and loss of *TFL1* function causes not only shoot-to-flower conversions but also earlier flowering, suggesting that *TFL1* acts also as a negative regulator of *LFY* (and other flower-meristem-identity genes) during the vegetative phase.

9.4 Floral patterning

Once individual flowers have been initiated, floral organs arise in four concentric rings termed whorls, with the fate of individual floral organs being controlled by a group of organ-identity or homeotic genes. Most of these genes are expressed in specific patterns, indicating that they respond to an underlying prepattern that is formed while the flower primordium is still undifferentiated. While many of the details of homeotic gene function still need to be filled in, a major problem was solved in 1991 when Bowman, Coen, Meyerowitz and Smyth proposed the so-called ABC model of homeotic gene action (Bowman *et al.*, 1991; Coen and Meyerowitz, 1991). Because it has been reviewed extensively before, the ABC model is only discussed briefly here, and then the possible connections between flower-meristem-identity and homeotic genes are explored.

9.4.1 Floral homeotic genes and the ABC model

Floral homeotic genes fall into three classes, A, B and C, each of which affects organs in two adjacent whorls. The original ABC model was based on four ABC genes from *Arabidopsis*, with three of these also represented by loss-of-function mutations in *Antirrhinum*. The four *Arabidopsis* genes are *AP2*, required for A function, *APETALA3 (AP3)* and *PISTILLATA (PI)*, required for B function, and *AG*, required for C function. Recently, several other genes that have overlapping or redundant roles in specifying A function have been described in *Arabidopsis*. These include *AP1*, which functions in both meristem and organ identity, *LEUNIG (LUG)* and possibly *AINTEGUMENTA (ANT)* (Bowman *et al*, 1993; Liu and Meyerowitz, 1995; Elliott *et al.*, 1996).

The ABC model has three components. First, it predicts that the particular combination of ABC gene activities in a given whorl specifies the type of organ in that whorl. Thus, in whorl one, A function alone causes sepals to develop, in whorl two, A plus B function causes petals to develop, in whorl three, B plus C function causes stamens to develop, and in whorl four, C function alone causes carpels to develop (Figures 9.1 and 9.3). The second prediction is that the A and C functions are mutually inhibitory, such that A function excludes C function from the outer two whorls, and vice versa. The third tenet of the ABC model predicts that without any homeotic function, leaves or leaf-like organs will develop (Bowman *et al*, 1991; Coen and Meyerowitz, 1991; Weigel and Meyerowitz, 1994).

The validity of the ABC model, which was based on the phenotype of single mutants, has been tested by examining the phenotypes of doubly and triply mutant strains, and by deliberately targeting homeotic gene function to other whorls. For example, if ectopic C function is responsible for the development of carpels and stamens in whorls one and two of A mutants, a C function mutation should affect these outer whorls in an A mutant background. This is indeed the case. While *ag* single mutants have defects only in the third and fourth whorls, where stamens and carpels are replaced by petals and sepals, *ag* mutations in an *ap2* background affect also the first and second whorls, where (ectopic) carpels are replaced by leaf-like structures and (ectopic) stamens are replaced by stamen/petal intermediates (Bowman *et al.*, 1991).

Genetics could not predict the level at which the spatial activity of homeotic genes is regulated, but studies of expression patterns showed that most of these genes are regulated at the transcriptional level. In wild-type flowers, RNA of the C function gene *AG* accumulates only in whorls three and four, while the overlap in RNA domains of B function genes *AP3* and *PI* is restricted to whorls two and three (Drews *et al.*, 1991; Jack *et al.*, 1992; Goto and Meyerowitz, 1994).

AG, *AP3* and *PI* are activated during stage 3 of flower development, around the time that the first organ primordia, those of sepals, start to emerge (Figure 9.3). As with *AP1*, all three genes encode transcription factors that bind DNA through a MADS domain (Yanofsky *et al.*, 1990; Jack *et al.*, 1992; Goto and Meyerowitz, 1994; Riechmann *et al.*, 1996; Riechmann and Meyerowitz, 1997). Another similarity is that misexpression of their RNAs in mutants or transgenic plants leads to homeotic transformations that confirm the predictions of the ABC model. For example, constitutive expression of *AG* in all four whorls causes carpels to arise in whorl one and stamens in whorl two, thus phenocopying an *ap2* mutant in which *AG* is similarly misexpressed (Drews *et al.*, 1991; Mizukami and Ma, 1992).

9.4.2 *Activation of floral homeotic genes*

While the ABC model deals with regulatory interactions among homeotic genes and with the effects of homeotic genes on (unknown) target genes, it does not say anything about the initial activation of homeotic genes. However, expression studies have confirmed that A function and C function genes mutually repress each other because *AG* RNA expression is derepressed in the periphery of *ap2* mutants and *AP1* RNA expression is derepressed in the center of *ag* mutants (Drews *et al.*, 1991; Gustafson-Brown *et al.*, 1994). Thus, the minimal requirement for a successful model describing floral patterning is that it explains the activation of B function genes in whorls two and three, and either the activation of A function genes in the periphery of the flower or the activation of C function genes in the center of the flower.

Since flower-meristem-identity genes control flower initiation, which is a prerequisite for the activation of homeotic genes such as *AG*, *AP3* and *PI*, they are likely candidates for upstream regulators of homeotic genes. This is at least formally true, as shown by analyzing homeotic gene expression in *lfy* mutants. Analysis of *lfy* mutations is, however, complicated by the fact that not all flowers

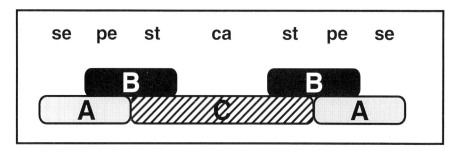

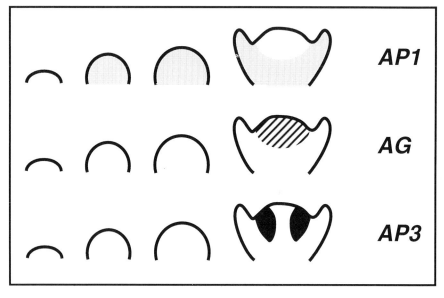

Figure 9.3 The ABC model of organ identity and the expression patterns of three representative ABC function genes, schematically drawn on cross-sections of stage 0 to stage 3 flowers.

are equally affected. Early arising flowers are converted into bracts with para-clades, but later arising flowers are replaced by structures that have only partial bract/paraclade character. *LFY* therefore appears to be an activator of all floral homeotic genes in early arising flowers, while the effects of *LFY* in later arising flowers are more limited. In strong *lfy* mutants, these abnormal flowers lack petals and stamens and consist mostly of sepals, carpels and bracts. The absence of petals and stamens has been traced back to a severe reduction in RNA levels of the B function genes *AP3* and *PI* (Weigel and Meyerowitz, 1993). In contrast to *AP3* and *PI*, the effects of *lfy* mutations on the expression of the C function gene *AG* are more complex. The appearance of *AG* RNA is delayed in *lfy* mutant flowers compared to wild-type, but there is also some ectopic expression, especially in the last flowers produced, which often consist mostly of carpel-like organs (Weigel *et al.*, 1992; Weigel and Meyerowitz, 1993). Thus, *LFY* has mostly a positive role in the regulation of B function genes but both positive and negative roles in the regulation of C function genes. These seemingly contradictory effects on C function genes could be explained through an additional role for *LFY* in regulating A function genes, which in turn repress C function genes. Double mutants indicate that both A and C function genes are at least partially active in an *lfy* mutant background (Huala and Sussex, 1992; Weigel *et al.*, 1992; Weigel and Meyerowitz, 1993).

While the phenotypic analyses and expression studies demonstrate that *LFY* is formally an upstream regulator of floral homeotic genes, it remains unknown how direct these interactions are. Since *LFY* is required for flower initiation, it is conceivable that the effects on homeotic gene expression are only an indirect consequence of earlier steps in flower development being disturbed. Furthermore, *LFY* RNA and protein are initially expressed uniformly in young flowers (Weigel *et al.*, 1992; Levin and Meyerowitz, 1995), yet downstream genes are activated in specific patterns. Thus, if flower-meristem-identity genes such as *LFY* were indeed directly involved in the regulation of homeotic gene expression, one would have to postulate that *LFY* is differentially active in different parts of the flower. That *LFY* activity is regulated would also follow from the observation that the flowers of *35S::LFY* plants are normally patterned, indicating that constitutive *LFY* expression does not cause indiscriminate activation of downstream homeotic genes throughout the flower (Weigel and Nilsson, 1995). Direct evidence for the activity of *LFY* being regulated has recently been obtained by examining the effects of another flower-meristem-identity gene, *UFO* (Lee *et al.*, 1997).

9.4.2.1 Regulation of B function homeotic genes

Although *ufo* and *lfy* mutant phenotypes are similar, *ufo* null mutations affect flower development less severely than *lfy* null mutations, and the *ufo* phenotype is completely masked by the stronger *lfy* phenotype in double mutants involving null alleles of both genes (Levin and Meyerowitz, 1995; Wilkinson and Haughn, 1995). The onset of *UFO* expression in flowers is later than that of *LFY* and its expression domain in young flowers is more limited than that of *LFY*, suggesting that *UFO* acts either downstream of, or together with, *LFY* (Ingram *et al*, 1995; Lee *et al.*, 1997).

The most prominent defect in *ufo* mutants is an apparent reduction in B function homeotic activity, as judged by partial conversion of petals into sepals and of stamens into carpels. An opposite phenotype is seen when *UFO* is expressed constitutively in transgenic plants (*35S::UFO*). In *35S::UFO* flowers the B function gene *AP3* is activated precociously and ectopically, with the consequence that supernumerary petals and stamens develop (Lee *et al.*, 1997). Importantly, all aspects of the *UFO* gain-of-function phenotype are suppressed in an *lfy* null background. This observation contrasts with the finding that a constitutively expressed *AP3* transgene can partially rescue the petal and stamen defects in *lfy* mutants (Jack *et al.*, 1994). The conclusion from these experiments has been that *UFO* is a partially dispensable co-regulator of *LFY*, since *LFY* can still activate *AP3* to a certain extent without *UFO* but *UFO* can activate *AP3* only in the presence of *LFY*. Thus, one way in which region-specific effects of a general regulator such as *LFY* could be achieved is through its interaction with co-regulators whose expression is more limited than that of the general regulator.

The question that arises from such an interpretation is how expression of co-regulators is regulated. The answer might come from the observation that *UFO* expression is not flower specific, but that it is expressed in a similar pattern in both shoot and stage-2-flower meristems (Lee *et al.*, 1997). Thus, it is possible that floral pattern results from the interaction of flower-specific, but not region-specific, factors (such as *LFY*) with region-specific, but not flower-specific, factors (such as *UFO*). In other words, to solve the problem of floral patterning, one might have to look to the general patterning of shoot meristems. Whether *UFO* itself has also a function, albeit redundant, in shoot meristems, or whether shoot expression of *UFO* is merely gratuitous, remains to be determined.

9.4.2.2 Regulation of C function genes

The traditional view of *AG* regulation is that flower-meristem-identity genes activate *AG* throughout the flower during stage 3, and that *AG* is specifically repressed in the outer whorls by A function homeotic genes such as *AP2* and *LUG*. An alternative model is that most A function genes are general repressors of *AG* during all stages of flower development, and that this repression is selectively overcome in the center of the flower during stage 3, through the interaction of general regulators with (unknown) region-specific regulators, in analogy with the *LFY/UFO* interaction.

There is at least some circumstantial evidence that regulation of C function genes involves general repressors as well as specific activators. First, both *AP2* and *LUG* have phenotypic effects throughout the flower, indicating that their activity is not restricted to the periphery of the flower. Second, by traditional criteria another *AG* repressor, *CURLY LEAF (CLF)*, would have been classified as an A function gene, since *clf* mutants display a floral phenotype similar to that of weak *ap2* mutants. In contrast to *AP2* and *LUG*, the more general function of *CLF* extends to its role in *AG* regulation and *CLF* is not required for repression of *AG* in flowers, but is required in vegetative tissue (Goodrich *et al.*, 1997). Since this was realized from the beginning, *CLF* was never termed an A function gene.

There are a few other observations that can only be explained if the ABC model is amended. For example, the ABC model postulates expansion of A function into the central whorls of C function mutants. Since *AP2* activity is not regulated at the transcriptional level, one cannot easily assess how its activity changes in an *ag* mutant. However, since *AP2* represses *AG* RNA expression in the outer whorls of wild-type, the ABC model would predict that *AP2* will repress *AG* RNA expression throughout the flower when its activity expands into whorls three and four of *ag* mutants. Thus, it was surprising when it was discovered that expression of (non-functional) *AG* RNA is normal in an apparent *ag* null mutant, i.e. *AG* RNA accumulates as in wild-type, in whorls three and four only (Gustafson-Brown *et al.*, 1994).

Another observation that has been largely ignored is that not only the spatial, but also the temporal aspects of *AG* RNA expression are changed in *ap2* and *lug* mutants. Besides being present in all whorls, *AG* RNA is also detected earlier in these mutants, during stage 2 instead of stage 3 (Drews *et al.*, 1991; Liu and Meyerowitz, 1995). In addition, there appears to be a general increase in levels of *AG* RNA in *ap2* and *lug* mutants, as observed by *in situ* hybridization or with a reporter gene (Liu and Meyerowitz, 1995; Sieburth and Meyerowitz, 1997).

These observations suggest that most A function genes are not simply repressors of *AG* in whorls one and two, but instead general factors that repress *AG* throughout the flower. During stage 3, this repression would be partially overcome and *AG* RNA accumulates specifically in whorls three and four, although still at lower levels than in the absence of A function genes. This could either be achieved through inactivation of *AG* repressors in the center or through selective activation of *AG* activators in the center of the flower. In this model, general factors (such as *LFY*) could only activate *AG* when either the repressors are absent or when their activity is enhanced by co-regulators in the center of the flower. While there is no evidence for the existence of region-specific co-regulators that interact with genes such as *LFY* in activating *AG*, several regulatory factors that are specifically expressed in the center of stage-2-flower meristems (as well as in shoot meristems) have been described, including *SHOOT MERISTEMLESS* and *CLAVATA1* (Long *et al.*, 1996; Clark *et al.*, 1997).

The revised model makes two predictions. First, *AG* RNA should be less abundant in the outer whorls of A function mutants than in the center. This is the case during early stages in both *ap2* and *lug* mutants. For example, in strong *ap2* mutants, ectopic RNA is initially lower in medial first-whorl organs, and absent from lateral first-whorl organs (Drews *et al.*, 1991). Similarly, *AG* RNA is observed in the center of *lug* flowers before it is detected in the outer whorls (Liu and Meyerowitz, 1995). Along with these correlative observations, there is direct genetic evidence for increased *AG* RNA levels in the center of *ap2* flowers. Apart from homeotic changes, the number of third-whorl organs is reduced in strong *ap2* mutants and this effect is reversed in an *ap2,ag* double mutant, indicating that *AP2* affects third-whorl organs through *AG* (Bowman *et al.*, 1991).

A second tenet of the revised model is that C function should override A function when both A and C functions are present. This prediction is compatible with the observation that constitutive expression of *AG* in transgenic plants leads to a close phenocopy of A function mutants (Mizukami and Ma, 1992). The main difference between *35S::AG* and strong *ap2* flowers is that all first-whorl organs are equally affected in *35S::AG* plants, confirming that ectopic expression of *AG* in these transgenic flowers is more uniform than in *ap2* flowers.

Despite the more general roles of most A function genes, *AP1* might be involved in specific repression of *AG*, especially in the second whorl (Bowman *et al.*, 1993; Weigel and Meyerowitz, 1993).

9.4.3 A preliminary model for floral patterning

The data discussed so far can be integrated in the following way to describe floral patterning:

1. Both B and C function homeotic genes are activated through the interaction of flower-specific, but not region-specific, factors with region-specific, but not flower-specific, factors.
2. The cardinal A function gene is *AP1*, whose patterned expression during stage 3 of flower development is a direct consequence of activation of the C function gene *AG*. That there is not an absolute requirement of *AP1* function for the formation of petals, which according to the ABC model requires the interaction of A and B function genes, is postulated to be owing to the redundant action of other, *AP1*-related, genes, such as *CAL* and possibly *AGL8*.
3. Most other A function genes, such as *AP2* and *LUG*, are not specific repressors of C function genes in the outer whorls, but general repressors of C function genes throughout the flower.
4. When both A and C functions are present, C function overrides A function, which explains also why constitutive expression of *AP1* in *35S::AP1* plants does not induce homeotic transformations in the central two whorls.

Of course, this is a preliminary and crude model that does not account for many of the details of homeotic gene expression patterns. For example, the *AP3* and *PI* expression domains are initially larger than prospective whorls two and three, although their overlap is confined to these two whorls. The *AP3* and *PI* expression domains become refined as floral development progresses through an auto-regulatory loop that requires the presence of both *AP3* and *PI*, and through the action of another factor, *SUPERMAN (SUP)*, whose expression itself is *AP3/PI*-dependent and which regulates cell proliferation in the center of the flower (Sakai *et al.*, 1995; Krizek and Meyerowitz, 1996). In relation to this, it is important to mention that regulation of cell proliferation as a patterning mechanism might be directly linked to the activity of flower-meristem-identity genes, as has been proposed for *UFO*, which shares an F-box with several cell-cycle regulators from animals and fungi (Samach *et al.*, 1997). In analogy with the dual functions of other F-box factors, it is conceivable that *UFO* might control floral fate by integrating activation of downstream target genes with cell proliferation.

9.5   Perspectives

Early flower development can be described in three consecutive phases: the global switch from vegetative to reproductive development, the initiation of individual flowers, and the elaboration of a specific organ pattern within the flower. The genes involved in these phases are flowering-time, flower-meristem-identity and organ-identity genes, and genetic studies suggest that the interaction between the different groups might be direct, at least in the genetic sense. The challenge for the years ahead is to determine whether genetic interactions translate into direct interactions at the molecular level or, if not, what the intermediary factors are.

Acknowledgments

Our work on flower development is supported by the National Science Foundation, the US Department of Agriculture, the Human Frontier Science Program Organization, the Samuel Roberts Noble Foundation, Epitope Inc. and ForBio Ltd. F.P. was supported by Bourse Lavoisier du Ministère Français des Affaires Etrangères and a Human Frontier Science Program Organization Fellowship, and I.L. by Parson and Aron Foundation fellowships.

References

Adamse, P., Jaspers, P.A.P.M., Bakker, J.A., Kendrick, R.E. and Koornneef, M. (1988) Photophysiology and phytochrome content of long-hypocotyl mutant and wild-type cucumber seedlings. *Plant Physiol.*, **87** 264-68.

Ahmad, M. and Cashmore, A.R. (1993) *HY4* gene of *A. thaliana* encodes a protein with characteristics of a blue-light photoreceptor. *Nature*, **366** 162-66.

Ahmad, M. and Cashmore, A.R. (1996) The *pef* mutants of *Arabidopsis thaliana* define lesions early in the phytochrome signaling pathway. *Plant J.*, **10** 1103-10.

Alvarez, J., Guli, C.L., Yu, X.-H. and Smyth, D.R. (1992) *terminal flower*: a gene affecting inflorescence development in *Arabidopsis thaliana*. *Plant J.*, **2** 103-16.

Aukerman, M.J. and Amasino, R.M. (1996) Molecular genetic analysis of flowering time in *Arabidopsis*. *Sem. Cell Dev. Biol.*, **7** 427-33.

Aukerman, M.J., Hirschfeld, M., Wester, L., Weaver, M., Clack, T., Amasino, R.M. and Sharrock, R.A. (1997) A deletion in the *PHYD* gene of the Arabidopsis Wassilewskija ecotype defines a role for phytochrome D in red/far-red light sensing. *Plant Cell*, **9** 1317-26.

Bagnall, D.J. (1993) Light quality and vernalization interact in controlling late flowering in *Arabidopsis* ecotypes and mutants. *Ann. Bot.*, **71** 75-83.

Bagnall, D.J., King, R.W., Whitelam, G.C., Boylan, M.T., Wagner, D. and Quail, P.H. (1995) Flowering responses to altered expression of phytochrome in mutants and transgenic lines of *Arabidopsis thaliana* (L.) Heynh. *Plant Physiol.*, **108** 1495-1503.

Bai, S. and Sung, Z.R. (1995) The role of *EMF1* in regulating the vegetative and reproductive transition in *Arabidopsis thaliana* (Brassicaceae). *Am. J. Bot.*, **82** 1095-1103.

Bai, C., Sen, P., Hofmann, K., Ma, L., Goebl, M., Harper, J.W. and Elledge, S.J. (1996) *SKP1* connects cell cycle regulators to the ubiquitin proteolysis machinery through a novel motif, the F-box. *Cell*, **86** 263-74.

Barral, Y., Jentsch, S. and Mann, C. (1995) G1 cyclin turnover and nutrient uptake are controlled by a common pathway in yeast. *Genes Dev.*, **9** 399-409.

Beall, F.D., Morgan, P.W., Mander, L.N., Miller, F.R. and Babb, K.H. (1991) Genetic regulation of development in *Sorghum bicolor*. V. The ma_3^R allele results in gibberellin enrichment. *Plant Physiol.*, **95** 116-25.

Bernier, G. (1988) The control of floral evocation and morphogenesis. *Ann. Rev. Plant Physiol. Plant Mol. Biol.*, **39** 175-219.

Blázquez, M., Soowal, L., Lee, I. and Weigel, D. (1997) *LEAFY* expression and flower initiation in *Arabidopsis*. *Development*, **124** 3835-44.

Bocchi, A., Lona, F. and Sachs, R.M. (1956) Photoperiodic induction of disbudded *Perilla* plants. *Plant Physiol.*, **31** 480-82.

Bowman, J.L., Smyth, D.R. and Meyerowitz, E.M. (1991) Genetic interactions among floral homeotic genes of *Arabidopsis*. *Development*, **112** 1-20.

Bowman, J.L., Alvarez, J., Weigel, D., Meyerowitz, E.M. and Smyth, D.R. (1993) Control of flower development in *Arabidopsis thaliana* by *APETALA1* and interacting genes. *Development*, **119** 721-43.

Bradley, D.J., Ratcliffe, O.J., Vincent, C., Carpenter, R. and Coen, E.S. (1997) Inflorescence commitment and architecture in *Arabidopsis*. *Science*, **275** 80-83.

Brown, J.A.M. and Klein, W.H. (1971) Photomorphogenesis in *Arabidopsis thaliana* (L.) Heynh. *Plant Physiol.*, **47** 393-99.

Carpenter, R., Copsey, L., Vincent, C., Doyle, S., Magrath, R. and Coen, E. (1995) Control of flower development and phyllotaxy by meristem identity genes in Antirrhinum. *Plant Cell*, **7** 2001-11.

Chailakhyan, M.K. (1936) On the hormonal theory of plant development. *Dolk. Acad. Sci. USSR*, **12** 443-47.

Chandler, J., Wilson, A. and Dean, C. (1996) *Arabidopsis* mutants showing an altered response to vernalization. *Plant J.*, **10** 637-44.

Chien, J.C. and Sussex, I.M. (1996) Differential regulation of trichome formation on the adaxial and abaxial leaf surfaces by gibberellins and photoperiod in *Arabidopsis thaliana* (L.) Heynh. *Plant Physiol.*, **111** 1321-28.

Childs, K.L., Lu, J.L., Mullet, J.E. and Morgan, P.W. (1995) Genetic regulation of development in *Sorghum bicolor*. X. Greatly attenuated photoperiod sensitivity in a phytochrome-deficient *Sorghum* possessing a biological clock but lacking a red light-high irradiance response. *Plant Physiol.*, **108** 345-51.

Chouard, P. (1960) Vernalization and its relations to dormancy. *Ann. Rev. Plant Physiol.*, **11** 191-238.

Clark, S.E., Williams, R.W. and Meyerowitz, E.M. (1997) The *CLAVATA1* gene encodes a putative receptor kinase that controls shoot and floral meristem size in Arabidopsis. *Cell*, **89** 575-85.

Clarke, J.H. and Dean, C. (1994) Mapping *FRI*, a locus controlling flowering time and vernalization response in *Arabidopsis thaliana*. *Mol. Gen. Genet.*, **242** 81-89.

Clarke, J.H., Mithen, R., Brown, J.K.M. and Dean, C. (1995) QTL analysis of flowering time in *Arabidopsis thaliana*. *Mol. Gen. Genet.*, **248** 278-86.

Coen, E.S. and Meyerowitz, E.M. (1991) The war of the whorls: genetic interactions controlling flower development. *Nature*, **353** 31-37.

Coupland, G. (1995) Genetic and environmental control of flowering time in *Arabidopsis*. *Trends Genet.*, **11** 393-97.

Devlin, P.F., Rood, S.B., Somers, D.E., Quail, P.H. and Whitelam, G.C. (1992) Photophysiology of the *elongated internode (ein)* mutant of *Brassica rapa*. *Plant Physiol.*, **100** 1442-47.

Doorenbos, J. and Wellensiek, S.J. (1959) Photoperiodic control of floral induction. *Ann. Rev. Plant Physiol.*, **10** 147-84.

Drews, G.N., Bowman, J.L. and Meyerowitz, E.M. (1991) Negative regulation of the Arabidopsis homeotic gene *AGAMOUS* by the *APETALA2* product. *Cell,* **65** 991-1002.

Elliott, R.C., Betzner, A.S., Huttner, E., Oakes, M.P., Tucker, W.Q.J., Gerentes, D., Perez, P. and Smyth, D.R. (1996) *AINTEGUMENTA,* an *APETALA2*-like gene of Arabidopsis with pleiotropic roles in ovule development and floral organ growth. *Plant Cell,* **8** 155-68.

Eskins, K. (1992) Light-quality effects on *Arabidopsis* development. Red, blue and far-red regulation of flowering and morphology. *Physiol. Plant.,* **86** 439-44.

Flick, J.S. and Johnston, M. (1991) *GRR1* of *Saccharomyces cerevisiae* is required for glucose repression and encodes a protein with leucine-rich repeats. *Mol. Cell. Biol.,* **11** 5101-12.

Garner, W.W. and Allard, H.A. (1920) Effect of the relative length of day and night and other factors of the environment on growth and reproduction in plants. *J. Agric. Res.,* **18** 553-606.

Gaßner, G. (1918) Beiträge zur physiologischen Charakteristik sommer- und winterannueller Gewächse, insbesondere der Getreidepflanzen. *Z. Bot.,* **10** 417-80.

Goodrich, J., Puangsomlee, P., Martin, M., Long, D., Meyerowitz, E.M. and Coupland, G. (1997) A Polycomb-group gene regulates homeotic gene expression in *Arabidopsis. Nature,* **386** 44-48.

Goto, K. and Meyerowitz, E.M. (1994) Function and regulation of the *Arabidopsis* floral homeotic gene *PISTILLATA. Genes Dev.,* **8** 1548-60.

Goto, K., Kumagai, T. and Koornneef, M. (1991a) Flowering responses to light breaks in photomorphogenic mutants of *Arabidopsis thaliana,* a long-day plant. *Physiol. Plant.,* **83** 209-15.

Goto, N., Katoh, N. and Kranz, A.R. (1991b) Morphogenesis of floral organs in *Arabidopsis*: predominant carpel formation of the pin-formed mutant. *Jpn. J. Genet.,* **66** 551-67.

Gustafson-Brown, C. (1996) Characterization of the *Arabidopsis* floral homeotic gene *APETALA1,* Ph.D. thesis, University of California, San Diego, La Jolla.

Gustafson-Brown, C., Savidge, B. and Yanofsky, M.F. (1994) Regulation of the Arabidopsis floral homeotic gene *APETALA1. Cell,* **76** 131-43.

Hempel, F.D. and Feldman, L.J. (1994) Bi-directional inflorescence development in *Arabidopsis thaliana*: acropetal initiation of flowers and basipetal initiation of paraclades. *Planta,* **192** 276-86.

Hempel, F.D., Weigel, D., Mandel, M.A., Ditta, G., Zambryski, P., Feldman, L.J. and Yanofsky, M.F. (1997) Floral determination and expression of floral regulatory genes in *Arabidopsis. Development,* **124** 3845-53.

Hicks, K.A., Millar, A.J., Carre, I.A., Somers, D.E., Straume, M., Meeks-Wagner, D.R. and Kay, S.A. (1996) Conditional circadian dysfunction of the *Arabidopsis early-flowering 3* mutant. *Science,* **274** 790-92.

Huala, E. and Sussex, I.M. (1992) *LEAFY* interacts with floral homeotic genes to regulate Arabidopsis floral development. *Plant Cell,* **4** 901-13.

Hussein, H.A.S. (1968) Genetic analysis of mutagen-induced flowering time variation in *Arabidopsis thaliana* (L.) Heynh. Ph.D. thesis, Agricultural University, Wageningen, The Netherlands.

Ingram, G.C., Goodrich, J., Wilkinson, M.D., Simon, R., Haughn, G.W. and Coen, E.S. (1995) Parallels between *UNUSUAL FLORAL ORGANS* and *FIMBRIATA,* genes controlling flower development in Arabidopsis and Antirrhinum. *Plant Cell,* **7** 1501-10.

Irish, V.F. and Sussex, I.M. (1990) Function of the *apetala-1* gene during *Arabidopsis* floral development. *Plant Cell,* **2** 741-51.

Ishioka, N., Tanimoto, S. and Harada, H. (1991) Flower-inducing activity of phloem exudates from *Pharbitis* cotyyledons exposed to various photoperiods. *Plant Cell Physiol.,* **32** 921-24.

Jack, T., Brockman, L.L. and Meyerowitz, E.M. (1992) The homeotic gene APETALA3 of *Arabidopsis thaliana* encodes a MADS-box and is expressed in petals and stamens. *Cell,* **68** 683-97.

Jack, T., Fox, G.L. and Meyerowitz, E.M. (1994) *Arabidopsis* homeotic gene APETALA3 ectopic expression: transcriptional and post-transcriptional regulation determine organ identity. *Cell,* **76** 703-16.

Jacobsen, S.E. and Olszewski, N.E. (1993) Mutations at the *SPINDLY* locus of Arabidopsis alter gibberellin signal transduction. *Plant Cell,* **5** 887-96.

Jofuku, K.D., den Boer, B.G.W., Van Montagu, M. and Okamuro, J.K. (1994) Control of Arabidopsis flower and seed development by the homeotic gene *APETALA2*. *Plant Cell*, **6** 1211-25.

Johnson, E., Bradley, M., Harberd, N.P. and Whitelam, G.C. (1994) Photoresponses of light-grown *phyA* mutants of *Arabidopsis*. *Plant Physiol.*, **105** 141-49.

Karlovska, V. (1974) Genotypic control of the speed of development in *Arabidopsis thaliana* (L.) Heynh. Lines obtained from natural populations. *Biol. Plant.*, **16** 107-17.

Kempin, S.A., Savidge, B. and Yanofsky, M.F. (1995) Molecular basis of the *cauliflower* phenotype in *Arabidopsis*. *Science*, **267** 522-25.

King, R.W. and Zeevaart, J.A.D. (1973) Floral stimulus movement in *Perilla* and flower inhibition caused by noninduced leaves. *Plant Physiol.*, **51** 727-38.

King, R.W. and Bagnall, D. (1996) Photoreceptors and the photoperiodic response controlling flowering of *Arabidopsis*. *Sem. Cell Dev. Biol.*, **7** 449-44.

Klucher, K.M., Chow, H., Reiser, L. and Fischer, R.L. (1996) The *AINTEGUMENTA* gene of *Arabidopsis* required for ovule and female gametophyte development is related to the floral homeotic gene *AP2*. *Plant Cell*, **8** 137-53.

Koornneef, M., Rolff, E. and Spruit, C.J.P. (1980) Genetic control of light-inhibited hypocotyl elongation in *Arabidopsis thaliana* (L.) Heynh. *Z. Pflanzenphysiol.*, **100** 147-60.

Koornneef, M., Hanhart, C.J. and van der Veen, J.H. (1991) A genetic and physiological analysis of late flowering mutants in *Arabidopsis thaliana*. *Mol. Gen. Genet.*, **229** 57-66.

Koornneef, M., Blankestijn-de Vries, H., Hanhart, C., Soppe, W. and Peeters, T. (1994) The phenotype of some late-flowering mutants is enhanced by a locus on chromosome 5 that is not effective in the Landsberg *erecta* phenotype. *Plant J.*, **6** 911-19.

Kowalski, S.P., Lan, T.H., Feldmann, K.A. and Paterson, A.H. (1994) QTL mapping of naturally occurring variation in flowering time of *Arabidopsis thaliana*. *Mol. Gen. Genet.*, **245** 548-55.

Kozaki, A., Takeba, G. and Tanaka, O. (1991) A polypeptide that induces flowering in *Lemna paucicostata* at a very low concentration. *Plant Physiol.*, **95** 1288-90.

Krizek, B.A. and Meyerowitz, E.M. (1996) The *Arabidopsis* homeotic genes *APETALA3* and *PISTILLATA* are sufficient to provide the B class organ identity function. *Development*, **122** 11-22.

Lang, A. (1965) Physiology of flower initiation, in *Encyclopedia of Plant Physiology* (ed. W. Ruhland), Springer, Berlin, pp. 1371-1536.

Lang, A., Chailakhyan, M.K. and Frolova, I.A. (1977) Promotion and inhibition of flower formation in a dayneutral plant in grafts with a short-day and a long-day plant. *Proc. Natl Acad. Sci. USA*, **74** 2412-16.

Langridge, J. (1957) Effect of day-length and gibberellic acid on the flowering of *Arabidopsis*. *Nature*, **180** 36-37.

Lee, I. and Amasino, R.M. (1995) Effect of vernalization, photoperiod, and light quality on the flowering phenotype of *Arabidopsis* plants containing the *FRIGIDA* gene. *Plant Physiol.*, **108** 157-62.

Lee, I., Bleecker, A. and Amasino, R. (1993) Analysis of naturally occurring late flowering in *Arabidopsis thaliana*. *Mol. Gen. Genet.*, **237** 171-76.

Lee, I., Aukerman, M.J., Gore, S.L., Lohman, K.N., Michaels, S.D., Weaver, L.M., John, M.C., Feldmann, K.A. and Amasino, R.M. (1994a) Isolation of *LUMINIDEPENDENS*: a gene involved in the control of flowering time in Arabidopsis. *Plant Cell*, **6** 75-83.

Lee, I., Michaels, S.D., Masshardt, A.S. and Amasino, R.M. (1994b) The late-flowering phenotype of *FRIGIDA* and mutations in *LUMINIDEPENDENS* is suppressed in the Landsberg *erecta* strain of *Arabidopsis*. *Plant J.*, **6** 903-909.

Lee, I., Wolfe, D.S., Nilsson, O. and Weigel, D. (1997) A *LEAFY* co-regulator encoded by *UNUSUAL FLORAL ORGANS*. *Curr. Biol.*, **7** 95-104.

Levin, J.Z. and Meyerowitz, E.M. (1995) *UFO*: an Arabidopsis gene involved in both floral meristem and floral organ development. *Plant Cell*, **7** 529-48.

Li, F.N. and Johnston, M. (1997) Grr1 of *Saccharomyces cerevisiae* is connected to the ubiquitin proteolysis machinery through Skp1: coupling glucose sensing to gene expression and the cell cycle. *EMBO J.*, **16** 5629-38.

Liljegren, S.J. and Yanofsky, M.F. (1996) Genetic control of shoot and flower meristem behavior. *Curr. Opin. Cell Biol.*, **8** 865-69.

Liu, Z. and Meyerowitz, E.M. (1995) *LEUNIG* regulates *AGAMOUS* expression in *Arabidopsis* flowers. *Development*, **121** 975-91.

Long, J.A., Moan, E.I., Medford, J.I. and Barton, M.K. (1996) A member of the KNOTTED class of homeodomain proteins encoded by the *STM* gene *Arabidopsis*. *Nature*, **379** 66-69.

Ma, H. (1994) The unfolding drama of flower development: recent results from genetic and molecular analyses. *Genes Dev.*, **8** 745-56.

Macknight, R., Bancroft, I., Lister, C., Page, T., Love, K., Schmidt, R., Westphal, L., Murphy, G., Sherson, S., Cobbett, C. and Dean, C. (1997) *FCA*, a gene controlling flowering time in Arabidopsis, encodes a protein containing RNA-binding domains. *Cell*, **89** 737-45.

Mandel, M.A. and Yanofsky, M.F. (1995a) The Arabidopsis AGL8 MADS box gene is expressed in inflorescence meristems and is negatively regulated by *APETALA1*. *Plant Cell*, **7** 1763-71.

Mandel, M.A. and Yanofsky, M.F. (1995b) A gene triggering flower development in *Arabidopsis*. *Nature*, **377** 522-24.

Mandel, M.A., Gustafson-Brown, C., Savidge, B. and Yanofsky, M.F. (1992) Molecular characterization of the *Arabidopsis* floral homeotic gene *APETALA1*. *Nature*, **360** 273-77.

Martínez-Zapater, J.M. and Somerville, C.R. (1990) Effect of light quality and vernalization on late-flowering mutants of *Arabidopsis thaliana*. *Plant Physiol.*, **92** 770-76.

Martínez-Zapater, J.M., Coupland, G., Dean, C. and Koornneef, M. (1994) The transition to flowering in *Arabidopsis*, in *Arabidopsis* (eds E.M. Meyerowitz and C.R. Somerville), Cold Spring Harbor Laboratory, Cold Spring Harbor, New York, pp. 403-33.

McKelvie, A.D. (1962) A list of mutant genes in *Arabidopsis thaliana* (L.) Heynh. *Radiat. Bot.*, **1** 233-41.

McNellis, T.W., von Arnim, A.G., Araki, T., Komeda, Y., Miséra, S. and Deng, X.-W. (1994) Genetic and molecular analysis of an allelic series of *cop*1 mutants suggests functional roles for the multiple protein domains. *Plant Cell*, **6** 487-500.

Millar, A.J., Straume, M., Chory, J., Chua, N.H. and Kay, S.A. (1995) The regulation of circadian period by phototransduction pathways in *Arabidopsis*. *Science*, **267** 1163-66.

Mizukami, Y. and Ma, H. (1992) Ectopic expression of the floral homeotic gene *AGAMOUS* in transgenic Arabidopsis plants alters floral organ identity. *Cell*, **71** 119-31.

Napp-Zinn, K. (1969) *Arabidopsis thaliana* (L.) Heynh, in *The Induction of Flowering: Some Case Histories* (ed. L.T. Evans), MacMillan, Melbourne, pp. 291-304.

Napp-Zinn, K. (1985) *Arabidopsis thaliana*, in *CRC Handbook of Flowering* (ed. H.A. Halevy), CRC Press, Boca Raton, pp. 492-503.

Nilsson, O. and Weigel, D. (1997) Modulating the timing of flowering. *Curr. Opin. Biotechnol.*, **8** 195-199.

Ohme-Takagi, M. and Shinshi, H. (1995) Ethylene-inducible DNA binding proteins that interact with an ethylene-responsive element. *Plant Cell*, **7** 173-82.

Okada, K. and Shimura, Y. (1995) Genetic analyses of signalling in flower development using *Arabidopsis*. *Plant Mol. Biol.*, **26** 1357-77.

Okamuro, J.K., den Boer, B.G.W. and Jofuku, K.D. (1993) Regulation of Arabidopsis flower development. *Plant Cell*, **5** 1183-93.

Okamuro, J.K., den Boer, B.G.W., Lotys-Prass, C., Szeto, W. and Jofuku, K.D. (1996) Flowers into shoots: photo and hormonal control of a meristem identity switch in *Arabidopsis*. *Proc. Natl Acad. Sci. USA*, **93** 13831-36.

Okamuro, J.K., Szeto, W., Lotys-Prass, C. and Jofuku, K.D. (1997a) Photo and hormonal control of meristem identity in the Arabidopsis flower mutants *apetala2* and *apetala1*. *Plant Cell*, **9** 37-47.

Okamuro, J.K., Caster, B., Villarroel, R., Van Montagu, M. and Jofuku, K.D. (1997b) The AP2 domain of APETALA2 defines a large new family of DNA binding proteins in *Arabidopsis*. *Proc. Natl Acad. Sci. USA*, **94** 7076-81.

Özcan, S. and Johnston, M. (1995) Three different regulatory mechanisms enable yeast hexose transporter (*HXT*) genes to be induced by different levels of glucose. *Mol. Cell. Biol.*, **15** 1564-72.

Paton, D.M. and Barber, H.N. (1955) Physiological genetics of *Pisum*. *Aust. J. Biol. Sci.*, **8** 231-40.

Peeters, A.J.M. and Koornneef, M. (1996) Genetic variation of flowering time in *Arabidopsis thaliana*. *Sem. Cell Dev. Biol.*, **7** 381-89.

Pepper, A.E. and Chory, J. (1997) Extragenic suppressors of the *Arabidopsis det1* mutant identify elements of flowering-time and light-response regulatory pathways. *Genetics*, **145** 1125-37.

Pharis, R.P. and King, R.W. (1985) Gibberellins and reproductive development in seed plants. *Ann. Rev. Plant Physiol. Plant Mol. Biol.*, **36** 517-68.

Purugganan, M.D., Rounsley, S.D., Schmidt, R.J. and Yanofsky, M.F. (1995) Molecular evolution of flower development: diversification of the plant MADS-box regulatory gene family. *Genetics*, **140** 345-56.

Putterill, J., Robson, F., Lee, K., Simon, R. and Coupland, G. (1995) The *CONSTANS* gene of Arabidopsis promotes flowering and encodes a protein showing similarities to zinc finger transcription factors. *Cell*, **80** 847-57.

Ray, A., Lang, J.D., Golden, T. and Ray, S. (1996) *SHORT INTEGUMENT (SIN1)*, a gene required for ovule development in *Arabidopsis*, also controls flowering time. *Development*, **122** 2631-38.

Redeí, G.P. (1962) Supervital mutants of *Arabidopsis*. *Genetics*, **47** 443-60.

Reed, J.W., Nagatani, A., Elich, T.D., Fagan, M. and Chory, J. (1994) Phytochrome A and phytochrome B have overlapping but distinct functions in *Arabidopsis* development. *Plant Physiol.*, **104** 1139-49.

Reed, J.W., Foster, K.R., Morgan, P.W. and Chory, J. (1996) Phytochrome B affects responsiveness to gibberellins in *Arabidopsis*. *Plant Physiol.*, **112** 337-42.

Riechmann, J.L. and Meyerowitz, E.M. (1997) Determination of floral organ identity by *Arabidopsis* MADS domain homeotic proteins *AP1*, *AP3*, *PI*, and *AG* is independent of their DNA-binding specificity. *Mol. Biol. Cell*, **8** 1243-59.

Riechmann, J.L., Krizek, B.A. and Meyerowitz, E.M. (1996) Dimerization specificity of *Arabidopsis* MADS domain homeotic proteins APETALA1, APETALA3, PISTILLATA, and AGAMOUS. *Proc. Natl Acad. Sci. USA*, **93** 4793-98.

Sakai, H., Medrano, L.J. and Meyerowitz, E.M. (1995) Role of *SUPERMAN* in maintaining *Arabidopsis* floral whorl boundaries. *Nature*, **378** 199-203.

Samach, A., Kohalmi, S., Haughn, G. and Crosby, W. (1997) UFO encounters of a floral kind. *Plant Physiol.*, **114** (suppl.), 60.

Sanda, S., John, M. and Amasino, R. (1997) Analysis of flowering time in ecotypes of *Arabidopsis thaliana*. *J. Hered.*, **88** 69-72.

Schultz, E.A. and Haughn, G.W. (1991) *LEAFY*, a homeotic gene that regulates inflorescence development in *Arabidopsis*. *Plant Cell*, **3** 771-81.

Schultz, E.A. and Haughn, G.W. (1993) Genetic analysis of the floral initiation process (FLIP) in *Arabidopsis*. *Development*, **119** 745-65.

Shannon, S. and Meeks-Wagner, D.R. (1991) A mutation in the Arabidopsis *TFL1* gene affects inflorescence meristem development. *Plant Cell*, **3** 877-92.

Shannon, S. and Meeks-Wagner, D.R. (1993) Genetic interactions that regulate inflorescence development in Arabidopsis. *Plant Cell*, **5** 639-55.

Sharrock, R.A. and Quail, P.H. (1989) Novel phytochrome sequences in *Arabidopsis thaliana*: structure, evolution, and differential expression of a plant regulatory photoreceptor family. *Genes Dev.*, **3** 1745-57.

Sieburth, L.E. and Meyerowitz, E.M. (1997) Molecular dissection of the *AGAMOUS* control region shows that *cis* elements for spatial regulation are located intragenically. *Plant Cell*, **9** 355-65.

Simon, R. and Coupland, G. (1996) *Arabidopsis* genes that regulate flowering time in response to day-length. *Sem. Cell Dev. Biol.*, **7** 419-25.

Simon, R., Igeño, M.I. and Coupland, G. (1996) Activation of floral meristem identity genes in *Arabidopsis*. *Nature*, **382** 59-62.

Smyth, D.R., Bowman, J.L. and Meyerowitz, E.M. (1990) Early flower development in *Arabidopsis*. *Plant Cell*, **2** 755-67.

Sun, T.P. and Kamiya, Y. (1994) The Arabidopsis *GA1* locus encodes the cyclase *ent*-kaurene synthetase A of gibberellin biosynthesis. *Plant Cell*, **6** 1509-18.

Sung, Z.R., Belachew, A., Shunong, B. and Bertrand-Garcia, R. (1992) *EMF*, an *Arabidopsis* gene required for vegetative shoot development. *Science*, **258** 1645-47.

Takimoto, A. and Kaihara, S. (1990) Production of the water-extractable flower-inducing substance(s) in *Lemna*. *Plant Cell Physiol.*, **31** 887-91.

Telfer, A., Bollman, K.M. and Poethig, R.S. (1997) Phase change and the regulation of trichome distribution in *Arabidopsis thaliana*. *Development*, **124** 637-44.

Theißen, G. and Saedler, H. (1995) MADS-box genes in plant ontogeny and phylogeny: Haeckel's 'biogenetic law' revisited. *Curr. Opin. Genet. Dev.*, **5** 628-39.

Tran Thanh Van, K., Toubart, P., Cousson, A., Darvill, A.G., Gollin, D.J., Chelf, P. and Albersheim, P. (1985) Manipulation of the morphogenetic pathways of tobacco explants by oligosaccharins. *Nature*, **314** 615-17.

Vetrilova, M. (1973) Genetic and physiological analysis of induced late mutants of *Arabidopsis thaliana* (L.) Heynh. *Biol. Plant.*, **15** 391-97.

Weigel, D. (1995) The genetics of flower development: from floral induction to ovule morphogenesis. *Ann. Rev. Genet.*, **29** 19-39.

Weigel, D. and Meyerowitz, E.M. (1993) Activation of floral homeotic genes in *Arabidopsis*. *Science*, **261** 1723-26.

Weigel, D. and Meyerowitz, E.M. (1994) The ABCs of floral homeotic genes. *Cell*, **78** 203-209.

Weigel, D. and Nilsson, O. (1995) A developmental switch sufficient for flower initiation in diverse plants. *Nature*, **377** 495-500.

Weigel, D., Alvarez, J., Smyth, D.R., Yanofsky, M.F. and Meyerowitz, E.M. (1992) *LEAFY* controls floral meristem identity in *Arabidopsis*. *Cell*, **69** 843-59.

Weller, J.L., Murfet, I.C. and Reid, J.B. (1997) Pea mutants with reduced sensitivity to far-red light define an important role for phytochrome A in day length detection. *Plant Physiol.*, **114** 1225-36.

Wilkinson, M.D. and Haughn, G.W. (1995) *UNUSUAL FLORAL ORGANS* controls meristem identity and organ primordia fate in Arabidopsis. *Plant Cell*, **7** 1485-99.

Wilson, A. and Dean, C. (1996) Analysis of the molecular basis of vernalization in *Arabidopsis thaliana*. *Sem. Cell Dev. Biol.*, **7** 435-40.

Wilson, R.N., Heckman, J.W. and Somerville, C.R. (1992) Gibberellin is required for flowering in *Arabidopsis thaliana* under short days. *Plant Physiol.*, **100** 403-408.

Yang, C.-H., Chen, L.-J. and Sung, Z.R. (1995) Genetic regulation of shoot development in *Arabidopsis*: role of the *EMF* genes. *Dev. Biol.*, **169** 421-35.

Yanofsky, M. (1995) Floral meristems to floral organs: genes controlling early events in *Arabidopsis* flower development. *Ann. Rev. Plant Physiol. Plant Mol. Biol.*, **46** 167-88.

Yanofsky, M.F., Ma, H., Bowman, J.L., Drews, G.N., Feldmann, K.A. and Meyerowitz, E.M. (1990) The protein encoded by the *Arabidopsis* homeotic gene *agamous* resembles transcription factors. *Nature*, **346** 35-39.

Zagotta, M.T., Hicks, K.A., Jacobs, C.I., Young, J.C., Hangarter, R.P. and Meeks-Wagner, D.R. (1996) The *Arabidopsis ELF3* gene regulates vegetative photomorphogenesis and the photoperiodic induction of flowering. *Plant J.*, **10** 691-702.

Zagotta, M.T., Shannon, S., Jacobs, C. and Meeks-Wagner, D.R. (1992) Early-flowering mutants of *Arabidopsis thaliana*. *Aust. J. Plant Physiol.*, **19** 411-18.

Zeevaart, J.A.D. (1958) Flower formation as studied by grafting. *Meded. Langbouwhogesch. Wageningen*, **58** 1-88.

Zeevaart, J.A.D. (1976) Physiology of flower formation. *Ann. Rev. Plant Physiol. Plant Mol. Biol.*, **27** 321-48.

Zeevaart, J.A.D. and Talón, M. (1992) Gibberellin mutants in *Arabidopsis thaliana*, in *Progress in Plant Growth Regulation* (eds C.M. Karssen, L.C. van Loon and D. Vreugdenhil), Kluwer, Dordrecht, pp. 34-42.

10   Light regulation and biological clocks

Garry C. Whitelam and Andrew J. Millar

10.1   Introduction

All living organisms have the ability to acquire information about their surround-
ings, via the perception of environmental signals, and they use this information
to modify their behaviour or development. As sessile organisms, plants cannot
choose their environment and so need to be especially plastic in their development in order to optimise growth in response to environmental changes. Plants
have evolved an array of exquisite sensory systems for monitoring their environment and initiating appropriate developmental responses. As photoautotrophs,
depending upon photosynthesis for their survival, plants are especially sensitive
to variations in the light environment. Plants monitor a range of light signals,
such as intensity, quality and direction and use this information to modulate
developmental responses such as seed germination, de-etiolation and seedling
establishment, the control of plant architecture and the onset of flowering.

Light signals, interacting with an endogenous circadian oscillator, also
provide plants with a means to monitor the length of the day (photoperiod).
Since evolution began, the Earth's rotation has imposed regular cycles of light
and temperature on the environment. This regularity allows plants and other
organisms to anticipate the light/dark cycle as well as simply to respond to
unpredictable changes in the light environment. Daily anticipation requires a
daily timing mechanism. Circadian rhythms (from *circa dies*, about a day) are
the most widespread adaptation to the demands of biological timing. These
ubiquitous 'biological clocks' have supported a boom in research on various
organisms (see Carré, 1996; Dunlap, 1996). Reviews on plants have concentrated
on molecular and genetic data, for which *Arabidopsis* is a central model
(McClung and Kay, 1994; Anderson and Kay, 1996; Beator and Kloppstech,
1996; Millar and Kay, 1997).

The various developmental responses of plants to signals from the light
environment are collectively referred to as photomorphogenesis and are dependent on the action of specialised signal-transducing photoreceptors.

10.2   Photoreceptor properties

The photomorphogenic responses of higher plants involve the action of at least
three distinct classes of signal-transducing photoreceptors: the phytochromes,

which absorb mainly in the red and far-red regions of the spectrum, a series of blue/UVA light-absorbing photoreceptors, and UVB absorbing photoreceptors. The phytochromes are the most extensively characterised of these photoreceptors. Some of the blue/UVA photoreceptors have been identified and characterised at the molecular level. The identity of any of the UVB photoreceptors remains elusive.

10.2.1 The phytochromes

The phytochromes are reversibly photochromic biliproteins that absorb maximally in the red (R) and far-red (FR) regions of the spectrum. Higher plants contain multiple discrete phytochrome species that make up a family of closely related photoreceptors, the apoproteins of which are encoded by a small family of divergent genes (see Quail, 1994). All of the higher plant phytochromes appear to share the same basic structure, consisting of a dimer of identical ~124 kDa polypeptides. Each monomer carries a single covalently linked linear tetrapyrrole chromophore (phytochromobilin), attached via a thioether bond to a conserved cysteine residue in the N-terminal globular domain of the protein (Furuya and Song, 1994). The more elongated, non-chromophorylated, C-terminal domain of the protein is involved in dimerization (Edgerton and Jones, 1992). Phytochromes can exist in either of two relatively stable isoforms: an R-light-absorbing form, Pr, with an absorption maximum at about 660 nm, or an FR-light-absorbing form, Pfr, with an absorption maximum at about 730 nm. The Pfr form of phytochrome is generally considered to be biologically active and Pr is considered to be inactive. The absorption spectra of Pr and Pfr show considerable overlap throughout the visible light spectrum, and so under almost all irradiation conditions phytochromes are present in an equilibrium mixture of the two forms.

Light-induced interconversions between Pr and Pfr involve a $Z-E$ isomerisation about the C15 double bond that links the C and D tetrapyrrole rings (see Terry et al., 1993). These isomerisations are accompanied, and stabilised, by reversible conformational changes throughout the protein moiety (Quail, 1991). It is reasonably assumed that these protein conformational changes account for the difference in activity between the Pr and Pfr forms of phytochrome.

The size of the phytochrome family varies among different plant species. In Arabidopsis thaliana, five apophytochrome encoding genes (PHYA to PHYE) have been characterised (Sharrock and Quail, 1989; Clack et al., 1994). The PHYA gene has been demonstrated to encode the apoprotein of a well-characterised light-labile phytochrome that predominates in etiolated seedlings. This phytochrome, previously referred to as type I phytochrome, is now called phytochrome A (phyA). Phytochrome A is rapidly depleted when etiolated tissues are exposed to light as a consequence of Pfr-induced proteolysis (Quail, 1994). Light exposure also induces a reduction in PHYA transcript abundance (Quail, 1994 and below). The other *PHY (B–E)* genes encode the apoproteins of lower abundance, more light-stable phytochromes, previously referred to as type II phytochromes (see Kendrick and Kronenberg, 1994). Counterparts of *PHYA*, *PHYB* and other *PHY* genes have been isolated from several other plant species (Quail, 1994).

The spatial and temporal expression patterns of the *Arabidopsis PHYA* and *PHYB* genes have been studied using reporter gene methods. Fusions between *PHY* promoter regions and the *Escherichia coli uidA (GUS)* gene have been introduced into *Arabidopsis*. Staining for GUS indicated significant activity of both the *PHYA* and *PHYB* promoters in most cells throughout seedling development (Somers and Quail, 1995). The promoters displayed similar spatial patterns of activity, although in dry seeds and pollen grains the *PHYB* promoter drove significantly greater expression than the *PHYA* promoter. In seedlings, the highest activities of both promoters were observed in the apical hook region and in roots. In mature plants promoter activity was highest in the aerial parts, and in leaves GUS expression was limited to chloroplast-containing cells. In shoots, the activity of both promoters was found to be strongly down-regulated by light, with the *PHYA* promoter showing a ten-fold down-regulation on transfer of etiolated seedlings to light and the *PHYB* promoter showing a two-fold down-regulation (Somers and Quail, 1995). The similar temporal and spatial patterns of expression of *PHYA* and *PHYB* suggests that differences in photoreceptor gene expression cannot account for the different physiological roles of these two phytochromes (see below).

Immunocytochemical methods have been used to establish that phyA is a cytoplasmic protein in several etiolated plant tissues. However, because of their lower abundance and the difficulties of obtaining specific antibodies, the intracellular localisation of the other phytochromes has not been extensively studied. Sakamoto and Nagatani (1996) have presented two lines of evidence suggesting that phyB from *Arabidopsis* is a nuclear protein. Since the derived amino acid sequence of PHYB (and the related PHYD), is known to contain sequences with homology to bipartite nuclear localisation signals, Sakamoto and Nagatani (1996) constructed translational fusions between the non-chromophorylated C-terminus of PHYB and GUS. The gene fusions were expressed in transgenic *Arabidopsis*. In cotyledon protoplasts and epidermal peels, GUS activity was observed to be localised in the nucleus. Furthermore, Sakamoto and Nagatani (1996) went on to show that immunochemically detectable PHYB protein was associated with isolated nuclei and that anti-PHYB monoclonal antibodies stained nuclei in immunocytochemical tests. The authors speculate that the different subcellular locations of phyA and phyB may be related to the differences in the physiological activities of these two phytochromes. However, this is not readily reconciled with the findings of Wagner *et al.* (1996), who created a chimeric phytochrome comprising a fusions between the N-terminal, chromophore-bearing half of phyA and the C-terminal half of phyB, as well as the reciprocal chimera. These chimeras were overexpressed in transgenic *Arabidopsis* and shown to be biologically active. Although their subcellular locations were not determined, the chimeric phyAB protein (i.e. the N-terminal half of phyA and the C-terminal half of phyB) had the same physiological properties as authentic phyA, while the chimeric phyBA protein had the same properties as authentic phyB (Wagner *et al.*, 1996). Furthermore, the phyAB chimera was found to display the same light lability as authentic phyA, while the phyBA chimera was light stable.

10.2.2 Blue/UVA photoreceptors

Although specific blue/UVA-light sensing systems have been studied for more than a century, the chemical nature of any of the blue (B) light photoreceptors remained elusive for many years. As a pun on the cryptic nature of these photo-receptors and the prevalence of B-light responses in cryptogamic (lower) plants, the term cryptochrome was coined for these receptors.

The cryptochromes of *Arabidopsis* include photoreceptors whose apoproteins are the products of the *CRY1* (*HY4*) and *CRY2* genes. The genes encode proteins with an N-terminal region that bears a strong resemblance to the chromophore-binding region of type I microbial DNA photolyases (Ahmad and Cashmore, 1993; Hoffman *et al.*, 1996), a class of flavoenzymes that catalyse light-dependent repair of cyclobutane-type pyrimidine dimers in DNA that has been damaged by UV light. Cryptochrome 1 (CRY1) and cryptochrome 2 (CRY2) differ in the sequence of their C-terminal regions; the C-terminus of CRY1 shows some similarity to smooth muscle tropomyosin (Ahmad and Cashmore, 1993), whereas the C-terminus of CRY2 shows some similarity to neuromodulin (Lin *et al.*, 1996).

The DNA photolyases typically contain two B-light absorbing chromophores, a light-harvesting chromophore, usually methenyltetrahydrofolate (a pterin) or a deazaflavin, bound at the N-terminus of the enzyme and a reduced flavin adenine dinucleotide ($FADH_2$), involved in catalysis and bound at the C-terminus of the enzyme. On absorption of a photon of light by the pterin (or deazaflavin) chromophore, the energy is transferred to the $FADH_2$, which in turn transfers an electron to a bound pyrimidine dimer, initiating bond rearrangement that splits the cyclobutane ring thus restoring the pyrimidines. It has been shown that the CRY1 apoprotein, when expressed in *Escherichia coli*, binds both FAD and methenyltetrahydrofolate (Malhorta *et al.*, 1995). However, the recombinant CRY1 protein has no photoreactivation activity. Interestingly, FAD was found to be stably bound as the green light-absorbing neutral radical flavosemiquinone (FADH·), a semi-reduced flavin (Lin *et al.*, 1995). This observation may explain some responsivity of CRY1 to green light. The nature of the cryptochrome molecule has led to the proposal of a novel method of signal transduction in B-light signalling, involving electron transfer (Malhorta *et al.*, 1995).

Although the cryptochromes have homology to type I photolyases they have no photolyase activity (Malhorta *et al.*, 1995). It has been shown that *Arabidopsis* possesses a gene (*PHR1*) with homology to type II microbial photolyase (Ahmad *et al.*, 1997). Type II photolyases also bind two chromophores and have been shown to be functionally interchangeable with type I photolyases in *E. coli*. The *Arabidopsis uvr2* mutant, which lacks cyclobutane pyrimidine dimer repair activity and is hypersensitive to low levels of UVB irradiation, has been shown to contain a deletion within the *PHR1* gene, providing direct evidence that *PHR1* encodes a photolyase (Landry *et al.*, 1997).

A further blue/UVA photoreceptor from *Arabidopsis* has been identified. The genetic analysis of B-light-mediated phototropism has led to the identification of several mutants that are specifically defective in this response (see Liscum and Hangarter, 1994; Liscum and Briggs, 1995). Mutants deficient in the CRY1 photoreceptor display normal B-light-mediated phototropic responses. However, some alleles of the *nph1* (=JK224) mutant are phototropic nulls, consistent with the notion that *NPH1* encodes the photoreceptor for phototropism (Liscum and Briggs, 1995). Furthermore, the levels of a 120 kDa plasma-membrane-associated protein that is phosphorylated within seconds of blue light irradiation and has previously been suggested to define the photoreceptor for phototropism, are reduced in all *nph1* alleles (Liscum and Briggs, 1995). The *NPH1* gene has recently been cloned and shown to encode a protein with homology to Ser/Thr kinases (W.R. Briggs, personal communication).

10.3 Photoreceptor functions

A focus of current photomorphogenesis research is the identification of the roles of individual photoreceptors. The most revealing approaches have been those involving the analysis of mutants that are deficient in one or more photoreceptor species, along with the study of transgenic plants that overexpress photoreceptor genes. Dark-grown *Arabidopsis* seedlings display a typical etiolated phenotype, characterised by extreme hypocotyl elongation growth, a hypocotyl hook, small folded cotyledons, absence of chlorophyll and lack of plastid development. In response to light signals, a series of growth and developmental changes are initiated that lead to de-etiolation and photomorphogenic seedlings development. Genetic screens for mutants that are insensitive to light with regard to photomorphogenic seedling development have proved especially powerful in the identification of photoreceptor mutants.

10.3.1 Phytochrome A functions

Mutants deficient in phyA (*phyA*) were initially isolated on the basis of their long hypocotyl phenotype following growth of etiolated seedlings under prolonged

FR light (Nagatani *et al.*, 1993; Parks and Quail, 1993; Whitelam *et al.*, 1993). In fact, etiolated *phyA* seedlings display a complete insensitivity to FR light with respect to a whole range of processes involved in seedling photomorphogenesis, including the inhibition of hypocotyl elongation (Figure 10.1), the opening and expansion of cotyledons, the synthesis of anthocyanin and the regulation of expression of several genes (e.g. Whitelam *et al.*, 1993; Johnson *et al.*, 1994; Barnes *et al.*, 1996). In contrast, *phyA* seedlings display more or less normal sensitivity to R light with respect to the inhibition of hypocotyl elongation. Since other photoreceptor mutants are not impaired in their responses to FR light, phyA is implicated as the sole photoreceptor mediating seedling responses to this region of the spectrum (see Quail *et al.*, 1995).

Two distinct physiological response modes have been identified for the responses of etiolated seedlings to FR light (Smith and Whitelam, 1990). The FR high irradiance response (HIR) is initiated by prolonged irradiation with FR light

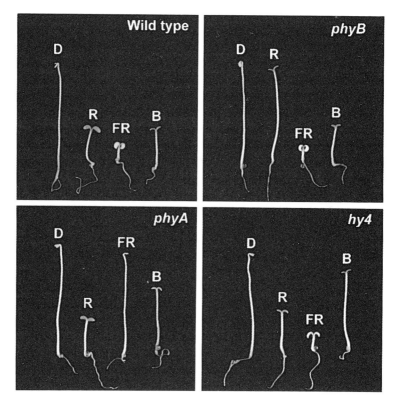

Figure 10.1 Photoreceptor mutants of *Arabidopsis*. Wild-type, *phyA, phyB* and CRY1-deficient, *hy4*, seedlings were grown in the dark (D), under continuous red light (R), under continuous far-red light (FR) or under continuous blue light (B) for four days.

of a relatively high fluence rate (0.1–$50 \, \mu\text{mol m}^{-2}\text{s}^{-1}$). The extent of the HIR is dependent on the fluence rate of the FR light (see Smith and Whitelam, 1990). In contrast, the very low fluence response (VLFR) is triggered by fluences of light as low as 10^{-9}mol m^{-2}. As a consequence, this response is fully saturated by very low concentrations of Pfr. Indeed, saturation of the VLFR can be achieved by pulses of FR light, so this response mode does not display R/FR reversibility (see Smith and Whitelam, 1990). Both HIR and VLFR (e.g. for the promotion of seed germination) modes are defective in *phyA* mutant seedlings (Botto *et al.*, 1996; Shinomura *et al.*, 1996).

Phytochrome A also plays a role in some seedlings responses to R light. For example, the induction of chalcone synthase (*CHS*) gene expression and anthocyanin accumulation under prolonged R light is essentially absent in *phyA* seedlings (Barnes *et al.,* 1996). Also, Parks *et al* (1996) have shown that whereas wild-type and phyB-deficient *Arabidopsis* seedlings show a marked enhancement of first positive phototropic curvature in response to a few seconds of R light irradiation, this response is absent in *phyA* seedlings.

Despite the light-lability of phyA, it nevertheless remains one of the most abundant phytochromes in the light-grown plant (Clack *et al.*, 1994). Furthermore, phyA continues to play a key role in the regulation of development in the fully de-etiolated light-grown plant. *Arabidopsis* is a facultative long-day plant and flowering is promoted as day length increases. The *phyA* mutant shows greatly reduced sensitivity to low fluence rate incandescent day extensions which induce flowering in wild-type plants (Johnson *et al.*, 1994). Similarly, *phyA* mutants are relatively insensitive to flower-inducing R light night breaks (Reed *et al.*, 1994). These observations clearly implicate phyA in day length perception in *Arabidopsis*. However, *phyA* mutants are not completely insensitive to day length and other photoreceptors are therefore presumed to play a role in day-length measurement.

The *phyA* mutation is partially dominant and so etiolated seedlings of the *PHYA/phyA* heterozygote display a hypocotyl length intermediate between that of the *PHYA/PHYA* and *phyA/phyA* homozygotes in FR light (Whitelam *et al.*, 1993). From this it can be concluded that no compensatory mechanism exists to regulate the levels of the photoreceptor. Such a situation would be expected where the degree of response itself is determined by the level of the active photoreceptor.

10.3.2 Phytochrome B functions

Etiolated phytochrome B-deficient (*phyB*) seedlings display an elongated hypocotyl, reduced cotyledon expansion and reduced pigmentation when grown under prolonged R (Figure 10.1) or prolonged white light but are indistinguishable from wild-type seedlings when grown under prolonged FR or B light (Koornneef *et al.*, 1980; Reed *et al*, 1993). This indicates that phyB plays a major role in seedling responses to R light but none in responses to FR light.

In addition to its role in seedling photomorphogenesis in R light, phyB plays a major role in the promotion of *Arabidopsis* seed germination (Shinomura *et al.*, 1994, 1996; Reed *et al.*, 1994). Germination of wild-type *Arabidopsis* seeds can be strongly promoted by a pulse of R light. The effect of the R light pulse can be nullified by a subsequent pulse of FR light; the classical hallmark of phyto-chrome action. This R/FR reversible response, the low fluence response (LFR) mode of phytochrome action, is defective in *phyB* seeds (Shinomura *et al.*, 1994, 1996; Reed *et al.*, 1994). It has also been demonstrated that the Pfr form of phyB stored in the dry seed can promote subsequent germination in darkness (McCormac *et al.*, 1993).

One of the most important aspects of photomorphogenesis in light-grown plants in the natural environment is the perception of alterations in the relative amounts of R and FR light and the initiation of the shade-avoidance syndrome of responses (see Smith, 1995). The light that is reflected from (or transmitted through) living vegetation is depleted in R (and B) light relative to FR light, because of selective absorption by chlorophyll, and consequently it has a reduced R:FR ratio (see Smith, 1995). Reductions in the R:FR ratio of incident light are perceived by photochromic phytochrome and translated into a change in equilib-rium between Pr and Pfr. This provides an unambiguous signal that indicates that potential competitors are nearby. Many competitive plants, including *Arabidop-sis*, respond to low R:FR ratio signals by increasing elongation internodes and/or petioles, reducing leaf growth, increasing apical dominance and accelerating flowering in an attempt to avoid being shaded (Smith, 1995). Phytochrome B-deficient mutants have elongated petioles, show increased apical dominance and are early flowering when grown under high R:FR ratio conditions (e.g. Whitelam and Smith, 1991; Devlin *et al.*, 1996). In addition to displaying a constitutive shade avoidance phenotype in the absence of low R:FR ratio signals, *phyB* seed-lings show greatly attenuated responses to a low R:FR ratio (e.g. Whitelam and Smith, 1991; Halliday *et al.*, 1994; Devlin *et al.*, 1996). In wild-type plants, shade avoidance responses can also be induced by pulses of FR light given at the end of each photoperiod. These responses are also greatly attenuated in *phyB* seedlings (Nagatani *et al.*, 1991; Devlin *et al.* 1996). Taken together, these find-ings implicate phyB as a major contributor to shade avoidance.

The *Arabidopsis phyB* mutant displays very marked early flowering follow-ing growth under either long- or short- day conditions, presumably as a conse-quence of its constitutive shade avoidance phenotype (e.g. Goto *et al.*, 1991; Halliday *et al.*, 1994). However, the *phyB* mutant is still responsive to day length, suggesting that phyB may not play a significant role in the detection of day length.

10.3.3 The functions of other phytochromes

The analysis of *phyA,phyB* double mutants has provided clear evidence that some responses to R and FR light are controlled by other members of the phytochrome family. For example, although already elongated and early flowering, *phyA,phyB* double mutants respond to low R:FR ratio signals, and end-of-day FR light treatments, by further promotion of elongation growth and earlier flowering (Devlin *et al.*, 1996). The responses of *phyA,phyB* seedlings to end-of-day FR light pulses show R/FR reversibility, clearly implicating the action of one or more other phytochromes (Devlin *et al.*, 1996). The steady-state level of transcripts of the *Athb-2* gene, which encodes a novel type of homeodomain protein, is also regulated in an R/FR reversible manner in *phyA,phyB* mutants (Carabelli *et al.*, 1996).

It has been established that certain accessions of the WS ecotype of *Arabidopsis* carry a 14 bp deletion within the coding region of the *PHYD* gene (R.A. Sharrock, personal communication and unpublished observations). This *phyD* mutation has been introgressed into other *Arabidopsis* ecotypes and combined with other *phy* mutations. Preliminary analyses suggest that phyD is partly responsible for the responses of *phyA,phyB* seedlings to low R:FR ratio signals (Whitelam, unpublished).

10.3.4 Cryptochrome 1 functions

The *hy4* mutant of *Arabidopsis* is deficient in CRY1 (Ahmad and Cashmore, 1993). Etiolated *hy4* seedlings display a reduced sensitivity to prolonged B/UVA light for several aspects of seedling photomorphogenesis, including the inhibition of hypocotyl elongation (Figure 10.1) and the promotion of cotyledon expansion (Koornneef *et al.*, 1980; Blum *et al.*, 1994; Ahmad *et al.*, 1995). In addition, *hy4* seedlings are defective in B-light-induced expression of the *CHS* gene and the accumulation of anthocyanin (Ahmad *et al.*, 1995; Jackson and Jenkins, 1995; Barnes *et al.*, 1996). However, etiolated *hy4* seedlings are not completely deficient in responses to B light, suggesting the involvement of other B light photoreceptors in de-etiolation responses. Of course, the phytochromes absorb in the B region of the spectrum and their action may contribute to seedlings responses to B light (e.g. Whitelam *et al.*, 1993). Nevertheless, there is evidence for the operation of specific B light photoreceptors, in addition to CRY1, mediating seedling de-etiolation. For example, B light induction of *CAB* gene expression still occurs in the *hy4* mutant. The time course for induction indicates that this response is not mediated by a phytochrome, possibly implicating the action of the CRY2 photoreceptor (Gao and Kaufman, 1994).

Apart from its role in seedlings establishment, the CRY1 photoreceptor has also been implicated in the perception of day length. Wild-type *Arabidopsis* seedlings respond more strongly to day extensions that contain B light than to B-light-depleted day extensions for the promotion of flowering (Mozley and Thomas, 1995). Furthermore, the *hy4* mutant displays an aberrant flowering time response. However, the effect of the *hy4* mutation on flowering time appears to be allele specific since both late-flowering (Bagnall *et al.*, 1996) and early-flowering (Zagotta *et al.*, 1996) alleles have been described.

10.4 Interactions and redundancy among photoreceptors

The photoreceptors responsible for light signalling in *Arabidopsis* do not act in isolation and there are many examples of interactions among them. For instance, Casal and Boccalandro (1995) have demonstrated a co-action between CRY1 and phyB during seedling establishment. A single pulse of R light has little effect on etiolated *Arabidopsis* seedlings. However, a pulse of R light can lead to an inhibition of hypocotyl elongation and opening of the cotyledons if it is preceded by a period of B light irradiation. This effect is observed in both wild-type and *phyA* seedlings, but it is impaired in both *phyB* and *hy4* mutant seedlings. It has been shown that the action of CRY1 in mediating photomorphogenic responses of etiolated seedlings to B light requires that phyA or phyB be present. Thus, despite possessing CRY1, the *phyA,phyB* double mutant was found to be essentially insensitive to B light with respect to the inhibition of hypocotyl elongation or the promotion of anthocyanin synthesis (Ahmad and Cashmore, 1997). Another type of interaction between phytochrome and a B/UVA photoreceptor is seen for phototropic curvature, which is enhanced by prior stimulation of phytochrome action by exposure of seedlings to R light (Parks *et al.*, 1996; Janoudi *et al.*, 1997).

Fuglevand *et al.* (1996) have reported a synergistic interaction between B/UVA and UVB photoreceptors for the induction of *CHS* gene expression in *Arabidopsis*. Here, the induction of *CHS* transcripts in seedlings exposed to mixtures of B and UVB light is greater than the sum of inductions by each waveband alone. The synergistic effect was also observed if B light irradiation preceded UVB irradiation, suggesting the participation of a stable B-light-induced signal. A similar synergism was observed between UVA and UVB irradiation, although UVA irradiation did not lead to production of a stable signal. Significantly, the effects of B light and UVA in amplifying the effects of UVB irradiation were still apparent in the CRY1-deficient *hy4* mutant (Fuglevand *et al.*, 1996), implicating the operation of some other photoreceptor(s).

Analysis of *phyA,phyB* double mutants has indicated that there is some functional redundancy between these two photoreceptors. For example, the induction of the *CAB* transcripts in etiolated seedlings by a pulse of R light occurs to wild type levels in either the monogenic *phyA* or the monogenic *phyB* mutant, but the induction is severely reduced in the *phyA,phyB* double mutant (Reed *et al.*, 1994). Thus, it seems that either phyA or phyB is capable of mediating the response and that in the absence of both of these phytochromes other members of the phytochrome family cannot produce this response. Similarly, deficiency of either phyA or phyB does not affect the total biomass of mature *Arabidopsis* plants whereas the biomass of the *phyA,phyB* double mutant is severely reduced (Devlin *et al.*, 1996). Finally, phyA and phyB display redundancy for the R-light-mediated agravitropism of *Arabidopsis* hypocotyls. The hypocotyls of etiolated *Arabidopsis* seedlings display strong negative gravitropism and R light treatment causes the hypocotyls to become agravitropic. This R light response is displayed by monogenic *phyA* or monogenic *phyB* mutants, but not by the *phyA,phyB* double mutant (Robson and Smith, 1996).

10.5 Photoreceptor signal transduction

The absorption of a photon of light by a photoreceptor has to be coupled to the modulation of gene expression or other changes in cell physiology in order to bring about the modifications of growth and development that constitute photo-morphogenesis. It is only relatively recently that some of the intermediates that act downstream of photoreceptors in the transduction of light signals have begun to be elucidated. Several laboratories have adopted the genetic approach in efforts to dissect light signal transduction pathways in *Arabidopsis*. Mutant screening strategies have tended to focus on the identification of two classes of mutants that show gross alterations of photomorphogenic development (see Table 10.1): (1) mutants that are insensitive to light, for example with respect to the photoregulation of hypocotyl growth, and therefore may define positively acting intermediates and (2) mutants that display a light-grown phenotype in the absence of light and so may define negatively acting intermediates.

10.5.1 Positive regulators

To date, almost all mutants isolated on the basis of an insensitivity to light, for example long hypocotyl mutants, have defined genes encoding the apoproteins of photoreceptors or proteins necessary for photoreceptor chromophore synthesis. However, there is a small number of exceptions. Loss-of-function mutations at the *HY5* locus lead to a light-insensitive phenotype characterised by a long hypo-cotyl for seedlings grown under R, FR, B and UVA conditions, but not UVB conditions (Koornneef *et al.*, 1980). The insensitivity to different wavebands of light suggests that the *HY5* gene product is required as a positive regulator in the transduction of light signals perceived by several different photoreceptors. Although *hy5* seedlings show reduced accumulation of anthocyanin in the light compared with wild-type seedlings, many other light-regulated phenotypes are normal in this mutant (Chory, 1992). This indicates that the *HY5* gene product plays a restricted role in a subset of responses that are regulated by light.

A mutant that displays a long hypocotyl phenotype selectively under R light has been isolated from a screen for chlorate-resistant mutants (Lin and Cheng, 1997). Since phyB is the principal photoreceptor mediating seedling response to R light, this mutant appears to have a selective defect in the phyB signalling pathway. The mutant, *cr88*, is resistant to chlorate because it is defective in the light-induction of *NR2*, one of the two *Arabidopsis* genes that encode nitrate

Table 10.1 Photomorphogenic mutants of *Arabidopsis*

Name	Phenotype
1. Mutants that define photoreceptor apoprotein genes	
phyA (*hy8*, *fre1*, *fhy2*)	Long hypocotyl in FR light, late flowering
phyB (*hy3*)	Long hypocotyl in R and white light, early flowering
phyD	Similar to *phyB*, but less severe
cry1 (*hy4*)	Long hypocotyl in B and white light
nph1? (*JK224*)	Hypocotyl is non-phototropic
2. Mutants defective in phytochrome chromophore synthesis	
hy1, *hy2*	Long hypocotyl in R, FR and white light, early flowering
3. Mutants that define positive regulators of light signalling	
hy5	Long hypocotyl in R, FR, B and white light
cr88	Chlorate-resistant, long hypocotyl in R light
pef1	Early flowering, long hypocotyl in R and FR light
pef2, *pef3*	Early flowering, long hypocotyl in R light
red1	Long hypocotyl in R light
fhy1	Long hypocotyl in FR light
fhy3	Long hypocotyl in FR light
elf3	Early flowering, long hypocotyl in R, FR and B light
4. Mutants that define negative regulators of light signalling a. Pleiotropic, seedling lethal mutants	
det1 (*fus2*), *cop1* (*fus1*), *cop9* (*fus7*), *fus4*, *fus5*, *fus6* (*cop11*), *fus8* (*cop8*), *fus9* (*cop10*), *fus11*, *fus12*	In the dark: short hypocotyl, open and expanded cotyledons, chloroplast development, expression of light-regulated genes
b. Less pleiotropic, non-lethal mutants	
det2	Similar to *det1*, brassinosteroid-deficient
cpd	Similar to *det1*, brassinosteroid-deficient
cop2 (*amp1*)	Similar to *det1*
shy1,*shy2*	Suppresses the long hypocotyl phenotype of *hy2*

reductase. The light-mediated induction of *CAB* and *RBCS* transcripts is also impaired in *cr88*, although the inductions of *NR1* and *NiR* are normal. In addition to the long hypocotyl phenotype and defective gene expression, light-grown *cr88* plants also display a pale-leaf phenotype and possess underdeveloped chloroplasts. These observations suggest that *CR88* may encode a signalling component involved in the light-regulation of a subset of genes, including some that are involved in chloroplast development. However, since *phyB* mutants do not display a similar pale-leaf phenotype, the *CR88* gene product cannot be specific to the phyB signalling pathway.

Mutants in which hypocotyl elongation growth shows a selective insensitivity to R light have also been described by Ahmad and Cashmore (1996). The *pef* mutants were selected on the basis of their early-flowering behaviour under both long-day and short-day conditions. In addition to early flowering, *pef2* and *pef3* mutants show a long hypocotyl phenotype under R light, but a normal inhibition of hypocotyl elongation in response to B, UVA and FR light. Neither mutant is allelic with previously described photoreceptor/chromophore mutants. Since *phyB* mutants are characterised by a long hypocotyl phenotype under R light (but not FR light) and early flowering, it is possible that *PEF2* and *PEF3* may encode proteins that act early in the signalling pathway from this photoreceptor. The *pef1* mutant, which is also not allelic to previously identified photoreceptor/ chromophore mutants, was found to display an elongated hypocotyl following growth under either R or FR light. This suggests that *pef1* may be impaired in the signalling pathways downstream of both phyA and phyB.

A third class of mutant that displays a selective insensitivity to R light has been described by Wagner *et al.* (1997). The *red1* mutant was selected from the M2 population derived following mutagenesis of a transgenic *Arabidopsis* line that overexpresses phyB. The transgenic overexpressors display a marked short hypocotyl phenotype under R light and the M2 population were screened for individuals that no longer show this hypersensitivity to R light. The *red1* mutation could be segregated away from the *PHYB* transgene and the mutant phenotype was also clearly displayed in non-transgenic wild-type plants. Light-grown *red1* mutants resembled the elongated phenotype of *phyB* mutants and, like *phyB* seedlings, showed attenuated responses to end-of-day FR light treatments (Wagner *et al.*, 1997). These observations suggest that the *RED1* gene product functions as an early intermediate in a phyB signal transduction pathway that regulates elongation growth. Significantly, Wagner *et al.* (1997) reported that *red1* mutants are not noticeably early flowering. This suggests that the early flowering of *phyB* mutants may be regulated by a separate pathway to that which regulates elongation growth.

Mutants that are selectively impaired in responses to FR light, and so may define components in the phyA signalling pathway, have also been described (Whitelam *et al.*, 1993). The *fhy1* and *fhy3* mutants are phenotypically similar to *phyA* mutants following growth under FR, R and white light. However, neither mutation is linked to the *PHYA* gene and the levels of immunochemically and spectrally detectable phyA are normal in the mutants. It has been suggested therefore that *FHY1* and *FHY3* encode positively acting downstream components of the phyA signal transduction pathway. Further analyses of the *fhy1* mutant have revealed that is defective in only a subset of phyA-mediated physiological responses, suggesting that the phyA signalling pathway may be branched (Johnson *et al.*, 1994). Barnes *et al.* (1996) have established that *fhy1* is impaired in the phyA-mediated regulation of only a subset of the genes that are phyA-regulated in wild-type seedlings. Thus, FR-light-mediated induction of *CHS* is deficient in *fhy1*, whereas the induction of *CAB* and *NR* is relatively unaffected by the mutation (Barnes *et al.*, 1996). Interestingly, the *CHS* and *CAB* genes had previously been proposed to define distinct branches of the phyA signalling pathway based on biochemical analyses performed in microinjected tomato hypocotyl cells (see Barnes *et al.*, 1997). The *FHY1* gene product may therefore provide a link between the genetic and biochemical dissections of phytochrome signalling pathways.

Screens for mutants that show long hypocotyls selectively under B light have failed to identify loci in addition to *HY4*, the gene encoding the apoprotein of CRY1. Nevertheless, it has been proposed that the product of the *ELF3* gene may play a particular role in some aspects of CRY1-mediated B light signal transduction (Zagotta *et al.*, 1996). The early-flowering *elf3* mutant is insensitive to day length with respect to flowering, a phenotype that appears to be related to circadian dysfunction in *elf3* (Hicks *et al.*, 1996 and below). In addition, *elf3* mutants display some defects in vegetative photomorphogenesis. Specifically, *elf3* mutants have long hypocotyls following growth under R, FR, UV and particularly B and green light (Zagotta *et al.*, 1996). This suggests a defect in signalling from several photoreceptors. However, for flowering time, the *elf3* mutation was found to be epistatic to an allele of *hy4*, and to act additively with *hy2*, a phytochrome chromophore-deficient mutant. Consequently it has been proposed that although the phytochrome-mediated pathway controlling flowering time is intact in *elf3*, the CRY1-dependent pathway for flowering is defective.

10.5.2 Negative regulators

Genetic screens for mutants that display photomorphogenic development in the dark, and so may define suppressers of photomorphogenesis, have identified at least 16 discrete loci, including *DEETIOLATED (DET)*, *CONSTITUTIVELY PHOTOMORPHOGENIC (COP)* and *FUSCA (FUS)*. The *det/cop/fus* mutants are all recessive and cause seedlings to display varying degrees of photomorphogenic development in the dark. For ten of these loci, *DET1 (= FUS2)*, *COP1 (= FUS1)*, *COP9 (= FUS7)*, *FUS4*, *FUS5*, *FUS6 (= COP11)*, *FUS8 (= COP8)*, *FUS9 (= COP10)*, *FUS11* and *FUS12*, severe mutations lead to an extremely pleiotropic photomorphogenic phenotype and seedling lethality (see von Arnim and Deng, 1996). The seedling lethal nature of these mutants suggests that the products of these *DET/COP/FUS* genes are not simply involved in repressing photomorphogenesis in the dark but are also involved in other essential cellular processes both in the light and in the dark (see Castle and Meinke, 1994; Mayer *et al.*, 1996). Dark-grown seedlings of severe alleles of any these ten pleiotropic mutations have very short hypocotyls and open and expanded cotyledons such that they are essentially indistinguishable from light-grown wild-type seedlings. In addition, these severe mutants accumulate significant amounts of antho-cyanins in the dark, as well as displaying chloroplast development and the con-stitutive or ectopic expression of several normally light-regulated genes (see von Arnim and Deng, 1996; Chamovitz and Deng, 1996). As might be expected, the phenotypes of double mutants carrying a photoreceptor mutation and a mutation of the pleiotropic *det/cop/fus* type indicate that the products of the *DET/COP/FUS* genes act downstream of the phytochromes and of CRY1.

Four of the *DET/COP/FUS* genes, *DET1*, *COP1*, *COP9* and *FUS6*, have been cloned and shown to encode novel nuclear-localised proteins (Deng *et al.*, 1992; Pepper *et al.*, 1994; Wei *et al.*, 1994; Castle and Meinke, 1994). The deduced amino acid sequence of COP1 reveals some interesting features. In particular, the C-terminus of the protein contains six repeats of the WD40 sequence motif (Deng *et al.*, 1992). This motif is found in several regulatory proteins that are involved in the repression of transcription (see von Arnim and Deng, 1996). In the case of yeast TUP1, the WD40 motif directs the protein to its target promoters by interacting with promoter-specific DNA-binding proteins (Komachi *et al.*, 1994), while the WD40 motif of Drosophila dTAF$_{II}$80, a subunit of TFIID, is proposed to directly contact the transcriptional machinery (Dynlacht *et al.*, 1993). The presence of WD40 domains in COP1 is consistent with its proposed role as an inhibitor of the transcription process.

Light has been reported to influence the subcellular location of COP1. von Arnim and Deng (1994) have determined the cellular location of a GUS-COP1 fusion protein expressed either transiently in onion bulb epidermal cells or stably in *Arabidopsis* hypocotyl cells. They showed that the fusion protein accumulates in the nucleus in cells maintained in the dark and that light treatment leads to a reduction in protein levels and a shift to a cytoplasmic localisation. The authors propose that in dark-grown seedlings COP1 acts in the nucleus as a repressor of photomorphogenesis and that in the light this repression is relieved by the depletion of COP1 from the nucleus via its translocation or degradation. This model seems adequate to explain to the activity of COP1 in relation to photomor-phogenesis. However, it is clear that since severe *cop1* alleles are lethal, COP1 must also perform essential functions in light-grown seedlings for which it presumably needs to be located in the nucleus.

The similar phenotypes displayed by each of the pleiotropic *det/cop/fus* mutants are consistent with the notion that the wild-type genes encode the subunits of a multi-protein complex. Significantly, cell fractionation studies have shown that COP9, a small hydrophilic protein, exists only as part of large, nuclear-localised protein complex comprising at least fifteen major polypeptides (Wei *et al.*, 1994; Chamovitz *et al.*, 1996; Staub *et al.*, 1996). Staub *et al.* (1996) have shown that the product of the *FUS6* gene co-fractionates with COP9 and that antibodies specific to either protein will co-precipitate the other. This, together with the observation that the COP9 complex is absent from *fus6* mutants, indicates that FUS6 is a component of the COP9 complex. The COP9 complex is also absent from *cop8* (*fus8*) mutants and so it has been suggested that the COP8 protein may form part of the complex or may be required for its assembly or stability (Wei *et al.*, 1994).

COP1 is not part of the COP9 complex, however Chamovitz *et al.* (1996) have shown that COP9 is required for the nuclear localisation of COP1 in the dark. Furthermore, although the nuclear localisation of the COP9 complex is not affected by light treatments, the complex is apparently larger in etiolated seedlings than in light-grown seedlings (Chamovitz and Deng, 1996). These findings have led to a model in which COP1 is proposed to interact with the COP9 complex in the dark to repress photomorphogenesis and where light is proposed to cause the dissociation of COP1 from the complex and then its export or degradation, leaving a smaller complex (Chamovitz and Deng, 1996).

The photomorphogenic phenotypes of dark-grown *det/cop/fus* mutants, and the reported effects of light on COP1 and the COP9 complex, support the view that the DET/COP/FUS proteins play a specific role in light signalling. However, several lines of evidence suggest that the DET/COP/FUS proteins function as general or global repressors. The seedling lethal nature of severe alleles of several of the mutants argues that the products of the wild-type genes are necessary for general development in both the light and the dark (Castle and Meinke, 1994). Nevertheless, it has been argued that because weak alleles of *cop1* and *det1*, which are not lethal, still show a photomorphogenic phenotype then perhaps DET/COP/FUS proteins are general repressors that play a particular role in light signalling (e.g. Chamovitz and Deng, 1996). However, Mayer *et al.* (1996) have shown that weak alleles of *cop1*, *det1* and *cop9* lead to inappropriate expression of several sets of genes and not just those that are normally light regulated. In addition, the finding that there appear to be human counterparts of DET1, COP1, COP9 and FUS6 supports the view that these proteins are not specific repressors of photomorphogenesis, but may instead form part of a diverse general regulatory mechanism (see von Arnim and Deng, 1996).

10.5.3 Dominant mutations that suppress photomorphogenesis

Two dominant mutations that cause photomorphogenic development in the dark have been isolated (Kim *et al.*, 1996). The *shy1* and *shy2* mutants were identified as suppressors of the elongated hypocotyl phenotype of the phytochrome chromophore-deficient mutant *hy2*. Both mutations are able to suppress the long hypocotyl phenotype of *hy2* by causing light-independent inhibition of hypocotyl elongation and by increasing seedling sensitivity to light. Intriguingly, the *shy1* mutant, which has only a mild photomorphogenic phenotype in the dark, is more effective in suppressing the long hypocotyl phenotype of *hy2* under R light than under FR light (Kim *et al.*, 1996). Thus, it has been suggested that SHY1 may be specifically involved in the phyB-mediated signalling pathway and play no role in the phyA-mediated pathway.

10.5.4 Biochemical analysis of blue light and UV signalling in Arabidopsis

Biochemical evidence for the participation of G-proteins, calcium, calmodulin and cGMP in phytochrome signal transduction pathways has been obtained from microinjection studies performed with single tomato hypocotyl cells (see Barnes *et al.*, 1997). These same signalling molecules have been implicated in phytochrome action by a series of pharmacological studies performed with photoautotrophic soybean cell suspension cultures that are responsive to light signals (Bowler and Chua, 1994). So far it has not been possible to apply the microinjection approach to the dissection of phytochrome signalling in *Arabidopsis*. However, Christie and Jenkins (1996) have reported the application of photomixotrophic *Arabidopsis* cell suspension cultures for the biochemical analysis of UVB and B/UVA signal transduction pathways. Such cell suspensions respond well to UVB and B/UVA irradiation by the induction of *CHS* gene expression. The involvement of calcium in these light effects was suggested by the observation that nifedipine, the voltage-dependent calcium channel blocker, completely inhibited the induction of *CHS* transcripts by either UVB or B/UVA (Christie and Jenkins, 1996). However, in contrast with the effects reported for microinjected tomato hypocotyl cells, an artificial elevation of intracellular calcium was not able to substitute for light treatment in *CHS* induction. The calmodulin antagonist W7 was found to inhibit the inductive effect of UVB light. However, W7 had no effect on *CHS* induction by B/UVA, providing evidence for distinct transduction pathways for these different light signals.

10.6 Plant growth regulators and light signalling

It has been known for many years that the application of plant growth regulators to light- or dark-grown seedlings can induce a range of responses identical to those initiated by photoreceptor action (see Chory *et al.*, 1995). In particular cytokinin application to seedlings grown in the dark can induce pleiotropic, 'photomorphogenic' development. The analysis of some of the less severe, nonlethal *cop/det* mutants has provided genetic evidence that cytokinins may play a role in mediating seedling responses to light signals. For example, the *det1* mutant phenotype can be copied by application of cytokinin to wild-type seedlings and *det1* seedlings show increased responsiveness to added cytokinins (Chory *et al.*, 1995). Additionally, the *altered meristem program (amp1)* mutant, which is known to be allelic with *cop2* (Wei and Deng, 1996), has been shown to be a cytokinin overproducer (Chinatkins *et al.*, 1996). Dark-grown *amp1* seedlings have short hypocotyls, expanded leaves and display a de-etiolated plastid morphology and increased abundance of transcripts of several light-regulated genes (Chinatkins *et al.*, 1996). From this, it has been inferred that cytokinins, or some cytokinin-dependent process, is involved in the regulation of de-etiolation. The possibility that light signals and cytokinins interact in the control of

Arabidopsis hypocotyl elongation has been assessed directly by Su and Howell (1995), who studies cytokinin-dose and light-fluence relationships in wild-type and photoreceptor mutant seedlings. These authors concluded that cytokinin effects were independent of light effects.

Other growth regulators have been implicated in light signalling. Most notably, the *det2* mutant and the *constitutive photomorphogenesis and dwarfism* (*cpd*) mutant have been shown to be defective in the biosynthesis of brassinosteroids (Li *et al.*, 1996; Szekeres *et al.*, 1996). Following growth in the dark, both of these mutants display a typical photomorphogenic phenotype, characterised by a short hypocotyl, open expanded cotyledons, leaf primordia and elevated expression of several normally light-regulated genes. The phenotypes of both mutants are ameliorated by application of the plant steroid brassinolide. The *DET2* and *CPD* genes have been cloned and shown to encode enzymes involved in brassinolide biosynthesis. The *DET2* gene encodes a protein with sequence similarity to mammalian steroid 5α-reductases, enzymes involved in steroid hormone synthesis (Li *et al.*, 1996, 1997). Furthermore, the DET2 protein, when expressed in human embryonic kidney 293 cells, catalyses the 5α-reduction of several animal steroid substrates (Li *et al.*, 1997). What is more, human type 1 or type 2 steroid 5α-reductase expressed in *det2* mutant plants can substitute for DET2 in brassinosteroid biosynthesis. This indicates that DET2 is an ortholog of the mammalian steroid 5α-reductases. These findings have led to the proposal that light may act by modulating the brassinosteroid signal transduction pathway. The *CPD* gene encodes a novel cytochrome P450 that shares homology with steroid hydroxylases (Szekeres *et al.*, 1996). The expression of several stress-related genes are altered in the *cpd* mutant, indicating that brassinosteroids do not only function in light-signalling pathways.

10.7 Circadian rhythms in *Arabidopsis*

The empirical regularities of circadian systems have resulted in a widely held model for the fundamentals of a circadian clock (Figure 10.2; Sweeney, 1987). The three major parts are (1) the input pathways, which transmit environmental signals (mainly light) to (2) the central oscillator, a negative feedback loop that generates a 24-hour rhythm, and (3) the output pathways, via which rhythmic signals from the oscillator are transmitted to the cell to control the overt, measurable rhythms. The period remains close to 24 hours at a wide range of constant temperatures, indicating that the oscillator is temperature-compensated.

The biological clock is very closely linked to light signalling (Figure 10.2) and the many advantages of *Arabidopsis* for phototransduction studies also apply to circadian rhythms. The clock has two principle interactions with light signalling. Firstly, the oscillator must be set to local time (entrained) in order to co-ordinate endogenous timing with the environmental day/night cycle (Johnson,

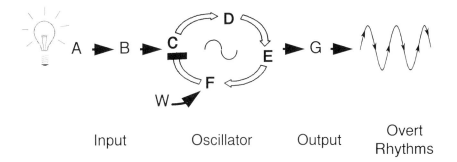

Input   Oscillator   Output   Overt
Rhythms

Figure 10.2 The major components of a circadian clock. The input pathway (A, B) from light (bulb) is shown, with the oscillator components (C, D, E, F) and an output pathway (G) to one overt rhythm. The negative feedback required for oscillation is shown between F and C. W represents an accessory component that is required to sustain the oscillator (see text).

1994). The circadian input pathways transduce light signals to the circadian oscillator, where the incoming signals reset the oscillator, changing either its phase or its period (Millar and Kay, 1997). The phase shifts are probably mediated by the rapid synthesis or destruction of an oscillator component (Dunlap, 1996). Secondly, the cellular targets of circadian regulation often exhibit a rapid (acute) response to light, which is directly regulated by phototransduction pathways. This provides a dual control by light and by the clock, but these forms of regulation are not independent: the circadian clock can rhythmically modulate or 'gate' the acute response to light (Millar and Kay, 1996).

10.7.1   Markers for circadian rhythms in Arabidopsis

Arabidopsis shares several of the classically described circadian markers with other plants, including rhythms in leaf movement (Engelmann *et al.*, 1992; Millar *et al.*, 1995a) and in the responsiveness of floral induction to FR light pulses (Deitzer, 1984). Transgenic plants expressing the calcium-dependent photoprotein, aequorin, exhibit a rhythm in bioluminescence (Johnson *et al.*, 1995). This suggests that the resting concentration of cytoplasmic free calcium is oscillating (about three-fold) under the control of the circadian clock. Presumably, the clock rhythmically modulates the mechanisms of calcium homeostasis: the implications of this rhythm for calcium-dependent signal transduction may be widespread.

One of the more recent rhythmic phenotypes to be developed in higher plants is the circadian control of gene transcription (Beator and Kloppstech, 1996). The family of genes that encode the chlorophyll a/b binding proteins (*CAB*s) of the photosynthetic light harvesting complexes is among the best characterised (McClung and Kay, 1994; Anderson and Kay, 1996). *CAB* genes in many plants exhibit rhythmic transcription, rising before dawn to peak around midday and falling off again in the afternoon. This rhythm persists in continuous light but the level of *CAB* expression damps rapidly in constant darkness, probably because of the loss of phytochrome activation (Millar and Kay, 1997). The firefly luciferase gene (*luc*) has provided a real-time reporter for the fine details of *CAB* promoter regulation. The bioluminescence pattern of living, transgenic plants carrying *cab::Luc* gene fusions recapitulates both circadian control and the other regulatory responses of *CAB* (Millar *et al.*, 1992). This marker enabled much of the *Arabidopsis* rhythm research discussed here, including the discovery of gated *CAB* induction (below); such quantitative reporters can be extremely powerful in studies of regulatory networks. The construction of new luciferase reporters may now mark specific circadian phases and tissues that otherwise lack visible, rhythmic markers, for example, markers for the night phases or for rhythms in roots.

Accordingly, there is considerable interest in identifying clock-controlled genes (*ccgs*) with novel expression patterns. Different members of the catalase gene family, for example, are expressed at different phases (Zhong and McClung, 1996). Cold and circadian-regulated (*CCR*) genes, encoding RNA-binding protein homologues, are maximally expressed at the end of the subjective day (Kreps and Simon, 1997). The period of *CCR* expression shows the same dependence as *CAB* on lighting conditions and the *timing of CAB expression 1* gene (*toc1*, see below), suggesting that *CCR* is probably regulated by the same circadian oscillator as *CAB* (Kreps and Simon, 1997).

The identification of a 'clock box' or CCRE (circadian-clock regulated element) in a *ccg* promoter has been complicated, in part because the *ccgs* are also light-regulated. The smallest promoter fragments that support circadian rhythmicity of downstream reporter genes will also confer light responsiveness (Anderson and Kay, 1995). Gel-shift assays show that the relevant 36 bp in the *Arabidopsis CAB2* promoter are the target for a number of DNA-binding complexes (Carré, 1996), not including a GT-1 related complex (Anderson and Kay, 1995; Teakle and Kay, 1995). The development of new model promoters, such as rubisco activase (Liu *et al.*, 1996), for comparative studies may now be a valuable approach. Other DNA-binding complexes are also coming to light, such as a novel *myb* relative, CCA-1 (Wang *et al.*, 1997).

10.8 Circadian gating

The classical rhythmic markers in higher plants, including stomatal aperture and the expression of *CAB* genes, have long been known to exhibit acute responses to light. The acute responses are not dependent on the circadian clock but are often modulated by the clock: both stomatal opening (Gorton *et al.*, 1993) and *CAB* induction are rhythmically responsive to light. *CAB* induction was tested in seedlings grown under light/dark cycles and transferred to a period of extended darkness (Millar and Kay, 1996). Maximal induction in response to white light coincided with the peaks in the circadian rhythm of *CAB* expression in darkness, but there was little or no response to light during the first night. A rhythmic signal from the circadian clock thus antagonises signalling from the phototransduction pathway to create a circadian 'gate'. The acute response to light is observed only at circadian phases when the gate is open. For comparison, a similar circadian rhythm in the responsiveness to light is widely accepted to be the basis for photoperiodic time measurement in plants (see King and Bagnall, 1996). The circadian clock may therefore have two types of output in the plant cell: a direct regulation, creating a circadian rhythm, and/or an indirect effect, modulating the response to a non-rhythmic signal. The rhythm of cytoplasmic calcium may point to a possible gating mechanism (Johnson *et al.*, 1995). Circadian gating may extend to many signalling pathways, 'fine-tuning' responses appropriately for different times of day.

10.9 The circadian oscillator

The molecular analysis of rhythm mutants in *Drosophila melanogaster* and *Neurospora crassa* has generated an attractive model (Dunlap, 1996), comprising the rhythmic transcription, translation, protein interaction and nuclear entry of the products of critical 'clock genes'. The negative feedback required to complete the oscillator cycle occurs when the 'clock proteins' suppress transcription from their cognate genes, by mechanisms as yet unknown. The model has withstood several tests, in the case of the genes *frequency (frq)* in Neurospora, *timeless (tim)* and *period (per)* in Drosophila, though its generality remains to be determined (Hall, 1996). The functions of these 'clock genes' are largely specific to circadian rhythms and to other timed processes (Hall, 1995), suggesting that the circadian oscillator is not generated only from the interactions of cellular 'housekeeping' components.

An analogous screen in *Arabidopsis thaliana* has uncovered more than twenty *timing of CAB expression (toc)* mutants. A short-period line, *toc1-1*, was characterised initially (Millar *et al.*, 1995b; Millar and Kay, 1997). The 21-hour period of *CAB* expression in this line is semi-dominant, a property it shares with all the period alleles of *per* and *frq*, among others. The leaf movement rhythm also has a shortened period in *toc1*, though there are none of the morphological phenotypes associated with the *det* or *hy* mutants (Millar *et al.*, 1995b). The 'clock genes' in other species were highlighted by their frequent recovery in clock mutant screens. Several alleles of *toc1* may have been recovered in the initial screen; this outstanding question should soon be resolved by ongoing work on their detailed characterisation and molecular cloning (S. Kay, personal communication).

10.10 Light input to the circadian oscillator

The photoreceptors involved in circadian entrainment vary greatly among species (Johnson, 1995), but many rhythms in many plants are reset by the phytochromes (Millar and Kay, 1997). The control of circadian period by constant illumination has been the most convenient assay (Millar and Kay, 1997) to identify input pathway components in *Arabidopsis*. Rhythmic *CAB* transcription exhibits approximately a 24.5 hour period in constant light but a 30 hour period in constant darkness (Millar *et al.*, 1995a). The rhythm persists with a 25 hour period in either constant R light or constant B light, indicating that the input pathways must perceive both wavebands. The phytochrome chromophore-deficient *hy1* mutant lengthens the circadian period in constant R light towards the period of the wild-type in darkness, identifying the phytochrome family as one of the R light photoreceptors for the circadian input pathway. A non-phytochrome, B light photoreceptor is also involved, because the period of *hy1* is slightly *shorter* than wild-type in B light (Millar *et al.*, 1995a).

Genetics is also beginning to identify components that affect the input pathways. The pleiotropic *det1* and *cop1* mutations include short circadian periods in their range of phenotypes (Millar *et al.*, 1995a; Millar and Kay, 1997). The brassinosteroid biosynthesis mutant *det2*, in contrast, has only a minor effect on the period of the *CAB* rhythm in darkness (Millar *et al.*, 1995b). A major challenge is now to determine the mechanism of the input pathway from phytochromes to the circadian oscillator, where it branches from the pathway responsible for acute light responses, and how it relates to biochemical models of the phototransduction pathway (Bowler and Chua, 1994).

10.10.1 Closer links between phototransduction and circadian rhythms

Given that photoperiodic time measurement depends on the circadian clock, some photoperiod-insensitive phenotypes may result from a primary defect in circadian rhythms. The circadian rhythms of the photoperiod-insensitive *elf3* mutant (Zagotta *et al.*, 1992) have been tested based on this premise. *CAB* expression and leaf movement are arrhythmic in *elf3* under constant light (Hicks *et al.*, 1996). *CAB* expression in *elf3* retains some circadian control in light/dark cycles, and in dark-adapted and dark-grown seedlings (Hicks *et al.*, 1996). The arrhythmic phenotypes are therefore conditional on the illumination of the plants. The function of the *ELF3* gene product is unclear: the light-dependent arrhythmicity suggests that it could affect the circadian input pathway, although other models are possible (Millar, 1998). If this is so, the input pathways can have an even greater influence on the circadian clock than indicated by the control of the circadian period by light.

Results from Neurospora suggests that phototransduction and circadian rhythms may not only be functionally related but also share evolutionary ancestors. Mutations of the two *white-collar* (*wc*) genes are unique in eliminating all known light responses in this filamentous fungus. Both turn out to encode zinc-finger proteins (see Crosthwaite *et al.*, 1997). The *wc* mutants are also arrhythmic for visible and molecular circadian markers, and contain very low levels of *frq* RNA (Crosthwaite *et al.*, 1997), suggesting that the WC proteins normally activate *frq* transcription. Formal analyses suggest that this type of positive input (represented by W in Figure 10.2) is required to *sustain* the oscillator, which would damp out if only the negative feedback mechanism (required for oscillation) was present. Most intriguingly, both WC proteins contain a PAS domain. One of the prototype PAS proteins is none other than PER (the 'P' of 'PAS'), which requires the PAS domain for protein–protein interaction with the TIM protein (Dunlap, 1996). PAS domains have also been identified in eukaryotic and prokaryotic photoreceptors, including phytochrome from the green alga *Mesotaenium* (Lagarias *et al.*, 1995). The link between phytochrome and the circadian clock may be much closer than previously suspected.

10.11 Conclusions

In the last few years our understanding of the perception and transduction of the light signals that lead to the control of seedling development have benefited enormously from the combination of genetic and biochemical studies. The genes encoding the apoproteins of several relevant *Arabidopsis* photoreceptors have now been cloned. *Arabidopsis* mutants that define at least three of the phytochromes, and at least two of the B light photoreceptors, have also been identified and characterised. As a consequence, a detailed picture of the roles of these different photoreceptors, as well as of the interactions among them, is now emerging.

Many of the messengers that act downstream of the photoreceptors in the photocontrol of gene expression and development have also been identified. Largely by virtue of the mutant screening strategies that have been adopted, many of the genetically defined messengers, such as the *DET/COP/FUS* gene products, act as negative regulators. Several of these genes have been cloned and their products shown to form a nuclear-localised complex that is a point of convergence for light signals and other signals that regulate plant development. Messengers that are specific to light signalling initiated by individual photoreceptors, and which will lie closer to the primary events triggered by photoreceptor action, are likely to act upstream of the DET/COP/FUS complex. The application of more subtle genetic screens has led to the identification of several mutants, such as *fhy1/fhy3* and *red1* that define photoreceptor-specific signalling components. The cloning of the genes defined by these mutations, and the identification of interacting partners for their products, as well as interacting partners for the photoreceptors themselves, will be crucial in dissecting the early steps of light signalling.

For the future, a major challenge will be to understand how light signals originating from the different photoreceptors are integrated together in the control of gene expression and development. Similarly, another future focus of research will be to determine the ways in which light-signalling mechanisms interact with other developmental signalling pathways, for example those mediated by the plant hormones, and with the biological clock in order to bring about precise and appropriate responses to regulatory light cues.

References

Ahmad, M. and Cashmore, A.R. (1993) *HY4* gene of *Arabidopsis thaliana* encodes a protein with characteristics of a blue-light photoreceptor. *Nature*, **306** 162-66.

Ahmad, M. and Cashmore, A.R. (1996) The *pef* mutants of *Arabidopsis thaliana* define lesions early in the phytochrome signaling pathway. *Plant J.*, **10** 1103-10.

Ahmad, M. and Cashmore, A.R. (1997) The blue-light receptor cryptochrome 1 shows functional dependence on phytochrome A and phytochrome B in *Arabidopsis thaliana. Plant J.*, **11** 421-27.

Ahmad M., Lin C. and Cashmore, A.R. (1995) Mutations throughout an *Arabidopsis* blue-light photoreceptor impair blue-light responsive anthocyanin accumulation and inhibition of hypocotyl elongation. *Plant J.*, **8** 653-58.

Ahmad, M., Jarillo, J.A., Klimczak, L.J., Landry, L.G., Peng, T., Last, R.L. and Cashmore, A.R. (1997) An enzyme similar to animal type II photolyases mediates photoreactivation in *Arabidopsis. Plant Cell*, **9** 199-207.

Anderson, S.L. and Kay, S.A. (1995) Functional dissection of circadian clock-regulated and phytochrome-regulated transcription of the *Arabidopsis CAB2* gene. *Proc. Natl Acad. Sci. USA*, **92** 1500-1504.

Anderson, S.L. and Kay, S.A. (1996) Illuminating the mechanism of the circadian clock in plants. *Trends Plant Sci.*, **1** 51-57.

Bagnall, D.J., King, R.W. and Hangarter, R.P. (1996) Blue-light promotion of flowering is absent in *hy4* mutants of *Arabidopsis. Planta*, **200** 268-80.

Barnes, S.A., Quaggio, R.B., Whitelam, G.C. and Chua, N.-H. (1996) *fhy1* defines a branch point in phytochrome A signal transduction pathways for gene expression. *Plant J.*, **10** 1155-61.

Barnes, S.A., McGrath, R.B. and Chua, N.-H. (1997) Light signal transduction in plants. *Trends Cell Biol.*, **7** 21-26.

Beator, J. and Kloppstech, K. (1996) Significance of circadian gene expression in higher plants. *Chronobiol. Int.*, **13** 319-39.

Blum, D.E., Neff, M.M. and Vanvolkenburgh, E. (1994) Light-stimulated cotyledon expansion in the *blu3* and *hy4* mutants of *Arabidopsis thaliana. Plant Physiol.*, **105** 1433-36.

Botto, J.F., Sánchez, R.A., Whitelam, G.C. and Casal, J.J. (1996) Phytochrome A mediates the promotion of seed germination by very low fluences of light and canopy shade-light in *Arabidopsis. Plant Physiol.*, **110** 439-44.

Bowler, C. and Chua, N.-H. (1994) Emerging themes of plant signal transduction. *Plant Cell*, **6** 1529-41.

Carabelli, M., Morelli, G., Whitelam, G. and Ruberti, I. (1996) Twilight-zone and canopy shade induction of the *Athb-2* homeobox gene in green plants. *Proc. Natl Acad. Sci. USA*, **93** 3530-35.

Carré, I.A. (1996) Biological timing in plants. *Sem. Cell Dev. Biol.*, **7** 775-80.

Casal, J.J. and Boccalandro, H. (1995) Co-action between phytochrome B and HY4 in *Arabidopsis thaliana. Planta*, **197** 213-18.

Castle, L.A and Meinke, D.W. (1994) A *FUSCA* gene of *Arabidopsis* encodes a novel protein essential for plant development. *Plant Cell*, **6** 25-41.

Chamovitz, D.A. and Deng, X.-W. (1996) Light signalling in plants. *Crit. Rev. Plant Sci.*, **15** 455-78.

Chamovitz, D.A., Wei, N., Osterlund, M.T., von Arnim, A.G., Staub, J.M., Matsui, M. and Deng, X.-W. (1996) The COP9 complex, a novel multisubunit nuclear regulator involved in the light control of plant developmental switch. *Cell*, **86** 115-21.

Chinatkins, A.N., Craig, S., Hocart, C.H., Dennis, E.S. and Chaudhury, A.M. (1996) Increased endogenous cytokinin in the *Arabidopsis amp1* mutant corresponds with de-etiolation responses. *Planta*, **198** 549-56.

Chory, J. (1992) A genetic model for light-regulated seedling development in *Arabidopsis*. *Development*, **115** 337-54.

Chory, J., Cook, R.K., Dixon, R., Elich, T., Li, H.M., Lopez, E., Mochizuki, N., Nagpal, P., Pepper, A., Poole, D. and Reed, J. (1995) Signal-transduction pathways controlling light-regulated development in *Arabidopsis*. *Phil. Trans. Roy. Soc. Lond. Ser. B*, **350** 59-65.

Christie, J.M. and Jenkins, G.I. (1996) Distinct UV-B and UV-A/blue light signal transduction pathways induce chalcone synthase gene expression in *Arabidopsis* cells. *Plant Cell*, **8** 1555-67.

Clack, E., Matthews, S. and Sharrock, R.A. (1994) The phytochrome apoprotein family in *Arabidopsis* is encoded by five genes: the sequences and expression of *PHYD* and *PHYE*. *Plant Mol. Biol.*, **25** 413-27.

Crosthwaite, S.K., Dunlap, J.C. and Loros, J.J. (1997) Neurospora *wc-1* and *wc-2*: Transcription, photoresponses, and the origins of circadian rhythmicity. *Science*, **276** 763-69.

Deitzer, G. (1984) Photoperiodic induction in long-day plants, in *Light and the Flowering Process* (eds D. Vince-Prue, B. Thomas and K.E. Cockshull), Academic Press, New York, pp. 51-63.

Deng, X.-W., Matsui, M., Wei, N., Wagner, D., Chu, A.M., Feldmann, K.A. and Quail, P.H. (1992) *COP1*, an *Arabidopsis* regulatory gene, encodes a protein with both a zinc-binding motif and a G_β homologous domain. *Cell*, **71** 791-801.

Devlin, P.F., Halliday, K.J., Harberd, N.P. and Whitelam, G.C. (1996) The rosette habit of *Arabidopsis thaliana* is dependent upon phytochrome action: novel phytochromes control internode elongation and flowering time. *Plant J.*, **10** 1127-34.

Dunlap, J.C. (1996) Genetic and molecular analysis of circadian rhythms. *Ann. Rev. Genet.*, **30** 579-602.

Dynlacht, B.D., Weinzierl, R.O.J., Admon, A. and Tijan, R. (1993) The dTAFII80 subunit of Drosophila TFIID contains β-transducin repeats. *Nature*, **363** 176-79.

Edgerton, M.D. and Jones, A.M. (1992) Localization of protein–protein interactions between the subunits of phytochrome. *Plant Cell*, **4** 161-71.

Engelmann, W., Simon, K. and Phen, C.J. (1992) Leaf movement rhythm in *Arabidopsis thaliana*. *Zeitschrift fur Naturforschung*, **47C** 925-28.

Fuglevand, G., Jackson, J.A. and Jenkins, G.I. (1996) UV-A, UV-B and blue light signal transduction pathways interact synergistically to regulate chalcone synthase gene expression in *Arabidopsis*. *Plant Cell*, **8** 2347-57.

Furuya, M. and Song, P.-S. (1994) Assembly and properties of holophytochrome, in *Photomorphogenesis in Plants*, 2nd ed. (eds R.E. Kendrick and G.H.M. Kronenberg), Kluwer Academic Publishers, Dordrecht, pp. 105-140.

Gao, J. and Kaufman, L.S. (1994) Blue light regulation of the *Arabidopsis thaliana CAB1* gene. *Plant Physiol.*, **104** 1251-57.

Gorton, H.L., Williams, W.E. and Assmann, S.M. (1993) Circadian rhythms in stomatal responsiveness to red and blue light. *Plant Physiol.*, **103** 399-406.

Goto, N., Kumagai, T. and Koornneef, M. (1991) Flowering responses to light-breaks in photomorphogenic mutants of *Arabidopsis thaliana*, a long-day plant. *Physiol. Plant.*, **83** 209-15.

Hall, J.C. (1995) Tripping along the trail to the molecular mechanisms of biological clocks. *Trends Neurosci.*, **18** 230-40.

Hall, J.C. (1996) Are cycling gene-products as internal zeitgebers no longer the zeitgeist of chrono-biology. *Neuron,* **17** 799-802.

Halliday, K., Koornneef, M. and Whitelam, G.C. (1994) Phytochrome B and at least one other phytochrome mediate the accelerated flowering response of *Arabidopsis thaliana* L. to low red/far-red ratio. *Plant Physiol.,* **104** 1311-15.

Hicks, K.A., Millar, A.J., Carré, I.A., Somers, D.E., Straume, M., Meeks-Wagner, D.R. and Kay, S.A. (1996) Conditional circadian dysfunction of the *Arabidopsis early-flowering 3* mutant. *Science,* **274** 790-92.

Hoffman, P.D., Batschauer, A. and Hays, J.B. (1996) *PHH1,* a novel gene from *Arabidopsis thaliana* that encodes a protein similar to plant blue light photoreceptors and microbial photolyases. *Mol. Gen. Genet.,* **253** 259-65.

Jackson, J.A. and Jenkins, G.I. (1995) Extension-growth responses and expression of flavonoid biosynthesis genes in the *Arabidopsis hy4* mutant. *Planta,* **197** 233-39.

Janoudi, A.K., Gordon, W.R., Wagner, D., Quail, P.H. and Poff, K.L. (1997) Multiple phytochromes are involved in red-light-induced enhancement of first-positive phototropism in *Arabidopsis thaliana. Plant Physiol.,* **113** 975-79.

Johnson, C.H. (1994) Illuminating the clock: circadian photobiology. *Sem. Cell Biol.,* **5** 355-62.

Johnson, C.H. (1995) Photobiology of circadian rhythms, in *CRC Handbook of Organic Photo-chemistry and Photobiology* (eds W.M. Horspool and P.-S. Song), CRC Press, Boca Raton, pp. 1602-10.

Johnson, C.H., Knight, M.R., Kondo, T., Masson, P., Sedbrook, J., Haley, A. and Trewavas, A. (1995) Circadian oscillations of cytosolic and chloroplastic free calcium in plants. *Science,* **269** 1863-1865.

Johnson, E., Bradley, M., Harberd, N.P. and Whitelam, G.C. (1994) Photoresponses of light-grown *phyA* mutants of *Arabidopsis. Plant Physiol.,* **105** 141-49.

Kendrick, R.E. and Kronenberg, G.H.M. (eds) (1994) *Photomorphogenesis in Plants,* 2nd edn, Kluwer Academic Publishers, Dordrecht.

Kim, B.C., Soh, M.S., Kang, B.J., Furuya, M. and Nam, H.G. (1996) Two dominant photomorpho-genic mutations of *Arabidopsis thaliana* identified as suppressor mutations of *hy2. Plant J.,* **9** 441-56.

King, R.W. and Bagnall, D.J. (1996) Photoreceptors and the photoperiodic response controlling flowering of *Arabidopsis. Sem. Cell Dev. Biol.,* **7** 449-54.

Komachi, K., Reed, M.J. and Johnson, A.D. (1994) The WD repaets of Tup1 interact with the homeodomain protein a2. *Genes Dev.,* **8** 2857-67.

Koornneef, M., Rolff, E. and Spruit, C.J.P. (1980) Genetic control of light-inhibited hypocotyl elongation in *Arabidopsis thaliana* L. Heynh. *Zeitschrift fur Pflanzenphysiologie,* **100** 147-60.

Kreps, J.A. and Simon, A.E. (1997) Environmental and genetic effects on circadian clock-regulated gene expression in *Arabidopsis. Plant Cell,* **9** 297-304.

Lagarias, D.M., Wu, S.H. and Lagarias, J.C. (1995) Atypical phytochrome gene structure in the green alga *Mesotaenium caldariorum. Plant Mol. Biol.,* **29** 1127-42.

Landry, L.G., Stapleton, A.E., Lim, J., Hoffman, P., Hays, J.B., Walbot, V. and Last, R.L. (1997) An *Arabidopsis* photolyase mutant is hypersensitive to ultraviolet-B radiation. *Proc. Natl Acad. Sci. USA,* **94** 328-32.

Li, J., Nagpal, P., Vitart, V., McNorris, T.C. and Chory, J. (1996) A role for brassinosteroids in light-dependent development in *Arabidopsis. Science,* **272** 398-401.

Li, J., Biswas, M.G., Chao, A., Russell, D.W. and Chory, J. (1997) Conservation of function between mammalian and plant steroid 5α-reductases. *Proc. Natl Acad. Sci. USA,* **94** 3554-59.

Lin, Y. and Cheng, C,-L. (1997) A chlorate-resistant mutant defective in the regulation of nitrate reductase gene expression in *Arabidopsis* defines a new *HY* locus. *Plant Cell,* **9** 21-35.

Lin, C., Robertson, D.E., Ahmad, M., Raibekas, A.A., Jorns, M.S., Dutton, P.L. and Cashmore, A.R. (1995) Association of the flavin adenine-dinucleotide with the *Arabidopsis* blue light receptor CRY1. *Science,* **269** 968-70.

Lin, C., Ahmad, M., Chan, J. and Cashmore, A.R. (1996) *CRY2*: a second member of the *Arabidopsis* cryptochrome gene family. *Plant Physiol.*, **110** 1047.

Liscum, E. and Briggs, W.R. (1995) Mutations in the *NPH1* locus of *Arabidopsis* disrupt the perception of phototropic stimuli. *Plant Cell*, **7** 473-85.

Liscum, E. and Hangarter, R.P. (1994) Mutational analysis of blue-light sensing in *Arabidopsis*. *Plant Cell Environ.*, **17** 639-48.

Liu, Z.R., Taub, C.C. and McClung, C.R. (1996) Identification of an *Arabidopsis thaliana* ribulose-1,5-bisphosphate carboxylase oxygenase activase (*rca*) minimal promoter regulated by light and the circadian clock. *Plant Physiol.*, **112** 43-51.

Malhorta, K., Kim, S.T., Batschauer, A., Dawut, L. and Sancar, A. (1995) Putative blue light photoreceptors from *Arabidopsis thaliana* and *Sinapis alba* with a high degree of sequence homology to DNA photolyase contain the two photolyase cofactors but lack DNA-repair activity. *Biochemistry*, **34** 6892-99.

Mayer, R., Raventos, D. and Chua, N.-H. (1996) *det1*, *cop1* and *cop9* mutations cause inappropriate expression of several gene sets. *Plant Cell*, **8** 1951-59.

McClung, C.R. and Kay, S.A. (1994) Circadian rhythms in the higher plant, *Arabidopsis thaliana*, in *Arabidopsis thaliana* (eds C.S. Somerville and E. Meyerowitz), Cold Spring Harbor Press, Cold Spring Harbor, New York, pp. 615-37.

McCormac, A.C., Smith, H. and Whitelam, G.C. (1993) Photoregulation of germination in seed of transgenic lines of tobacco and *Arabidopsis* which express an introduced cDNA encoding phytochrome A or phytochrome B. *Planta*, **191** 386-93.

Millar, A.J. (19978) Molecular intrigue between phototransduction and the circadian clock. *Ann. Bot.*, in press.

Millar, A.J. and Kay, S.A. (1996) Integration of circadian and phototransduction pathways in the network controlling *CAB* gene transcription in *Arabidopsis*. *Proc. Natl Acad. Sci. USA*, **93** 15491-96.

Millar, A.J. and Kay, S.A. (1997) The genetics of phototransduction and circadian rhythms in *Arabidopsis*. *BioEssays*, **19** 209-14.

Millar, A.J., Short, S.R., Chua, N.-H. and Kay, S.A. (1992) A novel circadian phenotype based on firefly luciferase expression in transgenic plants. *Plant Cell*, **4** 1075-87.

Millar, A.J., Straume, M., Chory, J., Chua, N.-H. and Kay, S.A. (1995a) The regulation of circadian period by phototransduction pathways in *Arabidopsis*. *Science*, **267** 1163-66.

Millar, A.J., Carré, I.A., Strayer, C.A., Chua, N.-H. and Kay, S.A. (1995b) Circadian clock mutants in *Arabidopsis* identified by luciferase imaging. *Science*, **267** 1161-63.

Mozley, D. and Thomas, B. (1995) Developmental and photobiological factors affecting photoperiodic induction in *Arabidopsis thaliana* Heynh. Landsberg *erecta*. *J. Exper. Bot.*, **46** 173-79.

Nagatani, A., Chory, J. and Furuya, M. (1991) Phytochrome B is not detectable in the *hy3* mutant of *Arabidopsis*, which is deficient in responding to end-of-day far-red light treatments. *Plant Cell Physiol.*, **32** 1119-22.

Nagatani, A., Reed, J.W. and Chory, J. (1993) Isolation and initial characterisation of *Arabidopsis* mutants that are deficient in phytochrome A. *Plant Physiol.*, **102** 269-77.

Parks, B.M. and Quail, P.H. (1993) *hy8*, a new class of *Arabidopsis* long hypocotyl mutants deficient in functional phytochrome A. *Plant Cell*, **5** 39-48.

Parks, B.M., Quail, P.H. and Hangarter, R.P. (1996) Phytochrome A regulates red-light induction of phototropic enhancement in *Arabidopsis*. *Plant Physiol.*, **110** 155-12.

Pepper, A., Delaney, T., Washburn, T., Poole, D. and Chory, J. (1994) *DET1*, A negative regulator of light-mediated development and gene expression in *Arabidopsis*, encodes a novel nuclear-localized protein. *Cell*, **78** 109-16.

Quail, P.H. (1991) Phytochrome. A light-activated molecular switch that regulates plant gene expression. *Ann. Rev. Gen.*, **25** 389-409.

Quail, P.H. (1994) Phytochrome genes and their expression, in *Photomorphogenesis in Plants*, 2nd edn (eds R.E. Kendrick and G.H.M. Kronenberg), Kluwer Academic Publishers, Dordrecht, pp. 71-104.

Quail, P.H., Boylan, M.T., Parks, B.M., Short, T.W., Xu, Y. and Wagner, D. (1995) Phytochromes: photosensory perception and signal transduction. *Science*, **268** 675-80.

Reed, J.W., Nagpal, P., Poole, D.S., Furuya, M. and Chory, J. (1993) Mutations in the gene for red/far-red light receptor phytochrome B alter cell elongation and physiological responses throughout *Arabidopsis* development. *Plant Cell*, **5** 147-57.

Reed, J.W., Nagatani, A., Elich, T.D., Fagan, M. and Chory, J. (1994) Phytochrome A and phytochrome B have overlapping but distinct functions in *Arabidopsis* development. *Plant Physiol.*, **104** 1139-49.

Robson, P.R.H. and Smith, H. (1996) Genetic and transgenic evidence that phytochrome A and phytochrome B act to modulate the gravitropic orientation of *Arabidopsis thaliana* hypocotyls. *Plant Physiol.*, **110** 211-16.

Sakamoto, K. and Nagatani, A. (1996) Nuclear localization activity of phytochrome B. *Plant J.*, **10** 859-68.

Sharrock, R.A. and Quail, P.H. (1989) Novel phytochrome sequences in *Arabidopsis thaliana*: structure, evolution, and differential expression of a plant regulatory photoreceptor family. *Genes Dev.*, **3** 1745-57.

Shinomura, T., Nagatani, A., Chory, J., and Furuya, M. (1994) The induction of seed-germination in *Arabidopsis thaliana* is regulated principally by phytochrome B and secondarily by phytochrome A. *Plant Physiol.*, **104** 363-71.

Shinomura, T., Nagatani, A., Hanzawa, H., Kubota, M., Watanabe, M. and Furuya, M. (1996) Action spectra for phytochrome A-specific and B-specific photoinduction of seed-germination in *Arabidopsis thaliana*. *Proc. Natl Acad. Sci. USA*, **93** 8129-33.

Smith, H. (1995) Physiological and ecological function within the phytochrome family. *Ann. Rev. Plant Physiol. Plant Mol. Biol.*, **46** 289-315.

Smith, H. and Whitelam, G.C. (1990) Phytochrome, a family of photoreceptors with multiple physiological roles. *Plant Cell Environ.*, **13** 695-707.

Somers, D.E. and Quail, P.H. (1995) Temporal and spatial expression patterns of *PHYA* and *PHYB* genes in *Arabidopsis*. *Plant J.*, **7** 413-27.

Staub, J.M., Wei, N. and Deng, X.-W. (1996) Evidence for FUS6 as a component of the nuclear-localized COP9 complex in *Arabidopsis*. *Plant Cell*, **8** 2047-56.

Su, W.P. and Howell, S.H. (1995) The effects of cytokinin and light on hypocotyl elongation in *Arabidopsis* seedlings are independent and additive. *Plant Physiol.*, **108** 1423-30.

Sweeney, B.M. (1987) *Rhythmic Phenomena in Plants*, Academic Press, San Diego.

Szekeres, M., Nemeth, K., Koncz-Kalman, Z., Mathur, J., Kauschmann, A., Altmann, T., Redei, G., Nagy, F., Schell, J. and Koncz, C. (1996) Brassinosteroids rescue the deficiency of CYP90, a cytochrome P450, controlling cell elongation and de-etiolation in *Arabidopsis*. *Cell*, **85** 171-82.

Teakle, G.R. and Kay, S.A. (1995) The GATA-binding protein CGF-1 is closely-related to GT-1. *Plant Mol. Biol.*, **29** 1253-66.

Terry, M.J., Wahleithner, J.A. and Lagarias, J.C. (1993) Biosynthesis of the plant photoreceptor phytochrome. *Arch. Biochem. Biophys.*, **306** 1-15.

von Arnim, A. and Deng, X.-W. (1994) Light inactivation of *Arabidopsis* photomorphogenic repressor COP1 involves a cell-specific regulation of its nucleocytoplasmic partitioning. *Cell*, **79** 1035-45.

von Arnim, A. and Deng, X.-W. (1996) A role for transcriptional repression during light control of plant development. *BioEssays*, **18** 905-10.

Wagner, D., Fairchild, C.D., Kuhn, R.M. and Quail, P.H. (1996) Chromophore-bearing NH_2-terminal domains of phytochrome A and phytochrome B determine their photosensory specificity and differential light lability. *Proc. Natl Acad. Sci. USA*, **93** 4011-15.

Wagner, D., Hoecker, U. and Quail, P.H. (1997) *RED1* is necessary for phytochrome B-mediated red light-specific signal transduction in *Arabidopsis*. *Plant Cell,* **9** 731-43.

Wang, Z.-Y., Kenigbuch, D., Sun, L., Haerl, E., Ong, M.S. and Tobin, E.M. (1997) A Myb-related transcription factor is involved in the phytochrome regulation of an *Arabidopsis Lhcb* gene. *Plant Cell,* **9** 491-507.

Wei, N. and Deng, X.-W. (1996) The role of the *COP/DET/FUS* genes in light control of *Arabidopsis* seedling development. *Plant Physiol.,* **112** 871-78.

Wei, N., Chamovitz, D.A. and Deng, X.-W. (1994) *Arabidopsis* COP9 is a component of a novel signalling complex mediating light control of development. *Cell,* **78** 117-24.

Whitelam, G.C. and Smith, H. (1991) Retention of phytochrome-mediated shade avoidance responses in phytochrome-deficient mutants of *Arabidopsis,* cucumber and tomato. *J. Plant Physiol.,* **39** 119-25.

Whitelam, G.C., Johnson, E., Peng, J., Carol, P., Anderson, M.L., Cowl, J.S. and Harberd, N.P. (1993) Phytochrome A null mutants of *Arabidopsis* display a wild-type phenotype in white light. *Plant Cell,* **5** 757-68.

Zagotta, M.T., Hicks, K.A., Jacobs, C.I., Young, J.C., Hangarter, R.P. and Meeks-Wagner, D.R. (1996) The *Arabidopsis ELF3* gene regulates vegetative photomorphogenesis and the photoperiodic induction of flowering. *Plant J.,* **10** 691-702.

Zagotta, M.T., Shannon, S., Jacobs, C. and Meeks-Wagner, D.R. (1992) Early-flowering mutants of *Arabidopsis thaliana. Aust. J. Plant Physiol.,* **19** 411-18.

Zhong, H.H. and McClung, C.R. (1996) The circadian clock gates expression of two *Arabidopsis* catalase genes to distinct and opposite circadian phases. *Mol. Gen. Genet.,* **251** 196-203.

11 Programmed cell death in plants

John Gray and Guri S. Johal

11.1 Introduction

> ...and in a field in which plant biologists might be expected to be first and foremost—
> programmed cell death—we find ourselves following the leads offered by mammalian,
> worm and fly biologists. So? Get curious. Pick a tough problem and go to work. (Joe
> Varner, *Plant Cell*, July 1995)

Joe Varner's spur to further investigation of the phenomenon known as programmed cell death has indeed been addressed in recent years by plant biologists. It is becoming clear that programmed cell death events in plants include some broad parallels with animal and invertebrate systems but also other events that are peculiar to plants (Johal *et al.*, 1995; Dangl *et al.*, 1996; Greenberg, 1996; Havel and Durzan, 1996a; Jones and Dangl, 1996a; McCabe and Pennell, 1996; Mittler and Lam, 1996; Greenberg, 1997; Pennell and Lamb, 1997).

The life cycles of multicellular organisms revolve through a continuum of stages involving one cell as zygote to trillions of cells in some adult organisms. Many cells are lost because of the gross disruption of cellular homeostasis by external physical and chemical agents, which if too severe result in breaking of the life cycle, i.e. death of the individual before reproduction. It took a long time to realize, however, that a successful life cycle can require the deliberate death of certain cells in an individual and that this is an energy-requiring process dynamically regulated by both developmental and environmental cues (Jacobson *et al.*, 1997). This process has been named programmed cell death (PCD) and a well-known example is the readsorption of tail tissue during the metamorphosis of a tadpole into a frog.

PCD is now envisaged to occur at many points in the life cycle of plants and these are summarized in Figure 11.1. The descriptions of these processes are still limited. The parallels and differences between PCD pathways and processes in plants and other organisms are only beginning to be defined. In this chapter the current paradigms and definitions for PCD that have been largely defined by animal studies are described. These concepts are then employed in the examination of studies that illuminate cell death events in plants. It has been hypothesized that PCD evolved prior to multicellularity and thus animals and plants may share components of PCD pathways (Ameisen, 1996; Jacobson *et al.*,

1997). Research has not yet established that this is so and we point to instances where mechanisms of PCD seem to have evolved uniquely in plants. The tools to reveal these mechanisms are being applied in model organisms such as *Arabidopsis* and maize, and possible avenues of future research are proposed.

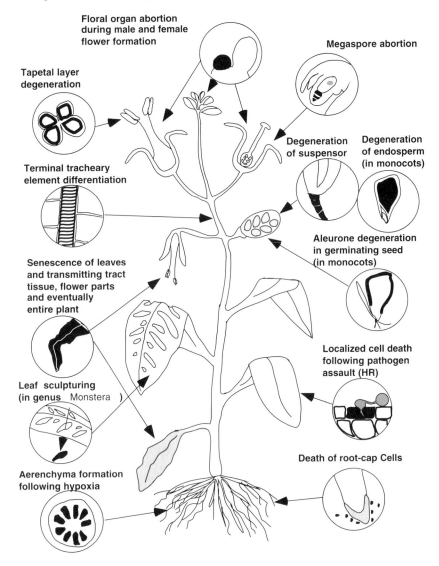

Floral organ abortion during male and female flower formation

Megaspore abortion

Tapetal layer degeneration

Terminal tracheary element differentiation

Degeneration of suspensor

Degeneration of endosperm (in monocots)

Aleurone degeneration in germinating seed (in monocots)

Senescence of leaves and transmitting tract tissue, flower parts and eventually entire plant

Localized cell death following pathogen assault (HR)

Leaf sculpturing (in genus Monstera **)**

Death of root-cap Cells

Aerenchyma formation following hypoxia

Figure 11.1 Putative PCD events in higher plants. This idealized schematic indicates the cells, organs and tissues in a plant that undergo cell death events during the normal developmental cycle. Dark areas indicate where PCD events are believed to occur. See text for details.

11.2 Definitions and emerging paradigms for PCD

In animals, cell deaths have been distinguished by whether they are caused by 'accident' or 'design' and named traumatic (accidental) or programmed cell death, respectively (Schwartzman and Cidlowski, 1993). The mechanisms involved in both cell death processes are largely but not completely exclusive. The term 'necrosis' has been used to describe traumatic cell death but is more correctly limited to the description of changes secondary to the initiation of cell death by any mechanism, including apoptosis (Majno and Joris, 1995).

11.2.1 Traumatic cell death

Traumatic events for the cell, including low pH, heating and freezing, or exposure to toxins such as azide or ethanol, cause traumatic cell death (TCD). The loss of control over cell volume owing to changes in cell permeability caused by either membrane phospholipid degradation, production of amphipathic lipids, damage to the cytoskeleton or generation of toxic oxygen species and free radicals, is thought to be the primary cellular cause of necrosis (Buja *et al.*, 1993). Cytoplasmic and mitochondrial swelling are typically the first microscopically observable events. Other cytoplasmic organelles become dissolute and 'necrotic' cells become separated from neighboring cells. Membrane permeability changes lead to increased intracellular sodium and calcium levels. Irreversible damage is thought to occur at the point when mitochondrial structure and function cannot be restored. Lysosomes rupture, releasing hydrolytic enzymes into the cytoplasm and leading to further instability of the nucleus, which swells, becomes darker (pycnosis) and ruptures (karyolysis). DNA is exposed following proteolysis of histones and is degraded by lysosomal nucleases into a continuous spectrum of sizes. Finally the entire cell ruptures and the release of chemotactic and other signals attracts inflammatory cells into the surrounding tissue (Buja *et al.*, 1993; Schwartzman and Cidlowski, 1993).

Whether a cell dies in a traumatic or a programmed fashion appears to be dependent on the concentration of the toxin and the ability of the cell to perceive and nullify that toxin. For example, $100\,\mu M$ inorganic mercuric chloride will induce TCD in cell monolayers but at $35\,\mu M$ of the same toxin cell death with apoptotic features occurs (Duncan-Achanzar *et al.*, 1996). In rat neurons, treatment with 3-nitropropionic acid (an inhibitor of succinate dehydrogenase) induces a receptor-mediated TCD and a receptor-independent apoptosis. The number of receptors in the cell influences which cell death event will occur (Pang and Geddes, 1997). Thus cells appear to monitor cellular damage and activate a PCD pathway at certain threshold levels of damage unless first they are overwhelmed by excessive toxin levels, in which case TCD will ensue.

11.2.2 Programmed cell death and apoptosis

'Spontaneous' cell death events that occur in scattered cells of healthy tissue were first defined using morphological criteria. Inhibitors of protein and RNA synthesis were later used to show that such cell death is an active process requiring the synthesis of new molecules. The term 'apoptosis' (from the Greek *apo* away from, *ptosis* falling) was originally used to refer these cells in which an intracellular death program with associated morphological changes was activated or inhibited by various environmental stimuli. Programmed cell death (PCD) includes any cell death event that is triggered by external or internal factors and actively mediated by an intracellular death program, even though all of the morphological characteristics of apoptosis may not be observed. PCD is used extensively in animal development for sculpting organs (e.g. elimination of interdigit tissue), deleting unwanted organs (e.g. tadpole tail), adjusting cell numbers (e.g. redundant neurons) and in eliminating unwanted or damaged cells (e.g. cancer cells) (Jacobson *et al.*, 1997).

11.2.2.1 Cellular features of PCD in animals

Several cellular changes are frequently observed in cells undergoing PCD although these should not be considered as typical of all PCD events (Schwartz *et al.*, 1993). Using light and electron microscopy it can be observed that cells undergoing PCD often exhibit cytoplasmic and nuclear condensation. Cell junctions are lost and cells are rapidly phagocytosed so that they often appear within other cells prior to exhibiting some of the further features of the cell death process (Jacobson *et al.*, 1997). Alterations in mitochondrial permeability linked to membrane potential disruption precede nuclear and plasma membrane changes. The nuclear chromatin coalesces into one or several clumps and the nucleus later fragments (karyohexis).

Chromatin degradation during PCD can occur in a non-random fashion owing to the cleaving of the internucleosomal DNA, which results in a 200 bp ladder detected in electrophoretically separated DNA samples (Schwartzman and Cidlowski, 1993). Histologically the 3'OH ends of degraded DNA can be detected by fluorescent end-labeling (known as a TUNEL assay). However, since this labeling technique will also label DNA cleaved non-specifically it cannot be used as a sole method for determining that apoptosis and oligonucleosomal laddering is taking place (Collins *et al.*, 1992). In contrast to TCD there is no apparent proteolysis of histones and other nuclear proteins at the time the DNA fragmentation becomes observable (Schwartzman and Cidlowski, 1993).

Cellular condensation is the result of loss of cellular contents as endoplasmic reticulum vesicles form and on fusing with the cell membrane empty their contents extracellularly (membrane blebbing). The cell breaks into a number of membrane-bound 'apoptotic bodies' containing nuclear fragments and intact organelles (Schwartzman and Cidlowski, 1993). The apoptotic bodies are usually rapidly phagocytosed and this process may be mediated by specific receptors.

Apoptotic bodies do not lyse and cellular contents are not released, and thus the evocation of an inflammatory response is avoided. The entire process can be very rapid (sometimes less than 1 hour long) and this leads to the belief that the full extent of apoptosis in development is underestimated (Jacobson *et al.*, 1997).

Although the above characteristics are most often used to define PCD events, it is thought that there is considerable variation in the subcellular events that can occur during PCD events (Schwartz *et al.*, 1993). During metamorphosis of the blow-fly (*Calliphara vomitoria*) the salivary glands cells are seen to vacuolate and swell (oncosis) rather than condense and shrink. A transient enhancement in autophagy and acid phosphatase activity followed by autolysis is observed. The mitochondria persist until the cell fragments but the nucleus does not show the distinct chromatin margination and blebbing that is typical of apoptosis (Bowen *et al.*, 1996). It has been suggested that a larger series of PCD descriptions should be used to reflect the extent to which the morphological and biochemical features of traumatic cell death, apoptosis and autophagy are observed in a particular case (Schulte-Hermann *et al.*, 1997).

11.2.2.2 Model for molecular program underlying PCD in animals

Tremendous advances have been made in the elucidation of the mechanisms of PCD in animals (Schwartzman and Cidlowski, 1993; Majno and Joris, 1995; Hale *et al.*, 1996; Miura and Yuan, 1996; Jacobson *et al.*, 1997; Jehn and Osborne, 1997). Figure 11.2 shows an idealized schematic of the signalling and execution steps known to function during animal PCD. This model serves as a framework for the types of molecules that we might expect to find underlying PCD programs in plants with the caveat that new paradigms particular to plants may have to be devised.

A key feature of PCD is that it is regulated by intracellular and extracellular signals. Both cell survival and cell death factors are known that bind to cell surface receptors and transduce a signal to an intracellular cell death or cell survival pathway (Figure 11.2) (Deshmukh and Johnson, 1997; Nagata, 1997). One regulator of apoptosis in animals is ceramide (N-acyl-sphingosine), which can induce apoptosis in various cell types either by a stress-activated protein kinase cascade or possibly by inhibiting a cell death suppressor (Pena *et al.*, 1997). Toxin-damaged, dysfunctional or harmful cells may be eliminated from an individual by intracellular monitoring and checkpoints during the cell cycle. Transcription-dependent and transcription-independent programs affect this process in different cell types (Griffiths *et al.*, 1997). A key component in this process is the p53 tumor suppressor protein—a transcription factor that regulates the expression of cell cycle, DNA repair and apoptosis genes (e.g. bax, see below) (Levine, 1997).

Key execution molecules of PCD are proteases named caspases, as evidenced by caspase inhibitors that inhibit PCD. Regulatory caspases (e.g. *ced3* of

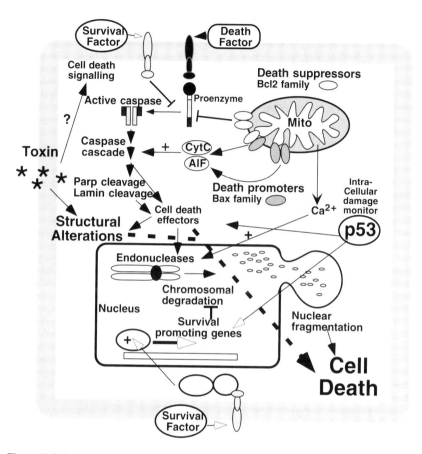

Figure 11.2 Summary model for molecular program(s) underlying PCD. This idealized cell is based on animal studies and indicates features of PCD and some of the known molecular components involved in the signalling and execution of this process. White or dark arrows indicate cell survival or cell death promoting factors respectively. A plethora of molecules serve to integrate external and internal signals to regulate proteins or genes involved in cell maintenance or degradation. Caspase proteases are prominent in cleaving enzymes to the inactive or active forms that cause degradation of the cell. The caspase cascade is regulated by the ratio of cell death suppressor to promoter proteins of the bcl-2 and bax families. Cell death effectors include molecules such as endonucleases which randomly or non-randomly degrade DNA, thus ensuring exit from the cell cycle and ultimately cell death. Toxins may cause a traumatic cell death event or, at low levels and by an unknown mechanism, may trigger a destructive program by the cell itself. For further details see text.

C.elegans) cleave other proteases into their heteromeric active forms via a caspase cascade (Figure 11.2). Caspases cleave several substrates such as poly(ADP)ribose polymerase (PARP), lamin, actin and DNA-dependent kinase. The cleaved substrates are thought to permit other destructive processes such as chromosomal degradation, resulting in an orderly cell death (Hale *et al.*, 1996). Members of the caspase family have not been identified in plants.

PCD-inhibiting (*bcl-2* family) and PCD-promoting (bax family) proteins have been identified whose dimerization ratio serves to enhance or antagonize the other's function (Figure 11.2). These appear to be pore-forming membrane proteins and may regulate membrane permeability. The formation of mitochondrial pores could contribute to apoptosis by the generation of free radicals, release of Ca^{2+} into the cytosol to activate endonucleases, and the release of mitochondrial proteins (cytochrome c and apoptosis-inducing factor (AIF)) that may activate caspases, which effect cell death. These pore proteins may also interact directly with caspases to regulate their activity (Figure 11.2) (Reed, 1997).

It is clear that in order to eliminate unwanted cells animals have a constitutively expressed intracellular death program involving a proteolytic cascade that is carefully regulated by both intracellular and extracellular signals. We may not find the same molecules in plant PCD programs but an equally ordered sequence of morphological changes is expected. Such changes require an underlying mechanism that is carefully regulated and utilizes identifiable regulatory and execution molecules.

11.3 Overview of cell death in plants

In recent years there has been a surge of interest in documenting PCD events during the life cycle of plants and in isolating the genes involved. PCD is thought to occur at many stages in the life cycle of plants (Figure 11.1) and eleven of these stages are discussed below. The mechanisms of PCD are beginning to be elucidated but (with one possible exception which will be discussed later) homologues of animal PCD-related genes have not been identified in plants. Some cell death descriptions are dependent on morphological and histological observations but in some areas, such as xylem differentiation, the *h*ypersensitive *r*esponse (HR) and senescence, there is also active pursuit of the molecular mechanisms involved. The use of molecular genetic analysis to identify the components of PCD has centered on genetic mutations in disease resistance and 'lesion mimic' genes. The study of these genes promises to reveal how plant cells integrate external and internal signals with the PCD machinery.

11.3.1 Megaspore abortion

Approximately 70% of plant species examined (including maize and *Arabidopsis*) exhibit a monokaryosporic pattern of embryo sac development (Reiser, 1993), although numerous variations in this process have been described (Russell, 1993). In this process a single megasporocyte gives rise to four megaspores via meiosis and three of these subsequently degenerate (Bell, 1996). The remaining megaspore gives rise to a mature megagametophyte by subsequent mitotic divisions. In *Marsilea* and *Arabidopsis* the tetrad of megaspores exhibits a tetrahedral symmetry within which the abortive megaspores are apparently randomly placed and no recognizable controlling gradient has been found. A possible nutritive advantage to the megaspore that lies closest to the chalaza has been proposed but this seems inadequate to explain why the remaining sporocytes degenerate so rapidly before the surviving megaspore begins to develop (Bell, 1996). In *Clintonia* a tetrasporic development occurs followed by degeneration of megaspore nucleii as opposed to cells, resulting in a functionally monosporic development (ex-tetra-monokaryoscopic) (Russell, 1993). This and other variations in megaspore development suggest that nuclear degeneration may precede cellular degeneration. The genetic mechanism underlying this process is not understood at the present time (Bell, 1996). Several sporophytic mutants exist in maize (Golubovskaya *et al.*, 1992) and *Arabidopsis* (Reiser, 1993) which may provide approaches to identifying the genes involved in megaspore formation. For instance, the female-sterile ovule mutant-3 (*ovm*3) of *Arabidopsis* fails to develop integuments around the developing megaspore and although meiosis is initiated normally, all products of meiosis degenerate. The sterile nature of such mutations, however, confounds current approaches to cloning the genes involved.

11.3.2 Anther/tapetal degeneration

The importance of male sterility in plant breeding practices has stimulated the study of anther development in plants (Goldberg *et al.*, 1993). During anther dehiscence, certain cell types, including the tapetum, undergo an ordered process of cell death in which many hydrolytic enzymes such as RNases, proteases, pectate lyases, polygalacturonases, and cellulases are induced. One marker for this process is a thiol endopeptidase gene whose mRNA accumulates in cells destined to degenerate just prior to their destruction (Goldberg *et al.*, 1993). In *Arabidopsis*, following completion of the exine of the pollen, the tapetal cell plastids develop membrane-bound inclusions with osmiophilic and electron-transparent regions. The plastids undergo ultrastructural changes that suggest breakdown of the inclusion membranes followed by release of their contents into the locule prior to the complete degeneration of the tapetal cells (Owen and Makaroff, 1995). There are several mutants that alter anther development, for example *late dehiscence* in *Arabidopsis*, which fails to dehisce in time and releases pollen after the stigma is no longer receptive (Goldberg *et al.*, 1993). In the *stamenless*2 mutation of tomato the degeneration of the tapetal layer is delayed and the pollen exine layer fails to develop, resulting in male sterility (Sawhney and Bhadula, 1988). The genetically controlled temporal, tissue and cell specific nature of cell death in anthers and the associated subcellular changes strongly suggest that PCD events occur during anther development. Further biochemical characterization is required and cloning of anther mutations should be fruitful in elucidating the regulatory and execution genes involved in these cell death events.

11.3.3 Floral organ abortion

In plants with perfect (hermaphroditic) flowers, such as *Arabidopsis*, there is no evidence that PCD is used to achieve appropriate cell numbers in stem cell populations. However, in plants with unisexual flowers either certain primordia are not formed or, as in the case of maize, both male and female primordia are formed and one of these will later degenerate by a process referred to as programmed organ death (POD) (DeLong *et al.*, 1993; Dellaporta and Calderon-Urrea, 1994). This deliberate degeneration of tissue may involve a PCD mechanism.

In the tassel of maize, gynoecial abortion occurs and stamens develop to maturity whereas in the ear gynoecia develop into fertile ovules and stamen primordia are aborted (Cheng *et al.*, 1983). Several mutations in the maize TASSELSEED (*TS*) genes result in bisexual tassel florets (*ts4, ts5* and *ts6*) or completely pistillate tassels (*ts1* and *ts2*). The *tasselseed 2* gene has been isolated and encodes a protein that bears homology to short-chain alcohol dehydro-genases and hydroxysteroid dehydrogenases from a variety of species and also to an RNA from tomato whose abundance is regulated by gibberellins (GAs) (DeLong *et al.*, 1993; Jacobsen and Olszewski, 1996). A second class of muta-tions in maize (dwarf mutants *d1, d2, d3* and *d5*) blocks specific steps in gibberellin biosynthesis. These mutants have perfect flowers in the ear owing to failure of the stamen to abort. Application of GA_3 can cause partial to complete staminate-to-pistillate conversion of the tassel when applied before the deter-mination of sexuality. Together these results indicate that tasselseed genes act antagonistically to the feminizing effects of GA possibly by direct inactivation or redirecting of biosynthesis (Dellaporta and Calderon-Urrea, 1994). *In situ* localization indicates that the tasselseed 2 RNA localizes to gynoecial cells prior to cell death. The cell death process itself exhibits vacuolation and loss of cyto-plasmic organelles (Cheng *et al.*, 1983) but a thorough microscopic examination of these events has not yet been conducted.

This process of organ abortion bears a strong resemblance to animal PCD where certain neurons are maintained only if they receive a sufficient supply of a survival factor (Deshmukh and Johnson, 1997). In this instance a steroid com-pound interacts with an underlying genetic program to determine the develop-mental fate of cells to further division or death, and would appear to comprise a clear instance of PCD in plants.

11.3.4 Suspensor degeneration

One of the first cell death events to occur following fertilization is the degrada-tion of the suspensor (earlier possible PCD events might be the degeneration of synergid and antipodal cells). The suspensor develops from the basal cell after the first cell division of the zygote. Massive suspensors develop in runner bean (*Phaseolus coccineus*) and nasturtium (*Tropaeolum majus*) which have permitted the development and degeneration of suspensors to be examined in considerable detail (Gärtner and Nagl, 1980; Singh *et al.*, 1980). In addition to positioning the embryo within the seed, the suspensor functions as an 'embryonic root' to transport essential nutrients from maternal tissue to the embryo as well as synthesizing growth factors such as gibberellin (Yeung and Meinke, 1993). After the embryo reaches the 'heart' stage of development, the suspensor begins to degenerate in an autophagous manner. Specialized organelles named 'plastolysosomes', which are a form of specialized leukoplast, develop prior to degeneration in *P. coccineus* and appear to envelope neighboring cytoplasm. In *T. majus* it is the mitochondria that develop into autophagous organelles (Gärtner and Nagl, 1980; Singh *et al.*, 1980; Jones and Dangl, 1996). Degradation probably occurs by hydrolysis owing to the detection of hydrolytic enzymes and of acid phosphatase activity in autolysing suspensors (Gärtner and Nagl, 1980; Singh *et al.*, 1980). While suspensor degradation appears to be dependent on intracellular active processes, other studies suggest that the embryo may play a role in regulating suspensor viability.

The developmental mutant *twin* of *Arabidopsis* causes the occasional production of a second embryo from the suspensor but cell death still occurs between the first and second embryos (Vernon and Meinke, 1994). In the abnormal suspensor mutants *sus1* and *sus2* abnormal embryo development precedes the enlargement of the suspensor, which acquires some characteristics normally restricted to the embryo (Schwartz *et al.*, 1994). It is hypothesized that the embryo fails to produce an inhibiting compound that serves to restrict suspensor growth (Yeung and Meinke, 1993; Meinke, 1995). In embryogenic suspension cultures, totipotent cells divide into asymmetric cell pairs: one goes on to establish an embryo but the second stops synthesizing DNA and dies. There is vesiculation of the cytoplasmic organelles and positive TUNEL staining in the dying cells. When dead cells are immunologically cell sorted they are found to contain DNA fragments of about 145 bp, which may represent the end products of DNA degradation during a PCD-type event (Havel and Durzan, 1996a; McCabe *et al.*, 1997; Pennell and Lamb, 1997). This system does not distinguish whether or not the dying cells are progenitors of suspensors but the observations indicate that there is positive and negative regulation of suspensor development by the embryo and allow for the possibility that a PCD event may be requisite for proper embryo development (Jones and Dangl, 1996; Pennell and Lamb, 1997). A somatic embryogenesis system for Norway Spruce also provided microscopic evidence and TUNEL staining evidence for apoptosis-like cell death events in degenerating egg equivalents and differentiating suspensors (Havel and Durzan, 1996b). It would be of value if this system could be extended to *Arabidopsis*, where somatic embryogenesis of suspensor mutants could be examined.

11.3.5 Terminal differentiation of aleurone and endosperm

The control by the embryo over terminal aleurone growth is well understood in cereals and initiates with the induction of hydrolytic enzyme genes in the aleurone by GA released from the embryo. Following degradation of the endosperm the aleurone cells die. This process can be completely blocked by a protein phosphatase inhibitor (okadaic acid), which also blocks cytosolic Ca^{2+} increases that occur following GA triggering of the aleurone cells. ABA can cause decreases in aleurone cell Ca^{2+} levels and appears to have opposing regulatory effects to GA (Kuo *et al.*, 1996). The dying aleurone cells exhibit features of apoptosis, namely cytoplasmic and nucleus condensation, and TUNEL staining (Kuo *et al.*, 1996; Wang *et al*, 1996a). Furthermore DNA laddering typical of apoptosis was observed when DNA from imbibed aleurone layers was examined. This observation was pronounced when cellulase was used to create protoplasts before DNA isolation, possibly suggesting that cell wall degradation is part of the triggering mechanism and may allow better synchronization of the apoptotic events in aleurone cells. ABA and okadaic acid were also found to counteract the DNA fragmentation in aleurone protoplasts (Wang *et al.*, 1996b). The association of the morphological and biochemical characteristics of apoptosis and a hormonal signalling system with aleurone cell death is strong evidence that PCD is occurring in this tissue.

In species that lack an aleurone layer, such as lettuce, the seed endosperm tissue is alive and active during germination. In other species the endosperm transports nutrients to embryonic leaves (cotyledons) which then serve as storage tissues for later embryonic development. The reversal of metabolism so that the endosperm serves as a source to the embryo is reminiscent of the processes of senescence in leaves and petals that export cell constituents in a highly co-ordinated manner prior to cell death, which may also be a form of PCD (discussed later). In cereals, endosperm storage cells die during normal kernel development. It has been shown that endosperm cell death in maize is accompanied by internucleosomal DNA fragmentation and that this process is more pronounced in the starch defective mutant *shrunken2* (*sh2*), which exhibits earlier and more rapid endosperm cell death than normal. Ethylene precursor levels and ethylene production rates are elevated in this mutant and an ethylene antagonist reduced the extent of DNA fragmentation. Furthermore, the application of ethylene to another defective endosperm mutant *sugary1* (*su1*) caused earlier endosperm cell death and more extensive DNA fragmentation (Young *et al.*, 1997). Thus it appears that selective PCD occurs within the kernel and that ethylene is involved in triggering this process. The regulation of ethylene production, possibly through developmentally regulated hormonal signals and the distribution of ethylene receptors in the endosperm, is deserving of closer examination. Because of the temporal and spatial separation of cell death in aleurone and endosperm and the plenitude of mutants available that affect these tissues, the maize kernel provides an attractive system for elucidating the interaction of the regulatory mechanisms that program such events.

11.3.6 Aerenchyma formation

Many plants can adapt to hypoxic (but not anoxic) soil conditions by the formation of intercellular spaces (lysigeny) in parenchymatous tissue and the resulting tissues are referred to as aerenchyma (Vartapetian and Jackson, 1997). These air spaces accelerate O_2 transfer from the shoot to root cells that are undergoing metabolic adjustment to fermentative conditions. Air spaces can arise from the creation of gaps between cells in growing tissue (schizogeny), such as is observed in the leaves of *Sagittaria lancifolia* (Schussler and Longstreth, 1996). In maize roots, only cortical cells interior to the hypodermis and exterior to the endodermis give rise to aerenchyma and in these cells there is complete lysis of tissue with the disappearance of all cell components including the cell wall within 24 hours of the initiation of cell death (Campbell and Drew, 1983). Ethylene regulates this process, as evidenced by increased ethylene in hypoxic maize roots and the induction of aerenchyma formation by ethylene even under completely aerobic conditions (Jackson *et al.*, 1985; He *et al.*, 1996). Increased cellulase activity, which precedes visible cell degradation, is used as a marker of aerenchyma formation although it has not been shown if cellulase activity is limited to the cells that disappear. Both cell death and cellulase activity are inhibited in hypoxic roots in the presence of inositol phospholipids, Ca^{2+}-calmodulin and protein kinases, but are promoted by reagents that activate G-proteins, increase cytosolic Ca^{2+} or inhibit protein phosphatases. Thus increases in intracellular Ca^{2+} levels are thought to mediate the ethylene signal that promotes aerenchyma formation during hypoxia (He *et al.*, 1996). The genes that are activated in response to hypoxia are those involved in glycolysis and related processes and a xyloglucan endo-transglycosylase has been isolated from maize whose expression is tightly linked to aerenchyma formation in an ethylene-dependent manner (Saab and Sachs, 1996). Other degradative enzymes associated with aerenchyma formation have yet to be identified. The tissue and cellular nature of cell death during aerenchyma formation and the role of ethylene in signalling these events indicate that aerenchyma formation could be a good example of extrinsically triggered PCD in plants. Plants cannot phagocytose but in the case of aerenchyma complete dissolution of cells occurs, effected in part by autophagocytosis. Further analyses of the sequence of molecular events will determine the extent to which this phenomenon can be considered a true PCD event. Elucidation of the genes involved will also assist the development of flood- and drought-resistant plants, which is a major objective for several major crops.

11.3.7 Differentiation of root cap cells

The root apical meristem produces a group of cells at the very root tip called root cap cells. The root cap cells are continuously moved laterally by new cap cells and die after five to six days. Root cap cell death occurs when plants are grown in water, indicating that it is not soil abrasion that is the cause of the death of these cells. In onion root cap cells, dying cells exhibit nuclear condensation and TUNEL staining (Wang *et al.*, 1996a). These observations suggest that PCD may occur in root cap cells. This hypothesis may be strengthened when one considers that the root cap cells have the same clonal origin as the suspensor which, as discussed above, appears to also undergo PCD. In *Arabidopsis*, following the first zygotic cell division, the apical cell gives rise to the embryo and the basal cell gives rise to the suspensor (Schiefelbein and Benfey, 1994). The topmost cell of the suspensor forms the hypophysis, which during embryogenesis divides to form a lens-shaped cell that is the progenitor of the part of the embryo that gives rise to the 'columella' root cap. All other cells of the embryonic root meristem are derived from the apical cell. This formation of the root meristem from cells of different clonal origin has given rise to the hypothesis that cell–cell interactions may co-ordinate this process (Schiefelbein and Benfey, 1994). Root hairs do not have root caps but lateral roots do and difficulty in obtaining lateral root mutants has prompted the interpretation that lateral root formation is regulated by the same genes as those that control embryonic root development or other organs (Schiefelbein and Benfey, 1994). This parallel between suspensor and root cap clonal origin suggests that it would be worth examining root development in suspensor mutants such as *twin* and *sus* discussed above. In addition, inhibitors of PCD signals might be easily utilized in root cap studies because of their easy accessibility (Pennell and Lamb, 1997).

11.3.8 Sculpturing of leaf form

PCD is not generally used for the development of leaf form but an exception is observed in the genus *Monstera* of the Araceae family. The mature leaf blades of *Monstera* species exhibit holes or lacunae giving a characteristic 'Swiss Cheese' appearance (Figure 11.1). The developmental origin of these lacunae has been examined using scanning electron microscopy and shown to occur as localized regions of cell death. After the major lateral veins are defined (leaf length >5 mm), small indentations appear on the lamina surface as the dying cells have less volume than those around them. Ultimately the dying tissue dries up and falls out, leaving a hole in the blade surface. The mesophyll cells in the lacuna border redifferentiate into secondary epidermis much like an abscission scar (Kaplan, 1984). There is no apparent external cause of these lacunae and the variations in pattern between species imply a genetic regulation.

In sphagnum moss (*Sphagnopsida. S.* magellanicum) mature branch leaves exhibit well-defined pores that are important for the water relations of the plant. Pores form first in cells near the apex of the leaves and occur through a process of autophagy and autolysis of pre-existing cells. Acid phosphatase activity is associated with degeneration of the protoplasm in these cells (Holcombe, 1984). The apparent programmed nature of cell death in the sculpturing of leaf form is reminiscent of similar sculpturing of body parts in animals and is deserving of further study.

11.3.9 Terminal differentiation of tracheary elements

The evolution of xylem elements was essential for the conduction of water in land plants and the terminal differentiation of these cells appears to fulfill all the criteria for PCD. Tracheary element differentiation has been extensively studied during normal development and following wounding or grafting (Fukuda, 1996, 1997; Savidge, 1996; Groover *et al.*, 1997).

Homeobox genes are thought to determine the cell fate of cambial initials from which xylem and phloem derive, and the differentiation of these elements is regulated by sucrose and the phytohormones auxin and cytokinin (Aloni, 1993). The study of xylem differentiation has benefited enormously from the ability to induce xylem element formation *in vitro* from *Zinnia* mesophyll cells by using an appropriate combination of the aforementioned compounds (Fukuda, 1997). The steps in this process are summarized in Figure 11.3 (Fukuda, 1996, 1997; Groover *et al.*, 1997).

This system has been used to identify the changes in cell wall composition, architecture and gene expression that are associated with stages of dedifferentiation, redifferentiation and differentiation of the mesophyll cells to xylem elements. *In vitro* this process can occur in four days.

Stage 1 *In vivo* young xylem elements differentiate from procambium tissue or dedifferentiate from parenchyma following wounding and exhibit dense cytoplasm containing Golgi apparati and rough endoplasmic reticulum. *In vitro*, following hormone induction, chloroplasts move away from the cell wall and lose photosynthetic capacity. There is increased expression of proteases and protein apparatus genes.

Stage 2 During the second stage of redifferentiation there are increases in TE-element specific gene expression, protein synthesis machinery and cell size with concommittant actin and tubulin changes. *ted2*, *ted3* and *ted4* (Demura, 1994) are expressed at this stage and one of these encodes a protein with homology to crystallin—a quinone oxido-reductase. It has been suggested that this gene may function to generate the oxidative cell death signals that are also associated with cell death in animals (Greenberg, 1996). Cells are committed to form xylem after 48 hours and inhibition of brassinosteroid biosynthesis or the use of cysteine proteinase inhibitors prevents further differentiation (Kuriyama and Fukuda, unpublished results, cited in Fukuda, 1997). Cell wall changes include new arabinogalactoproteins (AGP) in primary walls, which are retained

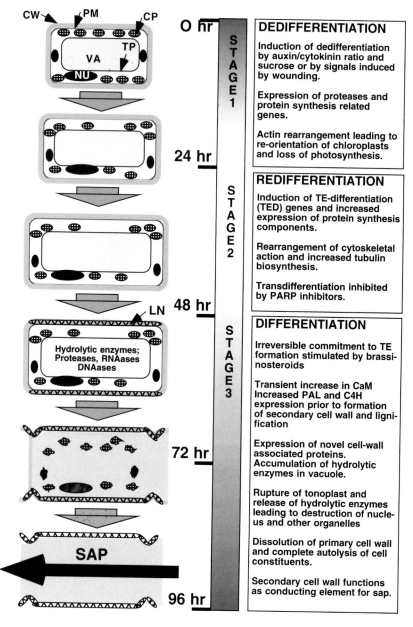

Figure 11.3 Stages in xylem cell differentiation: The major cellular events associated with the redifferentiation of mesophyll cells to xylem elements in *Zinnia* culture are indicated. CW, cell wall; PM, plasma membrane; CP, chloroplast; TP, tonoplast; NU, nucleus; VA, vacuole; LN, lignification; CaM, calmodulin; PAL, phenylalanine ammonia-lyase; C4H, cinnamic acid-4-hydroxylase. See text for details. (Modified from Fukuda, 1997.)

in the secondary thickenings of mature TEs, as well as a transient fucose-containing epitope that disappears abruptly 6 hours before any secondary thickenings are visible in the cells. In coleoptiles it appears that the related AGPs identify those cells that are committed to PCD (Stacey *et al.*, 1995).

Stage 3 The genes required for cell death and secondary wall formation accumulate after commitment to xylem differentiation. An S-adenosyl-L-methionine:trans-caffeoyl-coenzyme A 3-O-methyltransferase (CCoAOMT) has been implicated in the lignification of TEs *in vivo* and *in vitro* (Ye *et al.*, 1994; Beers and Freeman, 1997) as well as major phenylpropanoid genes. Hydrolytic enzymes, including several proteases (p48h-17, *ZCP4*, 30 and 60kD proteases), a Zn^{2+}-activated 43 kD nuclease (similar to a barley aleurone nuclease) and a 22 kD RNAase (*ZRNase1*) with possible vacuole targeting peptides, accumulate prior to autolysis of the cell (Thelen and Northcote, 1989; Ye and Varner, 1996) (Minami *et al.*, unpublished results, cited in Fukuda, 1997). TUNEL staining has been used to detect DNA degradation during xylem terminal differentiation (Mittler and Lam, 1995; Groover *et al.*, 1997) and it may be possible to determine if an oligonucleosomal ladder also is formed during this process using the *Zinnia* system.

The final step of xylem differentiation is when the nucleus, endoplasmic reticulum and organelles degenerate in a process named 'breakdown' or 'autolysis' or PCD (Esau *et al.*, 1963; Greenberg, 1996; Havel and Durzan, 1996a; Groover *et al.*, 1997). The controlled degradation of the cell does not appear to involve an oxidative burst (Groover *et al.*, 1997). Esau *et al.* (1963) described how nucleii of differentiating xylem vessels enlarge and become lobed with pores apparently missing. At later stages, cells exhibit little or no endoplasmic reticulum with the characteristic cellular component being a single-membraned vesicle smaller in size than a mitochondrion. Golgi apparati were not found in association with endoplasmic reticulum that is undergoing vesiculation and mitochondria degenerate and disappear completely. Finally the entire protoplast disappears so that only the lignified cell wall remains, through which water is conducted.

In summary, the differentiation of xylem through a combination of *in vivo* and *in vitro* studies exhibits all the hallmarks of PCD, i.e. (1) the fate of these cells appears to be homeotically determined in the laying down of cambium initials, (2) the underlying genetic determination is regulated by the phytohormones auxin and cytokinin, (3) the process of xylem differentiation is an active one involving a series of RNA and protein synthesis, cell wall formation and gradual degradation of intracellular contents, (4) the protoplasm of the xylem entirely disappears presumably being broken down and reabsorbed by surrounding cells, (5) the terminally differentiated xylem cell is dead and comprises the 'apoplastic' or non-living portion of the plant but its ability to conduct water is essential for the existence of land plants.

Only a few mutations affecting xylem differentiation may exist owing to the essential requirement of this process. The *irx* (irregular xylem) mutants of *Arabidopsis* exhibit decreased cellulose in their cell walls and xylem vessels can collapse, leading to a reduced stature of plants. However, the terminal differentiation of xylem is not affected (Turner and Somerville, 1997). In maize, a recessive mutant named *wil* (*wilted*) exhibits retarded differentiation of two large metaxylem elements in the vascular bundles of the stem although this improves with the maturation of the plant (Neuffer *et al.*, 1997). In addition to study of the *Zinnia* system it may be possible to isolate components of the TE PCD pathway by examining appropriate mutants (e.g. wilted or conditional phenotype) of *Arabidopsis* that can be discerned by microscopic examination.

11.3.10 Cell death in response to pathogen attack

In plants, rapid localized death of cells at an infection site is thought to comprise a determined disease-resistance mechanism by the host. This process is named the hypersensitive response (HR) and may be the primary cause of limiting pathogen spread, particularly in the case of obligate biotrophic fungi (Greenberg, 1996, 1997; Mittler and Lam, 1996; Morel and Dangl, 1997; Pennell and Lamb, 1997). In animals and bacteria many pathogens, particularly viruses, cause the host cells to commit suicide (Shen and Shenk, 1995; Jacobson *et al.*, 1997). Suicidal cell death may have been selected for in plants as an effective protection strategy from further infection for several reasons: dead cells will have limited plasmadesmatal connections through which viruses may propagate, dehydration of the dead cell may prevent bacterial survival and cellular disruption will cause the release of defense-related proteins and toxic metabolites into the apoplast and inhibit bacterial or fungal growth. Furthermore, the delaying of pathogen spread may also allow time for surrounding cells to mount more effective defense responses (Greenberg, 1997; Morel and Dangl, 1997; Pennell and Lamb, 1997).

In other cases, the HR may be a TCD event owing to cell disruption by the pathogen or a doublesafe resistance mechanism occurring after the arrest of pathogen growth, e.g. *Mlg*-mediated powdery mildew resistance in barley (Görg *et al.*, 1993). We outline here the morphological and molecular evidence that the HR is an active process genetically programmed by the plant cell. The HR can be induced by purified bacterial or fungal elicitors, active plant metabolism is required for such death to occur and the plant expresses a battery of defense-related genes and associated cellular changes (Croft *et al.*, 1990; He *et al.*, 1993; Nurnberger *et al.*, 1994; Levine *et al.*, 1996).

11.3.10.1 Cellular features of the hypersensitive response
Identifying the train of subcellular events that follows pathogen attack in plants is made difficult by the rapidity and asynchronicity with which cell death can occur (Freytag *et al.*, 1994). However, cells undergoing HR appear to exhibit

some features associated with PCD in animals, such as controlled DNA fragmentation (Collins *et al.*, 1992; Ryerson and Heath, 1996; Wang *et al.*, 1996a; Mittler *et al.*, 1997).

The application of an *Alternaria alternata* f sp *lycopersici* (AAL) toxin to tomato protoplasts or intact leaflets caused cell death and generation of a weak oligonucleosomal DNA ladder that was inhibited by cyclohexamide (Wang *et al.*, 1996a). Ladder intensity was enhanced by Ca^{2+} and inhibited by Zn^{2+}—a fact that parallels endonucleases that cause DNA fragmentation in animal apoptosis and those associated with xylem formation (Schwartzman and Cidlowski, 1993; Fukuda, 1997). Apoptotic bodies in AAL-treated protoplasts were reported by Wang *et al.* (1996a), and possibly also in intact soybean and *Arabidopsis* tissue (Levine *et al.*, 1996). AAL toxins are sphinganine analogues that inhibit the synthesis of ceramide, a regulator of apoptosis in animals (Merrill *et al.*, 1996; Pena *et al.*, 1997). However, a homologue of the suspected target of ceramide, $Bcl-X_L$, has not been found in plants and transgenic plants overexpressing mammalian $Bcl-X_L$ do not exhibit an altered death phenotype (Mittler *et al.*, 1996). Resistant cultivars of cowpea exhibited both oligonucleosomal laddering and TUNEL staining in cells containing fungal haustoria whereas susceptible cultivars that did not undergo HR did not (Ryerson and Heath, 1996). This chromosome degradation occurs early when some chloroplast damage has occurred but before evidence of further cellular disorganization.

Rapid DNA fragmentation but not an oligonucleosomal ladder is also observed in resistant tobacco plants infected with TMV (Mittler *et al.*, 1996). Condensation and vacuolization of the cytoplasm are observed with morphological changes first occurring in the chloroplast before nuclear changes are observed. DNA cleavage in the nucleus generates large 50kb fragments but condensation of the nucleus does not result in the formation of apoptotic bodies. Thus, whereas the genetic regulation and temporal rapidity of the cell death of cells resistant to TMV infection is indicative of a programmed cell death event the HR exhibits events that are similar but not typical of apoptosis in animals (Mittler and Lam, 1996, 1997).

11.3.10.2 Regulation of the hypersensitive response

Studies revealing components that regulate, execute and limit cell death during the HR and their possible interactions are discussed below and summarized in Figure 11.4. Efforts to isolate HR cell death regulatory and execution genes have focused on differential expression analysis and two major groups of plant genetic mutants, namely disease-resistance and lesion-mimic mutants. Disease-resistance mutants fail to mount an effective defense and lesion-mimic mutants produce disease-like lesions even in the absence of any pathogen (Walbot *et al.*, 1983; Greenberg and Ausubel, 1993; Dietrich *et al.*, 1994; Johal *et al.*, 1995; Dangl *et al.*, 1996; Hammond-Kosack and Jones, 1997; Kosslak *et al.*, 1997).

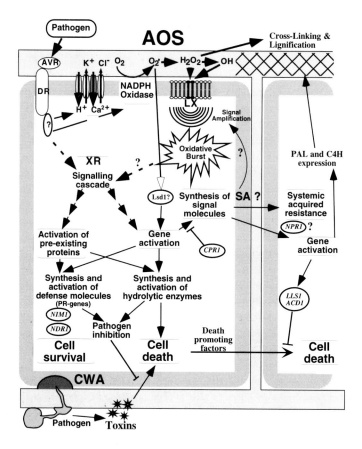

Figure 11.4 Features of HR in plants that may reveal an underlying cell death program. Signal molecules from pathogens (e.g. avirulence factor, AVR) bind to receptor proteins (disease-resistance genes, DR) in the membrane or within the cell and may trigger membrane changes such as ion fluxes (XR) and lipid peroxidation (LX), the production of activated oxygen species (AOS), or a kinase/ phosphatase cascade (dotted arrow). These signals acting alone or in concert activate pre-existing proteins and activate new gene expression (possibly mediated via NIM1 and NDR1) leading to proteins that protect the cell from the pathogen (e.g. callose deposition in the case of cell wall apposition formation (CWA), phytoalexin biosynthetic genes and oxidative stress protectant enzymes including PR (pathogenesis-related) genes). Alternatively, cell death effector genes (e.g. nucleases and proteases) are activated thus limiting pathogen spread but also killing the host cell (HR). HR is accompanied by the production of survival-promoting molecules (possibly salicylic acid, SA) that signal neighbouring cells to bolster defence responses through increased expression of phenyl-propanoid genes and lignification of cell walls. CPR1 and NPR1 may negatively and positively mediate this process respectively. Dying cells also release toxic or death-signalling molecules, which are removed by neighbouring cells (e.g. LLS1/ACD1). An oxidative burst that accompanies HR may enhance the rate of cell death either directly or possibly via redox responsive transcriptional regulators (e.g. LSD1). SA may potentiate this oxidative burst, which also produces H_2O_2. This may in turn act as a short-distance signalling molecule. Toxins released by the pathogen may cause TCD or at low levels trigger a PCD event, as is suggested for animal systems (see also Figure 11.2). For further details see text.

The initiation of a response to specific pathogens involves a recognition step involving direct interaction between a receptor and ligand (Figure 11.4). Several disease-resistance genes have been isolated, many of which encode receptor-like proteins (Hammond-Kosack and Jones, 1997). For instance, the Pto resistance protein from potato interacts with the AvrPto protein from the pathogen *Pseudomonas syringae* pv *tomato* (Scofield *et al.*, 1996; Tang *et al.*, 1996). Protein interaction studies reveal how proteins such as Pto effect a signalling cascade, regulating plant pathogenesis-related genes (Zhou *et al.*, 1997).

Some lesion-mimic mutations may result from disease-resistance genes that have gone awry (Walbot *et al.*, 1983; Johal *et al*, 1995). Some of these mutants exhibit altered gene expression and morphological features similar to those that occur during pathogen infection (Greenberg and Ausubel, 1993; Dietrich *et al.*, 1994). Recombination events within the complex maize rust-resistance (*Rp1*) locus lead to the creation of an allele (*Rp1-Kr1*) that exhibited lesion development in the absence of any biotic stimulus and the loss of all race specificity (Hu *et al*, 1996). It is postulated that the *Rp1-* gene encodes a receptor for a specific ligand and that *Rp1-Kr1* may either inappropriately recognize a ligand produced by the plant itself or aberrantly transduce a signal without the necessity of ligand binding (Hu *et al.*, 1996; Anderson *et al.*, 1997).

Similarly the *mlo* recessive resistance mutation of barley confers broad spectrum resistance to all known races of powdery mildew (*Erysiphe graminis* f. sp. hordei) and encodes a putative transmembrane protein (Büschges *et al.*, 1997). Under pathogen-free and axenic conditions, *mlo* plants grown at low temperature produce leaf lesions preceded by the formation of cell wall appositions (CWA), a characteristic response to fungal infections. However, the resistance provided by the absence of MLO does not utilize the HR and does not overlap genetically with the HR underlying race-specific resistance in barley (Peterhänsel *et al.*, 1997). It is hypothesized that in *mlo* plants either a sensor of cellular homeostasis becomes hypersensitized or cellular homeostasis itself is disrupted to magnify a signal that triggers the non-host resistance pathway. The identification of genes required for *mlo* resistance and CWA formation promises to further clarify how 'threshold responses' constitute part of the control of plant cell death and plant defense (Freialdenhoven *et al*, 1996; Morel and Dangl, 1997). The examples of *mlo* and *Rp1-Kr1* indicate that the normal wild-type plant encodes regulatory components that link cell homeostasis and survival with signals derived from the pathogen and that these exhibit fine control and specificity of response.

An active area of debate is whether or not the oxidative burst (Figure 11.4), which plays protective, oxidative cross-linking and cell–cell signalling roles in plant defence, is also necessary for triggering cell death during the HR (Low and Merida, 1996; Lamb and Dixon, 1997). Certain bacterial mutants are defective in HR elicitation due to mutations affecting secretion of virulence and avirulence factors. Using these mutants to infect tobacco cells in culture it was found that they induced H_2O_2 in a similar way to wild-type strains, but they did not induce cell collapse and death (Glazener *et al.*, 1996). Although the oxidative burst appears inadequate to trigger HR, it may enhance the amount of cell death that occurs. Using inhibitors to block the NADPH oxidase that generates the oxidative burst, the amount of cell death induced by an *avirulent* strain of *Pseudomonas* was decreased two-fold (Levine *et al.*, 1994; Greenberg, 1997; Lamb and Dixon, 1997). A sustained H_2O_2 release correlated with Ca^{2+} influx (the ion flux is indicated by XR in Fig. 11.4) in soybean cells and this was followed by cell shrinkage, plasma membrane blebbing, nuclear condensation and cell death (Levine *et al.*, 1996). It was also suggested that induction of the cellular protectant glutathione-S-transferase was Ca^{2+} dependent. The cell death that ensues in HR tobacco cells (genotype NN) infected with tobacco mosaic virus (TMV) can be inhibited at low partial oxygen pressures suggesting again that PCD can be separated from the activation of defense responses (Mittler *et al.*, 1996). It is not likely that superoxide or H_2O_2 are perceived through specific receptors but by their generation of other activated metabolites they may affect other redox-regulated systems.

Cells may have to integrate different sensors of cellular homeostasis before triggering an HR cell death pathway (Figure 11.4). Triggering of cell death may involve pre-existing molecules or require the induction of new gene expression. The LSD-1 protein from *Arabidopsis* may be involved in transducing such signals. *lsd-1* plants form spontaneous lesions in the absence of pathogens under long-day conditions, and extracellular superoxide but not hydrogen peroxide is both necessary and sufficient for *lsd1* lesion initiation and propagation (Dietrich *et al.*, 1994, 1997; Jabs *et al.*, 1996). The finding that *lsd1* encodes a putative novel zinc-finger protein suggests that LSD1 responds to upstream cell death effectors by activating genes that will protect the cell from further damage (Figure 11.4) (Dietrich *et al.*, 1997). The *lsd-1* phenotype is abolished in nahG transgenic plants in which salicylic acid (SA) cannot accumulate, suggesting that LSD-1 negatively regulates an amplification loop mediated by SA (Dangl *et al.*, 1996). The isolation of suppressors of lesion-mimic mutations such as *lsd-1* in *Arabidopsis* promises to uncover other cellular components regulating or effecting the cell death pathway (Dangl *et al.*, 1996).

11.3.10.3 Execution of the hypersensitive response

Expression analysis has revealed genes such as PR (pathogenesis related) genes which exhibit increased expression in response to pathogens, abiotic stresses and even during normal development (Stintzi *et al.*, 1993). Some of these genes have anti-pathogen (e.g. glucanases and chitinases) and degradative properties (e.g. ribonucleases) (Hammond-Kosack and Jones, 1996). So far, none of the PR proteins can be implicated directly in a programmed cell death pathway. In *Arabidopsis*, mutants such as *eds-1* and *ndr-1* re sent mutations in pre- and post-convergence steps in signalling pathways pre downstream of pathogen perception but their exact function is not yet known (Figure 11.4) (Morel and Dangl,

1997). Some enzymes may be expected to play a direct role in plant PCD such as proteases and endonucleases, which are key players in animal PCD. During infection of *NN* tobacco with tobacco mosaic virus (TMV) an increased nuclear localized nuclease activity can be detected, which is stimulated by Ca^{2+} and inhibited by Mg chelators. The specific association of these nucleases with HR cell death has not yet been established and some of these nucleases may function to recycle DNA after the plasma membrane is compromised. They may also may be secreted by neighbouring cells not undergoing the HR to degrade pathogen DNA (Mittler and Lam, 1997). Thus it remains to be shown that any nucleases or proteases are activated deliberately by the host cell to effect its own demise, such as is the case for caspases in animals (Figure 11.1).

HR cell death results in cell lysis and release of toxic molecules such as phenolics that may serve to propagate damage to neighbouring cells. Salicylic acid may act directly as a short-distance signal to enhance resistance in cells surrounding an HR cell or by possibly influencing H_2O_2 production by an undefined mechanism (Figure 11.4) (Rao *et al.*, 1997). SA may mediate the regulation of pathogenesis-related proteins via the *npr1-1* (*nonexpresser of PR genes*) gene, which may encode a transcriptional regulator (Cao *et al.*, 1997).

Cells surrounding an HR cell appear to sense nearby cell death and exhibit increased resistance to pathogen infection in the form of increased secondary wall reinforcement (lignification) and systemic acquired resistance (Figure 11.4) (Ryals *et al.*, 1996). It is possible that some compounds released from dying cells are toxic above a certain threshold level but below that threshold act as signals regulating cell survival functions (e.g. SA). Mutant maize *lls1* plants exhibit spreading type lesions that are triggered developmentally, by non-pathogens, or by physical wounding. The LLS1 lesion mimic gene has been isolated and may encode a phenolic degrading enzyme whose putative substrate has a signalling and/or toxic function (Figure 11.4) (Greenberg and Ausubel, 1993; Gray *et al.*, 1997) (Gray, J. *et al.*, unpublished data).

The genetic class of mutations known as lesion mimics has proved to yield novel genes that serve to maintain cell survival in the face of mechanical and pathogen attack. However, not all lesion mimics are expected to be involved specifically in regulating PCD. This is exemplified by *Les22* , a dominant maize lesion mimic that exhibits dense non-propagative necrotic lesions on heterozygous plants (Johal *et al.*, 1994). The *Les22* gene encodes the enzyme uroporphyrinogen decarboxylase (*uroD*), an enzyme in the haem biosynthetic pathway (Hu, G. *et al.*, unpublished). Inactivation of this gene results in the accumulation of porphyrin precursor molecules that become free radicals in the light and cause cell damage and death, leading to visible lesions. A similar pathology underlies the porphyria syndrome in humans. The *Les22* mutation thus gives rise to lesions because of the failure of cellular homeostasis owing to the inability to remove a toxic metabolite. This result provokes caution in the interpretation of transgenic plants in which expression of both plant and non-plant genes has given rise to aberrant or 'serendipitous' cell death phenotypes (reviewed in (Dangl *et al.*, 1996). It remains to be determined if these plants exhibit a train of events that truly parallels events during a normal HR response (Dangl *et al.*, 1996)

Verification of the deterministic nature of the 'cell death program' underlying HR will require an understanding of the complex sequence of interactions between the various signalling, execution and propagation molecules currently known (Figure 11.4) and awaiting isolation.

11.3.11 Senescence of plant parts or entire plants

Plants exhibit a wide variation in life spans but the completion of the plant life cycle often involves degeneration (senescence) of the entire plant body except for the seed. Plants provide an attractive system in which to evaluate models of senescence and its evolution and this phenomenon has been intensely studied (Noodén and Leopold, 1988; Noodén and Guiamét, 1996; Buchanan Wollaston, 1997; Gan and Amasino, 1997; Weaver *et al.*, 1997). First, the terms 'senescence' and 'ageing' are distinguished and then evidence for senescence as a PCD event in plants is set forth.

Senescence and ageing have been distinguished as active and passive processes in plants but these terms are often used interchangeably at the organismic level in animals (Noodén and Leopold, 1988; Noodén and Guiamét, 1996). There is, however, a specific meaning for cellular senescence in animals. Animal cells in culture undergo approximately 40 to 50 cell divisions before actively and permanently exiting the cell cycle. These cells are termed senescent but may remain metabolically active for a long period before dying (Afshari and Barrett, 1996). The exit from the cell cycle may be the result of cellular damage (DNA mutation, oxidative damage), a programmed molecular clock (e.g. telomere shortening) and regulated by yet to be identified genes (Afshari and Barrett, 1996). Ageing, in contrast, is considered a passive process caused by the detrimental action of external agents such as ionizing radiation, active ions and free radicals, and the damage is accumulative in time since protective mechanisms are never totally successful. Thus it may be seen that 'ageing' processes (e.g. DNA mutation) may cause a cell to become senescent (exit the cell cycle) but also that senescent cells may 'age' (e.g. undergo further mutation). Studies haven indicated that a major mediator of both processes is the p53 protein, which senses DNA damage and activates genes that can cause cell cycle arrest or genes that actively promote cell death (Levine, 1997).

Similar ageing processes must affect plants (e.g. in stored seeds) and a model for PCD in plant ontogeny based on internal and external factors pre-disposing scheduled or unscheduled entry into and exit from divisional cycles has been proposed (Havel and Durzan, 1996a). Some cell cycle control in plants parallels that in animals but only a few components have been identified to date (Ach *et al.*, 1997). The telomere hypothesis suggests that telomere length serves as a mitotic clock for timing cellular replicative life span and telomerase activation may be a conserved mechanism involved in allowing long-term proliferation capacity. Plant telomerase activity has been detected in differentiating plant tissue (Fitzgerald *et al.*, 1996). It would be expected, if the telomere hypothesis is true, that plant tumor tissues should exhibit reactivated telomerase expression and that a telomerase gene expressed behind a SAG promoter may inhibit senescence indefinitely.

Senescence in plants involves nutrient recycling and disposal of entire plants or obsolete organ systems and is often interrelated with reproduction. Plants that die after one or several reproductive cycles are said to exhibit monocarpic or polycarpic senescence, respectively (Noodén, 1980). It will be seen that senescence in plants is an actively regulated event that results in cell death, but whereas in animals unwanted dead cells are reabsorbed by surrounding cells, in plants senescent cells release their components by controlled autodigestion and export and finally by autolysis. It has been proposed that in plants senescence has evolved to optimize reproductive success by not investing energy in plant parts that are no longer of much use to the plant (Bleecker and Patterson, 1997).

11.3.11.1 Morphological and cellular features of senescence

Stages of senescence in a monocarpic plant (e.g. *Arabidopsis* or maize) have been defined using chlorophyll levels as a convenient visual marker (Figure 11.5) (Buchanan Wollaston, 1997). Young leaves (YG) acheive maximum chlorophyll levels and photosynthetic capacity before full leaf expansion. In *Arabidopsis* mature plants are green through flowering (MG1) until the first siliques form (MG2) at which point the first visible yellowing owing to degradation of chlorophyll is observed (SS1 through SS3). The initiation of senescence is regulated by the multiple factors discussed below. Chloroplasts exhibit swelling of thylakoids and the appearance of lipids and large plastoglobuli. The chloroplasts are removed from the cytoplasm by an autophagic process whereby they are enveloped in vacuole-like structures derived from the vacuole, endoplasmic reticulum or leucoplasts. Ribosomes begin to be degraded by SS3 and changes to mitochondria, nuclear and vacuolar membranes occur late during the final necrotic stage (SS4) (Noodén and Leopold, 1988; Noodén and Guiamét, 1996; Bleecker and Patterson, 1997; Buchanan Wollaston, 1997). In starved *Arabidopsis* suspension cells, which may resemble senescent cells, a clear generation of an oligonucleosomal ladder has been demonstrated, suggesting that the final stages of nuclear degradation occurs in an apoptotic fashion (Callard *et al.*, 1996). Cell death appears to occur when the tonoplast ruptures and all membrane integrity is lost, thus removing all possibility of cell recovery. Morphologically it appears that senescence begins as an autophagic process, such as is found in starving cells (Aubert *et al.*, 1996; Moriyasu and Ohsumi, 1996). However, nutrients thus gained are exported out of the leaf and the cells appear ultimately to die because they are unable to energetically maintain membrane integrity. The fact that cell walls are not degraded may reflect an energetic budget that would be too high.

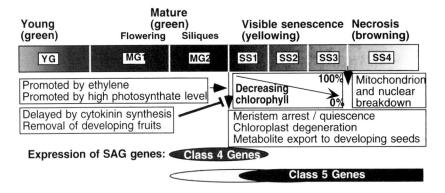

Figure 11.5 Stages of senescence in the life cycle of *Arabidopsis*. The advent of senescence in a monocarpic plant such as *Arabidopsis* is monitored most often by visible changes in chlorophyll content (shaded box, dark shading is high chlorophyll content). The different phases and SAG classes are designated as in Buchanan-Wollaston (1997). The onset of senescence is regulated by internal and external factors, resulting in a large plasticity in the progression of this event. Gene expression changes that accompany senescence are indicated by expanding and contracting bars in accordance with increasing or decreasing expression. See text for details.

11.3.11.2 Regulation of senescence

That senescence is a non-chaotic, active energy-requiring process is supported by several lines of evidence in plants (summarized in Figure 11.5) (Gan and Amasino, 1997). There is considerable evidence for the regulation of this process by hormones and developmental processes. Auxins, GAs and cytokinins have been observed to delay senescence (Figure 11.5) (Smart, 1994; Gan and Amasino, 1997). A cytokinin biosynthetic enzyme (from *Agrobacterium tumefaciens*) expressed in transgenic plants behind a senescence-specific promoter delays senescence indefinitely (Gan and Amasino, 1995). The isolation of a strong candidate for a cytokinin receptor in *Arabidopsis*, which belongs to the histidine-kinase family, brings the possibility that this aspect of cytokinin action will finally begin to be dissected (Kakimoto, 1996).

Ethylene, ABA and jasmonic acid promote senescence (Smart, 1994; Bleecker and Patterson, 1997; Gan and Amasino, 1997). Unfertilized carpels undergo a process of senescence that exhibits characteristics of apoptosis such as nuclear condensation and internucleosomal DNA fragmentation (DNA laddering). Ethylene action inhibitors delay this process and it is suggested that ethylene is the signal that promotes senescence and that signals generated during successful pollination override this signal (Orzáez and Granell, 1997a). Pollination itself may trigger ethylene production and cause petal/corolla senescence and an unknown factor may mediate this process (Tang and Woodsen, 1996; Woltering *et al*, 1997). Ethylene insensitive mutants or transgenic plants expressing antisense genes of ethylene biosynthesis enzymes exhibit delayed senescence and retarded loss of petals and anthers (Hamilton *et al.*, 1990; John *et al.*, 1995; Chao *et al.*, 1997). In *Arabidopsis*, proteins such as EIN3 (a possible transcriptional activating protein) act downstream of ethylene reception and it will be of interest to elucidate the genes that are activated during the ethylene response, which presumably will include senescence-associated genes (SAGs). When ein3 was overexpressed in *Arabidopsis*, the rosette leaves senesced immediately after the initiation of the inflorescence, and the petals and stamens also senesced earlier than in the wild-type (Chao *et al.*, 1997). Mutations such as *etr1*, *ein2* and *ein3* have proved very useful but no mutant has been found in which senescence completely failed to occur. This may reflect the leakiness of these mutations or that there are multiple avenues of senescence initiation (Gan and Amasino, 1997).

In monocarpic *Arabidopsis* the formation of siliques induces meristem arrest and chloroplast degeneration initiates in rosette leaves. This process is reversible, however, if developing siliques are removed, indicating that the meristem is quiescent rather than permanently arrested at least for some time (Bleecker and Patterson, 1997). This observation is reinforced by the observation that male-sterile plants such as *ms1-1* in *Arabidopsis* and *indeterminate* in maize continue to proliferate almost indefinitely (Hensel *et al.*, 1993; Colasanti and Sundaresan, 1995). The *proliferous* mutation of *Arabidopsis* is a single recessive mutation that disrupts proliferative arrest of meristems and twice as much seed as normal is formed without the loss of seed weight or viability (Bleecker and Patterson, 1997). Not only reproduction but also light dosage and photosynthate level appear to regulate senescence initiation (Noodén *et al.*, 1996; Buchanan Wollaston, 1997). Thus there are many regulatory influences that form a network to determine when a plant or plant part begins the senescence process leading to a certain 'plasticity' in the progression of cell death (Gan and Amasino, 1997).

11.3.11.2 Execution of senescence

Plant senescence is an active process requiring gene transcription and translation (Noodén and Leopold, 1988; Noodén and Guiamét, 1996; Buchanan Wollaston, 1997; Gan and Amasino, 1997). Studies of a 'stay green' mutant of *Festuca* indicate that non-enzymic, free-radical-mediated 'photobleaching' cannot account for the rate of chlorophyll loss from senescent tissues (Thomas and Matile, 1988) but involves activity of degradative enzymes (Vincentini *et al.*, 1995). Glyoxylate cycle, amino acid metabolism and RNAase activities have been documented to increase during senescence (Bleecker and Patterson, 1997). Many genes whose expression is altered during senescence have been isolated and these have been thoroughly reviewed recently and organized into ten classes according to temporal spacing of expression (Smart, 1994; Buchanan Wollaston, 1997; Gan and Amasino, 1997). These 'senescence-related' or 'senescence-associated' genes (SAGs) include several nucleases and proteases that may be involved in salvage pathways. Genes particularly attractive to further study are class 4 genes, which are expressed only around the time of the beginning of senescence, and class 5 genes, which are expressed in an increasing fashion from senescence initiation until cell death (Figure 11.5) (Buchanan Wollaston, 1997). Other genes of interest include SAG 15 (*ERD1*), which encodes a protein with homology to a regulatory subunit of a bacterial protease system that has been found in plant chloroplasts and could be involved in chloroplast turnover. SAG13 encodes a short-chain alcohol dehydrogenase related to the *tasselseed2* gene involved in floral primordia degeneration (discussed earlier). In *Arabidopsis* cell suspension cultures that are undergoing a loss of cell viability owing to nutrient exhaustion, the expression of three SAGs has been reported (Callard *et al.*, 1996). It is clear from these differentially expressed genes that senescence is not a passive process but requires the orderly progression of changes in gene expression. The cloning of senescence-associated genes will permit temporal, spatial and regulatory expression analysis of these genes during the senescence process (Buchanan Wollaston, 1997). The direct role of these genes in senescence can be further assessed through the analysis of transgenic plants with altered expression of these genes. One gene in particular is discussed here because of its association with cell death in animals.

Dad-1 (for *defender against apoptotic death*) is a gene which, when mutated, results under permissive conditions in a cell death event with apoptotic features in invertebrates and mammals (Nakashima *et al.*, 1993). Expression studies using the pea *dad-1* homologue show that *dad-1* mRNA declines dramatically on flower anthesis, disappearing in senescent petals, and is down-regulated by the plant hormone ethylene (Orzáez and Granell, 1997b). Rice and *Arabidopsis* DAD-1 cDNAs complement a hamster ts BN7 mutant cell line that contains a temperature-sensitive DAD-1 point mutation, proving that DAD-1 function is conserved across kingdoms (Gallois *et al.*, 1997; Tanaka *et al.*, 1997). The study of a homologue in yeast (Ost2) indicates that Dad-1 is part of a glycosylation complex for a range of soluble and membrane-bound glycoproteins (Kelleher and Gilmore, 1997). The DAD-1 protein caused cell death reminiscent of apoptosis in animals (Nakashima *et al.*, 1993) but it has not been shown that *dad-1* regulation is a key feature in normal PCD events during development. Thus, in plants it would appear that DAD-1 may function to maintain cell viability in cells, but it remains to be determined if the down-regulation of *dad-1* (such as in pea petals) is a common feature of senescence and whether it is a key regulatory step or merely a consequence of cell death already initiated.

Finally, in animal studies there is considerable evidence for genetic variation in life span and single gene mutations influencing longevity (gerontogenes) have been isolated in *C. elegans* and *Drosophila* (Finch and Tanzi, 1997). Homologues to these genes have not been identified in plants but there is evidence in plants for a genetic component for longevity as witnessed by the genetic variation in foliar senescence that has been exploited in crop plants such as alfalfa (Thomas and Smart, 1993; Van Oosterom *et al.*, 1996). Variations in these genes may also provide the basis for the occurrence of monocarpic versus polycarpic plants. In maize a recessive *pre-mature senescence* (*pre-1*) mutation has been described in which senescence begins at least two weeks prior to anthesis. Chlorosis, dehydration and death begin on the lowermost leaf tip and progress upwards in a fashion reminiscent of normal senescence (Neuffer *et al.*, 1997). Considerable efforts are being made to isolate defective senescence mutants of *Arabidopsis* (Buchanan Wollaston, 1997). The proposed existence of multiple regulatory cascades that initiate senescence may explain why such mutations that cause delayed senescence may be masked in mutant screens (Gan and Amasino, 1997).

11.4 Conclusions and forward perspectives

The attention that has been applied to studying cell death in plants has resulted in significant advances in distinguishing traumatic from programmed events and in isolating the molecular components of the latter. Xylem differentiation, the HR and senescence have received most attention from researchers, however an integrated picture linking regulatory and execution steps still remains to emerge for any PCD pathway in plants.

There are some parallels with PCD in animals, especially in that many cell death events are required for the completion of the life cycle and these cell death events are highly regulated by environmental and developmental signals, and require active transcription of novel execution genes. Although some degradative processes such as chromatin loss may be shared, the sequence of events in cellular removal has yet to reveal components that are shared across kingdoms and this may be largely owing to the existence of a cell wall in plants and the absence of phagocytosis. Autophagic processes appear to play a more important role in plants and the mechanisms involved deserve closer examination.

Arabidopsis has emerged as a widely utilized model organism for plant biology owing to its ease of mutant and suppressor screening, and the development of convenient transformation and tagging strategies. In the area of PCD research the lesion-mimic mutations and disease-resistance mutants have provided a rich source of germplasm from which to isolate components of PCD pathways. Determination of the epistatic effects of mutants and their extragenic suppressors in *Arabidopsis* promises to reveal how these pathways may parallel or intersect each other in the protection of cells from cell death. However, most of the putative PCD pathways described in this chapter have been documented and studied largely or exclusively in other systems. Some of these PCD events are not amenable in *Arabidopsis*, e.g. floral organ abortion in unisexual flowers, and the utility of different model systems is required to fully explore the PCD phenomenon in plants. The rich reservoir of maize mutants (e.g. lesion mimic and necrotic mutants) will continue to provide material for study but in all species a re-examination of existing mutants (especially those in flower and seed development) is deserved, with attention given to possible loss or gain of PCD phenotypes.

In addition to molecular genetic studies, extensive microscopic examination of the temporal sequence of subcellular events that occur during PCD is required. The development of PCD in isolated cell systems is also required so that isolated events in plants can be more easily studied when PCD events are induced in a synchronous fashion.

It is clear that Joe Varner's words, quoted at the beginning of this chapter, are still resonant today as we strive forward in research together with animal, worm and fly PCD researchers—there is a tough problem and much work to do!

Acknowledgements

We thank Gongshe Hu, Brent Buckner and Diane Janick-Buckner for critical reading of the manuscript. The preparation of this chapter was supported in part by a grant from Pioneer Hi-Bred Int. Inc. to G.S. Johal.

References

Ach, R.A., Taranto, P. and Gruissem, W. (1997) A conserved family of Wd-40 proteins binds to the retinoblastoma protein in both plants and animals. *Plant Cell*, **9** 1595-1606.

Afshari, C.A. and Barrett, J.C. (1996) Molecular genetics of *in vitro* cellular senescence, in *Cellular Aging and Cell Death* (eds N.J. Holbrook, G.R. Martin and R.A. Lockshin), Wiley-Liss, Inc., New York, pp. 109-21.

Aloni, R. (1993) The role of cytokinin in organised differentiation of vascular tissues. *Austr. J. Plant Physiol.*, **20** 601-608.

Ameisen, J.C. (1996) The origin of programmed cell death. *Science*, **272** 1278-79.

Anderson, P.A., Lawrence, G.J., Morrish, B.C., Ayliffe, M.A., Finnegan, E.J., and Ellis, J.G. (1997) Inactivation of the flax rust resistance gene *M* associated with loss of a repeated unit within the leucine-rich repeat coding region. *Plant Cell*, **9** 641-51.

Aubert, S., Gout, E., Bligny, R., Marty-Mazars, D., Barrieu, F., Alabouvette, J., Marty, F. and Douce, R. (1996) Ultrastructural and biochemical characterization of autophagy in higher plant cells subjected to carbon deprivation: control by the supply of mitochondria with respiratory substrates. *J. Cell Biol.*, **133** 1251-63.

Beers, E.P. and Freeman, T.B. (1997) Protease activity during tracheary element differentiation in *Zinnia* mesophyll cultures. *Plant Physiol.*, **113** 873-80.

Bell, P.R. (1996) Megaspore abortion: a consequence of selective apoptosis? *Int. J. Plant Sci.*, **157** 1-7.

Bleecker, A.B. and Patterson, S.E. (1997) Last exit: senescence, abscission, and meristem arrest in *Arabidopsis*. *Plant Cell*, **9** 1169-79.

Bowen, I.D., Mullarkey, K. and Morgan, S.M. (1996) Programmed cell death during metamorphosis in the blow-fly Calliphora vomitoria. *Microsc. Res. Tech.*, **34** 202-17.

Buchanan-Wollaston, V. (1997) The molecular biology of leaf senescence. *J. Exper. Bot.*, **48** 181-99.

Buja, L.M., Eigenbrodt, M.L. and Eigenbrodt, E.H. (1993) Apoptosis and necrosis. Basic types and mechanisms of cell death. *Arch. Pathol. Lab. Med.*, **117** 1208-14.

Büschges, R., Hollricher, K., Panstruga, R., Simons, G., Wolter, M., Frijters, A., van Daelen, R., van der Lee, T., Diergaarde, P., Groenendijk, J., Töpsch, S., Vos, P., Salamini, F. and Schulze-Lefert, P. (1997) The barley Mlo gene: a novel control element of plant pathogen resistance. *Cell*, **88** 695-705.

Callard, D., Axelos, M. and Mazzolini, L. (1996) Novel molecular markers for late phases of the growth cycle of *Arabidopsis* thaliana cell-suspension cultures are expressed during organ senescence. *Plant Physiology*, **112** 705-15.

Campbell, R. and Drew, M.C. (1983) Electron microscopy of gas space (aerenchyma) formation in adventitious roots of *Zea mays* L. subjected to oxygen shortage. *Planta*, **157** 350-57.

Cao, H., Glazebrook, J., Clarke, J. D., Volko, S. and Dong, X. (1997) The *Arabidopsis* NPR1 gene that controls systemic acquired resistance encodes a novel protein containing ankyrin repeats. *Cell*, **88** 57-63.

Chao, Q., Rothenberg, M., Solano, R., Roman, G., Terzaghi, W. and Ecker, J.R. (1997) Activation of the ethylene gas response pathway in *Arabidopsis* by the nuclear protein ETHYLENE-INSENSITIVE 3 and related proteins. *Cell*, **89** 1133-44.

Cheng, P. C., Greyson, R. I. and Walden, D. B. (1983) Organ initiation and the development of unisexual flowers in the tassel and ear of *Zea mays*. *Am. J. Bot.*, **70** 450-62.

Colasanti, J. and Sundaresan, V. (1995) Transposon tagging of the *indeterminate* gene. *Maize Gene. Coop. Newslett.*, **69** 35-36.

Collins, R.J., Harmon, B.V., Gobe, G.C. and Kerr, J.F. (1992) Internucleosomal DNA cleavage should not be the sole criterion for identifying apoptosis. *Int. J. Rad. Biol.*, **61** 451-53.

Croft, K.P.C., Voisey, C.R. and Slusarenko, A.J. (1990) Mechanism of hypersensitive cell collapse correlation of increased lipoxygenase activity with membrane damage in leaves of phaseolus-vulgaris L. inoculated with an avirulent race of pseudomonas-syringae pathovar phaseolicola. *Physiolog. Mol. Plant Pathol.*, **36** 49-62.

Dangl, J.L., Dietrich, R.A. and Richberg, M.H. (1996) Death don't have no mercy: cell death programs in plant-microbe interactions. *Plant Cell*, **8** 1793-1807.

Dellaporta, S.L. and Calderon-Urrea, A. (1994) The sex determination process in maize. *Science*, **266** 1501-1504.

DeLong, A., Calderon-Urrea, A. and Dellaporta, S.L. (1993) Sex determination gene tasselseed-2 of maize encodes a short-chain alcohol dehydrogenase required for stage-specific floral organ abortion. *Cell*, **74** 757-68.

Demura, T.F.H. (1994) Novel vascular cell-specific genes whose expression is regulated temporally and spatially during vascular system development. *Plant Cell*, **6** 967-81.

Deshmukh, M. and Johnson, E.M., Jr (1997) Programmed cell death in neurons: focus on the pathway of nerve growth factor deprivation-induced death of sympathetic neurons. *Mol. Pharmacol.*, **51** 897-906.

Dietrich, R.A., Delaney, T.P., Uknes, S.J., Ward, E.R., Ryals, J.A. and Dangl, J.L. (1994) *Arabidopsis* mutants simulating disease resistance response. *Cell*, **77** 565-77.

Dietrich, R.A., Richberg, M.H., Schmidt, R., Dean, C. and Dangl, J.L. (1997) A novel zinc finger protein is encoded by the *Arabidopsis* LSD1 gene and functions as a negative regulator of plant cell death. *Cell*, **88** 685-94.

Duncan-Achanzar, K.B., Jones, J.T., Burke, M.F., Carter, D.E. and Laird, H.E. (1996) Inorganic mercury chloride-induced apoptosis in the cultured porcine renal cell line LLC-PK1. *J. Pharmacol. Exper. Therap.*, **277** 1726-32.

Esau, K., Cheadle, V.I. and Risley, E.B. (1963) A view of ultrastructure of *Cucurbita* xylem. *Bot. Gaz.*, **124** 311-16.

Finch, C.E. and Tanzi, R.E. (1997) Genetics of aging. *Science*, **278** 407-11.

Fitzgerald, M.S., McKnight, T.D. and Shippen, D.E. (1996) Characterization and developmental patterns of telomerase expression in plants. *Proc. Natl Acad. Sci. USA*, **93** 14422-27.

Freialdenhoven, A., Peterhaensel, C., Kurth, J., Kreuzaler, F. and Schulze Lefert, P. (1996) Identification of genes required to the function of non-race-specific mlo resistance to powdery mildew in barley. *Plant Cell*, **8** 5-14.

Freytag, S., Arabatzis, N., Hahlbrock, K. and Schmeltzer, E. (1994) Reversible cytoplasmic rearrangements precede wall apposition, hypersensitive cell death and defense-related gene activation in potato/*Phytophthora infestans* interactions. *Planta*, **194** 123-35.

Fukuda, H. (1996) Xylogenesis: initiation, progression and cell death. *Ann. Rev. Plant Physiol. Plant Mol. Biol.*, **47** 299-325.

Fukuda, H. (1997) Tracheary element differentiation. *Plant Cell* **9**, 1147-56.

Gallois, P., Makishima, T., Hecht, V., Despres, B., Laudié, M., Nishimoto, T. and Cooke, R. (1997) An *Arabidopsis thaliana* cDNA complementing a hamster apoptosis suppressor mutant. *Plant J.*, **11** 1325-31.

Gan, S. and Amasino, R.M. (1995) Inhibition of leaf senescence by autoregulated production of cytokinin. *Science*, **270** 1986-88.

Gan, S. and Amasino, R.M. (1997) Making sense of senescence: Molecular genetic regulation and manipulation of leaf senescence. *Plant Physiology*, **113** 313-19.

Gärtner, P.J. and Nagl, W. (1980) Acid-phosphatase activity in plastids (plastolysosomes) of senescing embryo-suspensor cells. *Planta*, **149** 341-49.

Glazener, J.A., Orlandi, E.W. and Baker, C.J. (1996) The active oxygen response of cell suspensions to incompatible bacteria is not sufficient to cause hypersensitive cell death. *Plant Physiol.*, **110** 759-63.

Goldberg, R.B., Beals, T.P. and Sanders, P.M. (1993) Anther development: basic principles and practical applications. *Plant Cell*, **5** 1217-29.

Golubovskaya, I.N., Avalinka, N. and Sheridan, W. (1992) Effects of several meiotic mutants on female meiosis in maize. *Dev. Gen.*, **13** 411-24.

Görg, R., Hollricher, K. and Schulze-Lefert, P. (1993) Functional analysis and RFLP-mediated mapping of the *Mlg* resistance locus in barley. *Plant J.*, **3** 857-66.

Gray, J., Close, P.S., Briggs, S.P. and Johal, G.S. (1997) A novel suppressor of cell death in plants encoded by the Lls1 gene of maize. *Cell*, **89** 25-31.

Greenberg, J.T. (1996) Programmed cell death: a way of life for plants. *Proc. Natl Acad. Sci. USA*, **93** 12094-97.

Greenberg, J.T. (1997) Programmed cell death in plant-pathogen interactions. *Ann. Rev. Plant Physiol. Plant Mol. Biol.*, **48** 525-45.

Greenberg, J.T. and Ausubel, F.M. (1993) *Arabidopsis* mutants compromised for the control of cellular damage during pathogenesis and aging. *Plant J.*, **4**, 327-41.

Griffiths, S.D., Clarke, A.R., Healy, L.E., Ross, G., Ford, A.M., Hooper, M.L., Wyllie, A.H. and Greaves, M. (1997) Absence of p53 permits propagation of mutant cells following genotoxic damage. *Oncogene*, **14** 523-31.

Groover, A., DeWitt, N., Heidel, A. and Jones, A. (1997) Programmed cell death of plant tracheary elements differentiating *in vitro*. *Protoplasma*, **196** 197-211.

Hale, A.J., Smith, C.A., Sutherland, L.C., Stoneman, V.E.A., Longthorne, V.L., Culhane, A.C. and Williams, G.T. (1996) Apoptosis: molecular regulation of cell death. *Eur. J. Biochem.*, **236** 1-26.

Hamilton, A.J., Lycett, G.W. and Grierson, D. (1990) Antisense gene that inhibits synthesis of the hormone ethylene in transgenic plants. *Nature*, **346** 284-87.

Hammond-Kosack, K.E. and Jones, J.D.G. (1996) Resistance gene-dependent plant defense responses. *Plant Cell*, **8** 1773-91.

Hammond-Kosack, K.E. and Jones, J.D.G. (1997) Plant disease resistance genes. *Ann. Rev. Plant Physiol. Plant Mol. Biol.*, **48** 575-607.

Havel, L. and Durzan, D.J. (1996a) Apoptosis in plants. *Botanica Acta*, **109** 268-77.

Havel, L. and Durzan, D.J. (1996b) Apoptosis during diploid parthenogenesis and early somatic embryogenesis of Norway spruce. *Int. J. Plant Sci.*, **157** 8-16.

He, S.H., Huang, H.C. and Collmer, A. (1993) Pseudomonas syringae pv. syringae harpin Pss: a protein that is secreted via the Hrp pathway and elicits the hypersensitive response in plants. *Cell*, **73** 1255-66.

He, C.J., Morgan, P.W. and Drew, M.C. (1996) Transduction of an ethylene signal is required for cell death and lysis in the root cortex of maize during aerenchyma formation induced by hypoxia. *Plant Physiol.* **112**, 463-72.

Hensel, L.L., Grbic, V., Baumgarten, D.A. and Bleecker, A.B. (1993) Developmental and age-related processes that influence the longevity and senescence of photosynthetic tissues in *Arabidopsis*. *Plant Cell*, **5** 553-64.

Holcombe, J.W. (1984) Morphogenesis of branch leaves of sphagnum-magellanicum. *J. Hattori Bot. Lab.*, **57** 179-240.

Hu, G., Richter, T.E., Hulbert, S.H. and Pryor, T. (1996) Disease lesion mimicry caused by mutations in the rust resistance gene *rp*1. *Plant Cell*, **8** 1367-76.

Jabs, T., Dietrich, R.A. and Dangl, J.L. (1996) Initiation of runaway cell death in an *Arabidopsis* mutant by extracellular superoxide. *Science*, **273** 1853-56.

Jackson, M.B., Fenning, T.M., Drew, M.C. and Saker, L.R. (1985) Stimulation of ethylene production and gas-space aerenchyma formation in adventitious roots of zea-mays by small partial pressures of oxygen. *Planta*, **165** 486-92.

Jacobsen, S.E. and Olszewski, N.E. (1996) Gibberellins regulate the abundance of RNAs with sequence similarity to proteinase inhibitors, dioxygenases and dehydrogenases. *Planta*, **198** 78-86.

Jacobson, M.D., Weil, M. and Raff, M.C. (1997) Programmed cell death in animal development. *Cell*, **88** 347-54.

Jehn, B.M. and Osborne, B.A. (1997) Gene regulation associated with apoptosis. *Crit. Rev. Euk. Gene Expr.*, **7** 179-93.

Johal, G.S., Hulbert, S. and Briggs, S.P. (1995) Disease lesion mimics of maize: a model for cell death in plants. *BioEssays*, **17** 685-92.

Johal, G.S., Lee, E.A., Close, P.S., Coe, E.H., Neuffer, M.G. and Briggs, S.P. (1994) A tale of two mimics: transposon mutagenesis and characterization of two disease lesion mimic mutations of maize. *Maydica*, **39** 69-76.

John, I., Drake, R., Farrell, A., Cooper, W., Lee, P., Horton, P. and Grierson, D. (1995) Delayed leaf senescence in ethylene-deficient ACC-oxidase antisense tomato plants: molecular and physiological analysis. *Plant J.*, **7** 483-90.

Jones, A.M. and Dangl, J.L. (1996) Log Jam at the Styx: programmed cell death in plants. *Trends Plant Sci.*, **1** 114-19.

Kakimoto, T. (1996) CKI1, a histidine kinase homolog implicated in cytokinin signal transduction. *Science*, **274** 982-85.

Kaplan, D.R. (1984) Alternative modes of organogenesis in higher plants, in *Contemporary Problems in Plant Anatomy* (eds R.A. White and W.C. Dickison), Academic Press, Inc., Orlando, pp. 261-300.

Kelleher, D.J. and Gilmore, R. (1997) DAD1, the defender against apoptotic death, is a subunit of the mammalian oligosaccharyl transferase. *Proc. Natl Acad. Sci. USA*, **94** 4994-99.

Kosslak, R.M., Chamberlin, M.A., Palmer, R.G. and Bowen, B.A. (1997) Programmed cell death in the root cortex of soybean root necrosis mutants. *Plant J.*, **11** 729-45.

Kuo, A., Cappelluti, S., Cervantes-Cervantes, M., Rodriguez, M. and Bush, D.S. (1996) Okadaic acid, a protein phosphatase inhibitor, blocks calcium changes, gene expression, and cell death induced by gibberellin in wheat aleurone cells. *Plant Cell*, **8** 259-69.

Lamb, C. and Dixon, R.A. (1997) The oxidative burst in disease resistance. *Ann. Rev. Plant Physiol. Plant Mol. Biol.*, **48** 251-75.

Levine, A.J. (1997) p53, the cellular gatekeeper for growth and division. *Cell*, **88** 323-31.

Levine, A., Pennell, R.I., Alvarez, M.E., Palmer, R. and Lamb, C. (1996) Calcium-mediated apoptosis in a plant hypersensitive disease resistance response. *Curr. Biol.*, **6** 427-37.

Levine, A., Tenhaken, R., Dixon, R. and Lamb, C. (1994) H_2O_2 from the oxidative burst orchestrates the plant hypersensitive disease resistance response. *Cell*, **79** 583-93.

Low, P.S. and Merida, J.R. (1996) The oxidative burst in plant defense: function and signal transduction. *Physiol. Plant.*, **96** 533-42.

Majno, G. and Joris, I. (1995) Apoptosis, oncosis, and necrosis. An overview of cell death. *Am. J. Path.*, **146** 3-15.

McCabe, P.F. and Pennell, R.I. (1996) Apoptosis in plant cells *in vitro*, in *Techniques in Apoptosis*, (eds T.G. Koptter and S.J. Martin), Portland Press, London, pp. 301-26.

McCabe, P.F., Levine, A., Meijer, P.J., Tapon, N.A. and Pennell, R.I. (1997) A programmed cell death pathway activated in carrot cells cultured at low cell density. *Plant J.*, **12** 267-80.

Meinke, D.W. (1995) Molecular genetics of plant embryogenesis. *Ann. Rev. Plant Physiol. Plant Mol. Biol.*, **46** 369-94.

Merrill Jr, A. H., Liotta, D.C. and Riley, R.T. (1996) Fumonisins: fungal toxins that shed light on sphingolipid function. *Trends Cell Biol.*, **6** 218-23.

Mittler, R. and Lam, E. (1995) *In situ* detection of nDNA fragmentation during the differentiation of tracheary elements in higher plants. *Plant Physiol.*, **108** 489-93.

Mittler, R. and Lam, E. (1996) Sacrifice in the face of foes: pathogen-induced programmed cell death in plants. *Trends Microbiol.*, **4** 10-15.

Mittler, R. and Lam, E. (1997) Characterization of nuclease activities and DNA fragmentation induced upon hypersensitive response cell death and mechanical stress. *Plant Mol. Biol.*, **34** 209-21.

Mittler, R., Shulaev, V., Seskar, M. and Lam, E. (1996) Inhibition of programmed cell death in tobacco plants during a pathogen-induced hypersensitive response at low oxygen pressure. *Plant Cell* **8**, 1991-2001.

Mittler, R., Simon, L. and Lam, E. (1997) Pathogen-induced programmed cell death in tobacco. *J. Cell Sci.*, **110** 1333-44.

Miura, M. and Yuan, J. (1996) Mechanisms of programmed cell death in *Caenorhabditis elegans* and vertebrates. *Curr. Topics Dev. Biol.*, **32** 139-74.

Morel, J.-B. and Dangl, J.F. (1997) The hypersensitive response and the induction of cell death in plants. *Cell Death Diff.*, **4** 671-83.

Moriyasu, Y. and Ohsumi, Y. (1996) Autophagy in tobacco suspension-cultured cells in response to sucrose starvation. *Plant Physiol.*, **111** 1233-41.

Nagata, S. (1997) Apoptosis by death factor. *Cell*, **88** 355-65.

Nakashima, T., Sekiguchi, T., Kuraoka, A., Fukishima, K., Shibata, Y., Komiyama, S. and Nishimoto, T. (1993) Molecular cloning of a human cDNA encoding a novel protein, DAD1, whose defect causes apoptotic cell death in hamster BHK21 cells. *Mol. Cell. Biol.*, **13** 6367-74.

Neuffer, M.G., Coe, E.H. and Wessler, S.R. (1997) *Mutants of Maize*, Cold Spring Harbor Laboratory Press, Cold Spring Harbour, New York.

Noodén, L.D. (1980) Senescence in the whole plant, in *Senescence in Plants* (ed. K.V. Thimann), CRC Press, Boca Raton, pp. 276.

Noodén, L.D. and Leopold, A.C. (1988) *Senescence and Aging in Plants*, Academic Press, San Diego.

Noodén, L.D. and Guiamét, J.J. (1996) Genetic control of senescence and aging in plants, in *Handbook of the Biology of Aging* (eds E.L. Schneider and J.W. Rowe). Academic Press, Inc., San Diego, pp. 94-118.

Noodén, L.D., Hillsberg, J.W. and Schneider, M.J. (1996) Induction of leaf senescence in *Arabidopsis* thaliana by long days through a light-dosage effect. *Physiol. Plant.*, **96** 491-95.

Nurnberger, T., Nennstiel, D., Jabs, T., Sacks, W.R., Hahlbrock, K. and Scheel, D. (1994) High affinity binding of a fungal oligopeptide elicitor to parsley plasma membranes triggers multiple defense responses. *Cell*, **78** 449-60.

Orzáez, D. and Granell, A. (1997a) DNA fragmentation is regulated by ethylene during carpel senescence in *Pisum sativum. Plant J.*, **11** 137-44.

Orzáez, D. and Granell, A. (1997b) The plant homologue of the defender against apoptotic death gene is down-regulated during senescence of flower petals. *FEBS Lett.*, **404** 275-78.

Owen, H.A. and Makaroff, C.A. (1995) Ultrastructure of microsporogenesis and microgametogenesis in *Arabidopsis* thaliana (L.) Heynh. ecotype Wassilewskija (Brassicaceae). *Protoplasma*, **185** 7-21.

Pang, Z. and Geddes, J.W. (1997) Mechanisms of cell death induced by the mitochondrial toxin 3-nitropropionic acid: acute excitotoxic necrosis and delayed apoptosis. *J. Neurosci.*, **17** 3064-73.

Pena, L.A., Fuks, Z. and Kolesnick, R. (1997) Stress-induced apoptosis and the sphingomyelin pathway. *Biochem. Pharmacol.*, **53** 615-21.

Pennell, R.I. and Lamb, C. (1997) Programmed cell death in plants. *Plant Cell*, **9** 1157-68.

Peterhänsel, C., Freialdenhoven, A., Kurth, J., Kolsch, R. and Schulze-Lefert, P. (1997) Interaction analyses of genes required for resistance responses to powdery mildew in barley reveal distinct pathways leading to cell death. *Plant Cell*, **9** 1397-1409.

Rao, M.V., Paliyath, G., Ormrod, D.P., Murr, D.P. and Watkins, C.B. (1997) Influence of salicylic acid on H_2O_2 production, oxidative stress and H_2O_2-metabolizing enzymes. *Plant Physiol.*, **115** 137-49.

Reed, J.C. (1997) Double identity for proteins of the Bcl-2 family. *Nature*, **387** 773-76.

Reiser, L.F.R.L. (1993) The ovule and the embryo sac. *Plant Cell*, **5** 1291-1301.

Russell, S.D. (1993) The egg cell: development and role in fertilization and early embryogenesis. *Plant Cell*, **5** 1349-59.

Ryals, J.A., Neuenschwander, U.H., Willits, M.G., Molina, A., Steiner, H.-Y., and Hunt, M.D. (1996) Systemic acquired resistance. *Plant Cell*, **8** 1809-19.

Ryerson, D.E., and Heath, M.C. (1996) Cleavage of nuclear DNA into oligonucleosomal fragments during cell death induced by fungal infection or by abiotic treatments. *Plant Cell*, **8** 393-402.

Saab, I.N. and Sachs, M.M. (1996) A flooding-induced xyloglucan endo-transglycosylase homolog in maize is responsive to ethylene and associated with aerenchyma. *Plant Physiol.*, **112** 385-91.

Savidge, R.A. (1996) Xylogenesis, genetic and environmental regulation: a review. *IAWA J.*, **17** 269-310.

Sawhney, V.K. and Bhadula, S.K. (1988) Microsporogenesis in the normal and male-sterile stamenless-2 mutant of tomato lycopersicon-esculentum. *Can. J. Bot.*, **66** 2013-21.

Schiefelbein, J.W. and Benfey, P.N. (1994) Root development in *Arabidopsis*, in *Arabidopsis* (eds E.M. Meyerowitz and C.R. Somerville), Cold Spring Harbor Laboratory Press, Cold Spring Harbour, New York, pp. 335-53.

Schulte-Hermann, R., Bursch, W., Grasl-Kraupp, B., Marian, B., Torok, L., Kahl-Rainer, P. and Ellinger, A. (1997) Concepts of cell death and application to carcinogenesis. *Toxicol. Pathol.*, **25** 89-93.

Schussler, E.E. and Longstreth, D.J. (1996) Aerenchyma develops by cell lysis in roots and cell separation in leaf petioles in *Sagittaria lancifolia* (Alismataceae). *Am. J. Bot.*, **83** 1266-73.

Schwartz, L.M., Smith, S.W., Jones, M.E. and Osborne, B.A. (1993) Do all programmed cell deaths occur via apoptosis? *Proc. Natl Acad. Sci. USA*, **90** 980-84.

Schwartz, B.W., Yeung, E.C. and Meinke, D.W. (1994) Disruption of morphogenesis and transformation of the suspensor in abnormal suspensor mutants of *Arabidopsis. Development*, **120** 3235-45.

Schwartzman, R.A. and Cidlowski, J.A. (1993) Apoptosis: the biochemistry and molecular biology of programmed cell death. *Endocrine Rev.*, **14** 133-51.

Scofield, S.R., Tobias, C.M., Rathjen, J.P., Chang, J.H., Lavelle, D.T., Michelmore, R.W. and Staskawicz, B.J. (1996) Molecular basis of gene-for-gene specificity in bacterial speck disease of tomato. *Science*, **274** 2063-65.

Shen, Y. and Shenk, T. E. (1995) Viruses and apoptosis. *Curr. Biol.*, **5** 105-11.

Singh, M.B., Bhalla, P.L. and Malik, C.P. (1980) Activity of some hydrolytic enzymes in the autolysis of the embryo suspensor in *Tropaeolum majus* L. *Ann. Bot.*, **45** 523-27.

Smart, C.M. (1994) Tansley review no. 64 gene expression during leaf senescence. *New Phytol.*, **126** 419-48.

Stacey, N.J., Roberts, K., Carpita, N.C., Wells, B. and McCann, M.C. (1995) Dynamic changes in cell surface molecules are very early events in the differentiation of mesophyll cells from Zinnia elegans into tracheary elements. *Plant J.*, **8** 891-906.

Stintzi, A., Heitz, T., Prasad, V., Wiedemann-Merdinoglu, S., Kauffmann, S., Geoffroy, P., Legrand, M. and Fritig, B. (1993) Plant 'pathogenesis-related' proteins and their role in defense against pathogens. *Biochimie*, **75** 687-706.

Tanaka, Y., Makishima, T., Sasabe, M., Ichinose, Y., Shiraishi, T., Nishimoto, T. and Yamada, T. (1997) *dad-1*, A putative programmed cell death suppressor gene in rice. *Plant Cell Physiol.*, **38** 379-83.

Tang, X. and Woodsen, W.R. (1996) Temporal and spatial expression of 1-aminocyclopropane-1-carboxylate oxidase mRNA following pollination of immature and mature petunia flowers. *Plant Physiol.*, **112** 503-11.

Tang, X., Frederick, R.D., Zhou, J., Halterman, D.A., Jia, Y. and Martin, G.B. (1996) Initiation of plant disease resistance by physical interaction of AvrPto and Pto kinase. *Science*, **274** 2060-63.

Thelen, M.P. and Northcote, D.H. (1989) Identification and purification of a nuclease from zinnia-elegans L. a potential molecular marker for xylogenesis. *Planta*, **179** 189-95.

Thomas, H. and Matile, P. (1988) Photobleaching of chloroplast pigments in leaves of a non-yellowing mutant genotype of festuca-pratensis. *Phytochemistry*, **27** 345-48.

Thomas, H. and Smart, C.M. (1993) Crops that stay green. *Ann. App. Biol.*, **123** 193-219.

Turner, S.R. and Somerville, C.R. (1997) Collapsed xylem phenotype of *Arabidopsis* identifies mutants deficient in cellulose deposition in the secondary cell wall. *Plant Cell*, **9** 689-701.

Van Oosterom, E.J., Jayachandran, R. and Bidinger, F.R. (1996) Diallel analysis of the stay-green trait and its components in sorghum. *Crop Sci.*, **36** 549-55.

Vartapetian, B.B. and Jackson, M.B. (1997) Plant adaptations to anaerobic stress. *Ann. Bot.*, **79** 3-20.

Vernon, D.M. and Meinke, D.W. (1994) Embryogenic transformation of the suspensor in twin, a polyembryonic mutant of *Arabidopsis*. *Dev. Biol.*, **165** 566-73.

Vincentini, F., Hortensteiner, S., Schellenberg, M., Thomas, H. and Matile, P. (1995) Chlorophyll breakdown in senescent leaves: identification of the biochemical lesion in a stay-green genotype of Festuca pratensis Huds. *New Phytol.*, **129** 247-52.

Walbot, V., Hoisington, D.A. and Neuffer, M.G. (1983) Disease lesion mimic mutations, in *Genetic Engineering of Plants* (eds T. Kosuge, C.P. Meredith and A. Hollaender), Plenum Press, New York, pp. 431-42.

Wang, H., Li, J., Bostock, R.M. and Gilchrist, D. (1996a) Apoptosis: a functional paradigm for programmed cell death induced by a host-selective phytotoxin and invoked during development. *Plant Cell*, **8** 375-91.

Wang, M., Oppedijk, B.J., Lu, X., Van Duijn, B. and Schilperoort, R.A. (1996b) Apoptosis in barley aleurone during germination and its inhibition by abscisic acid. *Plant Mol. Biol.*, **32** 1125-34.

Weaver, L.M., Himelblau, E. and Amasino, R.M. (1997) Leaf senescence: gene expression and regulation, in *Genetic Engineering*, Plenum Press, New York, pp. 215-34.

Woltering, E.J., de Vrije, T., Harren, F. and Hoekstra, F.A. (1997) Pollination and stigma wounding: same response, different signal? *J. Exp. Bot.*, **48** 1027-33.

Ye, Z.H. and Varner, J.E. (1996) Induction of cysteine and serine proteases during xylogenesis in Zinnia elegans. *Plant Mol. Biol.*, **30** 1233-46.

Ye, Z.H., Kneusel, R.E., Matern, U. and Varner, J.E. (1994) An alternative methylation pathway in lignin biosynthesis in Zinnia. *Plant Cell*, **6** 1427-39.

Yeung, E.C. and Meinke, D.W. (1993) Embryogenesis in angiosperms: development of the suspensor. *Plant Cell*, **5** 1371-81.

Young, T.E., Gallie, D.R. and DeMasin, D.A. (1997) Ethylene-mediated programmed cell death during maize endosperm development of wild-type and *shrunken2* genotypes. *Plant Physiol.*, **115** 737-51.

Zhou, J., Tang, X. and Martin, G.B. (1997) The Pto kinase conferring resistance of tomato bacterial speck disease interacts with proteins that bind a *cis*-element of pathogenesis-related genes. *EMBO J.*, **16** 3207-18.

Appendix: Electronic *Arabidopsis* resources

Biological resources (seeds and DNA)
Arabidopsis Biological Resource Center http://aims.cps.msu.edu/aims/
Nottingham *Arabidopsis* Stock Centre http://nasc.nott.ac.uk/

DNA and sequencing
AtDB (*Arabidopsis thaliana* Database) http://genome-www.stanford.edu/Arabidopsis/
AGR (*Arabidopsis* Genome Resource) http://synteny.nott.ac.uk/agr/agr.html
Centre National de Sequenage http://www.genoscope.cns.fr/externe/arabidopsis/
    Arabidopsis.html
CSHL/Wash U Sequencing Group http://nucleus.cshl.org/protarab/
    http://genome.wustl.edu/gsc/
Kazusa Sequencing Group http://www.kazusa.or.jp/arabi/
MIPS, ESSA1-2 sequencing http://speedy.mips.biochem.mpg.de/
SPP Sequencing Group http://sequence-www.stanford.edu/ara/
    ArabidopsisSeqStanford.html
    http://cbil.humgen.upenn.edu/~atgc/ATGCUP.html
    http://pgec-genome.pw.usda.gov/
TIGR Sequencing Group http://www.tigr.org/tdb/at/at.html
University of Minnesota ESTs http://www.cbc.umn.edu/ResearchProjects/
    Arabidopsis/index.html

Genetic maps
Lister and Dean RI Map and Mapping Service http://nasc.nott.ac.uk/new_ri_map.html
AtDB (*Arabidopsis thaliana* Database) http://genome-www.stanford.edu/Arabidopsis/
AGR (*Arabidopsis* Genome Resource) http://synteny.nott.ac.uk/agr/agr.html

Physical maps
AGR (*Arabidopsis* Genome Resource) http://synteny.nott.ac.uk/agr/agr.html
AtDB (*Arabidopsis thaliana* Database) http://genome-www.stanford.edu/Arabidopsis/
ATCG (*Arabidopsis thaliana* Genome Center) http://cbil.humgen.upenn.edu/~atgc/ATGCUP.html
Toni Shaffner's lab http://www.lmb.uni-muenchen.de/groups/
    schaeffner/main.htm

TAMU BACs http://http.tamu.edu:8000/~creel/TOC.html
Goodman Laboratory http://weeds.mgh.harvard.edu/goodman.html

General
Arabidopsis Gene Registry (David Meinke) http://mutant.lse.okstate.edu/meinke.html
Electronic Newsletter, Weeds World http://nasc.nott.ac.uk:8300/

Index